ESSENTIALS OF STATISTICS FOR THE BEHAVIORAL SCIENCES

THIRD EDITION

ESSENTIALS
OF STATISTICS
FOR THE BEHAVIORAL
SCIENCES
THIRD EDITION

FREDERICK J GRAVETTER

State University of New York
College at Brockport

LARRY B. WALLNAU

State University of New York
College at Brockport

Brooks/Cole Publishing Company

 An International Thomson Publishing Company

Pacific Grove • Albany • Belmont • Bonn • Boston • Cincinnati • Detroit
Johannesburg • London • Madrid • Melbourne • Mexico City
New York • Paris • Singapore • Tokyo • Toronto • Washington

Sponsoring Editors: *Denis Ralling and Vicki Knight*
Marketing Team: *Alicia Barelli, Aaron Eden, and Jean Vevers Thompson*
Editorial Assistant: *Stephanie Andersen*
Production Coordinator: *Laurel Jackson*
Production Service: *The Clarinda Company*
Manuscript Editor: *Sherry Goldbecker*
Permissions Editor: *May Clark*
Design Editor: *Roy R. Neuhaus*
Cover Design: *Carolyn Deacy*
Cover Art: *Vasily Kandinsky*

Horizontal Blue (Wagerecht Blau), *December 1929.*
Watercolor, gouache, and blue ink on paper
24.2 × 31.7 cm (9 1/2 × 12 1/2 inches)
The Hilla von Rebay Foundation
The Solomon R. Guggenheim Museum, New York
Photography by David Heald © The Solomon R.
    Guggenheim Foundation, New York.
Typesetting: The Clarinda Company
Cover Printing: *Phoenix Color Corporation*
Printing and Binding: *Courier/Kendallville*

For more information, contact:

BROOKS/COLE PUBLISHING COMPANY
511 Forest Lodge Road
Pacific Grove, CA 93950
USA

International Thomson Publishing Europe
Berkshire House 168-173
High Holborn
London WC1V 7AA
England

Thomas Nelson Australia
102 Dodds Street
South Melbourne, 3205
Victoria, Australia

Nelson Canada
1120 Birchmount Road
Scarborough, Ontario
Canada M1K 5G4

International Thomson Editores
Seneca 53
Col. Polanco
11560 México, D.F., México

International Thomson Publishing GmbH
Königswinterer Strasse 418
53227 Bonn
Germany

International Thomson Publishing Asia
60 Albert St.
#15-01 Albert Complex
Singapore 189969

International Thomson Publishing Japan
Hirakawacho Kyowa Building, 3F
2-2-1 Hirakawacho
Chiyoda-ku, Tokyo 102
Japan

Printed in the United States of America

10   9   8   7   6   5   4   3   2

Library of Congress Cataloging-in-Publication Data

Gravetter, Frederick J.
    Essentials of statistics for the behavioral sciences / Frederick
J Gravetter, Larry B. Wallnau.—3rd ed.
        p.   cm.
    Includes bibliographical references (p.   ) and index.
    ISBN 0-534-35780-6
    1. Social sciences—Statistical methods.   I. Wallnau, Larry B.
II. Title.
HA29.G726   1998
300'.7'2—dc21
                                        98-20983
                                        CIP

To David W. Wallnau, my extraordinary mentor,
devoted father, and dear friend
—L. B. W.

CONTENTS IN BRIEF

CONTENTS

PREFACE

There are three kinds of lies: Lies, damned lies, and statistics

We have used this quote in previous editions because it is timeless as well as humorous. It is attributed by Mark Twain to Benjamin Disraeli and reflects a commonly held belief that statistics (or perhaps even statisticians) should not be trusted. Unfortunately, this mistrust does have at least some basis in reality. In this book, we shall see that statistical techniques are tools that we use to organize information and to make inferences from our data. Like any other tool, however, statistics can be misused, which can result in misleading, distorted, or incorrect conclusions. It is no small wonder, then, that we are sometimes skeptical when a statistician presents findings. However, if we understand the correct uses of statistical techniques, then we will recognize those situations in which statistical procedures have been incorrectly applied. We can decide which statistical reports are more believable. By understanding statistical techniques, we can examine someone else's results, understand how they were analyzed, and arrive at our own thoughtful conclusion about the study. Therefore, the goal of this book is to teach not only the methods of statistics, but also how to apply these methods appropriately. Finally, a certain amount of mistrust is healthy, that is, we should critically examine information and data before we accept its implications. As you will see, statistical techniques help us look at data with a critical eye and a questioning mind.

For those of you who are familiar with previous editions of *Essentials of Statistics for the Behavioral Sciences*, you will notice that some changes have been made. These changes are summarized in the section entitled "To the Instructor." In revising this text, our students have been foremost in our minds. Over the years, they have provided honest and useful feedback. Their hard work and perseverance has made our writing and teaching most rewarding. We sincerely thank them. For the students who are using this edition, please read the section of the preface entitled "To the Student."

ANCILLARIES

Ancillaries for this edition include the following:

- *Study Guide:* Contains chapter overviews, learning objectives, new terms and concepts, new formulas, step-by-step procedures for problem solving, study hints and cautions, self-tests, and review.

- *SPSS Manual:* Contains step-by-step instructions on how to use SPSS to carry out statistical analysis.
- *Minitab Manual:* Includes instructions on using Minitab software to perform the statistical analyses covered in the text.
- *Instructor's Manual:* Contains test items, as well as solutions to all problems included in the text.
- *Transparency Masters:* Include about 90 tables and figures taken directly from the text.

ACKNOWLEDGMENTS

Our friends at Brooks/Cole have made an enormous contribution to this book. We thank our editor, Vicki Knight, who has been most supportive and encouraging. Editorial assistant Stephanie Andersen, production coordinator Laurie Jackson, and ancillaries editor Faith Stoddard are very capable people who possess many professional and technical skills and have helped in this undertaking.

Special thanks go to Emily Autumn of Clarinda. We have previously worked with her on other projects, and once more, her work on production has been extraordinary. The people at Minitab, Inc., deserve accolades for their Author Assistance Program, which was very helpful for the preparation of the Minitab Manual. Roxy Peck was most helpful in checking computations and finding errors we had missed.

Reviewers play a very important role in the development of a manuscript. Accordingly, we offer our appreciation to the following colleagues for their thoughtful reviews: Suzanne Bousquet, Kean University; Gregory Burton, Seton Hall University; Maria Czyzewska, Southwest Texas State University; Mark B. Fineman, Southern Connecticut State Universtiy; James Gardner, Ozarks Technical Community College; Richard Grow, Weber State University; Joe W. Hatcher, Jr., Ripon College; Bonnie Kind, Worcester State College; Thomas Nelson, Adrian College; Kerri Pickel, Ball State University; Carol Pandey, Los Angeles Pierce College; Stephen Schepman, Central Washington University; Martha Spiker, University of Charleston; and Davin Youngclarke, California State University at Fresno.

A Special Note of Thanks   Our family members have endured weekends and evenings when we were not available because we were immersed in writing and proofreading. This endeavor would be impossible were it not for their support, encouragement, and patience. Heartfelt thanks go to Debbie, Justin, Melissa, Megan, JoAnn, and Nico.

TO THE INSTRUCTOR

Those of you familiar with the second edition of *Essentials of Statistics for the Behavioral Sciences* will notice several changes in the third edition. A general summary of the revisions follows:

- The end-of-chapter problem sets have been revised.
- Some learning checks have been revised, and more have been added to chapters where they were needed.
- Chapter 16 (Minitab) of the second edition is now an ancillary under separate cover (*The Minitab Manual*).
- A new chapter has been added to cover two-factor analysis of variance.

The following are examples of the specific and noteworthy revisions:

Chapter 1

- The section on the experimental method now includes new coverage of quasi-independent variables (subject variables).
- The discussion of theories and hypotheses now emphasizes the importance of developing *testable* hypotheses. The role of hypothesis tests in theory construction is discussed.
- The rules of summation notation have be reorganized, and information on order of operations is included.

Chapter 4

- A new section discusses the use of the standard deviation in descriptive statistics and the use of variance in inferential statistics.

Chapter 5

- The chapter now begins with an overview that "points the way" to inferential statistics. It briefly, and in simple terms, presents the interrelationship of issues addressed in Chapters 5, 6, and 7, which form the foundation for subsequent chapters on hypothesis testing.

Chapter 6

- The introduction begins with a new example that illustrates the notions of uncertainty and probability.
- A distinction is made between simple random samples and convenience samples.
- The *unit normal table* has been reorganized to make it easier for students to use when determining probabilities. The probabilities in the table columns now separate the tail of the distribution from the remaining body. (This change has been successful in *Statistics for the Behavioral Sciences, Fourth Edition.*) It also makes finding percentile ranks easier.

Chapter 7

- The standard error formula is developed in terms of variance, as well as of standard deviation. The inclusion of variance here facilitates the understanding of more complex standard error formulas in later chapters.
- The chapter ends with new discussions of the role of standard error in inferential statistics: (a) standard error and sampling error, (b) standard error as a measure of chance, and (c) standard error as a measure of reliability. These discussions provide a better transition to the hypothesis testing chapters that follow.

Chapter 8

- More attention is given to taking the student through the *entire* hypothesis test procedure early in the chapter, addressing other aspects of hypothesis testing later. This organizational change provides an uninterrupted flow of logic.

- Type I and Type II errors now are presented in the context of uncertainty in inferential statistics.
- A new section discusses concerns about the usefulness of hypothesis testing that have been argued among researchers recently.

Chapter 9

- The development of the t statistic has been modified to incorporate minor changes in the formula for standard error (Chapter 7); that is, the standard error formula is now expressed in terms of variance.

Chapter 12

- This chapter has been greatly streamlined. Rather than develop the estimation procedures and formulas that correspond to the test situations in Chapters 8 through 11, a general model for estimation of μ is developed from which all others are derived.

Chapter 13

- Explanations of between treatment and within treatment variability have been revised.
- A new box on sources of error variability has been included.
- The explanation of the F ratio was rewritten.
- A new section, "A Conceptual View of ANOVA," examines the components of the F ratio using several small data sets.

Chapter 14

- This chapter is entirely new to this book, covering two-factor analysis of variance.

Chapter 15

- There is coverage of the role of outriders in measuring correlation.
- A new section provides coverage of the Spearman correlation.

TO THE STUDENT   There is a common (and usually unfair) belief that visits to the dentist will be associated with fear and pain, even though dentists perform a service of great benefit to us. Although you initially may have similar fears and anxieties about this course, we could argue that a statistics course also performs a beneficial service. This is evident when one considers that our world has become information-laden and information-dependent. The media informs us of the latest findings on nutrition and health, global warming, economic trends, aging and memory, effects of television violence on children, success or failure of new welfare programs, and so on. All these data-gathering efforts provide an enormous and unmanageable amount of information. Enter the statisticians, who use statistical procedures to analyze, organize, and interpret vast amounts of data. Having a basic understanding of a variety of statistical procedures will help you to understand these findings, to examine the data critically, and to question the statisticians about what they have done.

What about the fear of taking statistics? One way to deal with the fear is to get plenty of practice. You will notice that this book provides you with a number of opportunities to repeat the techniques you will be learning, in the form of learning checks, examples, demonstrations, and end-of-chapter problems. We encourage you to take advantage of these opportunities. Also, we encourage you to read the text rather than just memorize the formulas. We have taken great pains to present each statistical procedure in a conceptual context that explains why the procedure was developed and when it should be used. If you read this material and gain an understanding of the basic concepts underlying a statistical formula, you will find that learning the formula and how to use it will be much easier. In the following section, "Study Hints," we provide advice that we give our own students. Ask your instructor for advice as well, we are sure other instructors will have ideas of their own.

Study Hints   You may find some of these tips helpful, as our own students have reported.

- You will learn (and remember) much more if you study for short periods several times per week rather than try to condense all of your studying into one long session. For example, it is far more effective to study half an hour every night than to have a single $3\frac{1}{2}$-hour study session once a week. We cannot even work on *writing* this book without frequent rest breaks.

- Do some work before class. Keep a little ahead of the instructor by reading the appropriate sections before they are presented in class. Although you may not fully understand what you read, you will have a general idea of the topic, which will make the lecture easier to follow. Also, you can identify material that is particularly confusing and then be sure the topic is clarified in class.

- Pay attention and think during class. Although this advice seems obvious, often it is not practiced. Many students spend so much time trying to write down every example presented or every word spoken by the instructor that they do not actually understand and process what is being said. Check with your instructor. There may not be a need to copy every example presented in class, especially if there are many examples like it in the text. Sometimes, we tell our students to put their pens and pencils down for a moment and just listen.

- Test yourself regularly. Do not wait until the end of the chapter or the end of the week to check your knowledge. After each lecture, work some of the end-of-chapter problems, and do the Learning Checks. Review the Demonstration Problems, and be sure you can define the Key Terms. If you are having trouble, get your questions answered *immediately* (re-read the section, go to your instructor, or ask questions in class). By doing so, you will be able to move ahead to new material.

- Do not kid yourself! Avoid denial. Many students observe their instructor solve problems in class and think to themselves, "This looks easy, I understand it." Do you really understand it? Can you really do the problem on your own without having to leaf through the pages of a chapter? Although there is nothing wrong with using examples in the text as models for solving problems, you should try working a problem with your book closed to test your level of mastery.

· We realize that many students are embarassed to ask for help. It is our biggest challenge as instructors. You must find a way to overcome this aversion. Perhaps contacting the instructor directly would be a good starting point, if asking questions in class is too anxiety-provoking. You could be pleasantly surprised to find that your instructor does not yell, scold, or bite! Also, your instructor might know of another student who can offer assistance. Peer tutoring can be very helpful.

Over the years, our students in our classes have given us many helpful suggestions. We learn from them. If you have any suggestions or comments about this book, you can send a note to us at the Department of Psychology, SUNY College at Brockport, 350 New Campus Drive, Brockport, NY 14420. Also, we can be reached by email at fgravett@po.brockport.edu and lwallnau@po.brockport.edu.

Frederick J Gravetter
Larry B. Wallnau

CHAPTER 1

INTRODUCTION TO STATISTICS

CONTENTS

1.1 STATISTICS, SCIENCE, AND OBSERVATIONS

The procedure is actually quite simple. First you arrange things into different groups depending on their makeup. Of course, one pile may be sufficient, depending on how much there is to do. If you have to go somewhere else due to lack of facilities, that is the next step; otherwise you are pretty well set. It is important not to overdo any particular endeavor. That is, it is better to do too few things at once than too many. In the short run this may not seem important, but complications from doing too many can easily arise. A mistake can be expensive as well. The manipulation of the appropriate mechanisms should be self-explanatory, and we need not dwell on it here. At first the whole procedure will seem complicated. Soon, however, it will become just another facet of life. It is difficult to foresee any end to the necessity for this task in the immediate future, but then one never can tell.*

The preceding paragraph was adapted from a psychology experiment reported by Bransford and Johnson (1972). If you have not read the paragraph yet, go back and read it now.

You probably find the paragraph a little confusing, and most of you probably think it is describing some obscure statistical procedure. Actually, this paragraph describes the everyday task of doing laundry. Now that you know the topic of the paragraph, try reading it again—it should make sense now.

Why did we begin a statistics textbook with a paragraph about washing clothes? Our goal is to demonstrate the importance of context—when not in the proper context, even the simplest material can appear difficult and confusing. In the Bransford and Johnson experiment, people who knew the topic before reading the paragraph were able to recall 73% more than people who did not know that it was about doing laundry. When you have the appropriate background, it is much easier to fit new material into your memory and to recall it later. As you work through the topics in this book, remember that all statistical methods were developed to serve a purpose. The purpose for each statistical procedure provides a background or context for the details of the formulas and calculations. If you understand why a new procedure is needed, you will find it much easier to learn the procedure.

The objectives for this first chapter are to provide an introduction to the topic of statistics and to give you some background for the rest of the book. We will discuss the role of statistics within the general field of scientific inquiry, and we will introduce some of the vocabulary and notation that are necessary for the statistical methods that follow.

Incidentally, we cannot promise that statistics will be as easy as washing clothes. But if you begin each new topic within the proper context, you should eliminate some unnecessary confusion.

DEFINITIONS OF STATISTICS   Why study statistics? is a question countless students ask. One simple answer is that statistics have become a common part of everyday life and therefore deserve some attention. A quick glance at the newspaper yields statistics that deal with crime rates, birth rates, average income, average snowfall, and so on. By a common definition, therefore, statistics consist of facts and figures. These statistics generally are

---

*Bransford, J. D., and Johnson, M. K. (1972). Contextual prerequisites for understanding: Some investigations of comprehension and recall. *Journal of Verbal Learning and Verbal Behavior, 11,* 717–726. Copyright by Academic Press. Reprinted by permission.

informative and time saving because they condense large quantities of information into a few simple figures or statements. For example, the average snowfall in Chicago during the month of January is based on many observations made over many years. Few people would be interested in seeing a complete list of day-by-day snowfall amounts for the past 50 years. Even fewer people would be able to make much sense of all those numbers at a quick glance. But nearly everyone can understand and appreciate the meaning of an average.

In this book, however, we will use another definition of the term *statistics*. When researchers use the word *statistics,* they are referring to a set of mathematical procedures that are used to organize, summarize, and interpret information.

DEFINITION

Caution: The term *statistics* (plural) is used as a general reference for the entire set of statistical procedures. Later, we will use the term *statistic* (singular) to refer to a specific type of statistical method.

The term *statistics* refers to a set of methods and rules for organizing, summarizing, and interpreting information.

Statistical procedures help ensure that the information or observations are presented and interpreted in an accurate and informative way. In somewhat grandiose terms, statistics help researchers to bring order out of chaos. In addition, statistics provide researchers with a set of standardized techniques that are recognized and understood throughout the scientific community. Thus, the statistical methods used by one researcher will be familiar to other researchers, who can accurately interpret the statistical analyses with a full understanding of how the analysis was done and what the results signify. Although facts and figures can be important, this book will focus on the methods and procedures of statistics.

STATISTICS AND SCIENCE

These observations should be public, in the sense that others are able to repeat the observations using the same methods to see if the same findings will be obtained.

It is frequently said that science is *empirical.* That is, scientific investigation is based on making observations. Statistical methods enable researchers to describe and analyze the observations they have made. Thus, statistical methods are tools for science. We might think of science as consisting of methods for making observations and of statistics as consisting of methods for analyzing them.

On one mission of the space shuttle *Columbia,* so many measurements and observations were relayed to computers on earth that scientists were hard pressed to convey, in terms the public could grasp, how much information had been gathered. One individual made a few quick computations on a pocket calculator and determined that if all the data from the mission were printed on pages, they would pile up as high as the Washington Monument. Such an enormous amount of scientific observation is unmanageable in this crude form. To interpret the data, many months of work have to be done by many people to statistically analyze them. Statistical methods serve scientific investigation by organizing and interpreting data.

1.2 POPULATIONS AND SAMPLES

WHAT ARE THEY?

Scientific research typically begins with a general question about a specific group (or groups) of individuals. For example, a researcher may be interested in the effect of divorce on the self-esteem of preteen children. Or a researcher may want to examine the attitudes toward abortion of men versus those of women. In the first example, the researcher is interested in the group of *preteen children*. In the second example, the researcher wants to compare the group of *men* to the group of *women*. In statistical terminology, the entire group that a researcher wishes to study is called a *population*.

DEFINITION

A *population* is the set of all individuals of interest in a particular study.

As you can well imagine, a population can be quite large—for example, the number of women on the planet Earth. A researcher might be more specific, limiting the population to women who are registered voters in the United States. Perhaps the investigator would like to study the population consisting of women who are heads of state. Populations can obviously vary in size from extremely large to very small, depending on how the investigator defines the population. The population being studied should always be identified by the researcher. In addition, the population need not consist of people—it could be a population of rats, corporations, parts produced in a factory, or anything else an investigator wants to study. In practice, populations are typically very large, such as the population of fourth-grade children in the United States or the population of small businesses.

Because populations are typically very large, it usually is impossible for a researcher to examine every single individual in the population of interest. Therefore, researchers typically select a smaller, more manageable group from the population and limit their studies to the individuals in the selected group. In statistical terms, a set of individuals selected from a population is called a *sample*. A sample is intended to be representative of its population, and a sample should always be identified in terms of the population from which it was selected.

DEFINITION

A *sample* is a set of individuals selected from a population, usually intended to represent the population in a research study.

Just as we saw with populations, samples can vary in size. For example, one study might examine a sample of only 10 children in a preschool program, and another study might use a sample of over 1000 people representing the population of a major city.

Before we move on, there is one additional point that we should make about samples and populations. Thus far we have defined populations and samples in terms of *individuals*. For example, we have discussed a population of preteen children and a sample of preschool children. You should be forewarned, however, that we will also refer to populations or samples of *scores*. The change from individuals to scores is usually very straightforward because research typically involves measuring each individual and recording a score (measurement) for that individual. Thus, each sample (or population) of individuals produces a corresponding sample (or population) of scores. Occasionally, a set of scores is called a *statistical population* or a *statistical sample* to differentiate it from a population or a sample of individuals.

PARAMETERS AND STATISTICS

When describing data, it is necessary to distinguish whether the data come from a population or a sample. Any characteristic of a population, for example, its average, is called a population *parameter*. On the other hand, a characteristic of a sample is called a *statistic*. The average of the scores for a sample is a statistic. The range of scores for a sample is another type of statistic. As we will see later, statisticians frequently use different symbols for a parameter and a statistic. By using different symbols, we can readily tell if a characteristic, such as an average, is describing a population or a sample.

DEFINITIONS

A *parameter* is a value, usually a numerical value, that describes a population. A parameter may be obtained from a single measurement, or it may be derived from a set of measurements from the population.

A *statistic* is a value, usually a numerical value, that describes a sample. A statistic may be obtained from a single measurement, or it may be derived from a set of measurements from the sample.

DESCRIPTIVE AND INFERENTIAL STATISTICAL METHODS

The task of answering a research question begins by gathering information. In science, information is gathered by making observations and recording measurements for the individuals being studied. The measurement or observation obtained for each individual is called a *datum,* or, more commonly, a *score* or *raw score.* The complete set of scores or measurements is called the *data set* or simply the *data.* After data are obtained, statistical methods are used to organize and interpret the data.

DEFINITIONS

Data (plural) are measurements or observations. A *data set* is a collection of measurements or observations. A *datum* (singular) is a single measurement or observation and is commonly called a *score* or *raw score.*

Although researchers have developed a variety of different statistical procedures to organize and interpret data, these different procedures can be classified into two general categories. The first category, *descriptive statistics,* consists of statistical procedures that are used to simplify and summarize data.

DEFINITION

Descriptive statistics are statistical procedures that are used to summarize, organize, and simplify data.

Descriptive statistics are techniques that take raw scores and summarize them in a form that is more manageable. There are many descriptive procedures, but a common technique is to compute an average. Note that even if the data set has hundreds of scores, the average of those scores provides a single descriptive value for the entire set. Other descriptive techniques, including tables and graphs, will be covered in the next several chapters.

The second general category of statistical techniques is called *inferential statistics.* Inferential statistics are methods that use sample data to make general statements about a population.

DEFINITION

Inferential statistics consist of techniques that allow us to study samples and then make generalizations about the populations from which they were selected.

It usually is not possible to make measurements of everyone in the population. Because populations are typically too large to observe every individual, a sample is selected. By analyzing the results from the sample, we hope to make general statements about the population. Typically, researchers use sample statistics as the basis for drawing conclusions about population parameters.

One problem with using samples, however, is that a sample provides only limited information about the population. It gives us just a "glimpse" of the population. Of great importance, however, is the notion that a sample should be *representative* of its population. That is, the general characteristics of the sample should be consistent with the characteristics of the population. The method of selecting a sample (as we shall see in Chapter 6) helps ensure that a sample is representative. Even so, a sample is not expected to give a perfectly accurate picture of the whole population. There usually is some discrepancy between a sample statistic and the corresponding population parameter. This discrepancy is called *sampling error,* and it creates another problem to be addressed by inferential statistics.

DEFINITION   *Sampling error* is the discrepancy, or amount of error, that exists between a sample statistic and the corresponding population parameter.

The concept of sampling error will be discussed in more detail in Chapter 7, but, for now, you should realize that a statistic obtained from a sample generally will not be identical to the corresponding population parameter. One common example of sampling error is the error associated with a sample proportion. For example, in newspaper articles reporting results from political polls, you frequently find statements such as this:

Candidate Brown leads the poll with 51 percent of the vote. Candidate Jones has 42 percent approval, and the remaining 7 percent are undecided. This poll was taken from a sample of registered voters and has a margin of error of plus-or-minus 4 percentage points.

The "margin of error" is the sampling error. A sample of voters was selected, and 51 percent of the individuals in the sample expressed a preference for candidate Brown. Thus, the 51 percent is a sample statistic. Because the sample is only a small part of the total population of voters, you should not expect the sample value to be exactly equal to the population parameter. A statistic always has some "margin of error," which is defined as sampling error.

The following example shows the general stages of a research study and demonstrates how descriptive statistics and inferential statistics are used to organize and interpret the data. At the end of the example, notice how sampling error can affect the interpretation of experimental results, and consider why inferential statistical methods are needed to deal with this problem.

EXAMPLE 1.1   Figure 1.1 shows an overview of a research situation and the role that descriptive and inferential statistics play. In this example, a researcher is examining the development of language by comparing vocabulary skills for 4-year-old girls with those for 4-year-old boys.

The first step in the research process is to obtain a sample from the population of 4-year-old girls and a similar sample of 4-year-old boys. Each child is given a vocabulary test, and the scores from the test make up the data for this study.

Next, descriptive statistics are used to simplify the pages of data. The researcher could use a graph to display the data, or the researcher could simply calculate the average score for the boys and the average score for the girls. Notice that the descriptive methods produce an organized summary of the data so that it is easy to see the results of the research study. In this example, the girls' scores are generally higher than the boys', with the girls averaging 58 and the boys averaging 51 on the test.

After the researcher has described the sample data, the next step is to interpret the outcome; that is, the researcher must use the sample data to reach a general conclusion about the population. This is the job of inferential statistics. The results of this research study, for example, seem to indicate that girls have higher

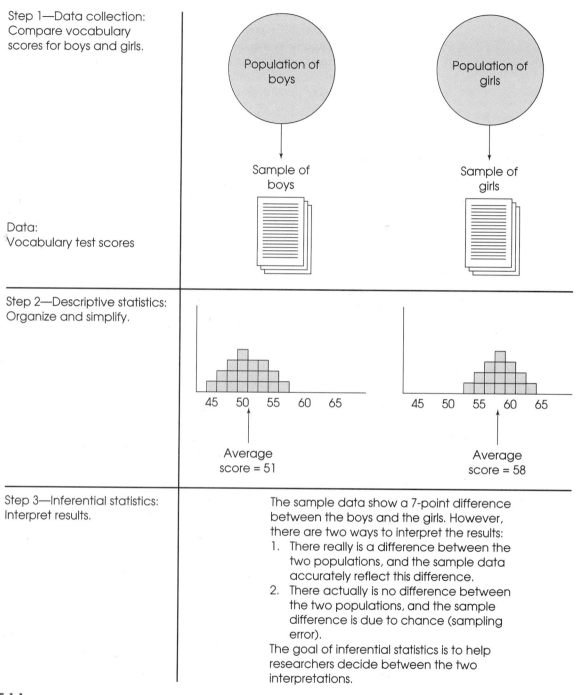

Step 1—Data collection: Compare vocabulary scores for boys and girls.

Population of boys

Population of girls

Sample of boys

Sample of girls

Data: Vocabulary test scores

Step 2—Descriptive statistics: Organize and simplify.

45 50 55 60 65

45 50 55 60 65

Average score = 51

Average score = 58

Step 3—Inferential statistics: Interpret results.

The sample data show a 7-point difference between the boys and the girls. However, there are two ways to interpret the results:
1. There really is a difference between the two populations, and the sample data accurately reflect this difference.
2. There actually is no difference between the two populations, and the sample difference is due to chance (sampling error).

The goal of inferential statistics is to help researchers decide between the two interpretations.

FIGURE 1.1

The role of statistics in experimental research

vocabulary scores than boys at age 4. Specifically, the girls scored an average of 7 points higher than the boys (an average of 58 versus an average of 51). However, you must remember that the averages are only *sample statistics* and there is a margin of error (sampling error) associated with each one.

Suppose, for example, that the researcher had selected a different sample of boys to participate in the research study. Although the original sample had a average score of 51 (7 points lower than the girls), a new sample would almost certainly produce a different average. A new sample of boys might even score higher than the girls. The point of this argument is that there are differences from one sample to another and no single sample will give you a perfectly accurate picture of the population. This is the general idea behind the concept of sampling error. For this research study, inferential statistics would be used to decide between the following two interpretations of the data:

1. The 7-point difference between the two samples is more than chance and represents a real difference in vocabulary skills between boys and girls. If the researcher continued to select new samples of boys and girls, the results would probably show a consistent pattern of higher scores for the girls.

2. The 7-point difference between the two samples is simply due to chance (sampling error) and does not indicate any real difference in vocabulary skills between boys and girls. After all, two samples are not expected to be exactly identical, and a 7-point difference is within the normal margin of error. If additional samples of boys and girls were selected, there would not be any consistent pattern of results; sometimes the girls would score higher, and sometimes the boys would score higher.

Notice that sampling error is an important factor in the interpretation of the research results. Computing the magnitude of sampling error will be a fundamental part of inferential statistics.

Inferential statistics are a crucial part of conducting experiments. Suppose a researcher would like to test the effectiveness of a new therapy program for depression. It would be too costly and time consuming to test the program on all depressed individuals. Therefore, the treatment is tested on a sample of depressed patients. Of course, the researcher would like to see these people overcome their misery, but keep in mind that one is not interested just in this sample of individuals. The investigator would like to generalize the findings to the entire population. If the treatment program is effective for the sample of depressed people, it would be great to be able to state with confidence that it will also work for others. It is important to note that inferential statistical methods will allow meaningful generalizations only if the individuals in the sample are representative of the population. One way to ensure that the sample is representative is to use *random selection*. Although there are several formally defined techniques for random sampling, one basic requirement is that all individuals in the population have the same chance of being selected. There will be more to say about this topic in later chapters.

D E F I N I T I O N   *Random selection,* or *random sampling,* is a process for obtaining a sample from a population that requires that every individual in the population have the same chance of being selected for the sample. A sample obtained by random selection is called a *random sample.*

LEARNING CHECK

1. What is a *population,* and what is a *sample?*

2. A characteristic that describes a population, such as the population average, is called a(n) _____.

3. The relationship between a population and a parameter is the same as the relationship between a sample and a(n) _____.

4. Statistical techniques are classified into two general categories. What are the two categories called, and what is the general purpose for the techniques in each category?

ANSWERS

1. The population is the entire set of individuals of interest for a particular research study. The sample is the specific set of individuals selected to participate in the study. The sample is selected from the population and is expected to be representative of the population.

2. parameter

3. statistic

4. The two categories are descriptive statistics and inferential statistics. Descriptive techniques are intended to organize, simplify, and summarize data. Inferential techniques use sample data to reach general conclusions about populations.

1.3 THE SCIENTIFIC METHOD AND THE DESIGN OF EXPERIMENTS

OBJECTIVITY

As noted earlier, science is empirical in that knowledge is acquired by observation. Another important aspect of scientific inquiry is that it should be *objective.* That is, theoretical biases of the researcher should not be allowed to influence the findings. Usually when a study is conducted, the investigator has a hunch about how it will turn out. This hunch, actually a prediction about the outcome of the study, is typically based on a theory that the researcher has. It is important that scientists conduct their studies in a way that will prevent these hunches or biases from influencing the outcome of the research. Experimenter bias can operate very subtly. Rosenthal and Fode (1963) had student volunteers act as experimenters in a learning study. The students were given rats to train in a maze. Half of the students were led to believe that their rats were "maze-bright," and the remainder were told their rats were "maze-dull." In reality, they all received the same kind of rat. Nevertheless, the data showed real differences in the rats' performance for the two groups of experimenters. Somehow, the students' expectations influenced the outcome of the experiment. Apparently, there were differences between the groups in how the students handled the rats, and the differences accounted for the effect. For a detailed look at experimenter bias, you might read the review by Rosenthal (1963).

RELATIONSHIPS BETWEEN VARIABLES

Science attempts to discover orderliness in the universe. Even people of ancient civilizations noted regularity in the world around them—the change of seasons, changes in the moon's phases, changes in the tides—and they were able to make many observations to document these orderly changes. Something that can change or have different values is called a *variable*.

DEFINITION

A *variable* is a characteristic or condition that changes or has different values for different individuals.

Variables are often identified by letters (usually *X* or *Y*), rather than a specific number. For example, the variable *height* could be identified by the letter *X*, and *shoe size* could be identified by *Y*. It is reasonable to expect a consistent, orderly relationship between these two variables: As *X* changes, *Y* also changes in a predictable way.

A value that does not change or vary is called a *constant*. For example, an instructor may adjust the exam scores for a class by adding 4 points to each student's score. Because every individual gets the same 4 points, this value is a constant.

DEFINITION

A *constant* is a characteristic or condition that does not vary but is the same for every individual.

A constant is often identified by its numerical value, such as 4, or by the letter *C*. Adding a constant to each score, for example, could be represented by the expression $X + C$.

Science involves a search for relationships between variables. For example, there is a relationship between the amount of rainfall and crop growth. Rainfall is one of the variables. It varies from year to year and season to season. Crop growth is the other variable. Some years, the cornstalks seem short and stunted; other years, they are tall and full. When there is very little rainfall, the crops are short and shriveled. When rain is ample, the crops show vigorous growth. Note that in order to document the relationship, one must make observations—that is, measurements of the amount of rainfall and size of the crops.

THE CORRELATIONAL METHOD

The simplest way to look for relationships between variables is to make observations of the two variables as they exist naturally for a set of individuals. This is called the *observational* or *correlational method.*

DEFINITION

With the *correlational method,* two variables are observed to see if there is a relationship.

Suppose a researcher wants to examine whether or not a relationship exists between length of time in an executive position and assertiveness. A large sample of executives takes a personality test designed to measure assertiveness. Also, the investigator determines how long each person has served in an executive-level job. Suppose the investigator found that there is a relationship between the two variables—that the longer a person had an executive position, the more assertive that person tended to be. Naturally, one might jump to the conclusion that being an executive for a long time makes a person more assertive. The problem with the correlational method is that it provides no information about cause-and-effect relationships. An equally plausible explanation for the relationship is that assertive people choose to stay or survive longer in executive positions than less assertive individuals.

FIGURE 1.2

Volunteers are randomly assigned to one of two treatment groups: 70° room or 90° room. After memorizing a list of words, subjects are tested by having them write down as many words as possible from the list. A difference between the groups in performance is attributed to the treatment—the temperature of the room.

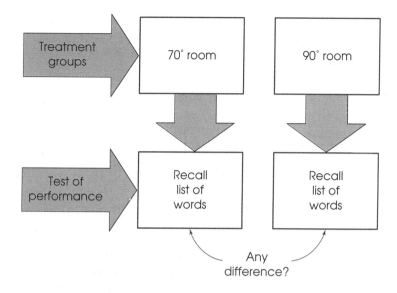

To establish a cause-and-effect relationship, it is necessary to exert a much greater level of control over the variables being studied. This is accomplished by the experimental method. We shall see that the experimental method is a highly structured, systematic approach to the study of relationships between variables.

THE EXPERIMENTAL METHOD

The goal of the *experimental method* is to establish a cause-and-effect relationship between two variables. That is, the method is intended to show that changes in one variable are *caused* by changes in the other variable. To accomplish this goal, the experimental method has two distinguishing characteristics:

1. The researcher *manipulates* one of the variables and observes the second variable to determine whether or not the manipulation causes changes to occur.

2. The researcher must exercise some *control* over the research situation to ensure that other, extraneous variables do not influence the relationship being examined.

In more complex experiments, a researcher may systematically manipulate more than one variable and may observe more than one variable. Here we are considering the simplest case, where only one variable is manipulated and only one variable is observed.

To demonstrate these two characteristics, consider an experiment where a researcher is examining the effects of room temperature on memory performance. The purpose of the experiment is to determine whether changes in room temperature *cause* changes in memory performance.

The researcher manipulates temperature by creating two or more different treatment conditions. For example, our researcher could set the temperature at 70 degrees for one condition and then change the temperature to 90 degrees for a second condition. The experiment would consist of observing the memory performance for a group of individuals (often called *subjects* or *participants*) in the 70-degree room and comparing their scores with those of another group that is tested in the 90-degree room. The structure of this experiment is shown in Figure 1.2.

To be able to say that differences in memory performance are caused by temperature, the researcher must rule out any other possible explanations for the difference. That is, any other variables that might affect memory performance must be controlled. Researchers typically use two basic techniques to control variables. First, the researcher could use *random assignment* so that each subject has an equal chance of being assigned to each of the treatment conditions. Random assignment helps assure that the subjects in one treatment condition are not substantially different from the subjects in another treatment condition. For example, a researcher should not assign all the young subjects to one condition and all the old subjects to another. If this were done, the researcher could not be sure that the difference in memory performance was caused by temperature; instead, it could be caused by age differences. Second, the experimental method requires that the treatment conditions be identical except for the one variable that is being manipulated. This is accomplished by *controlling* or *holding constant* any other variables that might influence performance. For example, the researcher should test both groups of subjects at the same time of day, in the same room, with the same instructions, and so on. Again, the goal is to eliminate the contribution of all other variables, except temperature, that might account for the difference in memory performance.

DEFINITION

In the *experimental method,* one variable is manipulated while changes are observed in another variable. To establish a cause-and-effect relationship between the two variables, an experiment attempts to eliminate or minimize the effect of all other variables by using *random assignment* and by *controlling* or *holding constant* other variables that might influence the results.

THE INDEPENDENT AND DEPENDENT VARIABLES

Specific names are used for the two variables that are studied by the experimental method. The variable that is manipulated by the experimenter is called the *independent variable.* It can be identified as the treatment conditions to which subjects are assigned. For the example in Figure 1.2, the temperature of the room is the independent variable. The variable that is observed to assess a possible effect of the manipulation is the *dependent variable.*

DEFINITIONS

The *independent variable* is the variable that is manipulated by the researcher. In behavioral research, the independent variable usually consists of the two (or more) treatment conditions to which subjects are exposed.

The *dependent variable* is the one that is observed for changes in order to assess the effect of the treatment.

In psychological research, the dependent variable is typically a measurement or score obtained for each subject. For the temperature experiment (Figure 1.2), the dependent variable is the number of words recalled on the learning test. Differences between groups in performance on the dependent variable suggest that the manipulation had an effect. That is, changes in the dependent variable *depend* on the independent variable.

Often we can identify one condition of the independent variable that receives no treatment. It is used for comparison purposes and is called the *control group.* The group that does receive the treatment is the *experimental group.*

DEFINITIONS

A *control group* is a condition of the independent variable that does not receive the experimental treatment. Typically, a control group either receives no treatment or receives a neutral, placebo treatment. The purpose of a control group is to provide a baseline for comparison with the experimental group.

An *experimental group* does receive an experimental treatment.

Note that the independent variable always consists of at least two values. (Something must have at least two different values before you can say that it is "variable.") For the temperature experiment (Figure 1.2), the independent variable is the temperature of the room, 70 degrees versus 90 degrees. For an experiment with an experimental group and a control group, the independent variable would be treatment versus no treatment.

THE QUASI-EXPERIMENTAL METHOD

As the name implies, the *quasi-experimental* method concerns research studies that are almost, but not quite, real experiments. You should recall that in a true experiment, the researcher manipulates one variable in order to create treatment conditions that can be compared. The manipulated variable is called the independent variable. In quasi-experimental research, there is no manipulation of an independent variable. Instead, the research compares groups that are defined by a naturally-occurring, nonmanipulated variable that is usually a *subject variable* or a *time variable*.

A *subject variable* is a characteristic such as age or gender that varies from one subject to another. For example, a researcher might want to compare personality scores for a group of males to those of a group of females. Notice that this study is comparing two groups (like an experiment) but the groups were not created by manipulating a variable. Specifically, a researcher cannot take one group of subjects and make them into males, then make a second group into females.

A *time variable* simply involves comparing individuals at different points in time. For example, a researcher may measure depression before therapy and then again after therapy. This study is comparing two groups of scores (before versus after) but the two groups were not created by manipulating a variable. Specifically, the researcher cannot control or manipulate the passage of time.

Research studies using the quasi-experimental method are similar to those using the experimental method because they both involve comparing one group of scores with another group of scores. In an experiment, the groups are created when the researcher manipulates the independent variable. For example, a researcher manipulates temperature to create a 70-degree room for one group and a 90-degree room for another group. In a quasi-experimental study, the researcher simply uses a nonmanipulated variable to define the groups. The variable defining the groups is called a *quasi-independent variable*. For example, a researcher compares vocabulary scores for a group of 4-year-old girls versus a group of 4-year-old boys. For this example, subject gender (male/female) is the quasi-independent variable.

DEFINITIONS

Instead of using an independent variable to create treatment conditions, a *quasi-experimental* research study uses a nonmanipulated variable to define the conditions that are being compared. The nonmanipulated variable is usually a subject variable (such as male versus female) or a time variable (such as before treatment versus after treatment).

The nonmanipulated variable that defines the conditions is called a *quasi-independent variable*.

THEORIES AND HYPOTHESES

Theories are a very important part of psychological research. A psychological theory typically consists of a number of statements about the underlying mechanisms of behavior. Theories are important in that they help organize and unify many observations. They may try to account for very large areas of psychology, such as

FIGURE 1.3

Theories should generate many testable hypotheses. When the data from a research study are consistent with the hypothesis, the theory is strengthened. When the hypothesis is falsified by the data, the theory must be revised.

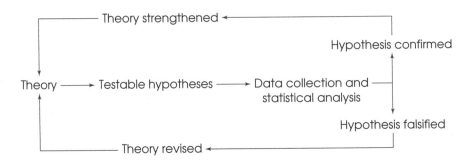

a general theory of learning. However, they may be more specific, such as a theory of the mechanisms involved just in avoidance learning. A theory is especially useful if it directs and promotes future research and provides specific predictions for the outcomes of that research.

These predictions, known as *hypotheses* (singular: hypothesis), make statements about the results that will be obtained from research studies. Simply stated, a hypothesis makes a prediction about the relationship between variables.

DEFINITION

A *hypothesis* is a prediction about the relationship between variables. In the context of an experiment, a hypothesis makes a prediction about how the manipulation of the independent variable will affect the dependent variable. In other kinds of research, a hypothesis predicts how a variable will behave under different circumstances or at different points in time.

Whenever an investigator designs a research study, there is a specific hypothesis that is addressed. It can be said that the hypothesis is being "tested" by the research. A theory can help research only if it generates *testable* hypotheses. That is, the hypothesis should have the potential to be disconfirmed by the data collected in the study. If the data are contrary to the hypothesis, then the hypothesis has been falsified. The researcher must go back to the theory, revise it to account for the new findings, and then generate new testable hypotheses. Often the researcher will test competing hypotheses in an experiment. If the hypothesis that supports the theory is confirmed by the data, then the theory is strengthened (Figure 1.3).

Some theories do not produce an abundance of testable hypotheses. Rather, they simply provide interpretations for existing observations. Suppose, for example, a Freudian theorist states that John F. Kennedy began to escalate the American presence in the Vietnam War because the rational part of his personality (the Ego) was too weak to control the impulsive urges of the irrational Id. This statement, like most from psychoanalytic theory, cannot be tested. It is an after-the-fact explanation rather than a prediction. What if President Kennedy had not been assassinated and was reelected for a second term? What if he then removed the troups from Vietnam? Hypotheses should make *predictions* that are later tested, not merely explain observations that have already been made.

CONSTRUCTS AND
OPERATIONAL
DEFINITIONS

Theories contain hypothetical concepts, which help describe the mechanisms that underlie behavioral phenomena. These concepts are called *constructs,* and they cannot be observed because they are hypothetical. For example, intelligence, personality types, and motives are hypothetical constructs. They are used in theories to organize observations in terms of underlying mechanisms. If constructs are hypo-

FIGURE 1.4

The constructs frustration and aggression are operationally defined so that a hypothesis can be tested.

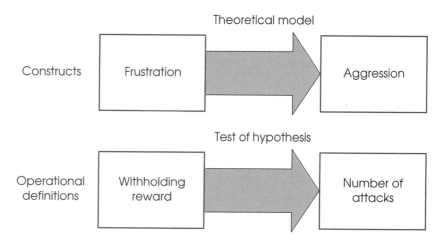

Test of Frustration-Aggression Hypothesis

thetical and cannot be observed, then how can they possibly be studied? The answer is that we have to *define* the construct so that it can be studied. An *operational definition* defines a construct in terms of an observable and measurable response. This definition should include the process and operations involved in making the observations. For example, an operational definition for emotionality might be stated as the amount of increase in heart rate after a person is insulted. An operational definition of intelligence might be the score on the Wechsler Adult Intelligence Scale.

DEFINITIONS

Constructs are hypothetical concepts that are used in theories to organize observations in terms of underlying mechanisms.

An *operational definition* defines a construct in terms of specific operations or procedures and the measurements that result from them. Thus, an operational definition consists of two components: First, it describes a set of operations or procedures for measuring a construct. Second, it defines the construct in terms of the resulting measurements.

Theories typically organize constructs into a model of behavior. That is, the theory contains statements about how constructs are interrelated with respect to behavior. With a useful theory, one can generate testable hypotheses by using operational definitions of the constructs and then doing an experiment. Consider the theory that frustration and aggression are related (Dollard et al., 1939). To test this hypothesis, we will need operational definitions for frustration and aggression. For example, frustration can be operationally defined as having a reward withheld when one is expected. This can be accomplished by training rats to press a level in a Skinner box for a food reward. After the response is learned, the food reward is withheld. Aggression can be operationally defined as the number of times the rats attack a plastic model of a rat that is in the Skinner box. The hypothesis, based on the theory, is that when food is withheld (frustration), animals will increase attacks (aggression). This example is depicted in Figure 1.4.

LEARNING CHECK   1. A researcher observes that elderly people who take regular doses of an anti-inflammatory drug tend to have a lower risk of Alzheimer's disease than their peers who do not take the drug. Is this study an example of the correlational method or the experimental method?

2. What are the two elements that are necessary for a research study to be an experiment?

3. The results from an experiment indicate that increasing the amount of indoor lighting during the winter months results in significantly lower levels of depression. Identify the independent variable and the dependent variable for this study.

ANSWERS   1. This is a correlational study. The researcher is simply observing the variables.

2. First, the researcher must manipulate one of the two variables being studied. Second, all other variables that might influence the results must be controlled.

3. The independent variable is the amount of indoor lighting, and the dependent variable is a measure of depression for each individual.

1.4   SCALES OF MEASUREMENT

WHAT IS A MEASUREMENT?   It should be obvious by now that data collection requires that we make measurements of our observations. Measurement involves either categorizing events (qualitative measurement) or using numbers to characterize the size of the event (quantitative measurement). Several types of scales are associated with measurements. The distinctions among the scales are important because they underscore the limitations of certain types of measurements and because certain statistical procedures are appropriate for data collected on some scales but not on others. If you were interested in people's heights, for example, you could measure a group of individuals by simply classifying them into three categories: tall, medium, and short. However, this simple classification would not tell you much about the actual heights of the individuals, and these measurements would not give you enough information to calculate an average height for the group. Although the simple classification would be adequate for some purposes, you would need more sophisticated measurements before you could answer more detailed questions. In this section, we examine four different scales of measurement, beginning with the simplest and moving to the most sophisticated.

NOMINAL SCALE   A *nominal scale* of measurement labels observations so that they fall into different categories.

DEFINITION   A *nominal scale* consists of a set of categories that have different names. Measurements on a nominal scale label and categorize observations but do not make any quantitative distinctions between observations.

The word *nominal* means "having to do with names." Measurements that are made on this scale involve naming things. For example, if we wish to know the sex of a person responding to a questionnaire, it would be measured on a nominal scale consisting of two categories. A researcher observing the behavior of a group of infant monkeys might categorize responses as playing, grooming, feeding, acting aggressively, or showing submissiveness. Again, this instance typifies a nominal scale of measurement. The nominal scale consists of qualitative distinctions.

If two individuals are measured on a nominal scale, it is possible to determine whether the two measurements are the same or different. However, if the measurements are different, you cannot determine how big the difference is, and you cannot say that one measurement is "more" or "less" than the other. A nominal scale consists of *qualitative* differences. It does not provide any information about *quantitative* differences between individuals.

Although the categories on a nominal scale are not quantitative values, they are occasionally represented by numbers. For example, the rooms or offices in a building may be identified by numbers. You should realize that the room numbers are simply names and do not reflect any quantative information. Room 109 is not necessarily bigger than room 100 and certainly not 9 points bigger. It also is fairly common to use numerical values as a code for nominal categories when data are entered into computer programs. For example, the data from a survey may code males with a 0 and females with a 1. Again, the numerical values are simply names and do not represent any quantitative difference.

ORDINAL SCALE In an *ordinal scale* of measurement, the categories that make up the scale not only have separate names (as in a nominal scale), but also are ranked in terms of magnitude. Thus, as the word *ordinal* implies, observations measured on an ordinal scale are categorized and arranged in rank order.

DEFINITION An *ordinal scale* consists of a set of categories that are organized in an ordered sequence. Measurements on an ordinal scale rank observations in terms of size or magnitude.

For example, a job supervisor is asked to rank employees in terms of how well they perform their work. The resulting data will tell us whom the supervisor considers the best worker, the second best, and so on. However, the data provide no information about the amount that the workers differ in job performance. The data may reveal that Jan, who is ranked second, is viewed as doing better work than Joe, who is ranked third. However, the data do not reveal *how much* better. This is a limitation of measurements on an ordinal scale. Thus, an ordinal scale provides information about the direction of difference between two measurements, but it does not reveal the magnitude of the difference.

INTERVAL AND RATIO SCALES An *interval scale* of measurement consists of an ordered set of categories (like an ordinal scale), with the additional requirement that the categories form a series of intervals that are all exactly the same size. The additional feature of equal-sized intervals makes it possible to compute distances between values on an interval scale. On a ruler, for example, a 1-inch interval is the same size at every location on the ruler, and a 4-inch distance is exactly the same size no matter where it is measured on the ruler. Thus, an interval scale allows you to measure differences in the size or amount of events. However, an interval scale does not have an absolute zero point that indicates complete absence of the variable being measured. Because there is no absolute zero point on an interval scale, ratios of values are not meaningful. (For example, you cannot compare two values by claiming that one is "twice as large" as another. See Example 1.2.)

A *ratio scale* of measurement has all the features of an interval scale but adds an absolute zero point. That is, on a ratio scale, a value of zero indicates *none* (a complete absence) of the variable being measured. The advantage of an absolute zero is that ratios of numbers on the scale reflect ratios of magnitude for the variable being measured. The distinction between an interval and a ratio scale is demonstrated in Example 1.2.

DEFINITIONS

An *interval scale* consists of ordered categories where all of the categories are intervals of exactly the same size. In an interval scale, equal differences between numbers on the scale reflect equal differences in magnitude. However, ratios of magnitudes are not meaningful.

A *ratio scale* is an interval scale with the additional feature of an absolute zero point. In a ratio scale, ratios of numbers reflect ratios of magnitude.

EXAMPLE 1.2

A researcher obtains measurements of height for a group of 8-year-old boys. Initially, the researcher simply records each child's height in inches, obtaining values such as 44, 51, 49, and so on. These initial measurements constitute a ratio scale. A value of zero represents no height (absolute zero). Also, it is possible to use these measurements to form ratios. For example, a child who is 80 inches tall is twice as tall as a 40-inch-tall child.

Now suppose the researcher converts the initial measurements into a new scale by calculating the difference between each child's actual height and the average height for this age group. A child who is 1 inch taller than average now gets a score of +1; a child 4 inches taller than average gets a score of +4. Similarly, a child who is 2 inches shorter than average gets a score of −2. The new scores constitute an interval scale of measurement. A score of zero no longer indicates an absence of height; now it simply means average height.

Notice that both sets of scores involve measurement in inches, and you can compute differences, or intervals, on either scale. For example, there is a 6-inch difference in height between two boys who measure 57 and 51 inches tall on the first scale. Likewise, there is a 6-inch difference between two boys who measure +9 and +3 on the second scale. However, you should also notice that ratio comparisons are not possible on the second scale. For example, a boy who measures +9 is *not* three times as tall as a boy who measures +3.

For most statistical applications, the distinction between an interval scale and a ratio scale is not particularly important. As we have seen, the process of measuring inches on a ruler can produce either interval scores or ratio scores depending on how the ruler is used. If you begin measuring at ground level (starting at zero) the ruler will produce ratio scores. On the other hand, measuring distances from an arbitrary (non-zero) point produces interval scores. However, the values from either scale allow basic arithmetic operations that permit us to calculate differences between scores, to sum scores, and to calculate average scores. For this reason, most dependent variables we will encounter are measured on either an interval or a ratio scale. Finally, you should realize that the distinction between different scales of measurement is often unclear when considering specific measurements. For example, the scores resulting from an IQ test are usually treated as measurements on an interval scale, but many researchers believe that IQ scores are more accurately described as ordinal data. An IQ score of 105 is clearly greater than a score of 100, but there is some question concerning *how much* difference in intelligence is reflected in the 5-point difference between these two scores.

1.5 DISCRETE AND CONTINUOUS VARIABLES

WHAT ARE THEY, AND HOW DO THEY DIFFER?

The variables in a study can be characterized by the type of values that can be assigned to them. A *discrete variable* consists of separate, indivisible categories. For this type of variable, there are no intermediate values between two adjacent categories. Consider the values displayed when dice are rolled. Between neighboring values—for example, seven dots and eight dots—no other values can ever be observed.

DEFINITION

A *discrete variable* consists of separate, indivisible categories. No values can exist between two neighboring categories.

A discrete variable is typically restricted to whole countable numbers—for example, the number of children in a family or the number of students attending class. If you observe class attendance from day to day, you may find 18 students one day and 19 students the next day. However, it is impossible ever to observe a value between 18 and 19. A discrete variable may also consist of observations that differ qualitatively. For example, a psychologist observing patients may classify some as having panic disorders, others as having dissociative disorders, and some as having psychotic disorders. The type of disorder is a discrete variable because there are distinct and finite categories that can be observed.

On the other hand, many variables are not discrete. Variables such as time, height, and weight are not limited to a fixed set of separate, indivisible categories. You can measure time, for example, in hours, minutes, seconds, or fractions of seconds. These variables are called *continuous* because they can be divided into an infinite number of fractional parts.

DEFINITION

For a *continuous variable,* there are an infinite number of possible values that fall between any two observed values. A continuous variable is divisible into an infinite number of fractional parts.

For example, subjects are given problems to solve, and a researcher records the amount of time it takes them to find the solutions. One person may take 31 seconds to solve the problems, whereas another may take 32 seconds. Between these two values, it is possible to find any fractional amount—$31\frac{1}{2}$, $31\frac{1}{4}$, $31\frac{1}{10}$—provided the measuring instrument is sufficiently accurate. Time is a continuous variable. A continuous variable can be pictured as a number line that is continuous. That is, there

FIGURE 1.5

Representation of time has a continuous number line. Note that there are an infinite number of possible values with no gaps in the line.

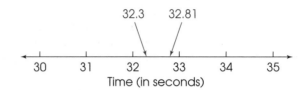

Time (in seconds)

are an infinite number of points on the line without any gaps or separations between neighboring points (see Figure 1.5).

LEARNING CHECK

1. An instructor records the order in which students complete their tests—that is, the first to finish, the second to finish, and so on. A(n) _____ scale of measurement is used in this instance.

2. The Scholastic Aptitude Test (SAT) most likely measures aptitude on a(n) _____ scale.

3. In a study on perception of facial expressions, subjects must classify the emotions displayed in photographs of people as anger, sadness, joy, disgust, fear, or surprise. Emotional expression is measured on a(n) _____ scale.

4. A researcher studies the factors that determine how many children couples decide to have. The variable, number of children, is a _____ (discrete/continuous) variable.

5. An investigator studies how concept-formation ability changes with age. Age is a _____ (discrete/continuous) variable.

ANSWERS 1. ordinal   2. interval   3. nominal   4. discrete   5. continuous

1.6 STATISTICAL NOTATION

Measurements of behavior usually will provide data composed of numerical values. These numbers form the basis of the computations that are done for statistical analyses. There is a standardized notation system for statistical procedures, and it is used to identify terms in equations and mathematical operations. Some general mathematical operations, notation, and basic algebra are outlined in the review section of Appendix A. There is also a skills assessment exam (page A-2) to help you determine whether you need the basic mathematics review. Here, we will introduce some statistical notation that is used throughout this book. In subsequent chapters, additional notation will be introduced as it is needed.

SCORES   Making observations of a dependent variable in a study will typically yield values or scores for each subject. Raw scores are the original, unchanged scores obtained

X
37
35
35
30
25
17
16

in the study. Scores for a particular variable are represented by the letter X. For example, if performance in your statistics course is measured by tests and you obtain a 35 on the first test, then we could state that $X = 35$. A set of scores can be presented in a column that is headed by X. For example, quiz scores from your class might be listed as shown in the margin.

When observations are made for two variables, there will be two scores for each subject. The data can be presented as two lists labeled X and Y for the two variables. For example, observations of people's height in inches (variable X) and weight in pounds (variable Y) can be presented in the following manner. Each X, Y pair represents the observations made of a single subject.

X	Y
72	165
68	151
67	160
68	146
70	160
66	133

It is also useful to specify how many scores are in a set. The number of scores in a data set is represented by the letter N. For populations, we will use an uppercase N, and for samples, we will use a lowercase n. (Throughout the book, notational differences are used to distinguish between samples and populations.) For the height and weight data, $N = 6$ for both variables.

SUMMATION NOTATION   Many of the computations required in statistics will involve adding a set of scores. Because this procedure is used so frequently, a special notation is used to refer to the sum of a set of scores. The Greek letter *sigma*, or Σ, is used to stand for summation. The expression ΣX means to add all the scores for variable X. The summation sign Σ can be read as "the sum of." Thus, ΣX is read "the sum of the scores." For the following set of quiz scores,

10, 6, 7, 4

$\Sigma X = 27$ and $N = 4$.

To use summation notation correctly, you should keep in mind the following two points:

1. The summation sign Σ is always followed by a symbol or mathematical expression. The symbol or expression identifies exactly which values are to be summed. To compute ΣX, for example, the symbol following the summation sign is X, and the task is to find the sum of the X values. On the other hand, to compute $\Sigma(X - 1)^2$, the summation sign is followed by a relatively complex mathematical expression, so your first task is to calculate all of the $(X - 1)^2$ values and then sum the results.

2. The summation process is often included with several other mathematical operations, such as multiplication or squaring. To obtain the correct answer,

it is essential that the different operations be done in the correct sequence. Following is a list showing the correct *order of operations* for performing mathematical operations. Most of this list should be familiar, but you should note that we have inserted the summation process as the fourth operation in the list.

More information on the order of operations for mathematics is available in the Math Review appendix, page A-3.

Order of Mathematical Operations

1. Any calculation contained within parentheses is done first.

2. Squaring (or raising to other exponents) is done second.

3. Multiplying and/or dividing is done third. A series of multiplication and/or division operations should be done in order from left to right.

4. Summation using the Σ notation is done next.

5. Finally, any other addition and/or subtraction is done.

The following examples demonstrate how summation notation will be used in most of the calculations and formulas we will present in this book.

EXAMPLE 1.3   A set of four scores consists of values 3, 1, 7, and 4. We will compute ΣX, ΣX^2, and $(\Sigma X)^2$ for these scores. To help demonstrate the calculations, we have constructed a *computational table* showing the original scores (the X values) and the squared scores (the X^2 values).

X	X^2
3	9
1	1
7	49
4	16

The first calculation, ΣX, does not include any parentheses, squaring, or multiplication, so we go directly to the summation operation. The X values are listed in the first column of the table, and we simply sum the values in this column:

$$\Sigma X = 3 + 1 + 7 + 4 = 15$$

To compute ΣX^2, the correct order of operations is to square the scores first and then sum the squared values. The squared scores are listed in the second column of the table, so we simply sum these values (see Box 1.1).

$$\Sigma X^2 = 9 + 1 + 49 + 16 = 75$$

The final calculation, $(\Sigma X)^2$, includes parentheses, so the first step is to perform the calculation inside the parentheses. Thus, we first find ΣX and then square this sum. Earlier, we computed $\Sigma X = 15$, so

$$(\Sigma X)^2 = (15)^2 = 225$$

1.1 COMPUTING ΣX^2 WITH A CALCULATOR

THE SUM of squared scores, ΣX^2, is a common expression in many statistical calculations. The following steps outline the most efficient procedure for using a typical, inexpensive hand calculator to find this sum. We assume that your calculator has one memory where you can store and retrieve information. *Caution:* The following instructions work for most calculators but not for every single model. If you encounter trouble, don't panic—check your manual or talk with your instructor.

1. Clear the calculator memory. You may press the memory-clear key (usually MC) or simply turn the calculator off and then back on.

2. Enter the first score.

3. Press the multiply key (×); then press the equals key (=). The squared score should appear in the display.

(Note that you do not need to enter a number twice to square it. Just follow the sequence: number-times-equals.)

4. Put the squared value into the calculator memory. For most calculators, you press the key labeled M+.

5. Enter the next score, square it, and add it to memory (steps 2, 3, and 4). (Note that you do not need to clear the display between scores.)

6. Continue this process for the full set of scores. Then retrieve the total, ΣX^2, from memory by pressing the memory-recall key (usually labeled MR).

Check this procedure with a simple set of scores, such as 1, 2, 3. You should find $\Sigma X^2 = 14$.

EXAMPLE 1.4   We will use the same set of four scores from Example 1.3 and compute $\Sigma(X - 1)$ and $\Sigma(X - 1)^2$. The following computational table will help demonstrate the calculations.

X	$(X - 1)$	$(X - 1)^2$	
3	2	4	The first column lists the
1	0	0	original scores, the second
7	6	36	column lists the $(X - 1)$ values, and the third column
4	3	9	shows the $(X - 1)^2$ values.

To compute $\Sigma(X - 1)$, the first step is to perform the operation inside the parentheses. Thus, we begin by subtracting 1 point from each of the X values. The resulting values are listed in the middle column of the table. The next step is to sum the $(X - 1)$ values.

$$\Sigma(X - 1) = 2 + 0 + 6 + 3 = 11$$

The calculation of $\Sigma(X - 1)^2$ begins within the parentheses by subtracting 1 point from each X value. The next step is to square the resulting values. In the table, the squared numbers are listed in the third column. Finally, you add the squared numbers to obtain

$$\Sigma(X - 1)^2 = 4 + 0 + 36 + 9 = 49$$

Notice that this calculation requires squaring before summing. A common mistake is to sum the $(X - 1)$ values and then square the total. Be careful!

EXAMPLE 1.5   In both of the preceding examples, and in many other situations, the summation operation is last step in the calculation. According to the order of operations, parentheses, exponents, and multiplication all come before summation. However, there are situations where extra addition and subtraction is completed after the summation. For this example, we will use the same scores that appeared in the previous two examples, and we will compute $\Sigma X - 1$.

With no parentheses, exponents, or multiplication, the first step is the summation. Thus, we begin by computing ΣX. Earlier, we found $\Sigma X = 15$. The next step is to subtract 1 point from the total. For these data,

$$\Sigma X - 1 = 15 - 1 = 14$$

EXAMPLE 1.6   For this example, each individual has two scores. The first score is identified as X, and the second score is Y. With the help of the following computational table, we will compute ΣX, ΣY, and ΣXY.

X	Y	XY
3	5	15
1	3	3
7	4	28
4	2	8

To find ΣX, simply sum the values in the X column.

$$\Sigma X = 3 + 1 + 7 + 4 = 15$$

Similarly, ΣY is the sum of the Y values.

$$\Sigma Y = 5 + 3 + 4 + 2 = 14$$

To compute ΣXY, the first step is to multiply X times Y for each individual. The resulting products (XY values) are listed in the third column of the table. Finally, we sum the products to obtain

$$\Sigma XY = 15 + 3 + 28 + 8 = 54$$

LEARNING CHECK   For the data in the margin, find each value requested.

X
4
3
5
2

1. ΣX 2. ΣX^2 3. $(\Sigma X)^2$ 4. $\Sigma X - 2$ 5. $\Sigma(X - 2)$

6. $\Sigma(X - 2)^2$ 7. N

ANSWERS   1. 14   2. 54   3. 196   4. 12   5. 6   6. 14   7. 4

A WORD ABOUT COMPUTERS   Long before desktop computers became commonplace at home, they were routinely used in statistics. The computer can make short work of the statistical analysis of large data sets and can eliminate the tedium that goes along with complex and repetitive computations. As the use of computers became widespread, so did the availability of statistical software packages. The typical software package contains a variety of specialized programs, each one capable of performing a specific type of statistical procedure. The user enters the data to be analyzed and then specifies the analysis to be used with one or more software commands.

There are many useful and popular statistical software packages, all differing in their ease of operation, command structure, sophistication, and breadth of coverage of statistical analyses. Two software manuals have been prepared to accompany this book.

Minitab manual   Minitab is a statistical package that is widely available on large computers as well as in DOS, Windows, and Macintosh versions for desktop computers. It is easy to master—or, in computer terms, "user friendly"—and it serves as a good model for many other statistics programs. The manual presents instruction on how to use Minitab and includes demonstrations of Minitab as it is used for many of the specific examples contained in other chapters throughout this book. As you work through the book, you will find computer symbols in the margins (see marginal note) with references to the manual. There you will find instructions on how to use Minitab to perform the statistical procedure that is discussed in the text.

This computer symbol will be used throughout the book to direct you to additional information or examples in the Minitab and SPSS computer manuals.

SPSS manual   This software is a sophisticated package of programs capable of performing many of the statistical analyses covered in this text and many more-advanced procedures as well. It commonly is used by researchers in the social and behavioral sciences. Again, computer symbols are used to reference the manual.

While we are on the topic of computers and statistics, the authors cannot resist an editorial comment. Many students have approached us with this question: If computers are so good at doing statistical calculations, why not just teach us how to use the computer and forget about teaching us statistics? This question usually is followed with an accurate observation: If I should get a job where I need to use statistics, I probably will just use a computer anyway.

Although there are many ways to respond to these comments, we will try to limit ourselves to one or two general ideas. The purpose of this book is to help you gain an understanding of statistics. Notice that we used the word *understanding*. Although computers can do statistics with incredible speed and accuracy, they are not capable of exercising any judgment or interpreting the material that you feed into them (input) or the results that they produce (output). Too often, students (and researchers) rely on computers to perform difficult calculations and have no idea of what the computer is actually doing. In some fields, this lack of understanding is acceptable—we are all allowed to use the telephone even if we do not understand how it works. But in scientific research, it is critical that you have a complete and accurate understanding of your data. Usually, this requires an understanding of the statistical techniques used to summarize and interpret data. So use your computer and enjoy it, but do not rely on it for statistical expertise.

SUMMARY

1. The term *statistics* is used to refer to methods for organizing, summarizing, and interpreting data.

2. Scientific questions usually concern a population that is the entire set of individuals that one wishes to study. Usually, populations are so large that it is impossible to examine every individual, so most research is conducted with samples. A sample is a group selected from a population, usually for purposes of a research study.

3. Statistical methods can be classified into two broad categories: descriptive statistics, which organize and summarize data, and inferential statistics, which use sample data to draw inferences about populations.

4. The correlational method looks for interrelationships between variables but cannot determine the cause-and-effect nature of the relationship. The experimental method is able to establish causes and effects in a relationship.

5. In the experimental method, one variable (the independent variable) is manipulated, and another variable (the dependent variable) is observed for changes that may occur as a result of the manipulation. All other variables are controlled or held constant.

6. A hypothesis is a prediction about the relationship between variables. Hypotheses are usually derived from theories. Experiments basically involve the test of a hypothesis.

7. Constructs are hypothetical concepts used in theories to describe the mechanisms of behavior. Because they are hypothetical, they cannot be observed. Constructs are studied by providing operational definitions for them. An operational definition defines a construct in terms of an observable and measurable response or event.

8. A measurement scale consists of a set of categories that are used to classify individuals. A nominal scale consists of categories that differ only in name and are not differentiated in terms of magnitude or direction. In an ordinal scale, the categories are differentiated in terms of direction, forming an ordered series. An interval scale consists of an ordered series of categories where the categories are all equal-sized intervals. With an interval scale, it is possible to differentiate direction and magnitude (or distance) between categories. Finally, a ratio scale is an interval scale where the zero point indicates none of the variable being measured. With a ratio scale, ratios of measurements reflect ratios of magnitude.

9. A discrete variable is one that can have only a finite number of values between any two values. It typically consists of whole numbers that vary in countable steps. A continuous variable can have an infinite number of values between any two values.

10. The letter X is used to represent scores for a variable. If a second variable is used, Y represents its scores. The letter N is used as the symbol for the number of scores in a set. The Greek letter *sigma* (Σ) is used to stand for summation. Therefore, the expression ΣX is read "the sum of the scores."

KEY TERMS

statistics	random selection	quasi-independent variable	nominal scale
population	variable	control group	ordinal scale
sample	constant	experimental group	interval scale
population parameter	correlational method	theory	ratio scale
sample statistic	experimental method	hypothesis	discrete variable
descriptive statistics	quasi-experimental method	construct	continuous variable
raw score	independent variable	operational definition	computational table
inferential statistics	dependent variable		

FOCUS ON PROBLEM SOLVING

1. It may help to simplify summation notation if you observe that the summation sign is always followed by a symbol (or symbolic expression)—for example, ΣX or $\Sigma(X + 3)$. This symbol specifies which values you are to add. If you use the symbol

as a column heading and list all the appropriate values in the column, your task is simply to add up the numbers in the column. To find $\Sigma(X + 3)$, for example, start a column headed with $(X + 3)$ next to the column of Xs. List all the $(X + 3)$ values; then find the total for the column.

2. To use summation notation correctly, you must be careful of two other factors:
   a. When you are determining the "symbol" that follows the summation sign, remember that everything within parentheses is part of the same symbol, for example, $\Sigma(X + 3)$, and a string of multiplied values is considered to be a single symbol, for example, ΣXY.
   b. Often it is necessary to use several intermediate columns before you can reach the column of values specified by a particular symbol. To compute $\Sigma(X - 1)^2$, for example, you will need three columns: first, the column of original scores, X values; second, a column of $(X - 1)$ values; and, third, a column of squared $(X - 1)$ values. It is the third column, headed by $(X - 1)^2$, that you should total.

DEMONSTRATION 1.1

SUMMATION NOTATION

A set of data consists of the following scores:

$$7, \quad 3, \quad 9, \quad 5, \quad 4$$

For these data, find the following values:
a. ΣX b. $(\Sigma X)^2$ c. ΣX^2 d. $\Sigma X + 5$ e. $\Sigma(X - 2)$

Compute ΣX. To compute ΣX, we simply add all of the scores in the group. For these data, we obtain

$$\Sigma X = 7 + 3 + 9 + 5 + 4 = 28$$

Compute $(\Sigma X)^2$. The key to determining the value of $(\Sigma X)^2$ is the presence of parentheses. The rule is to perform the operations that are inside the parentheses first.
   Step 1: Find the sum of the scores, ΣX.
   Step 2: Square the total.
   We have already determined that ΣX is 28. Squaring this total, we obtain

$$(\Sigma X)^2 = (28)^2 = 784$$

Compute ΣX^2. Calculating the sum of the squared scores, ΣX^2, involves two steps.
   Step 1: Square each score.
   Step 2: Sum the squared values.
   These steps are most easily accomplished by constructing a computational table. The first column has the heading X and lists the scores. The second column is labeled X^2 and contains the squared value for each score. For this example, the table is shown in the margin.
   To find the value for ΣX^2, we sum the X^2 column.

X	X^2
7	49
3	9
9	81
5	25
4	16

$$\Sigma X^2 = 49 + 9 + 81 + 25 + 16 = 180$$

Compute $\Sigma X + 5$. In this expression, there are no parentheses. Thus, the summation sign is applied only to the X values.
   Step 1: Find the sum of X.
   Step 2: Add the constant 5 to the total from step 1.

Earlier we found that the sum of the scores is 28. For $\Sigma X + 5$, we obtain the following.

$$\Sigma X + 5 = 28 + 5 = 33$$

Compute $\Sigma(X - 2)$. The summation sign is followed by an expression with parentheses. In this case, $X - 2$ is treated as a single expression, and the summation sign applies to the $(X - 2)$ values.

X	X − 2
7	5
3	1
9	7
5	3
4	2

Step 1: Subtract 2 from every score.
Step 2: Sum these new values.
This problem can be done by using a computational table with two columns, headed X and $X - 2$, respectively.
To determine the value for $\Sigma(X - 2)$, we sum the $X - 2$ column.

$$\Sigma(X - 2) = 5 + 1 + 7 + 3 + 2 = 18$$

——— **DEMONSTRATION 1.2** —————————————————————————

SUMMATION NOTATION WITH TWO VARIABLES

X	Y
5	8
2	10
3	11
7	2

The data to the left consist of pairs of scores (X and Y) for four individuals: Determine the values for the following expressions:

  a. $\Sigma X \Sigma Y$ **b.** ΣXY

Compute $\Sigma X \Sigma Y$. This expression indicates that we should multiply the sum of X by the sum of Y.
   Step 1: Find the sum of X.
   Step 2: Find the sum of Y.
   Step 3: Multiply the results of steps 1 and 2.
First, we find the sum of X.

$$\Sigma X = 5 + 2 + 3 + 7 = 17$$

Next, we compute the sum of Y.

$$\Sigma Y = 8 + 10 + 11 + 2 = 31$$

Finally, we multiply these two totals.

$$\Sigma X \Sigma Y = 17(31) = 527$$

Compute ΣXY. Now we are asked to find the sum of the products of X and Y.
   Step 1: Find the XY products.
   Step 2: Sum the products.
The computations are facilitated by using a third column labeled XY.

X	Y	XY
5	8	40
2	10	20
3	11	33
7	2	14

For these data, the sum of the XY products is

$$\Sigma XY = 40 + 20 + 33 + 14 = 107$$

PROBLEMS

*1. Describe the general purpose for descriptive statistics and for inferential statistics.

2. Define the terms *population* and *sample,* and explain how each of these is involved in the process of scientific research.

3. Identify the basic characteristics that distinguish the experimental method from the correlational method.

4. Define the concept of *sampling error.* Be sure that your definition includes the concepts of *statistic* and *parameter.*

5. In general, research attempts to establish and explain relationships between variables. What terms are used to identify the two variables in an experiment? Define each term.

6. A researcher reports that individuals who survived one heart attack and were given daily doses of aspirin were significantly less likely to suffer a second heart attack than survivors who did not take aspirin. For this study, identify the independent variable and the dependent variable.

7. A developmental psychologist conducts a research study comparing vocabulary skills for 5-year-old boys and 5-year-old girls. Is this study an *experiment?* Explain why or why not.

8. A researcher studying sensory processes manipulates the loudness of a buzzer and measures how quickly subjects respond to the sound at different levels of intensity. Identify the independent and dependent variables for this study.

9. A researcher would like to evaluate the claim that large doses of vitamin C can help prevent the common cold. One group of subjects is given a large dose of the vitamin (500 mg per day), and a second group is given a placebo (sugar pill). The researcher records whether or not each individual experiences a severe cold during the three-month winter season.
   a. Identify the dependent variable for this study.
   b. Is the dependent variable discrete or continuous?
   c. What scale of measurement (nominal, ordinal, interval, or ratio) is used to measure the dependent variable?
   d. What research method is being used (experimental or correlational)?

10. A questionnaire asks individuals to report age, sex, eye color, and height. For each of the four variables,
   a. Classify the variable as either discrete or continuous.
   b. Identify the scale of measurement that probably would be used.

11. Define and differentiate a discrete variable and a continuous variable.

12. A professor records the number of absences for each student in a class of $N = 20$.
   a. Is the professor measuring a discrete or a continuous variable?
   b. What scale of measurement is being used?

13. A researcher is comparing two brands of medication intended to reduce stomach acid. Before eating a very spicy meal, one group of people is given brand X, and another group is given brand Y. Two hours after eating, each person is asked to rate his or her level of indigestion using the following scale: no stomach upset, mild indigestion, moderate indigestion, severe indigestion.
   a. What is the independent variable for this study, and what scale of measurement is used for this variable?
   b. What is the dependent variable for this study, and what scale of measurement is used for this variable?

14. A researcher studying the effects of environment on mood asks subjects to sit alone in a waiting room for 15 minutes at the beginning of an experiment. Half of the subjects are assigned to a room with dark blue walls, and the other half are assigned to a room with bright yellow walls. After 15 minutes in the waiting room, each subject is brought into the lab and given a mood-assessment questionnaire.
   a. Identify the independent and dependent variables for this study.
   b. What scale of measurement is used for the independent variable?

15. A researcher records the number of errors each subject makes while performing a standardized problem-solving task.
   a. Is the researcher measuring a discrete or a continuous variable?
   b. What scale of measurement is being used?

16. Suppose two individuals are assigned to different categories on a scale of measurement. If the measurement scale is nominal, then the only information provided about the two individuals is that they are different.
   a. What additional information about the difference between the two individuals would be provided if the measurement scale were ordinal instead of nominal?
   b. What additional information about the two individuals would be provided by an interval scale versus an ordinal scale?
   c. What additional information would be provided by ratio scale measurements versus interval scale measurements?

*Solutions for odd-numbered problems are provided in Appendix C.

17. Define and differentiate a construct and an operational definition.

18. For the following scores, find the value of each expression:

a. ΣX
b. ΣX^2
c. $(\Sigma X)^2$
d. $\Sigma(X - 1)$

X
4
2
6
1

19. Two scores, X and Y, are recorded for each of $n = 4$ subjects. For these scores, find the value of each expression:

a. ΣX
b. ΣY
c. ΣXY

Subject	X	Y
A	3	2
B	4	1
C	2	3
D	6	4

20. Use summation notation to express each of the following calculations:

a. Square each score, and find the sum of the squared values.
b. Find the sum of the scores, and then square the sum.

21. Use summation notation to express each of the following calculations:

a. Sum the scores, and then add 3 points to the total.
b. Subtract 2 points from each score, and square the result. Then sum the squared values.
c. Square each score, and sum the squared values. Then subtract 10 points from the sum.

22. Each of the following summation expressions requires a specific sequence of mathematical operations. For each expression, state in words the sequence of operations necessary to perform the specified calculations. For example, the expression $\Sigma(X + 1)$ instructs you to add 1 point to each score and then sum the resulting values.

a. $\Sigma X - 4$
b. ΣX^2
c. $\Sigma(X + 4)^2$

23. For the following set of scores, find the value of each expression:

a. ΣX
b. ΣX^2
c. $\Sigma(X + 1)$
d. $\Sigma(X + 1)^2$

X
1
3
0
5

24. For the following set of scores, find the value of each expression:

a. ΣX
b. ΣX^2
c. $\Sigma(X + 3)$

X
3
-2
0
-1
-4

25. For the following set of scores, find the value of each expression:

a. ΣX
b. ΣY
c. ΣXY

X	Y
-2	4
0	5
3	2
-4	3

26. For the following set of scores, find the value of each expression:

a. ΣX^2
b. $(\Sigma X)^2$
c. $\Sigma(X - 5)$
d. $\Sigma(X - 5)^2$

X
7
4
10
3
1

CHAPTER 2

FREQUENCY DISTRIBUTIONS

TOOLS YOU WILL NEED

The following items are considered essential background material for this chapter. If you doubt your knowledge of any of these items, you should review the appropriate chapter or section before proceeding.

- Proportions (math review, Appendix A)
  - Fractions
  - Decimals
  - Percentages
- Scales of measurement (Chapter 1): nominal, ordinal, interval, and ratio
- Continuous and discrete variables (Chapter 1)

CONTENTS

2.1 INTRODUCTION

When a researcher finishes the data collection phase of an experiment, the results usually consist of pages of numbers. The immediate problem for the researcher is to organize the scores into some comprehensible form so that any trends in the data can be seen easily and communicated to others. This is the job of descriptive statistics: to simplify the organization and presentation of data. One of the most common procedures for organizing a set of data is to place the scores in a frequency distribution.

DEFINITION   A *frequency distribution* is an organized tabulation of the number of individual scores located in each category on the scale of measurement.

A frequency distribution takes a disorganized set of scores and places them in order from highest to lowest, grouping together all individual scores that are the same. If the highest score is $X = 10$, for example, the frequency distribution groups together all the 10s, then all the 9s, then the 8s, and so on. Thus, a frequency distribution allows the researcher to see "at a glance" the entire set of scores. It shows whether the scores are generally high or low and whether they are concentrated in one area or spread out across the entire scale and generally provides an organized picture of the data. In addition to providing a picture of the entire set of scores, a frequency distribution allows the researcher to see the location of any individual score relative to all of the other scores in the set.

A frequency distribution can be structured either as a table or as a graph, but in either case, the distribution presents the same two elements:

1. The set of categories that make up the original measurement scale
2. A record of the frequency, or number of individuals in each category

Thus, a frequency distribution presents a picture of how the individual scores are distributed on the measurement scale—hence the name *frequency distribution*.

2.2 FREQUENCY DISTRIBUTION TABLES

It is customary to list scores from highest to lowest, but this is an arbitrary arrangement. Many computer programs will list scores from lowest to highest.

The simplest frequency distribution table presents the measurement scale by listing the measurement categories (X values) in a column from highest to lowest. Beside each X value, we indicate the frequency, or the number of times the score occurred in the data. It is customary to use an X as the column heading for the scores and an f as the column heading for the frequencies. An example of a frequency distribution table follows.

EXAMPLE 2.1   The following set of $N = 20$ scores was obtained from a 10-point statistics quiz. We will organize these scores by constructing a frequency distribution table. Scores:

8, 9, 8, 7, 10, 9, 6, 4, 9, 8

7, 8, 10, 9, 8, 6, 9, 7, 8, 8

X	f
10	2
9	5
8	7
7	3
6	2
5	0
4	1

1. The highest score is $X = 10$, and the lowest score is $X = 4$. Therefore, the first column of the table will list the scale of measurement (X values) from 10 down to 4. Notice that all of the possible values are listed in the table. For example, no one had a score of $X = 5$, but this value is included.

2. The frequency associated with each score is recorded in the second column. For example, two people had scores of $X = 6$, so there is a 2 in the f column beside $X = 6$.

Because the table organizes the scores, it is possible to see very quickly the general quiz results. For example, there were only two perfect scores, but most of the class had high grades (8s and 9s). With one exception (the score of $X = 4$), it appears that the class has learned the material fairly well.

Notice that the X values in a frequency distribution table represent the scale of measurement, *not* the actual set of scores. For example, the X column shows only one value for $X = 10$, but the frequency column indicates that there are actually two values of $X = 10$. Also, the X column shows a value of $X = 5$, but the frequency column indicates that there are actually no scores of $X = 5$.

You also should notice that the frequencies can be used to find the total number of scores in the distribution. By adding up the frequencies, you will obtain the total number of individuals:

$$\Sigma f = N$$

OBTAINING ΣX FROM A FREQUENCY DISTRIBUTION TABLE

There may be times when you need to compute the sum of the scores, ΣX, or perform other computations for a set of scores that has been organized into a frequency distribution table. Such calculations can present a problem because many students tend to disregard the frequencies and use only the values listed in the X column of the table. However, this practice is incorrect because it ignores the information provided by the frequency (f) column. To calculate ΣX from a frequency distribution table, you must use both the X and the f columns.

Consider the frequency distribution table for Example 2.1. It tells us that the distribution has two 10s, five 9s, seven 8s, and so on. Therefore, you can obtain the total for X by first reconstructing the original distribution and then adding all the X values (compute ΣX): $10 + 10 + 9 + 9 + 9 + 9 + 9 + 8 + 8 + 8 + 8 + 8 + 8 + 8 + \ldots$ For the data in Example 2.1, $\Sigma X = 158$. Try it yourself.

An alternative way to get ΣX from a frequency distribution table is to multiply each X value by its frequency and then add these products. This sum may be expressed in symbols as ΣfX. The computation is summarized as follows for the data in Example 2.1:

X	f	f X	
10	2	20	(the two 10s total 20)
9	5	45	(the five 9s total 45)
8	7	56	(the seven 8s total 56)
7	3	21	(the three 7s total 21)
6	2	12	(the two 6s total 12)
5	0	0	(there are no 5s)
4	1	4	(the one 4 totals 4)

$$\Sigma fX = 158$$

In using either method to find ΣX, by reconstructing the distribution or by computing ΣfX, the important point is that you must use the information given in the frequency column.

PROPORTIONS AND PERCENTAGES

In addition to the two basic columns of a frequency distribution, there are other measures that describe the distribution of scores and can be incorporated into the table. The two most common are proportion and percentage.

Proportion measures the fraction of the total group that is associated with each score. In Example 2.1, there were two individuals with $X = 6$. Thus, 2 out of 20 people had $X = 6$, so the proportion would be $2/20 = 0.10$. In general, the proportion associated with each score is

$$\text{proportion} = p = \frac{f}{N}$$

Because proportions describe the frequency (f) in relation to the total number (N), they often are called *relative frequencies*. Although proportions can be expressed as fractions (for example, 2/20), they more commonly appear as decimals. A column of proportions, headed with a p, can be added to the basic frequency distribution table (see Example 2.2).

In addition to using frequencies (f) and proportions (p), researchers often describe a distribution of scores with percentages. For example, an instructor might describe the results of an exam by saying that 15% of the class earned *A*s, 23% *B*s, and so on. To compute the percentage associated with each score, you first find the proportion (p) and then multiply by 100:

$$\text{percentage} = p(100) = \frac{f}{N}(100)$$

Percentages can be included in a frequency distribution table by adding a column headed with % (see Example 2.2).

EXAMPLE 2.2

The frequency distribution table from Example 2.1 is repeated here. This time we have added columns showing the proportion (p) and the percentage (%) associated with each score.

X	f	$p = f/N$	$\% = p(100)$
10	2	$2/20 = 0.10$	10%
9	5	$5/20 = 0.25$	25%
8	7	$7/20 = 0.35$	35%
7	3	$3/20 = 0.15$	15%
6	2	$2/20 = 0.10$	10%
5	0	$0/20 = 0$	0%
4	1	$1/20 = 0.05$	5%

GROUPED FREQUENCY DISTRIBUTION TABLES

When the scores are whole numbers, the total number of rows for a regular table can be obtained by finding the difference between the highest and the lowest scores and adding 1:

$$\text{rows} = \text{highest} - \text{lowest} + 1$$

When a set of data covers a wide range of values, it is unreasonable to list all the individual scores in a frequency distribution table. For example, a set of exam scores ranges from a low of $X = 41$ to a high of $X = 96$. These scores cover a range of over 50 points.

If we were to list all the individual scores, it would take 56 rows to complete the frequency distribution table. Although this would organize and simplify the data, the table would be long and cumbersome. Additional simplification would be desirable. This is accomplished by dividing the range of scores into intervals and then listing these intervals in the frequency distribution table. For example, we could construct a table showing the number of students who had scores in the 90s, the number with scores in the 80s, and so on. The result is called a *grouped frequency distribution table* because we are presenting groups of scores, rather than individual values. The groups, or intervals, are called *class intervals.*

There are several rules that help guide you in the construction of a grouped frequency distribution table. These rules should be considered as guidelines, rather than absolute requirements, but they do help produce a simple, well-organized, and easily understood table.

RULE 1 The grouped frequency distribution table should have about 10 class intervals. If a table has many more than 10 intervals, it becomes cumbersome and defeats the purpose of a frequency distribution table. On the other hand, if you have too few intervals, you begin to lose information about the distribution of the scores. At the extreme, with only 1 interval, the table would not tell you anything about how the scores are distributed. Remember, the purpose of a frequency distribution is to help you see the data. With too few or too many intervals, the table will not provide a clear picture. You should note that 10 intervals is a general guide. If you are constructing a table on a blackboard, for example, you probably will want only 5 or 6 intervals. If the table is to be printed in a scientific report, you may want 12 or 15 intervals. In each case, your goal is to present a table that is easy to see and understand.

RULE 2 The width of each interval should be a relatively simple number. For example, 2, 5, 10, or 20 would be a good choice for the interval width. Notice that it is easy to count by 5s or 10s. These numbers are easy to understand and make it possible for someone to see quickly how you have divided the range.

RULE 3 Each class interval should start with a score that is a multiple of the width. If you decide to use an interval width of 10 points, for example, you should use a multiple of 10 as the lowest score in each interval. Thus, the bottom interval might start with 30, and successive intervals would start with 40, 50, and so on as you move up the scale. The goal is to make it easier for someone to understand how the table was constructed.

RULE 4 All intervals should be the same width. They should cover the range of scores completely, with no gaps and no overlaps, so that any particular score belongs in exactly 1 interval.

The application of these rules is demonstrated in Example 2.3.

EXAMPLE 2.3   An instructor has obtained the set of $N = 25$ exam scores shown here. To help organize these scores, we will place them in a frequency distribution table. Scores:

$$82, \quad 75, \quad 88, \quad 93, \quad 53, \quad 84, \quad 87, \quad 58, \quad 72, \quad 94, \quad 69, \quad 84, \quad 61$$

$$91, \quad 64, \quad 87, \quad 84, \quad 70, \quad 76, \quad 89, \quad 75, \quad 80, \quad 73, \quad 78, \quad 60$$

The first step is to examine the range of scores. For these data, the smallest score is $X = 53$, and the largest score is $X = 94$, so 42 rows would be needed for a table. Because it would require 42 rows to list each X value in a frequency distribution table, we will have to group the scores into class intervals.

The best method for finding a good interval width is a trial-and-error approach that uses rules 1 and 2 simultaneously. According to rule 1, we want about 10 intervals; according to rule 2, we want the interval width to be a simple number. For this example, the scores cover a range of 42 points, so we will try several different interval widths to see how many intervals are needed to cover this range. For example, if we try a width of 2, how many intervals would it take to cover the range of scores? With each interval only 2 points wide, we would need 21 intervals to cover the range. This is too many. What about an interval width of 5? What about a width of 10? The following table shows how many intervals would be needed for these possible widths:

To find the number of intervals, divide the range by the width and round up any fractions.

Width	Number of intervals needed to cover a range of 42 values	
2	21	(too many)
5	9	(OK)
10	5	(too few)

Notice that an interval width of 5 will result in about 10 intervals, which is exactly what we want.

The next step is to actually identify the intervals. In this example, we have decided to use an interval width of 5 points. According to rule 3, each of the class intervals should start with a multiple of 5. Because the lowest score for these data is $X = 53$, the bottom interval would start at 50 and go to 54. Notice that the lowest score, $X = 50$, is a multiple of 5. Also notice that the interval contains five values (50, 51, 52, 53, 54), so it does have a width of 5 points. The next interval would start at 55 (also a multiple of 5) and go to 59. The complete frequency distribution table showing all of the class intervals is presented in Table 2.1.

Once the class intervals are listed, you complete the table by adding a column of frequencies. The values in the frequency column indicate the number of individuals whose scores are located in that class interval. For this example, there were three students with scores in the 60–64 interval, so the frequency for this class interval is $f = 3$ (see Table 2.1).

TABLE 2.1

A grouped frequency distribution table showing the data from Example 2.3. The original scores range from a high of $X = 94$ to a low of $X = 53$. This range has been divided into 9 intervals, with each interval exactly 5 points wide. The frequency column (f) lists the number of individuals with scores in each of the class intervals.

X	f
90–94	3
85–89	4
80–84	5
75–79	4
70–74	3
65–69	1
60–64	3
55–59	1
50–54	1

CONTINUOUS VARIABLES AND REAL LIMITS

You should recall that a continuous variable has an infinite number of possible values. Therefore, it may be represented by a number line that is continuous and contains an infinite number of points. However, when a continuous variable is measured, we typically do not assign our observation to a single point on the line. In practice, we assign the observation to an *interval* on the number line. For example, if you are measuring body weight to the nearest pound, weights of $X = 150.3$ and $X = 149.6$ pounds would both be assigned a value of $X = 150$ (Figure 2.1). Note that a score of $X = 150$ is not a single point on the number line but instead represents an interval on the line.

To differentiate scores of 149, 150, and 151 we must use boundaries that are located exactly halfway between adjacent scores. Thus, a score of 150 actually corresponds to an interval bounded by 149.5 and 150.5. Any measurement that falls within this interval will be assigned a value of $X = 150$. The boundaries that form the interval are called the *real limits* of the interval. For this example, 149.5 is the *lower real limit,* and 150.5 is the *upper real limit* of the interval that corresponds to $X = 150$.* Thus, when a frequency distribution table shows that five people all have a weight of $X = 150$ pounds, you should realize that the five people do not necessarily have *exactly* the same weight; instead, they are all located in the same interval between 149.5 pounds and 150.5 pounds.

DEFINITION

Real limits are the boundaries of intervals for scores that are represented on a continuous number line. The real limit separating two adjacent scores is located exactly halfway between the scores. Each score has two real limits. The *upper real limit* is at the top of the interval, and the *lower real limit* is at the bottom.

Technical Note: It is important to distinguish between the *real limits* for an interval and the process of rounding scores. The real limits form boundaries that define an interval on a continuous scale. For example, the real limits 149.5 and 150.5 define an interval that is identified by the score $X = 150$. The real limits, however, are not necessarily a part of the interval. In this example, 149.5 is the lower real limit of the interval. If you are using a rounding process that causes 149.5 to be rounded up, then a measurement of 149.5 would be rounded to 150 and would be included in the 149.5–150.5 interval. On the other hand, if your rounding process causes 149.5 to be rounded down to 149, then this value would not be included in the interval. In general, the question of whether an upper real limit or a lower real limit belongs in an interval is determined by the rounding process that you have adopted.

FIGURE 2.1

When measuring weight to the nearest whole pound, 149.6 and 150.3 are assigned the value of 150 (top). Any value in the interval between 149.5 and 150.5 will be given the value of 150.

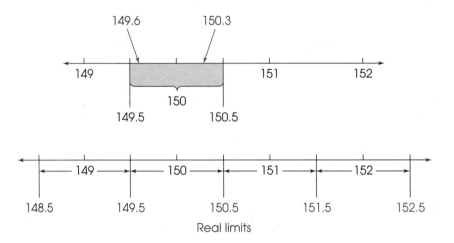

In Figure 2.1, note that neighboring X values share a real limit. For example, the score of $X = 150$ has an upper real limit of 150.5. The value of 150.5 is also the lower real limit for $X = 151$. Thus, on a continuous number line, such as Figure 2.1, there are no gaps between adjacent intervals.

The concept of real limits applies to any measurement of a continuous variable, even when the score categories are not whole numbers. For example, if you were measuring weight to the nearest 1/10 of a pound, the measurement categories would be 150.0, 150.1, 150.2, and so on. Each of these categories represents an interval on the scale that is bounded by real limits. For example, a score of $X = 150.1$ indicates that the actual weight is in an interval bounded by a lower real limit of 150.05 and an upper real limit of 150.15. Remember, the real limits are always halfway between adjacent categories.

Finally, you should realize that the concept of real limits also applies to the class intervals of a grouped frequency distribution table. For example, a class interval of 40–49 contains scores from $X = 40$ to $X = 49$. These values are called the *apparent limits* of the interval because it appears that they form the upper and lower boundaries for the class interval. But $X = 40$ is actually an interval from 39.5 to 40.5. Similarly, $X = 49$ is an interval from 48.5 to 49.5. Therefore, the real limits of the interval are 39.5 (the lower real limit) and 49.5 (the upper real limit). Notice that the next higher class interval would be 50–59, which has a lower real limit of 49.5. Thus, the two intervals meet at the real limit 49.5, so there are no gaps in the scale. You also should notice that the width of each class interval becomes easier to understand when you consider the real limits of an interval. For example, the interval 50–54 has real limits of 49.5 and 54.5. The distance between these two real limits (5 points) is the width of the interval.

Real limits will be used later for constructing graphs and for various calculations with continuous scales. For now, however, you should realize that real limits are a necessity whenever measurements are made of a continuous variable.

LEARNING CHECK   **1.** Place the following scores in a frequency distribution table. Scores:

2, 3, 1, 2, 5, 4, 5, 5, 1, 4, 2, 2

2. A set of scores ranges from a high of $X = 142$ to a low of $X = 65$.

    a. Explain why it would not be reasonable to display these scores in a *regular* frequency distribution table.

    b. Determine what interval width would be most appropriate for a grouped frequency distribution for this set of scores.

    c. For the grouped table, what are the apparent limits for the bottom interval, and what are the real limits for the bottom interval?

3. Find the values for N and ΣX for the population of scores in the following frequency distribution table:

X	f
5	1
4	3
3	5
2	2
1	2

ANSWERS **1.**

X	f
5	3
4	2
3	1
2	4
1	2

2. a. It would require a table with 78 rows to list all the individual score values. This is too many rows.

    b. An interval width of 10 points would probably be best. With a width of 10, it would require 8 intervals to cover the range of scores. In some circumstances, an interval width of 5 points (requiring 16 intervals) might be appropriate.

    c. With a width of 10, the apparent limits of the bottom interval would be 60–69. The real limits would be 59.5–69.5.

3. There are $N = 13$ scores summarized in the table ($\Sigma f = 13$). The 13 scores sum to $\Sigma X = 38$.

2.3 FREQUENCY DISTRIBUTION GRAPHS

A frequency distribution graph is basically a picture of the information available in a frequency distribution table. We will consider several different types of graphs, but all start with two perpendicular lines called axes. The horizontal line is called the X-axis, or the abscissa. The vertical line is called the Y-axis, or the ordinate. The scale of measurement (X values) is listed along the X-axis in increasing value from left to right. The frequencies are listed on the Y-axis in increasing value from bottom to top. As a general rule, the point where the two axes intersect should have a value

FIGURE 2.2

An example of a frequency distribution histogram. The same set of data is presented in a frequency distribution table and in a histogram.

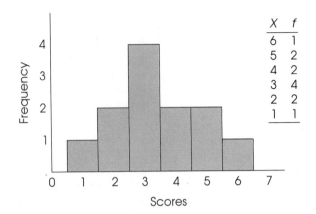

X	f
6	1
5	2
4	2
3	4
2	2
1	1

of zero for both the scores and the frequencies. A final general rule is that the graph should be constructed so that its height (Y-axis) is approximately two-thirds to three-quarters of its length (X-axis). Violating these guidelines can result in graphs that give a misleading picture of the data (see Box 2.1).

HISTOGRAMS AND BAR GRAPHS

The first type of graph we will consider is called either a histogram or a bar graph. For this type of graph, you simply draw a bar above each X value so that the height of the bar corresponds to the frequency of the score. As you will see, the choice of using a histogram or a bar graph is determined by the scale of measurement.

Histograms   When a frequency distribution graph is showing data from an interval or a ratio scale, the bars are drawn so that adjacent bars touch each other. The touching bars produce a continuous figure, which emphasizes the continuity of the variable. This type of frequency distribution graph is called a histogram. An example of a histogram is presented in Figure 2.2.

DEFINITION

For a *histogram,* a vertical bar is drawn above each score so that

1. The height of the bar corresponds to the frequency.
2. The width of the bar extends to the real limits of the score.

A histogram is used when the data are measured on an interval or a ratio scale.

When data have been grouped into class intervals, you can construct a frequency distribution histogram by drawing a bar above each interval so that the width of the bar extends to the real limits of the interval. This process is demonstrated in Figure 2.3.

Bar graphs   When you are presenting the frequency distribution for data from a nominal or an ordinal scale, the graph is constructed so that there is some space between the bars. In this case, the separate bars emphasize that the scale consists of separate, distinct categories. The resulting graph is called a bar graph. An example of a frequency distribution bar graph is given in Figure 2.4.

FIGURE 2.3

An example of a frequency distribution histogram for grouped data. The same set of data is presented in a grouped frequency distribution table and in a histogram.

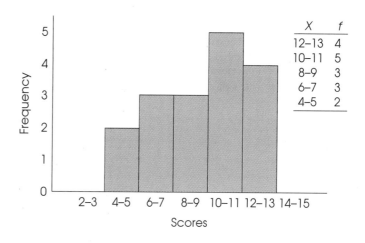

X	f
12–13	4
10–11	5
8–9	3
6–7	3
4–5	2

DEFINITION

For a *bar graph*, a vertical bar is drawn above each score (or category) so that

1. The height of the bar corresponds to the frequency.
2. There is a space separating each bar from the next.

A bar graph is used when the data are measured on a nominal or an ordinal scale.

FREQUENCY DISTRIBUTION POLYGONS

Instead of a histogram, many researchers prefer to display a frequency distribution using a polygon.

DEFINITION

In a *frequency distribution polygon*, a single dot is drawn above each score so that

1. The dot is centered above the score.
2. The vertical location (height) of the dot corresponds to the frequency.

A continuous line is then drawn connecting these dots. The graph is completed by drawing a line down to the *X*-axis (zero frequency) at a point just beyond each end of the range of scores.

FIGURE 2.4

A bar graph showing the distribution of personality types in a sample of college students. Because personality type is a discrete variable measured on a nominal scale, the graph is drawn with space between the bars.

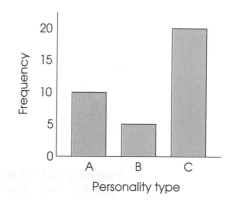

FIGURE 2.5

An example of a frequency distribution polygon. The same set of data is presented in a frequency distribution table and in a polygon. Note that these data are shown in a histogram in Figure 2.2.

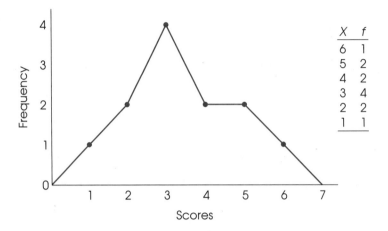

X	f
6	1
5	2
4	2
3	4
2	2
1	1

As with a histogram, the frequency distribution polygon is intended for use with interval or ratio scales. An example of a polygon is shown in Figure 2.5. A polygon also can be used with data that have been grouped into class intervals. In this case, you position the dots directly above the midpoint of each class interval. The midpoint can be found by averaging the apparent limits of the interval or by averaging the real limits of the interval. For example, a class interval of 40–49 would have a midpoint of 44.5.

$$\text{apparent limits: } \frac{40 + 49}{2} = \frac{89}{2} = 44.5$$

$$\text{real limits: } \frac{39.5 + 49.5}{2} = \frac{89}{2} = 44.5$$

An example of a frequency distribution polygon with grouped data is shown in Figure 2.6.

RELATIVE FREQUENCIES AND SMOOTH CURVES

Often it is impossible to construct a frequency distribution for a population because there are simply too many individuals for a researcher to obtain measurements and frequencies for the entire group. In this case, it is customary to draw a frequency distribution graph, showing *relative frequencies* (proportions) on the vertical axis. For example, a researcher may know that a particular species of animal has three times as many females as males in the population. This fact could be displayed in a bar graph by simply making the bar above "female" three times as tall as the bar above "male." Notice that the actual frequencies are unknown but that the relative frequencies of males and females can still be presented in a graph.

It also is possible to use a polygon to show relative frequencies for scores in a population. In this case, it is customary to draw a smooth curve instead of the series of straight lines that normally appears in a polygon. The smooth curve indicates that you are not connecting a series of dots (real frequencies) but rather are showing a distribution that is not limited to one specific set of data. One commonly occurring population distribution is the normal curve. The word *normal* refers to a specific shape that can be precisely defined by an equation. Less precisely, we can describe

FIGURE 2.6

An example of a frequency distribution polygon for grouped data. The same set of data is presented in a grouped frequency distribution table and in a polygon. Note that these data are shown in a histogram in Figure 2.3.

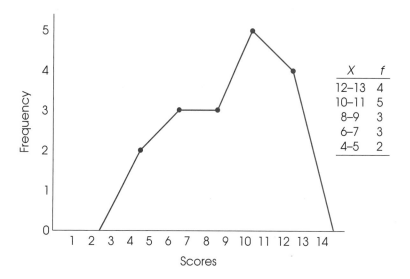

X	f
12–13	4
10–11	5
8–9	3
6–7	3
4–5	2

a normal distribution as being symmetrical, with the greatest frequency in the middle and relatively smaller frequencies as you move toward either extreme. A good example of a normal distribution is the population distribution for IQ scores shown in Figure 2.7. Because normal-shaped distributions occur commonly and because this shape is mathematically guaranteed in certain situations, it will receive extensive attention throughout this book.

In the future, we will be referring to *distributions of scores.* Whenever the term *distribution* appears, you should conjure up an image of a frequency distribution graph. The graph provides a picture showing exactly where the individual scores are located. To make this concept more concrete, you might find it useful to think of the graph as showing a pile of individuals. In Figure 2.7, for example, the pile is highest at an IQ score of around 100 because most people have "average" IQs. There are only a few individuals piled up with IQ scores above 130; it must be lonely at the top.

FIGURE 2.7

The population distribution of IQ scores: an example of a normal distribution.

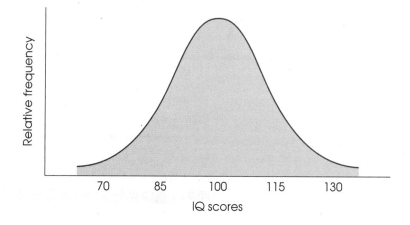

2.1   THE USE AND MISUSE OF GRAPHS

ALTHOUGH GRAPHS are intended to provide an accurate picture of a set of data, they can be used to exaggerate or misrepresent a set of scores. These misrepresentations generally result from failing to follow the basic rules for graph construction. The following example demonstrates how the same set of data can be presented in two entirely different ways by manipulating the structure of a graph.

For the past several years, the city has kept records of the number of major felonies. The data are summarized as follows:

Year	Number of major felonies
1990	218
1991	225
1992	229

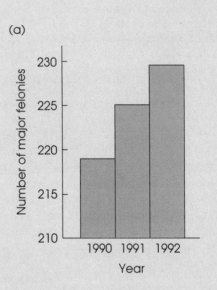

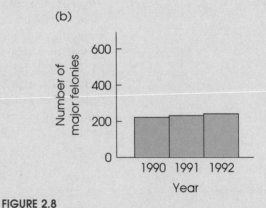

These same data are shown in two different graphs in Figure 2.8. In the first graph, we have exaggerated the height, and we started numbering the Y-axis at 210 rather than at zero. As a result, the graph seems to indicate a rapid rise in the crime rate over the three-year period. In the second graph, we have stretched out the X-axis and used zero as the starting point for the Y-axis. The result is a graph that shows little change in the crime rate over the three-year period.

Which graph is correct? The answer is that neither one is very good. Remember that the purpose of a graph is to provide an accurate display of the data. Figure 2.8(a) exaggerates the differences between years, while the graph in Figure 2.8(b) conceals the differences. Some compromise is needed. You also should note that in some cases a graph may not be the best way to display information. For these data, for example, showing the numbers in a table would be better than either graph.

FIGURE 2.8

Two graphs showing the number of major felonies in a city over a 3-year period. Both graphs are showing exactly the same data. However, the first graph gives the appearance that the crime rate is high and rising rapidly. The second graph gives the impression that the crime rate is low and has not changed over the 3-year period.

2.4   THE SHAPE OF A FREQUENCY DISTRIBUTION

Rather than drawing a complete frequency distribution graph, researchers often simply describe a distribution by listing its characteristics. There are three characteristics that completely describe any distribution: shape, central tendency, and variability. In simple terms, central tendency measures where the center of the distribution is located. Variability tells whether the scores are spread over a wide range or are clustered

FIGURE 2.9

Examples of different shapes for distributions.

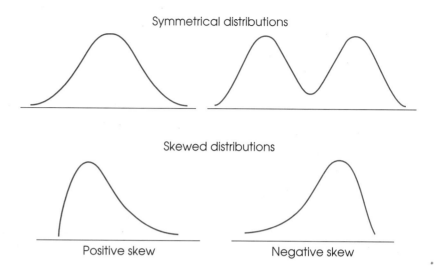

together. Central tendency and variability will be covered in detail in Chapters 3 and 4. Technically, the shape of a distribution is defined by an equation that prescribes the exact relationship between each X and Y value on the graph. However, we will rely on a few less-precise terms that will serve to describe the shape of most distributions.

Nearly all distributions can be classified as being either symmetrical or skewed.

DEFINITIONS

In a *symmetrical distribution,* it is possible to draw a vertical line through the middle so that one side of the distribution is a mirror image of the other (see Figure 2.9).

In a *skewed distribution,* the scores tend to pile up toward one end of the scale and taper off gradually at the other end (see Figure 2.9).

The section where the scores taper off toward one end of a distribution is called the *tail* of the distribution.

A skewed distribution with the tail on the right-hand side is said to be *positively skewed* because the tail points toward the positive (above-zero) end of the X-axis. If the tail points to the left, the distribution is said to be *negatively skewed* (see Figure 2.9).

For a very difficult exam, most scores will tend to be low, with only a few individuals earning high scores. This will produce a positively skewed distribution. Similarly, a very easy exam will tend to produce a negatively skewed distribution, with most of the students earning high scores and only a few with low values.

LEARNING CHECK

1. Sketch a frequency distribution histogram and a frequency distribution polygon for the data in the following table:

X	f
5	4
4	6
3	3
2	1
1	1

FIGURE 2.10

Answers to Learning Check exercise 1.

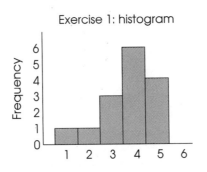

Exercise 1: histogram

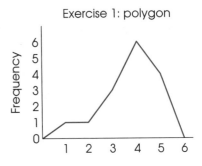

Exercise 1: polygon

2. Describe the shape of the distribution in exercise 1.

3. What type of graph would be appropriate to show the number of gold medals, silver medals, and bronze medals won by the United States during the 1998 Olympics?

4. What shape would you expect for the distribution of salaries for all employees of a major industry?

ANSWERS
1. The graphs are shown in Figure 2.10.

2. The distribution is negatively skewed.

3. A bar graph is appropriate for ordinal data.

4. The distribution probably would be positively skewed, with most employees earning an average salary and a relatively small number of top executives with very large salaries.

SUMMARY

1. The goal of descriptive statistics is to simplify the organization and presentation of data. One descriptive technique is to place the data in a frequency distribution table or graph that shows exactly how many individuals (or scores) are located in each category on the scale of measurement.

2. A frequency distribution table lists the categories that make up the scale of measurement (the X values) in one column. Beside each X value, in a second column, is the frequency or number of individuals in that category. The table may include a proportion column showing the relative frequency for each category:

$$\text{proportion} = p = \frac{f}{n}$$

And the table may include a percentage column showing the percentage associated with each X value:

$$\text{percentage} = p(100) = \frac{f}{n}(100)$$

3. It is recommended that a frequency distribution table have a maximum of 10 to 15 rows to keep it simple. If the scores cover a range that is wider than this suggested maximum, it is customary to divide the range into sections called class intervals. These intervals are then listed in the frequency distribution table along with the frequency or number of individuals with scores in each interval. The result is called a grouped frequency distribution. The guidelines for constructing a grouped frequency distribution table are as follows:
   a. There should be about 10 intervals.
   b. The width of each interval should be a simple number (e.g., 2, 5, or 10).
   c. The bottom score in each interval should be a multiple of the width.
   d. All intervals should be the same width, and they should cover the range of scores with no gaps.

4. For a continuous variable, each score corresponds to an interval on the scale. The boundaries that separate intervals are called real limits. The real limits are located exactly halfway between adjacent scores.

5. A frequency distribution graph lists scores on the horizontal axis and frequencies on the vertical axis. The type of graph used to display a distribution depends on the scale of measurement used. For interval or ratio scales, you

should use a histogram or a polygon. For a histogram, a bar is drawn above each score so that the height of the bar corresponds to the frequency. Each bar extends to the real limits of the score, so that adjacent bars touch. For a polygon, a dot is placed above the midpoint of each score or class interval so that the height of the dot corresponds to the frequency; then lines are drawn to connect the dots. Bar graphs are used with nominal or ordinal scales. Bar graphs are similar to histograms except that gaps are left between adjacent bars.

6. Shape is one of the basic characteristics used to describe a distribution of scores. Most distributions can be classified as either symmetrical or skewed. A skewed distribution that tails off to the right is said to be positively skewed. If it tails off to the left, it is negatively skewed.

KEY TERMS

frequency distribution	upper real limit	bar graph	positively skewed distribution
grouped frequency distribution	lower real limit	polygon	
class interval	apparent limit	relative frequency	negatively skewed distribution
real limit	histogram	symmetrical distribution	

FOCUS ON PROBLEM SOLVING

1. The reason for constructing frequency distributions is to put a disorganized set of raw data into a comprehensible, organized format. Because several different types of frequency distribution tables and graphs are available, one problem is deciding which type should be used. Tables have the advantage of being easier to construct, but graphs generally give a better picture of the data and are easier to understand.

   To help you decide exactly which type of frequency distribution is best, consider the following points:
   a. What is the range of scores? With a wide range, you will need to group the scores into class intervals.
   b. What is the scale of measurement? With an interval or a ratio scale, you can use a polygon or a histogram. With a nominal or an ordinal scale, you must use a bar graph.

2. In setting up class intervals, a common mistake is to determine the interval width by finding the difference between the apparent limits. This is incorrect! To determine the interval width, you must find the difference between the *real limits*. For example, the width for the interval 70–79 is 10 because its real limits are 69.5 and 79.5, and

$$\text{width} = \text{upper real limit} - \text{lower real limit}$$
$$= 79.5 - 69.5 = 10$$

Resist the temptation to state the width is 9. You must use the real limits.

DEMONSTRATION 2.1

A GROUPED FREQUENCY DISTRIBUTION TABLE

For the following set of $N = 20$ scores, construct a grouped frequency distribution table. Use an interval width of 5 points, and include columns for f and p. Scores:

14, 8, 27, 16, 10, 22, 9, 13, 16, 12

10, 9, 15, 17, 6, 14, 11, 18, 14, 11

X
25–29
20–24
15–19
10–14
5–9

STEP 1 Set up the class intervals.
The largest score in this distribution is $X = 27$, and the lowest is $X = 6$. Therefore, a frequency distribution table for these data would have 22 rows and would be too large. A grouped frequency distribution table would be better. We have asked specifically for an interval width of 5 points, and the resulting table (in the margin) will have five rows. Remember, the interval width is determined by the real limits of the interval. For example, the class interval 25–29 has an upper real limit of 29.5 and a lower real limit of 24.5. The difference between these two values is the width of the interval, namely, 5.

STEP 2 Determine the frequencies for each interval.
Examine the scores, and count how many fall into the class interval of 25–29. Cross out each score that you have already counted. Record the frequency for this class interval. Now repeat this process for the remaining intervals. The result is the following table:

X	f	
25–29	1	(the score $X = 27$)
20–24	1	($X = 22$)
15–19	5	(the scores $X = 16, 16, 15, 17,$ and 18)
10–14	9	($X = 14, 10, 13, 12, 10, 14, 11, 14,$ and 11)
5–9	4	($X = 8, 9, 9,$ and 6)

STEP 3 Compute the proportions.
The proportion (p) of scores contained in an interval is determined by dividing the frequency (f) of that interval by the number of scores (N) in the distribution. Thus, for each interval, we must compute the following:

$$p = \frac{f}{N}$$

This is demonstrated in the following table:

X	f	p
25–29	1	$f/N = 1/20 = 0.05$
20–24	1	$f/N = 1/20 = 0.05$
15–19	5	$f/N = 5/20 = 0.25$
10–14	9	$f/N = 9/20 = 0.45$
5–9	4	$f/N = 4/20 = 0.20$

PROBLEMS

1. An instructor obtained the following set of quiz scores for a class of 20 students:

3, 1, 1, 2, 5, 4, 4, 5, 3, 3

2, 3, 4, 3, 3, 4, 3, 2, 5, 3

a. Place the scores in a frequency distribution table including columns for proportion and percentage.
b. Using your table, provide the following information about the class:
   (1) Identify the highest and lowest scores for the class.

(2) What score was obtained by more students than any other?

(3) If a score of $X = 2$ or lower was considered failing, how many students failed the quiz? What percentage of the class failed?

2. Under what circumstances should you use a bar graph instead of a histogram to display a frequency distribution?

3. Under what circumstances should you use a grouped frequency distribution instead of a regular frequency distribution?

4. The following are reading comprehension scores for a third-grade class of 18 students:

5, 3, 5, 4, 5, 5, 4, 5, 2

4, 5, 3, 5, 4, 5, 5, 3, 5

a. Place the scores in a frequency distribution table.
b. Sketch a histogram showing the distribution.
c. Using your graph, provide the following information:
(1) Describe the shape of the distribution.
(2) If a score of $X = 3$ is considered normal for third-graders, how would you describe the general reading level for this class?

5. An instructor obtained the following set of scores from a 10-point quiz for a class of 26 students:

9, 2, 3, 8, 10, 9, 9, 2, 1,   2, 9, 8, 2

5, 2, 9, 9,   3, 2, 5, 7, 2, 10, 1, 2, 9

a. Place the scores in a frequency distribution table.
b. Sketch a histogram showing the distribution.
c. Using your graph, provide the following information:
(1) Describe the shape of the distribution.
(2) As a whole, how did the class do on the quiz? Were most scores high or low? Was the quiz easy or hard?

6. A set of scores has been organized into the following frequency distribution table. Find each of the following values for the original set of scores:
a. N
b. ΣX

X	f
5	4
4	3
3	3
2	0
1	2

7. Find each value requested for the set of scores summarized in the following table:
a. N
b. ΣX
c. ΣX^2

X	f
5	2
4	3
3	4
2	1
1	1

8. Three sets of data are described as follows:

Set I: $N = 40$, highest score is $X = 93$, lowest score is $X = 52$

Set II: $N = 62$, highest score is $X = 35$, lowest score is $X = 26$

Set III: $N = 25$, highest score is $X = 890$, lowest score is $X = 230$

For each set of data, identify whether a regular table or a grouped table should be used, and if a grouped table is necessary, identify the interval width that would be most appropriate.

9. Place the following set of scores in a grouped frequency distribution table using
a. An interval width of 2.
b. An interval width of 5.

23, 12, 16, 16, 17

20, 14, 21, 18, 24

18, 21, 22, 27, 21

22, 23, 21, 30, 27

10. For the following scores, construct a frequency distribution table using
a. An interval width of 5.
b. An interval width of 10.

64, 75, 50, 67, 86, 66, 62, 64, 71, 47

57, 74, 63, 67, 56, 65, 70, 87, 48, 50

41, 66, 73, 60, 63, 45, 78, 68, 53, 75

11. For each of the following sets of scores:

		Set I						Set II		
3	5	4	2	1		3	9	14	10	
3	3	2	3	4		6	4	9	1	4

a. Construct a frequency distribution table for each set of data.

b. Sketch a histogram showing each frequency distribution.

c. Describe each distribution using the following characteristics:

  (1) What is the shape of the distribution?

  (2) What score best identifies the center (average) for the distribution?

  (3) Are the scores clustered together around a central point, or are the scores spread out across the scale?

12. For the following set of scores:

5, 6, 2, 3, 6, 5, 6, 4, 1, 5, 6, 3, 4

a. Construct a frequency distribution table.

b. Sketch a polygon showing the distribution.

c. Describe the distribution using the following characteristics:

  (1) What is the shape of the distribution?

  (2) What score best identifies the center (average) for the distribution?

  (3) Are the scores clustered together, or are they spread out across the scale?

13. For the following set of scores:

4, 6, 9, 5, 3, 8, 9, 4, 2, 5, 10, 7, 4, 9, 8, 3

a. Construct a frequency distribution table.

b. Sketch a polygon showing the distribution.

c. Describe the distribution using the following characteristics:

  (1) What is the shape of the distribution?

  (2) What score best identifies the center (average) for the distribution?

  (3) Are the scores clustered together, or are they spread out across the scale?

14. For the following set of quiz scores:

3, 5, 4, 6, 2, 3, 4, 1, 4, 3

7, 7, 3, 4, 5, 8, 2, 4, 7, 10

a. Construct a frequency distribution table to organize the scores.

b. Draw a frequency distribution histogram for these data.

15. Construct a grouped frequency distribution table to organize the following set of scores:

206, 350, 590, 473, 450, 483

112, 380, 584, 620, 743, 816

685, 592, 712, 727, 686, 592

542, 490, 684, 491, 520, 380

16. A psychologist would like to examine the effects of diet on intelligence. Two groups of rats are selected, with 12 rats in each group. One group is fed the regular diet of Rat Chow, whereas the second group has special vitamins and minerals added to their food. After six months, each rat is tested on a discrimination problem. The psychologist records the number of errors each animal makes before it solves the problem. The data from this experiment are as follows:

regular diet scores: 13, 11, 12, 13, 11, 9
12, 10, 12, 14, 10, 12

special diet scores: 9, 8, 7, 8, 9, 10
7, 8, 9, 6, 8, 10

a. Identify the independent variable and the dependent variable for this experiment.

b. Sketch a frequency distribution polygon for the group of rats with the regular diet. On the same graph (in a different color), sketch the distribution for the rats with the special diet.

c. From looking at your graph, would you say that the special diet had any effect on intelligence? Explain your answer.

17. Sketch a frequency distribution histogram for the following data:

X	f
15	2
14	5
13	6
12	3
11	2
10	1

18. The following data are quiz scores from two different sections of an introductory statistics course:

Section I			Section II		
9	6	8	4	7	8
10	8	3	6	3	7
7	8	8	4	6	5
7	5	10	10	3	6
9	6	7	7	4	6

a. Organize the scores from each section in a frequency distribution histogram.
b. Describe the general differences between the two distributions.

19. Place the following set of scores in a frequency distribution table:

1, 3, 1, 1, 4, 1, 4, 5, 6, 2, 1, 1, 5

1, 3, 2, 1, 6, 2, 4, 5, 2, 3, 2, 3

Compute the proportion and the percentage of individuals with each score. From your frequency distribution table, you should be able to identify the shape of this distribution.

CHAPTER 3

CENTRAL TENDENCY

TOOLS YOU WILL NEED

The following items are considered essential background material for this chapter. If you doubt your knowledge of any of these items, you should review the appropriate chapter or section before proceeding.

- Summation notation (Chapter 1)
- Frequency distributions (Chapter 2)

CONTENTS

3.1 INTRODUCTION

The goal in measuring central tendency is to describe a group of individuals (more accurately, their scores) with a single measurement. Ideally, the value we use to describe the group will be the single value that is most representative of all the individuals.

DEFINITION

Central tendency is a statistical measure that identifies a single score as representative for an entire distribution. The goal of central tendency is to find the single score that is most typical or most representative of the entire group.

Usually we want to choose a value in the middle of the distribution because central scores are often the most representative. In everyday language, the goal of central tendency is to find the "average" or "typical" individual. This average value can then be used to provide a simple description of the entire population or sample. For example, archeological discoveries indicate that the average height for men in the ancient Roman city of Pompeii was 5 feet 7 inches. Obviously, not all the men were exactly 5 feet 7 inches, but this average value provides a general description of the population. Measures of central tendency also are useful for making comparisons between groups of individuals or between sets of figures. For example, suppose that weather data indicate that during the month of December, Seattle averages only 2 hours of sunshine per day, whereas Miami averages over 6 hours. The point of these examples is to demonstrate the great advantage of being able to describe a large set of data with a single, representative number. Central tendency characterizes what is typical for a large population and in doing so makes large amounts of data more digestable. Statisticians sometimes use the expression "number crunching" to illustrate this aspect of data description. That is, we take a distribution consisting of many scores and "crunch" them down to a single value that describes them all.

Unfortunately, there is no single standard procedure for determining central tendency. The problem is that no single measure will always produce a central, representative value in every situation. The three distributions shown in Figure 3.1 should help demonstrate this fact. Before we discuss the three distributions, take a moment to look at the figure, and try to identify the "center" or the "most representative score" for each distribution.

1. The first distribution [Figure 3.1(a)] is symmetrical, with the scores forming a distinct pile centered around $X = 5$. For this type of distribution, it is easy to identify the "center," and most people would agree that the value $X = 5$ is an appropriate measure of central tendency.

2. In the second distribution [Figure 3.1(b)], however, problems begin to appear. Now the scores form a negatively skewed distribution, piling up at the high end of the scale, around $X = 8$, but tapering off to the left all the way down to $X = 1$. Where is the "center" in this case? Some people might select $X = 8$ as the center because more individuals had this score than any other single value. However, $X = 8$ is clearly not in the middle of the distribution. In fact, the majority of the scores (10 out of 16) have values less

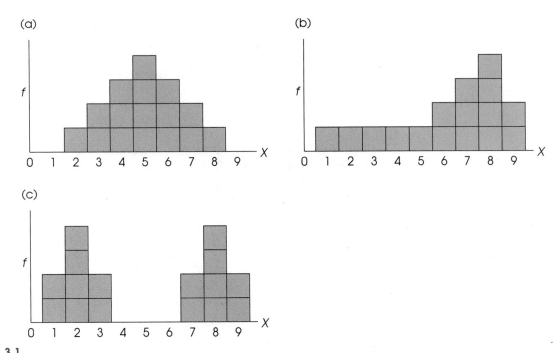

FIGURE 3.1

Three distributions demonstrating the difficulty of defining central tendency. In each case, try to locate the "center" of the distribution.

than 8, so it seems reasonable that the "center" should be defined by a value that is less than 8.

3. Now consider the third distribution [Figure 3.1(c)]. Again, the distribution is symmetrical, but now there are two distinct piles of scores. Because the distribution is symmetrical, with $X = 5$ as the midpoint, you may choose $X = 5$ as the "center." However, none of the scores is located at $X = 5$ (or even close), so this value is not particularly good as a representative score. On the other hand, because there are two separate piles of scores, with one group centered at $X = 2$ and the other centered at $X = 8$, it is tempting to say that this distribution has two centers. But can there be two centers?

Clearly, there are problems defining the "center" of a distribution. Occasionally, you will find a nice, neat distribution like the one shown in Figure 3.1(a), where everyone will agree on the center. But you should realize that other distributions are possible and that there may be different opinions concerning the definition of the center. To deal with these problems, statisticians have developed three different methods for measuring central tendency: the mean, the median, and the mode. They are computed differently and have different characteristics. To decide which of the three measures is best for any particular distribution, you should keep in mind that the general purpose of central tendency is to find the single most representative score. Each of the three measures we will present has been developed to work best in a specific situation. We will examine this issue in more detail after we define the three measures.

3.2 THE MEAN

The mean, commonly known as the arithmetic average, is computed by adding all the scores in the distribution and dividing by the number of scores. The mean for a population will be identified by the Greek letter mu, μ (pronounced "myoo"), and the mean for a sample will be identified by $\overline{X}$ (read "x-bar").

DEFINITION   The *mean* for a distribution is the sum of the scores divided by the number of scores.

The formula for the population mean is

$$\mu = \frac{\Sigma X}{N} \tag{3.1}$$

 First, sum all the scores in the population, and then divide by N. For a sample, the computation is done the same way, but the formula uses symbols that signify sample values:

$$\text{sample mean} = \overline{X} = \frac{\Sigma X}{n} \tag{3.2}$$

In general, we will use Greek letters to identify characteristics of a population and letters of our own alphabet to stand for sample values. If a mean is identified with the symbol $\overline{X}$, you should realize that we are dealing with a sample. Also note that n is used as the symbol for the number of scores in the sample.

EXAMPLE 3.1   For a population of $N = 4$ scores,

3,   7,   4,   6

the mean is

$$\mu = \frac{\Sigma X}{N} = \frac{20}{4} = 5$$

Although the procedure of adding the scores and dividing by the number provides a useful definition of the mean, there are two alternative definitions that may give you a better understanding of this important measure of central tendency.

The first alternative is to think of the mean as the amount each individual would get if the total (ΣX) were divided equally among all the individuals (N) in the distribution. This somewhat socialistic viewpoint is particularly useful in problems where you know the mean and must find the total. Consider the following example.

EXAMPLE 3.2   A group of six students decided to earn some extra money one weekend picking vegetables at a local farm. At the end of the weekend, the students discov-

FIGURE 3.2

The frequency distribution shown as a seesaw balanced at the mean.

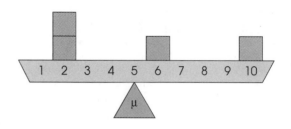

Based on Weinberg, G. H., Schumaker, J. A., & Oltman, D. (1981). *Statistics: An Intuitive Approach* (p. 14). Belmont: Wadsworth, Inc.

ered that their average income was $\mu = \$30$. If they decide to pool their money for a party, how much will they have?

You do not know how much money each student earned. But you do know that the mean is $30. This is the amount that each student would have if the total were divided equally. For each of six students to have $30, you must start with $6 \times \$30 = \$180$. The total, ΣX, is $180. To check this answer, use the formula for the mean:

$$\mu = \frac{\Sigma X}{N} = \frac{\$180}{6} = \$30$$

The second alternative definition of the mean is to describe it as a balance point for a distribution. Consider the population consisting of $N = 4$ scores (2, 2, 6, 10). For this population, $\Sigma X = 20$ and $N = 4$, so $\mu = \frac{20}{4} = 5$.

Imagine that the frequency distribution histogram for this population is drawn so that the X-axis, or number line, is a seesaw and the scores are boxes of equal weight that are placed on the seesaw (see Figure 3.2). If the seesaw is positioned so that it pivots at the value equal to the mean, it will be balanced and will rest level.

The reason the seesaw is balanced over the mean becomes clear when we measure the distance of each box (score) from the mean:

Score	Distance from the mean
$X = 2$	3 points below the mean
$X = 2$	3 points below the mean
$X = 6$	1 point above the mean
$X = 10$	5 points above the mean

Notice that the mean balances the distances. That is, the total distance below the mean is the same as the total distance above the mean:

3 points below + 3 points below = 6 points below

1 point above + 5 points above = 6 points above

THE WEIGHTED MEAN   Often it is necessary to combine two sets of scores and then find the overall mean for the combined group. For example, an instructor teaching two sections of introductory psychology obtains an average quiz score of $\overline{X} = 6$ for the 12 students in one section and an average of $\overline{X} = 7$ for the 8 students in the other section. If the two sections are combined, what is the mean for the total group?

The solution to this problem is straightforward if you remember the definition of the mean:

$$\overline{X} = \frac{\Sigma X}{n}$$

To find the overall mean, we must find two values: the total number of students for the combined group (n) and the overall sum of scores for the combined group (ΣX). Finding the number of students is easy. If there are $n = 12$ in one group and $n = 8$ in the other, then there must be $n = 12 + 8 = 20$ in the combined group. To find the sum of scores for the combined group, we will use the same method: Find the sum for one group, then find the sum for the other, and then add the two sums together.

We know that the first section has $n = 12$ and $\overline{X} = 6$. To obtain this mean, the total (ΣX) is divided equally among the $n = 12$ individuals. If each person has $X = 6$ and there are 12 people, then the total must be $\Sigma X = 12(6) = 72$. This same result can be obtained using the equation for the mean.

$$\overline{X} = \frac{\Sigma X}{n}$$

$$6 = \frac{\Sigma X}{12}$$

$$12(6) = \Sigma X$$

$$72 = \Sigma X$$

The second section has $n = 8$ and $\overline{X} = 7$, so ΣX must be equal to 56. When these two groups are combined, the sum of all 20 scores will be

$$72 + 56 = 128$$

Finally, the mean for the combined group is

$$\overline{X} = \frac{\Sigma X}{n} = \frac{128}{20} = 6.4$$

Notice that this value is not obtained by simply averaging the two means. (If we had simply averaged $\overline{X} = 6$ and $\overline{X} = 7$, we would have obtained a mean of 6.5.) Because the samples are not the same size, one will make a larger contribution to the total group and therefore will carry more weight in determining the overall mean. For this reason, the overall mean we have calculated is called the *weighted mean*. In this example, the overall mean of $\overline{X} = 6.4$ is closer to the value of $\overline{X} = 6$ (the larger sample) than it is to $\overline{X} = 7$ (the smaller sample).

In summary, when two samples are combined, the weighted mean is obtained as follows:

STEP 1   Determine the grand total of all the scores in both samples. This sum is obtained by adding the sum of the scores for the first sample (ΣX_1) and the sum of the scores for the second sample (ΣX_2).

TABLE 3.1

Statistics quiz scores for a section of $n = 8$ students

Quiz score (X)	f	fX
10	1	10
9	2	18
8	4	32
7	0	0
6	1	6

STEP 2 Determine the total number of scores in both samples. This value is obtained by adding the number in the first sample (n_1) and the number in the second sample (n_2).

STEP 3 Divide the sum of all the scores (step 1) by the total number of scores (step 2). Expressed as an equation,

$$\text{weighted mean} = \frac{\text{combined sum}}{\text{combined } n} = \frac{\Sigma X_1 + \Sigma X_2}{n_1 + n_2}$$

COMPUTING THE MEAN FROM A FREQUENCY DISTRIBUTION TABLE

Table 3.1 shows the scores on a quiz for a section of statistics students. Instead of listing all of the individual scores, these data are organized into a frequency distribution table. To compute the mean for this sample, you must use all the information in the table, the f values as well as the X values.

To find the mean for this sample, we will need the sum of the scores (ΣX) and the number of scores (n). The number n can be found by summing the frequencies:

$$n = \Sigma f = 8$$

It is very common for people to make mistakes when determining ΣX from a frequency distribution table. Often the column labeled X is summed, while the frequency column is ignored. Be sure to use the information in the f column when determining ΣX. (See Chapter 2, p. 33.)

Note that there is one 10, two 9s, four 8s, and one 6 for a total of $n = 8$ scores. To find ΣX, you must be careful to add all eight scores:

$$\Sigma X = 10 + 9 + 9 + 8 + 8 + 8 + 8 + 6 = 66$$

This sum also can be found by multiplying each score by its frequency and then adding up the results. This is done in the third column (fX) in Table 3.1. Note, for example, that the two 9s contribute 18 to the total.

Once you have found ΣX and n, you compute the mean as usual:

$$\overline{X} = \frac{\Sigma X}{n} = \frac{\Sigma fX}{\Sigma f} = \frac{66}{8} = 8.25$$

LEARNING CHECK

1. Compute the mean for the sample of scores shown in the following frequency distribution table:

X	f
4	2
3	4
2	3
1	1

2. Two samples were obtained from a population. For sample A, $n = 8$ and $\overline{X} = 14$. For sample B, $n = 20$ and $\overline{X} = 6$. If the two samples are combined, will the overall mean be closer to 14 than to 6, closer to 6 than to 14, or halfway between 6 and 14? Explain your answer.

3. A sample of $n = 20$ scores has a mean of $\overline{X} = 5$. What is ΣX for this sample?

ANSWERS 1. $\overline{X} = \frac{27}{10} = 2.7$

2. The mean will be closer to 6. The larger sample will carry more weight in the combined group.

3. $\Sigma X = 100$

CHARACTERISTICS OF THE MEAN

The mean has many characteristics that will be important in future discussions. In general, these characteristics result from the fact that every score in the distribution contributes to the value of the mean. Specifically, every score must be added into the total in order to compute the mean. Three of the more important characteristics will now be discussed.

Changing a score or introducing a new score   Changing the value of any score, or adding a new score to the distribution, will change the mean. For example, the quiz scores for a psychology lab section consist of

$$9, \quad 8, \quad 7, \quad 5, \quad 1$$

The mean for this sample is

$$\overline{X} = \frac{\Sigma X}{n} = \frac{30}{5} = 6.00$$

Suppose the student who received the score of $X = 1$ returned a few days later and explained that she was ill on the day of the quiz. In fact, she went straight to the infirmary after class and was admitted for two days with the flu. Out of the goodness of the instructor's heart, the student was given a makeup quiz, and she received an 8. By having changed her score from 1 to 8, the distribution now consists of

$$9, \quad 8, \quad 7, \quad 5, \quad 8$$

The new mean is

$$\overline{X} = \frac{\Sigma X}{n} = \frac{37}{5} = 7.40$$

Changing a single score in this sample has given us a different mean.

In general, the mean is determined by two values: ΣX and N (or n). Whenever either of these values is changed, the mean also will be changed. In the preceding example, the value of one score was changed. This produced a change in the total (ΣX) and therefore changed the mean. If you add a new score (or take away a score), you will change both ΣX and n, and you must compute the new mean using the changed values. The following example demonstrates how the mean is affected when a new score is added to an existing sample.

EXAMPLE 3.3   We begin with a sample of $n = 4$ scores with a mean of $\overline{X} = 7$. A new score, $X = 12$, is added to this sample. The problem is to find the mean for the new set of scores.

To find the mean, we must first determine the values for n (the number of scores) and ΣX (the sum of the scores). Finding n is easy: We started with four scores and added one more, so the new value for n is 5 ($n = 5$).

To find the sum for the new sample, you simply begin with the total (ΣX) for the original sample and then add in the value of the new score. The original set of $n = 4$ scores had a mean of $\overline{X} = 7$. Therefore, the original total must be $\Sigma X = 28$. Adding a score of $X = 12$ brings the new total to $28 + 12 = 40$.

Finally, the new mean is computed using the values for n and ΣX for the new set

$$\overline{X} = \frac{\Sigma X}{n} = \frac{40}{5} = 8$$

The entire process can be summarized as follows:

Original sample	New sample adding $X = 12$
$n = 4$	$n = 5$
$\Sigma X = 28$	$\Sigma X = 40$
$\overline{X} = {}^{28}\!/_{4} = 7$	$\overline{X} = {}^{40}\!/_{5} = 8$

Adding or subtracting a constant from each score   If a constant value is added to every score in a distribution, the same constant will be added to the mean. Similarly, if you subtract a constant from every score, the same constant will be subtracted from the mean.

Consider the feeding scores for a sample of $n = 6$ rats. See Table 3.2. These scores are the amounts of food (in grams) they ate during a 24-hour testing session. Note that $\Sigma X = 26$ for $n = 6$ rats, so $\overline{X} = 4.33$. On the following day, each rat is given an experimental drug that reduces appetite. Suppose this drug has the effect of reducing the meal size by 2 grams for each rat. Note that the effect of the drug is to subtract a constant (2 points) from each rat's feeding score. The new distribution is shown in Table 3.3. Now $\Sigma X = 14$ and n is still 6, so the mean amount of food consumed is $\overline{X} = 2.33$. Subtracting 2 points from each score has changed the mean by the same constant, from $\overline{X} = 4.33$ to $\overline{X} = 2.33$. (It is important to note that

TABLE 3.2

Amount of food (in grams) consumed during baseline session

Rat's identification	Amount (X)	
A	6	$\Sigma X = 26$
B	3	
C	5	$n = 6$
D	3	
E	4	$\overline{X} = 4.33$
F	5	

TABLE 3.3

Amount of food (in grams) consumed after drug injections

Rat's identification	Baseline score minus constant	Drug score (X)	
A	6 − 2	4	$\Sigma X = 14$
B	3 − 2	1	
C	5 − 2	3	$n = 6$
D	3 − 2	1	
E	4 − 2	2	$\overline{X} = 2.33$
F	5 − 2	3	

experimental effects are practically never so simple as the adding or subtracting of a constant. Nonetheless, the principle of this characteristic of the mean is important and will be addressed in later chapters when we are using statistics to evaluate the effects of experimental manipulations.)

Multiplying or dividing each score by a constant   If every score in a distribution is multiplied by (or divided by) a constant value, the mean will be changed in the same way.

Multiplying (or dividing) each score by a constant value is a common method for changing the unit of measurement. To change a set of measurements from minutes to seconds, for example, you multiply by 60; to change from inches to feet, you divide by 12. Table 3.4 shows how a sample of $n = 5$ scores originally measured in yards would be transformed into a set of scores measured in feet. The first column shows the original scores with total $\Sigma X = 50$ and $\overline{X} = 10$ yards. In the second column, each of the original values has been multiplied by 3 (to change from yards to feet), and the resulting values total $\Sigma X = 150$, with $\overline{X} = 30$ feet. Multiplying each score by 3 has also caused the mean to be multiplied by 3. You should realize, however, that although the numerical values for the individual scores and the sample mean have been multiplied, the actual measurements are not changed.

LEARNING CHECK

1. **a.** Compute the mean for the following sample of scores:

   6, 1, 8, 0, 5

   b. Add 4 points to each score, and then compute the mean.

   c. Multiply each of the original scores by 5, and then compute the mean.

2. After every score in a distribution is multiplied by 3, the mean is calculated to be $\overline{X} = 60$. What was the mean for the original distribution?

TABLE 3.4

Measurement of five pieces of wood

Original measurement in yards	Conversion to feet (Multiply by 3)	
10	30	$\Sigma X = 50$
9	27	$\overline{X} = 10$ yards
12	36	
8	24	$\Sigma X = 150$
11	33	$\overline{X} = 30$ feet

3. A sample of $n = 5$ scores has a mean of $\overline{X} = 8$. If a new score of $X = 2$ is added to the sample, what value would be obtained for the new sample mean?

ANSWERS   1. a. $\overline{X} = \frac{20}{5} = 4$ b. $\overline{X} = \frac{40}{5} = 8$ c. $\overline{X} = \frac{100}{5} = 20$

2. The original mean was $\overline{X} = 20$.

3. The original sample has $n = 5$ and $\Sigma X = 40$. With the new score, the sample has $n = 6$ and $\Sigma X = 42$. The new mean is $\frac{42}{6} = 7$.

3.3 THE MEDIAN

The second measure of central tendency we will consider is called the *median*. The median is the score that divides a distribution exactly in half. Exactly one-half of the scores are less than or equal to the median, and exactly one-half are greater than or equal to the median.

DEFINITION

The *median* is the score that divides a distribution exactly in half. Exactly 50 percent of the individuals in a distribution have scores at or below the median.

Earlier, when we introduced the mean, specific symbols and notation were used to identify the mean and to differentiate a sample mean and a population mean. For the median, however, there are no symbols or notation. Instead, the median is simply identified by the word *median*. In addition, the definition and the computations for the median are identical for a sample and for a population.

By midpoint of the distribution we mean that the area in the graph is divided into two equal parts. We are *not* locating the midpoint between the highest and lowest X values.

The goal of the median is to determine the precise midpoint of a distribution. The commonsense goal is demonstrated in the following three examples. The three examples are intended to cover all of the different types of data sets you are likely to encounter.

METHOD 1: WHEN *N* IS AN ODD NUMBER

With an odd number of scores, you list the scores in order (lowest to highest), and the median is the middle score in the list. Consider the following set of $N = 5$ scores, which have been listed in order:

3,   5,   8,   10,   11

The middle score is $X = 8$, so the median is equal to 8.0. In a graph, the median divides the space or area of the graph in half (Figure 3.3). The amount of area above the median consists of $2\frac{1}{2}$ "boxes," the same as the area below the median (shaded portion).

METHOD 2: WHEN *N* IS AN EVEN NUMBER

With an even number of scores in the distribution, you list the scores in order (lowest to highest) and then locate the median by finding the point halfway between the middle two scores. Consider the following population:

3,   3,   4,   5,   7,   8

FIGURE 3.3

The median divides the area in the graph exactly in half.

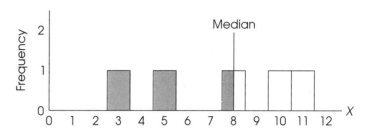

Now we select the middle pair of scores (4 and 5), add them together, and divide by 2:

$$\text{median} = \frac{4+5}{2} = \frac{9}{2} = 4.5$$

In terms of a graph, we see again that the median divides the area of the distribution exactly in half (Figure 3.4). There are three scores (or boxes) above the median and three below the median.

METHOD 3: WHEN THERE ARE SEVERAL SCORES WITH THE SAME VALUE IN THE MIDDLE OF THE DISTRIBUTION

In most cases, one of the two methods already outlined will provide you with a reasonable value for the median. However, when you have more than one individual at the median, these simple procedures may oversimplify the computations. Consider the following set of scores:

1,   2,   2,   3,   4,   4,   4,   4,   4,   5

There are 10 scores (an even number), so you normally would use method 2 and average the middle pair to determine the median. By this method, the median would be 4.

In many ways, this is a perfectly legitimate value for the median. However, when you look closely at the distribution of scores [see Figure 3.5(a)], you probably get the clear impression that $X = 4$ is not in the middle. The problem comes from the tendency to interpret the score of 4 as meaning exactly 4.00 instead of meaning an interval from 3.5 to 4.5. The simple method of computing the median has determined that the value we want is located in this interval. To locate the median with greater precision, it is necessary to use a process called *interpolation*. Although it is

FIGURE 3.4

The median divides the area in the graph exactly in half.

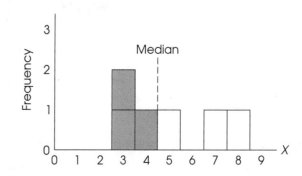

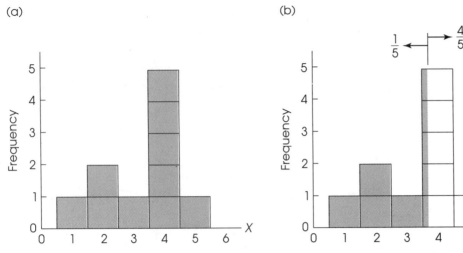

FIGURE 3.5

A distribution with several scores clustered at the median.

possible to describe the interpolation in general mathematical terms, we will use a simple graphic presentation to demonstrate how this procedure can be used to find the median.

EXAMPLE 3.4 For this example, we will use the data shown in Figure 3.5(a). Notice that the figure shows a population of $N = 10$ scores, with each score represented by a box in the histogram. To find the median, we must locate the position of a vertical line that will divide the distribution exactly in half, with 5 boxes on the left-hand side and 5 boxes on the right-hand side.

To begin the process, start at the left-hand side of the distribution and move up the scale of measurement (the X-axis), counting boxes as you go along. By the time you reach a value of 3.5 on the X-axis, you will have gathered a total of 4 boxes, so that only 1 more box is needed to give you exactly 50% of the distribution. The problem is that there are 5 boxes in the next interval. The solution is to take only a fraction of the next interval so that the fraction gives you the equivalent of 1 whole box. The following formula will determine the value for the fraction:

$$\text{fraction} = \frac{\text{number of boxes needed to reach 50\%}}{\text{number of boxes in the interval}}$$

For this example, we need 1 more box out of the 5 boxes in the interval, which gives us a fraction of $\frac{1}{5}$. Thus, the median is located exactly $\frac{1}{5}$ of the way into the interval containing the scores identified as $X = 4$. To find the exact location of the median, start at the lower real limit of the interval, which is 3.5, and add $\frac{1}{5}$ of the interval width. For this example, the interval is 1 point wide, so the median is

$$3.50 + \left(\tfrac{1}{5}\right)(1) = 3.50 + .20 = 3.70$$

This solution is shown in Figure 3.5(b). In general, once you have located the interval containing the tied scores and the median, the interpolation process can be expressed in the following formula, using the fraction we defined earlier:

$$\text{median} = \begin{array}{c} \text{lower real limit} \\ \text{of interval} \\ \text{with tied scores} \end{array} + (\text{fraction})(\text{interval width})$$

The interpolation process demonstrated in Example 3.4 can be summarized in the following four steps, which can be generalized to any situation where several scores are tied at the median.

STEP 1   Count the number of scores (boxes in the graph) below the tied value.

STEP 2   Find the number of additional scores (boxes) needed to make exactly one-half of the total distribution.

STEP 3   Form a fraction:

$$\frac{\text{number of boxes needed (step 2)}}{\text{number of tied boxes}}$$

STEP 4   Multiply the fraction (step 3) by the interval width, and add the result to the lower real limit of the interval containing the tied scores.

LEARNING CHECK   1. Find the median for each distribution of scores:
   a. 3, 10, 8, 4, 10, 7, 6
   b. 13, 8, 10, 11, 12, 10
   c. 3, 4, 3, 2, 1, 3, 2, 4

2. A distribution can have more than one median. (True or false?)

3. If you have a score of 52 on an 80-point exam, then you definitely scored above the median. (True or false?)

ANSWERS   1. a. The median is $X = 7$.
   b. The median is $X = 10.5$.
   c. The median is $X = 2.83$ (by interpolation).

2. False

3. False. The value of the median would depend on where the scores are located.

3.4 THE MODE

The final measure of central tendency that we will consider is called the mode. In its common usage, the word *mode* means "the customary fashion" or "a popular style." The statistical definition is similar in that the mode is the most common observation among a group of scores.

DEFINITION

In a frequency distribution, the *mode* is the score or category that has the greatest frequency.

In a frequency distribution graph, the greatest frequency will appear as the tallest part of the figure. To find the mode, you simply identify the score located directly beneath the highest point in the distribution.

As with the median, there are no symbols or special notation used to identify the mode or to differentiate between a sample mode and a population mode. In addition, the definition of the mode is the same for either a population or a sample distribution.

The mode is a useful measure of central tendency because it can be used to determine the typical or average value for any scale of measurement, including a nominal scale (see Chapter 1). Consider, for example, the data shown in Table 3.5. These data were obtained by asking a sample of 100 students to name their favorite restaurants in town. For these data, the mode is Luigi's, the restaurant (score) that was named most frequently as a favorite place. Although we can identify a modal response for these data, you should notice that it would be impossible to compute a mean or a median. For example, you cannot add the scores to determine a mean (How much is 5 College Grills plus 42 Luigi's?). Also, the restaurants do not form any natural order. Although they are listed alphabetically in this example, they could be listed in any order. Thus, it is impossible to list them in order and determine a 50% point (median). In general, the mode is the only measure of central tendency that can be used with data from a nominal scale of measurement.

Although a distribution will have only one mean and only one median, it is possible to have more than one mode. Specifically, it is possible to have two or more scores that have the same highest frequency. In a frequency distribution graph, the different modes will correspond to distinct, equally high peaks. A distribution with two modes is said to be *bimodal,* and a distribution with more than two modes is called *multimodal.* Occasionally, a distribution with several equally high points is said to have no mode.

Technically, the mode is the score with the absolute highest frequency. However, the term *mode* is often used more casually to refer to scores with relatively high frequencies, that is, scores that correspond to peaks in a distribution even though the peaks are not the absolute highest points. For example, Figure 3.6 shows the number of fish caught at various times during the day. There are two distinct peaks in this distribution, one at 6 A.M. and one at 6 P.M. Each of these values is a mode in

TABLE 3.5

Favorite restaurants named by a sample of $n = 100$ students

Caution: The mode is always a score or category, not a frequency. For this example, the mode is Luigi's, not $f = 42$.

Restaurant	f
College Grill	5
George & Harry's	16
Luigi's	42
Oasis Diner	18
Roxbury Inn	7
Sutter's Mill	12

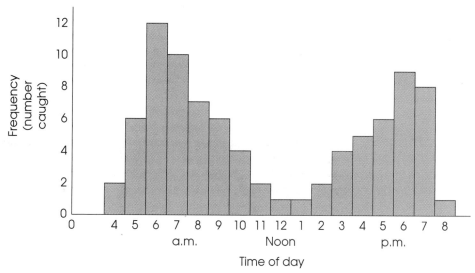

FIGURE 3.6

The relationship between time of day and number of fish caught.

the distribution. Close inspection of the graph reveals that the two modes do not have identical frequencies. Twelve fish were caught at 6 A.M., and 9 were caught at 6 P.M. Nonetheless, both of these points are called modes. The taller peak is called the *major mode*, and the shorter one is the *minor mode*.

LEARNING CHECK 1. Find the mode for the set of scores shown in the following frequency distribution table:

X	f
5	2
4	6
3	4
2	2
1	1

2. In a recent survey comparing picture quality for three brands of color televisions, 63 people preferred brand A, 29 people preferred brand B, and 58 people preferred brand C. What is the mode for this distribution?

3. It is possible for a distribution to have more than one mode. (True or false?)

ANSWERS 1. The mode is $X = 4$.

2. The mode is brand A.

3. True

3.5 SELECTING A MEASURE OF CENTRAL TENDENCY

How do you decide which measure of central tendency to use? The answer to this question depends on several factors. Before we discuss these factors, however, it should be noted that with many sets of data it is possible to compute two or even three different measures of central tendency. Often the three measures will produce similar results, but there are situations where they will be very different (see Section 3.6). Also, it should be noted that the mean is most often the preferred measure of central tendency. Because the mean uses every score in the distribution, it usually is a good representative value. Remember, the goal of central tendency is to find the single value that best represents the entire distribution. Besides being a good representative, the mean has the added advantage of being closely related to variance and standard deviation, the most common measures of variability (Chapter 4). This relationship makes the mean a valuable measure for purposes of inferential statistics. For these reasons, and others, the mean generally is considered to be the best of the three measures of central tendency. But there are specific situations where it is impossible to compute a mean or where the mean is not particularly representative. It is in these situations that the mode and the median are used.

WHEN TO USE THE MODE

The mode has two distinct advantages over the mean. First, it is easy to compute. Second, it can be used with any scale of measurement (nominal, ordinal, interval, ratio; see Chapter 1).

It is a bit misleading to say that the mode is easy to calculate because actually no calculation is required. When the scores are arranged in a frequency distribution, you identify the mode simply by finding the score with the greatest frequency. Because the value of the mode can be determined "at a glance," it is often included as a supplementary measure along with the mean or median as a no-cost extra. The value of the mode (or modes) in this situation is to give an indication of the shape of the distribution as well as a measure of central tendency. For example, if you are told that a set of exam scores has a mean of 72 and a mode of 80, you should have a better picture of the distribution than would be available from the mean alone (see Section 3.6).

The fact that the mode can be used with any scale of measurement makes it a very flexible value. When scores are measured on a nominal scale, it is impossible or meaningless to calculate either a mean or a median, so the mode is the only way to describe central tendency. Consider the frequency distribution shown in Figure 3.7. These data were obtained by recording the academic major for each student in a psychology lab section. Notice that "academic major" forms a nominal scale that simply classifies individuals into discrete categories.

You cannot compute a mean for these data because it is impossible to determine ΣX. (How much is one biologist plus six psychologists?) Also note that there is no natural ordering for the four categories in this distribution. It is a purely arbitrary decision to place biology on the scale before psychology. Because it is impossible to specify any order for the scores, you cannot determine a median. The mode, on the other hand, provides a very good measure of central tendency. The mode for this sample is psychology. This category describes the typical, or most representative, academic major for the sample.

Because the mode identifies the most typical case, it often produces a more sensible measure of central tendency. The mean, for example, will generate conclusions

FIGURE 3.7

Major field of study for $n = 9$ students enrolled in an experimental psychology laboratory section.

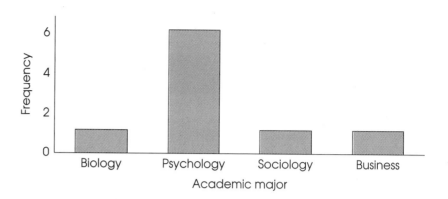

such as "the average family has 2.4 children and a house with 5.33 rooms." Many people would feel more comfortable saying "the typical, or modal, family has 2 children and a house with 5 rooms" (see Box 3.1).

WHEN TO USE THE MEDIAN

There are four specific situations where the median serves as a valuable alternative to the mean. These occur when (1) there are a few extreme scores in the distribution, (2) some scores have undetermined values, (3) there is an open-ended distribution, and (4) the data are measured on an ordinal scale.

Extreme scores or skewed distributions   When a distribution has a few extreme scores (scores that are very different in value from most of the others) then the mean will not be a good representative of the majority of the distribution. The problem comes from the fact that one or two extreme values can have a large influence and cause the mean to be displaced. In this situation, the fact that the mean uses all of the scores equally can be a disadvantage. For example, suppose a sample of $n = 10$ rats is tested in a T-maze for food reward. The animals must choose the correct arm of the T (right or left) to find the food in the goal box. The experimenter records the number of errors each rat makes before it solves the maze. Hypothetical data are presented in Figure 3.8.

FIGURE 3.8

Frequency distribution of errors committed before reaching learning criterion.

Notice that the graph in Figure 3.8 shows two *breaks* in the *X*-axis. Rather than listing all the scores from 0 to 100, the graph jumps directly to the first score, which is $X = 10$, and then jumps directly from $X = 15$ to $X = 100$. The breaks shown in the *X*-axis are the conventional way of notifying the reader that some values have been omitted.

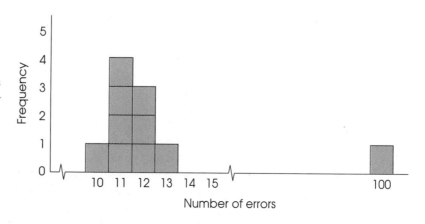

3.1   WHAT IS THE "AVERAGE"?

THE WORD *average* is used in everyday speech to describe what is typical and commonplace. Statistically, averages are measured by the mean, median, or mode. Agencies of the U.S. government frequently characterize demographic data, such as average income or average age, with the median. Winners of elections are determined by the mode, the most frequent choice. Scholastic Aptitude Test (SAT) scores for large groups of students usually are described with the mean. The important thing to remember about these measures of central tendency (or the "average") is that they describe and summarize a group of individuals rather than any single person. In fact, the "average person" may not actually exist.

Figure 3.9 shows data that were gathered from a sam-

FIGURE 3.10

Frequency distribution showing the age at which each infant in a sample of $n = 6$ uttered his or her first word.

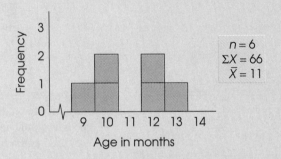

ple of $n = 9$ infants. The age at which each infant said his/her first word was recorded by the parents. Note that the mean for this group is $\overline{X} = 11$ months and that there are three infants who uttered their first intelligible words at this age.

Now look at the results of a different study, using a sample of $n = 6$ infants (Figure 3.10). Note that the mean for this group is also $\overline{X} = 11$ months but that the "average infant" does not exist in this sample.

The mean describes the group, not a single individual. It is for this reason that we find humor in statements such as "the average American family has 2.4 children," fully knowing that we never will encounter this family.

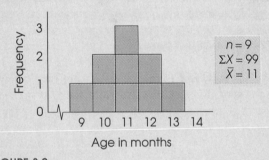

FIGURE 3.9

Frequency distribution showing the age at which each infant in a sample of $n = 9$ uttered his or her first word.

The mean for this sample is

$$\overline{X} = \frac{\Sigma X}{n} = \frac{203}{10} = 20.3$$

Notice that the mean is not very representative of any score in this distribution. Most of the scores are clustered between 10 and 13. The extreme score of $X = 100$ (a slow learner) inflates the value of ΣX and distorts the mean.

The median, on the other hand, is not easily affected by extreme scores. For this sample, $n = 10$, so there should be five scores on either side of the median. The median is 11.50. Notice that this is a very representative value. Also note that the median would be unchanged even if the slow learner made 1000 errors instead of only 100. The median commonly is used when reporting the average value for a skewed distribution. For example, the distribution of personal incomes is very skewed, with a small segment of the population earning incomes that are astronomical. These extreme values distort the mean, so that it is not very representative of the salaries that most of us earn. As in the previous example, the median is the preferred measure of central tendency when extreme scores exist.

Undetermined values   Occasionally, you will encounter a situation where an individual has an unknown or undetermined score. In psychology, this often occurs in learning experiments where you are measuring the number of errors an individual makes when solving a particular problem or the amount of time required for an individual to solve a particular problem. For example, suppose a sample of $n = 6$ people is asked to assemble a wooden puzzle as quickly as possible. The experimenter records how long (in minutes) it takes each individual to arrange all the pieces to complete the puzzle. Table 3.6 presents the outcome of this experiment.

Notice that person 6 never completed the puzzle. After an hour, this person still showed no sign of solving the puzzle, so the experimenter stopped him or her. This person has an undetermined score. (There are two important points to be noted. First, the experimenter should not throw out this individual's score. The whole purpose for using a sample is to gain a picture of the population, and this individual tells us that part of the population cannot solve the puzzle. Second, this person should not be given a score of $X = 60$ minutes. Even though the experimenter stopped the individual after 1 hour, the person did not finish the puzzle. The score that is recorded is the amount of time needed to finish. For this individual, we do not know how long this would be.)

It is impossible to compute the mean for these data because of the undetermined value. We cannot calculate the ΣX part of the formula for the mean. However, it is possible to compute the median. For these data, the median is 12.5. Three scores are below the median, and three scores (including the undetermined value) are above the median.

Open-ended distributions   A distribution is said to be *open-ended* when there is no upper limit (or lower limit) for one of the categories. The table in the margin provides an example of an open-ended distribution, showing the number of children in each family for a sample of $n = 20$ households. The top category in this distribution shows that three of the families have "5 or more" children. This is an open-ended category. Notice that it is impossible to compute a mean for these data because you cannot find ΣX (the total number of children for all 20 families). However, you can find the median. For these data, the median is 1.5 (exactly 50% of the families have fewer than 1.5 children).

Number of children (X)	f
5 or more	3
4	2
3	2
2	3
1	6
0	4

Ordinal scale   Many researchers believe that it is not appropriate to use the mean to describe central tendency for ordinal data. When scores are measured on an

TABLE 3.6

Amount of time to complete puzzle

Person	Time (min.)
1	8
2	11
3	12
4	13
5	17
6	Never finished

ordinal scale, the median is always appropriate and is usually the preferred measure of central tendency. The following example demonstrates that although it is possible to compute a "mean rank," the resulting value can be misleading.

EXAMPLE 3.5   Three children held a basketball competition to see who could hit the most baskets in 10 attempts. The contest was held twice; the results are shown in the following table:

	First contest			Second contest	
Child	Rank	Number of baskets	Child	Rank	Number of baskets
A	1st	10	C	1st	7
B	2nd	4	B	2nd	6
C	3rd	2	A	3rd	5

According to the data, child A finished first in one contest and third in the other, for a "mean rank" of 2. Child C also finished first one time and third one time and also has a mean rank of 2. Although these two children are identical in terms of mean rank, they are different in terms of the total number of baskets: Child A hit a total of 15 baskets, and child C hit only 9. The mean rank does not reflect this difference.

IN THE LITERATURE:
REPORTING THE MEAN

The mean is commonly used in behavioral science research when reporting the results of a study. Means may be reported with a verbal description of the results, in tables, and in graphs. Typically, the researcher is describing a relationship between one or more independent variables and a dependent variable. You should recall from Chapter 1 that the dependent variable is the score obtained for each subject and that the independent variable distinguishes the different treatment conditions or groups used in the experiment. The mean is often computed for the scores obtained for the dependent variable and reported for each treatment or group.

In reporting the results, many behavioral science journals use guidelines adopted by the American Psychological Association (APA), as outlined in the *Publication Manual of the American Psychological Association* (1994). We will refer to the APA manual from time to time in describing how data and research results are reported in the scientific literature. The APA style typically uses the letter M as the symbol for the sample mean. Thus, a study might state

The treatment group showed fewer errors ($M = 2.56$) on the task than the control group ($M = 11.76$).

When there are many means to report, tables with headings provide an organized and more easily understood presentation. Table 3.7 illustrates this point.

TABLE 3.7

The mean number of errors made on the task for treatment and control groups according to gender

	Treatment	Control
Females	1.45	8.36
Males	3.83	14.77

In addition to verbal descriptions and tables, graphs can be used to show relationships between the variables that have been studied. For example, a researcher testing a new diet drug might compare several dosages by measuring the amount of food that animals consume at each dose level. Figure 3.11 shows hypothetical data from this experiment. Notice that the dose levels (the independent variable) are on the X-axis and that food consumption (the dependent variable) is on the Y-axis. The points in the graph represent the mean food consumption for the groups at each dose level. The points are then connected with straight lines, and the resulting graph is called a *line graph*. A line graph is used when the variable on the X-axis (typically, the independent variable) is measured on an interval or a ratio scale. An alternative to the line graph is a histogram. For this example, the histogram would show a bar above each drug dose, with the height of each bar corresponding to the mean food consumption for that group, and leaving no space between adjacent bars.

Figure 3.12 shows a *bar graph* comparing four different means. A bar graph is used when the variable on the X-axis (the independent variable) is measured on a nominal or an ordinal scale.

When constructing graphs of any type, you should recall the basic rules we introduced in Chapter 2:

1. The height of a graph should be approximately two-thirds to three-quarters of its length.

2. Normally, you start numbering both the X-axis and the Y-axis with zero at the point where the two axes intersect. However, when a value of zero is part of the data, it is common to move the zero point away from the intersection so that the graph does not overlap the axes (see Figure 3.11).

More important, you should remember that the purpose of a graph is to give an accurate representation of the information in a set of data. Box 2.1 in Chapter 2 demonstrates what can happen when these basic principles are ignored.   ❏

FIGURE 3.11

The relationship between an independent variable (drug dose) and a dependent variable (food consumption). Because drug dose is a continuous variable, a continuous line is used to connect the different dose levels.

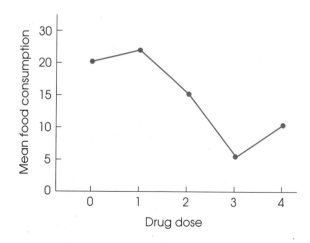

FIGURE 3.12

The relationship between an independent variable (brand of pain reliever) and a dependent variable (pain tolerance). The graph uses separate bars because the brand of pain reliever is measured on a nominal scale.

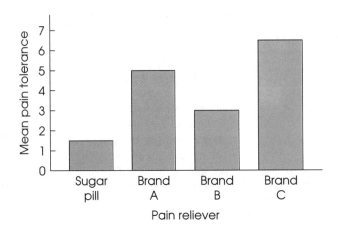

LEARNING CHECK

1. In a study on stress, a researcher exposed groups of rats to signaled shock, unsignaled shock, or no shock. The size (in millimeters) of ulcers for each group was then determined. Construct a graph of the following data:

	No shock	Signaled shock	Unsignaled shock
Mean size of ulcers	0	3	7

2. A psychologist studied the effect of sleep deprivation on mood. Groups of clinically depressed subjects were deprived of sleep for 0, 1, 2, or 3 nights. After deprivation, the amount of depression was measured with a depression inventory. Construct a graph of the following data:

Nights of deprivation	Mean depression score
0	22
1	17
2	9
3	7

ANSWERS   1. and 2. The graphs are shown in Figure 3.13.

3.6 CENTRAL TENDENCY AND THE SHAPE OF THE DISTRIBUTION

We have identified three different measures of central tendency, and often a researcher will calculate all three for a single set of data. Because the mean, the median, and the mode are all trying to measure the same thing (central tendency), it is reasonable to expect that these three values should be related. In fact, there are some

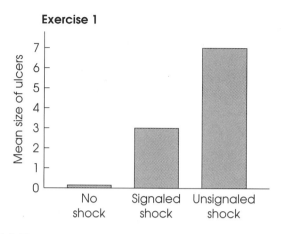

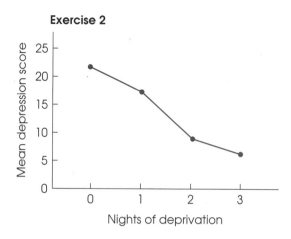

FIGURE 3.13

Answers to Learning Check exercises 1 and 2.

consistent and predictable relationships among the three measures of central tendency. Specifically, there are situations where all three measures will have exactly the same value. On the other hand, there are situations where the three measures are guaranteed to be different. In part, the relationships among the mean, median, and mode are determined by the shape of the distribution. We will consider two general types of distributions.

SYMMETRICAL DISTRIBUTIONS   For a *symmetrical distribution,* the right-hand side of the graph will be a mirror image of the left-hand side. By definition, the median will be exactly at the center of a symmetrical distribution because exactly half of the area in the graph will be on either side of the center. The mean also will be exactly at the center of a symmetrical distribution because each individual score in the distribution has a corresponding score on the other side (the mirror image), so that the average of these two values is exactly in the middle. Because all the scores can be paired in this way, the overall average will be exactly at the middle. For any symmetrical distribution, the mean and the median will be the same [Figure 3.14(a)].

If a symmetrical distribution has only one mode, then it must be exactly at the center, so that all three measures of central tendency will have the same value [see Figure 3.14(a)]. On the other hand, a bimodal distribution that is symmetrical

FIGURE 3.14

Measures of central tendency for three symmetrical distributions: (a) normal, (b) bimodal, and (c) rectangular.

(a)

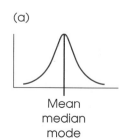

Mean
median
mode

(b)

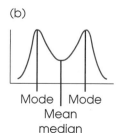

Mode │ Mode
Mean
median

(c)

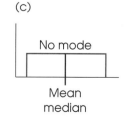

No mode

Mean
median

FIGURE 3.15

Measures of central tendency for skewed distributions.

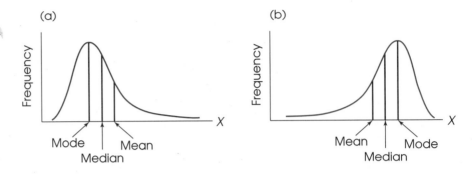

[Figure 3.14(b)] will have the mean and the median together in the center with the modes on each side. A rectangular distribution [Figure 3.14(c)] has no mode because all X values occur with the same frequency. Still, the mean and the median will be in the center of the distribution and equivalent in value.

SKEWED DISTRIBUTIONS

Distributions are not always symmetrical. In fact, quite often they are lopsided, or *skewed.* For example, Figure 3.15(a) shows a *positively skewed distribution.* In this distribution, the peak (highest frequency) is on the left-hand side. This is the position of the mode. If you examine Figure 3.15(a) carefully, it should be clear that the vertical line drawn at the mode does not divide the distribution into two equal parts. In order to have exactly 50% of the distribution on each side, the median must be located to the right of the mode. Finally, the mean will be located to the right of median because it is influenced most by extreme scores and will be displaced farthest to the right by the scores in the tail. Therefore, in a positively skewed distribution, the mean will have the largest value, followed by the median and then the mode [see Figure 3.15(a)].

Negatively skewed distributions are lopsided in the opposite direction, with the scores piling up on the right-hand side and the tail tapering off to the left. The grades on an easy exam, for example, will tend to form a negatively skewed distribution [see Figure 3.15(b)]. For a distribution with negative skew, the mode is on the right-hand side (with the peak), while the mean is displaced on the left by the extreme scores in the tail. As before, the median is located between the mean and the mode. In order from highest value to lowest value, the three measures of central tendency will be the mode, the median, and the mean.

SUMMARY

1. The purpose of central tendency is to determine the single value that best represents the entire distribution of scores. The three standard measures of central tendency are the mode, the median, and the mean.

2. The mean is the arithmetic average. It is computed by summing all the scores and then dividing by the number of scores. Conceptually, the mean is obtained by dividing the total (ΣX) equally among the number of individuals (N

or n). Although the calculation is the same for a population or a sample mean, a population mean is identified by the symbol μ, and a sample mean is identified by $\overline{X}$.

3. Changing any score in the distribution will cause the mean to be changed. When a constant value is added to (or subtracted from) every score in a distribution, the same constant value is added to (or subtracted from) the mean. If every score is multiplied by a constant, the mean will be

multiplied by the same constant. In nearly all circumstances, the mean is the best representative value and is the preferred measure of central tendency.

4. The median is the value that divides a distribution exactly in half. The median is the preferred measure of central tendency when a distribution has a few extreme scores that displace the value of the mean. The median also is used when there are undetermined (infinite) scores that make it impossible to compute a mean.

5. The mode is the most frequently occurring score in a distribution. It is easily located by finding the peak in a frequency distribution graph. For data measured on a nominal scale, the mode is the appropriate measure of central tendency. It is possible for a distribution to have more than one mode.

6. For symmetrical distributions, the mean will equal the median. If there is only one mode, then it will have the same value, too.

7. For skewed distributions, the mode will be located toward the side where the scores pile up, and the mean will be pulled toward the extreme scores in the tail. The median will be located between these two values.

KEY TERMS

central tendency	mode	bimodal distribution	skewed distribution
mean	major mode	multimodal distribution	positive skew
weighted mean	minor mode	symmetrical distribution	negative skew
median			

FOCUS ON PROBLEM SOLVING

1. Because there are three different measures of central tendency, your first problem is to decide which one is best for your specific set of data. Usually the mean is the preferred measure, but the median may provide a more representative value if you are working with a skewed distribution. With data measured on a nominal scale, you must use the mode.

2. Although the three measures of central tendency appear to be very simple to calculate, there is always a chance for errors. The most common sources of error are listed next.

   a. Many students find it very difficult to compute the mean for data presented in a frequency distribution table. They tend to ignore the frequencies in the table and simply average the score values listed in the X column. You must use the frequencies and the scores! Remember, the number of scores is found by $N = \Sigma f$, and the sum of all N scores is found by ΣfX.

   b. The median is the midpoint of the distribution of scores, not the midpoint of the scale of measurement. For a 100-point test, for example, many students incorrectly assume that the median must be $X = 50$. To find the median, you must have the *complete set* of individual scores. The median separates the individuals into two equal-sized groups.

   c. The most common error with the mode is for students to report the highest frequency in a distribution rather than the score with the highest frequency. Remember, the purpose of central tendency is to find the most representative score. Therefore, for the following data, the mode is $X = 3$, not $f = 8$.

X	f
4	3
3	8
2	5
1	2

DEMONSTRATION 3.1

COMPUTING MEASURES OF CENTRAL TENDENCY

For the following sample data, find the mean, median, and mode. Scores:

$$5, \quad 6, \quad 9, \quad 11, \quad 5, \quad 11, \quad 8, \quad 14, \quad 2, \quad 11$$

Compute the mean. Calculating the mean involves two steps:

1. Obtain the sum of the scores, ΣX.
2. Divide the sum by the number of scores, n.

For these data, the sum of the scores is as follows:

$$\Sigma X = 5 + 6 + 9 + 11 + 5 + 11 + 8 + 14 + 2 + 11 = 82$$

We can also observe that $n = 10$. Therefore, the mean of this sample is obtained by

$$\overline{X} = \frac{\Sigma X}{n} = \frac{82}{10} = 8.2$$

Find the median. The median divides the distribution in half, in that half of the scores are above or equal to the median and half are below or equal to it. In this demonstration, $n = 10$. Thus, the median should be a value that has 5 scores above it and 5 scores below it.

1. Arrange the scores in order.

$$2, \quad 5, \quad 5, \quad 6, \quad 8, \quad 9, \quad 11, \quad 11, \quad 11, \quad 14$$

2. With an even number of scores, locate the midpoint between the middle two scores. The middle scores are $X = 8$ and $X = 9$. The median is the midpoint between 8 and 9.

$$\text{median} = \frac{8 + 9}{2} = \frac{17}{2} = 8.5$$

Find the mode. The mode is the X value that has the highest frequency. Looking at the data, we can readily determine that $X = 11$ is the score that occurs most frequently.

DEMONSTRATION 3.2

COMPUTING THE MEAN FROM A FREQUENCY DISTRIBUTION TABLE

Compute the mean for the data in the following table:

X	f
6	1
5	0
4	3
3	3
2	2

To compute the mean from a frequency distribution table, you must use the information in *both* the X and the f columns.

STEP 1 Multiply each X value by its frequency.
You can create a third column labeled fX. For these data,

fX
6
0
12
9
4

STEP 2 Find the sum of the fX values.

$$\Sigma fX = 6 + 0 + 12 + 9 + 4 = 31$$

STEP 3 Find n for these data.
Remember, $n = \Sigma f$.

$$n = \Sigma f = 1 + 0 + 3 + 3 + 2 = 9$$

STEP 4 Divide the sum of fX by n.

$$\bar{X} = \frac{\Sigma fX}{n} = \frac{31}{9} = 3.44$$

PROBLEMS

1. Explain the general purpose for measuring central tendency.

2. Explain why there is more than one method for measuring central tendency. Why not use a single, standardized method for determining the "average" score?

3. Explain what is meant by each of the following statements:
a. The mean is the *balance point* of the distribution.
b. The median is the *midpoint* of the distribution.

4. Identify the circumstances where the median rather than the mean is the preferred measure of central tendency.

5. Under what circumstances will the mean, the median, and the mode all have the same value?

6. Under what circumstances is the mode the preferred measure of central tendency?

7. Explain why the mean is often not a good measure of central tendency for a skewed distribution.

8. Find the mean, median, and mode for the following sample of scores:

2, 5, 1, 4, 2, 3, 3, 2

9. For the following set of scores:

9, 7, 9, 8, 10, 8, 1, 9, 8, 9

a. Sketch a frequency distribution histogram.
b. Find the mean, median, and mode, and locate these values in your graph.

10. The following frequency distribution summarizes the number of absences for each student in a class of $n = 20$.

Number of absences (X)	f
5 or more	3
4	4
3	3
2	6
1	3
0	1

a. Find the mode for this distribution.

b. Find the median number of absences for the class.

c. Explain why you cannot compute the mean number of absences using the data provided in the table.

11. For the set of scores shown in the following frequency distribution table:

X	f
5	1
4	2
3	4
2	2
1	1

a. Sketch a histogram showing the distribution, and locate the median in your sketch.

b. Compute the mean for this set of scores.

c. Now suppose the score $X = 5$ is changed to $X = 45$. How does this change affect the mean and the median? Use your sketch to find the new median, and then compute the new mean.

12. For the set of scores presented in the following frequency distribution:

X	f
5	10
4	6
3	2
2	1
1	1

a. Find the mean, the median, and the mode.

b. Based on your answers from part a, identify the shape of the distribution.

13. A population of $N = 50$ scores has a mean of $\mu = 26$. What is ΣX for this population?

14. A sample has a mean of $\overline{X} = 5$, and $\Sigma X = 320$. How many scores are in this sample? ($n = $?)

15. A sample has a mean of $\overline{X} = 20$.

a. If each X value is multiplied by 6, then what is the value of the new mean?

b. If 5 points are added to each of the original X values, then what is the value of the new mean?

16. A sample of $n = 8$ scores has a mean of $\overline{X} = 12$. A new score of $X = 4$ is added to the sample.

a. What is ΣX for the original sample?

b. What is ΣX for the new sample?

c. What is n for the new sample?

d. What is the mean for the new sample?

17. A sample has a mean of $\overline{X} = 40$.

a. If a new score of $X = 50$ is added to the sample, what will happen to the sample mean (increase, decrease, stay the same)? Explain your answer.

b. If a new score of $X = 30$ is added to the sample, what will happen to the sample mean (increase, decrease, stay the same)? Explain your answer.

c. If a new score of $X = 40$ is added to the sample, what will happen to the sample mean (increase, decrease, stay the same)? Explain your answer.

18. A sample of $n = 5$ scores has a mean of $\overline{X} = 10$.

a. If a new score of $X = 4$ is added to the sample, what is the value of the new sample mean?

b. If one of the scores, $X = 18$, is removed from the original sample, what is the value of the new sample mean?

c. If one of the scores in the original sample is changed from $X = 5$ to $X = 20$, what is the value of the new sample mean?

19. A sample of $n = 5$ scores has a mean of $\overline{X} = 21$. One new score is added to the sample, and the mean for the resulting new sample is $\overline{X} = 25$. Find the value of the new score. (*Hint:* Find ΣX for the original sample and ΣX for the new sample.)

20. A sample of $n = 10$ scores has a mean of $\overline{X} = 23$. One score is removed from the sample, and the mean for the remaining scores is $\overline{X} = 25$. Find the value of the score that was removed.

21. One sample of $n = 3$ scores has a mean of $\overline{X} = 4$. A second sample of $n = 7$ scores has a mean of $\overline{X} = 10$. If these two samples are combined, what value will be obtained for the mean of the combined sample?

22. For each of the following situations, identify the measure of central tendency (mean, median, mode) that would provide the best description of the "average" score.

a. A researcher records each individual's favorite TV show for a sample of $n = 50$ children who are 6 years of age.

b. A researcher records how much weight is gained or lost for each client during a 6-week diet program.

c. A researcher studying motivation asks subjects to search through a newspaper for the word *discipline*. The researcher records how long (in minutes) each subject works at the task before finding the word or giving up. For a sample of $n = 20$ people, the mean time is $\overline{X} = 29$ minutes, the median is 17 minutes, and the mode is 15 minutes.

23. A distribution of exam scores has a mean of 71 and a median of 79. Is it more likely that this distribution is symmetrical, positively skewed, or negatively skewed?

24. On a standardized reading achievement test, the nation-wide average for seventh-grade children is $\mu = 7.0$. A seventh-grade teacher is interested in comparing class reading scores with the national average. The scores for the 16 students in this class are as follows:

8, 6, 5, 10, 5, 6, 8, 9

7, 6, 9, 5, 14, 4, 7, 6

a. Find the mean and median reading scores for this class.
b. If the mean is used to define the class average, how does this class compare with the national norm?
c. If the median is used to define the class average, how does this class compare with the national norm?

25. A researcher evaluated the taste of four leading brands of instant coffee by having a sample of individuals taste each coffee and then rate its flavor on a scale from 1 to 5 (1 = very bad and 5 = excellent). The results from this study are summarized as follows:

Coffee	Average rating
Brand A	2.5
Brand B	4.1
Brand C	3.2
Brand D	3.6

a. Identify the independent variable and the dependent variable for this study.
b. What scale of measurement was used for the independent variable (nominal, ordinal, interval, or ratio)?
c. If the researcher used a graph to show the obtained relationship between the independent variable and the dependent variable, what kind of graph would be appropriate (line graph, histogram, bar graph)?
d. Sketch a graph showing the results of this experiment.

26. A researcher examined the effect of amount of relaxation training on insomnia. Four treatment groups were used. Subjects received relaxation training for 2, 4, or 8 sessions. A control group received no training (0 sessions). Following training, the researcher measured how long it took the subjects to fall asleep. The average time for each group is presented in the following table:

Training sessions	Mean time (in minutes)
0	72
2	58
4	31
8	14

Present these data in a graph.

CHAPTER 4

VARIABILITY

TOOLS YOU WILL NEED

The following items are considered essential background material for this chapter. If you doubt your knowledge of any of these items, you should review the appropriate chapter or section before proceeding.

- Summation notation (Chapter 1)
- Central tendency (Chapter 3)
  - Mean
  - Median

CONTENTS

4.1   **INTRODUCTION**

The term *variability* has much the same meaning in statistics as it has in everyday language; to say that things are variable means that they are not all the same. In statistics, our goal is to measure the amount of variability for a particular set of scores, a distribution. In simple terms, if the scores in a distribution are all the same, then there is no variability. If there are small differences between scores, then the variability is small, and if there are big differences between scores, then the variability is large.

DEFINITION   *Variability* provides a quantitative measure of the degree to which scores in a distribution are spread out or clustered together.

Figure 4.1 shows two distributions of familiar values: Figure 4.1(a) shows the distribution of adult heights (in inches), and Figure 4.1(b) shows the distribution of adult weights (in pounds). You should notice that the two distributions differ in terms of central tendency. The mean adult height is 68 inches (5 feet and 8 inches), and the mean adult weight is 150 pounds. In addition, you should notice that the distributions differ in terms of variability. For example, most adult heights are clustered close together, within 5 or 6 inches of the mean. On the other hand, adult weights are spread over a much wider range. In the weight distribution, it is not unusual to find individuals who are located more than 30 pounds away from the mean, and it would not be surprising to find two individuals whose weights differ by more than 30 or 40 pounds. The purpose for measuring variability is to obtain an objective measure of how the scores are spread out in a distribution. In general, a good measure of variability will serve two purposes:

1. Variability describes the distribution. Specifically, it tells whether the scores are clustered close together or are spread out over a large distance. Usually variability is defined in terms of *distance*. It tells how much distance to expect between one score and another or how much distance to expect between an individual score and the mean. For example, we know that most adults' heights are clustered close together within 5 or 6 inches of the average. Although more-extreme heights exist, they are relatively rare.

2. Variability measures how well an individual score (or group of scores) represents the entire distribution. This aspect of variability is very important for inferential statistics where relatively small samples are used to answer questions about populations. For example, suppose that you selected a sample of one person to represent the entire population. Because most adult heights are within a few inches of the population average (the distances are small), there is a very good chance that you would select someone whose height is within 6 inches of the population mean. On the other hand, the scores are much more spread out (the distances are greater) in the distribution of adult weights. In this case, you probably would *not* obtain someone whose weight was within 6 pounds of the population mean. Thus, variability provides information about how much error to expect when you are using a sample to represent a population.

(a) The distribution of adult heights (in inches)

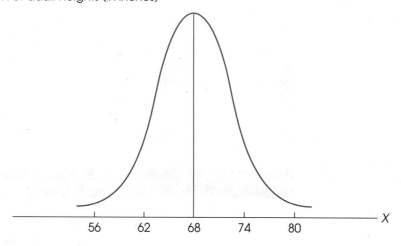

(b) The distribution of adult weights (in pounds)

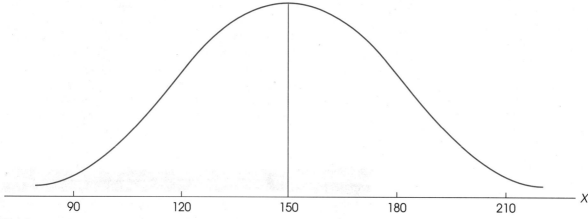

FIGURE 4.1

Population distributions of adult heights and adult weights.

Note: For simplicity, we have omitted the vertical axis for these graphs. As always, the height of any point on the curve indicates the relative frequency for that particular score.

In this chapter, we will consider three different measures of variability: the range, the interquartile range, and the standard deviation. Of these three, the standard deviation (and the related measure of variance) is by far the most important.

4.2 THE RANGE

The range is the distance between the largest score (X_{max}) and the smallest score in the distribution (X_{min}). In determining this distance, you must also take into account the real limits of the maximum and minimum X values. The range therefore is computed as the difference between the upper real limit (URL) for X_{max} and the lower real limit (LRL) for X_{min}.

$$\text{range} = \text{URL } X_{max} - \text{LRL } X_{min}$$

DEFINITION

The *range* is the difference between the upper real limit of the largest (maximum) X value and the lower real limit of the smallest (minimum) X value.

For example, consider the following data:

3, 7, 12, 8, 5, 10

For these data, $X_{max} = 12$, with an upper real limit of 12.5, and $X_{min} = 3$, with a lower real limit of 2.5. Thus, the range equals

When the distribution consists of whole numbers, the range also can be obtained as follows:

$$\text{range} = \text{highest } X - \text{lowest } X + 1$$

$$\begin{aligned} \text{range} &= \text{URL } X_{max} - \text{LRL } X_{min} \\ &= 12.5 - 2.5 = 10 \end{aligned}$$

The range is perhaps the most obvious way of describing how spread out the scores are—simply find the distance between the maximum and the minimum scores. The problem with using the range as a measure of variability is that it is completely determined by the two extreme values and ignores the other scores in the distribution. For example, the following two distributions have exactly the same range—10 points in each case. However, the scores in the first distribution are clustered together at one end of the range, whereas the scores in the second distribution are spread out over the entire range.

Distribution 1: 1, 8, 9, 9, 10, 10

Distribution 2: 1, 2, 4, 6, 8, 10

If, for example, these were scores on a 10-point quiz for two different class sections, there are clear differences between the two sections. Nearly all the students in the first section have mastered the material, but there is a wide range of abilities for students in the second section. A good measure of variability should show this difference.

Because the range does not consider all the scores in the distribution, it often does not give an accurate description of the variability for the entire distribution. For this reason, the range is considered to be a crude and unreliable measure of variability.

4.3 THE INTERQUARTILE RANGE AND SEMI-INTERQUARTILE RANGE

In Chapter 3, we defined the median as the score that divides a distribution exactly in half. In a similar way, a distribution can be divided into four equal parts using quartiles. By definition, the first quartile ($Q1$) is the score that separates the lower 25% of the distribution from the rest. The second quartile ($Q2$) is the score that has exactly two quarters, or 50%, of the distribution below it. Notice that the second

FIGURE 4.2

Frequency distribution for a population of
$N = 16$ scores. The first quartile is
$Q1 = 4.5$. The third quartile is $Q3 = 8.0$.
The interquartile range is 3.5 points. Note
that the third quartile ($Q3$) divides the
two boxes at $X = 8$ exactly in half so that
a total of 4 boxes is above $Q3$ and 12
boxes are below it.

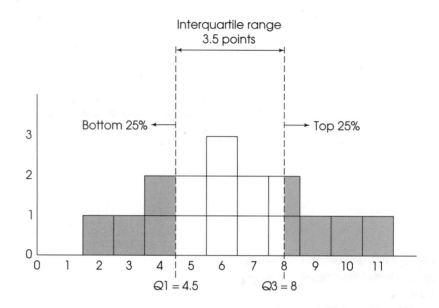

quartile and the median are the same. Finally, the third quartile ($Q3$) is the score that
divides the bottom three-quarters of the distribution from the top quarter.

The simplest method for finding the values of $Q1$, $Q2$, and $Q3$ is to construct a
frequency distribution histogram where each score is represented by a box (see Figure 4.2). Then determine exactly how many boxes are needed to make up exactly
one-quarter of the whole set. With a total of $n = 16$ scores (boxes), for example,
one-quarter is exactly 4 boxes. In this case, the first quartile separates the lowest 4
boxes (25%) from the rest, the second quartile has exactly 8 boxes (50%) on each
side, and the third quartile separates the lowest 12 boxes (75%) from the highest 4
boxes. Usually you can simply count boxes or fractions of boxes to locate $Q1$, $Q2$,
and $Q3$.

After the quartiles have been determined, the *interquartile range* is defined as the
distance between the first quartile and the third quartile. The interquartile range effectively ignores the top 25% and the bottom 25% of the distribution and measures
the range covered by the middle 50% of the distribution.

DEFINITION   The *interquartile range* is the distance between the first quartile and the
third quartile:

interquartile range = $Q3 - Q1$

When the interquartile range is used to describe variability, it commonly is transformed into the *semi-interquartile range*. As the name implies, the semi-interquartile range is simply one-half of the interquartile range. Conceptually, the
semi-interquartile range measures the distance from the middle of the distribution to
the boundaries that define the middle 50%.

DEFINITION   The *semi-interquartile range* is one-half of the interquartile range:

$$\text{semi-interquartile range} = \frac{Q3 - Q1}{2}$$

EXAMPLE 4.1   Figure 4.2 shows a frequency distribution histogram for a set of 16 scores. For this distribution, the first quartile is $Q1 = 4.5$. Exactly 25% of the scores (4 out of 16) are located below $X = 4.5$. Similarly, the third quartile is $Q3 = 8.0$. Note that this value separates the bottom 75% of the distribution (12 out of 16 scores) from the top 25%. For this set of scores, the interquartile range is

$$Q3 - Q1 = 8.0 - 4.5 = 3.5$$

The semi-interquartile range is simply one-half of this distance:

$$\text{semi-interquartile range} = \frac{3.5}{2} = 1.75$$

Because the semi-interquartile range is derived from the middle 50% of a distribution, it is less likely to be influenced by extreme scores and therefore gives a better and more stable measure of variability than the range. Nevertheless, the semi-interquartile range does not take into account the actual distances between individual scores, so it does not give a complete picture of how scattered or clustered the scores are. Like the range, the semi-interquartile range is considered to be a somewhat crude measure of variability. In Chapter 6, we will introduce another method to determine the semi-interquartile range for normal distributions.

LEARNING CHECK   1. For the following data, find the range and the semi-interquartile range:

3,   4,   5,   7,   9,   10,   11,   13

2. Consider the distribution of exercise 1, except replace the score of 13 with a score of 100. What are the new values for the range and the semi-interquartile range? In comparing the answer to the one of the previous exercise, what can you conclude about these measures of variability?

ANSWERS   1. Range = URL X_{max} − LRL X_{min} = 13.5 − 2.5 = 11; semi-interquartile range = $(Q3 - Q1)/2 = (10.5 - 4.5)/2 = 3$.

2. Range = 98; semi-interquartile range = 3. The range is greatly affected by extreme scores in the distribution.

4.4 STANDARD DEVIATION AND VARIANCE FOR A POPULATION

The standard deviation is the most commonly used and the most important measure of variability. Standard deviation uses the mean of the distribution as a reference point and measures variability by considering the distance between each score and the mean. It determines whether the scores are generally near or far from the mean. That is, are the scores clustered together or scattered? In simple terms, the standard deviation approximates the average distance from the mean.

Although the concept of standard deviation is straightforward, the actual equations will appear complex. Therefore, we will begin by looking at the logic that leads to these equations. If you remember that our goal is to measure the standard, or typical, distance from the mean, then this logic and the equations that follow should be easier to remember.

STEP 1   The first step in finding the standard distance from the mean is to determine the deviation, or distance from the mean, for each individual score. By definition, the deviation for each score is the difference between the score and the mean.

DEFINITION   *Deviation* is distance and direction from the mean:

$$\text{deviation score} = X - \mu$$

A deviation score occasionally is identified by a lowercase letter x.

For a distribution of scores with $\mu = 50$, if your score is $X = 53$, then your deviation score is

$$X - \mu = 53 - 50 = 3$$

If your score is $X = 45$, then your deviation score is

$$X - \mu = 45 - 50 = -5$$

Notice that there are two parts to a deviation score: the sign ($+$ or $-$) and the number. The sign tells the direction from the mean, that is, whether the score is located above ($+$) or below ($-$) the mean. The number gives the actual distance from the mean. For example, a deviation score of -6 corresponds to a score that is below the mean by 6 points.

STEP 2   Because our goal is to compute a measure of the standard distance from the mean, the obvious next step is to calculate the mean of the deviation scores. To compute this mean, you first add up the deviation scores and then divide by N. This process is demonstrated in the following example.

EXAMPLE 4.2   We start with the following set of $N = 4$ scores. These scores add up to $\Sigma X = 12$, so the mean is $\mu = \frac{12}{4} = 3$. For each score, we have computed the deviation.

X	$X - \mu$
8	+5
1	−2
3	0
0	−3
	$0 = \Sigma(X - \mu)$

Notice that the deviation scores add up to zero. This should not be surprising if you remember that the mean serves as a balance point for the distribution. The distances above the mean are equal to the distances below the mean (see page 56). Logically, the deviation scores must *always* add up to zero.

Because the mean deviation is always zero, it is of no value as a measure of variability. It is zero whether the scores are grouped together or are all scattered out. The mean deviation score provides no information about variability. (You should note, however, that the constant value of zero can be useful in other ways. Whenever you are working with deviation scores, you can check your calculations by making sure that the deviation scores add up to zero.)

STEP 3 The average of the deviation scores will not work as a measure of variability because it is always zero. Clearly, this problem results from the positive and negative values canceling each other out. The solution is to get rid of the signs (+ and −). The standard procedure for accomplishing this is to square each deviation score. Using these squared values, you then compute the mean squared deviation, which is called *variance*.

DEFINITION   *Population variance* = mean squared deviation. Variance is the mean of the squared deviation scores.

Note that the process of squaring deviation scores does more than simply get rid of plus and minus signs. It results in a measure of variability based on *squared* distances. Although variance is valuable for some of the *inferential* statistical methods covered later, the mean squared distance is not the best *descriptive* measure for variability.

STEP 4 Remember that our goal is to compute a measure of the standard distance from the mean. Variance, the mean squared deviation, is not exactly what we want. The final step simply makes a correction for having squared all the distances. The new measure, the standard deviation, is the square root of the variance.

DEFINITION   *Standard deviation* = $\sqrt{\text{variance}}$.

Technically, standard deviation is the square root of the mean squared deviation. But, conceptually, standard deviation is easier to understand if you think of it as describing the typical distance of scores from the mean (that is, the typical $X - \mu$). As the name implies, standard deviation measures the standard, or typical, deviation score.

The concept of standard deviation (or variance) is the same for a sample as for a population. However, the details of the calculations differ slightly, depending on whether you have data for a sample or a complete population. Therefore, we will first consider the formulas for measures of population variability.

SUM OF SQUARED DEVIATIONS (SS) Variance, you should recall, is defined as the mean squared deviation. This mean is computed exactly the same way you compute any mean: First, find the sum, and then divide by the number of scores:

$$\text{variance} = \text{mean squared deviation} = \frac{\text{sum of squared deviations}}{\text{number of scores}}$$

The value in the numerator of this equation, the sum of the squared deviations, is a basic component of variability, and we will focus on it. To simplify things, it is identified by the notation SS (for sum of squared deviations), and it generally is referred to as the *sum of squares.*

DEFINITION

SS, or *sum of squares,* is the sum of the squared deviation scores.

You will need to know two formulas in order to compute *SS*. These formulas are algebraically equivalent (they always produce the same answer), but they look different and are used in different situations.

The first of these formulas is called the definitional formula because the terms in the formula literally define the process of adding up the squared deviations:

$$\text{Definitional formula:} \quad SS = \Sigma(X - \mu)^2 \tag{4.1}$$

Note that the formula directs you to square each deviation score, $(X - \mu)^2$, and then add them. The result is the sum of the squared deviations, or *SS*. Following is an example using this formula.

EXAMPLE 4.3   We will compute *SS* for the following set of $N = 4$ scores. These scores have a sum of $\Sigma X = 8$, so the mean is $\mu = \frac{8}{4} = 2$. For each score, we have computed the deviation and the squared deviation in a table. The squared deviations add up to $SS = 22$.

Caution: The definitional formula requires that you first square the deviations and then add them. Note that we have done these computations in a table. Do not confuse this table with a frequency distribution table.

X	$X - \mu$	$(X - \mu)^2$	
1	-1	1	$\Sigma X = 8$
0	-2	4	$\mu = 2$
6	$+4$	16	
1	-1	$\underline{1}$	
		$22 = \Sigma(X - \mu)^2$	

The second formula for *SS* is called the computational formula (or the machine formula) because it works directly with the scores (X values) and therefore is generally easier to use for calculations, especially with an electronic calculator:

$$\text{Computational formula:} \quad SS = \Sigma X^2 - \frac{(\Sigma X)^2}{N} \tag{4.2}$$

The first part of this formula directs you to square each score and then add the squared values (ΣX^2). The second part requires you to add the scores (ΣX), square this total, and then divide the result by N (see Box 4.1). The use of this formula is shown in Example 4.4 with the same set of scores we used for the definitional formula.

4.1 COMPUTING *SS* WITH A CALCULATOR

THE COMPUTATIONAL formula for *SS* is intended to simplify calculations, especially when you are using an electronic calculator. The following steps outline the most efficient procedure for using a typical, inexpensive hand calculator to find *SS*. (We assume that your calculator has one memory, where you can store and retrieve information.) The computational formula for *SS* is presented here for easy reference.

$$SS = \Sigma X^2 - \frac{(\Sigma X)^2}{N}$$

STEP 1: The first term in the computational formula is ΣX^2. The procedure for finding this sum is described in Box 1.1 on page 23. Once you have calculated ΣX^2, write this sum on a piece of paper so you do not lose it. Leave ΣX^2 in the calculator memory, and go to the next step.

STEP 2: Now you must find the sum of the scores, ΣX. We assume that you can add a set of numbers with your calculator—just be sure to press the equals key (=) after the last score. Write this total on your paper.

(*Note:* You may want to clear the calculator display before you begin this process. It is not necessary, but you may feel more comfortable starting with zero.)

STEP 3: Now you are ready to plug the sums into the formula. Your calculator should still have the sum of the scores, ΣX, in the display. If not, enter this value.

1. With ΣX in the display, you can compute $(\Sigma X)^2$ simply by pressing the multiply key (×) and then the equals key (=).

2. Now you must divide the squared sum by N, the number of scores. Assuming that you have counted the number of scores, just press the divide key (÷), enter N, and then press the equals key.

Your calculator display now shows $(\Sigma X)^2/N$. Write this number on your paper.

3. Finally, you subtract the value of your calculator display from ΣX^2, which is in the calculator memory. You can do this by simply pressing the memory subtract key (usually M−). The value for *SS* is now in memory, and you can retrieve it by pressing the memory recall key (MR).

Try the whole procedure with a simple set of scores such as 1, 2, 3. You should obtain *SS* = 2 for these scores.

We asked you to write values at several steps during the calculation in case you make a mistake at some point. If you have written the values for ΣX^2, ΣX, and so on, you should be able to compute *SS* easily even if the contents of memory are lost.

EXAMPLE 4.4 The computational formula is used to calculate *SS* for the same set of $N = 4$ scores we used in Example 4.3. First, compute ΣX. Then square each score, and compute ΣX^2. These two values are used in the formula.

This table is used to organize the steps of the computations. It lists each individual score. Do not attempt to do these computations with a frequency distribution table.

X	X^2
1	1
0	0
6	36
1	1

$\Sigma X = 8$
$\Sigma X^2 = 38$

$$SS = \Sigma X^2 - \frac{(\Sigma X)^2}{N}$$

$$= 38 - \frac{(8)^2}{4}$$

$$= 38 - \frac{64}{4}$$

$$= 38 - 16$$

$$= 22$$

Notice that the two formulas produce exactly the same value for *SS*. Although the formulas look different, they are in fact equivalent. The definitional formula should be very easy to learn if you simply remember that *SS* stands for the sum of the squared deviations. If you use notation to write out "the sum of" (Σ) "squared deviations" $(X - \mu)^2$, then you have the definitional formula. Unfortunately, the terms in the computational formula do not translate directly into "sum of squared deviations," so you simply need to memorize this formula.

The definitional formula for *SS* is the most direct way of calculating the sum of squares, but it can be awkward to use for most sets of data. In particular, if the mean is not a whole number, then the deviation scores will all be fractions or decimals, and the calculations become difficult. In addition, calculations with decimals or fractions introduce the opportunity for rounding error, which makes the results less accurate. For these reasons, the computational formula is used most of the time. If you have a small group of scores and the mean is a whole number, then the definitional formula is fine; otherwise, use the computational formula.

FORMULAS FOR POPULATION STANDARD DEVIATION AND VARIANCE

With the definition and calculation of *SS* behind you, the equations for variance and standard deviation become relatively simple. Remember, variance is defined as the mean squared deviation. The mean is the sum divided by *N*, so the equation for variance is

$$\text{variance} = \frac{SS}{N}$$

In the same way that sum of squares, or *SS*, is used to refer to the sum of squared deviations, the term *mean square*, or *MS*, is often used to refer to variance, which is the mean squared deviation.

Standard deviation is the square root of variance, so the equation for standard deviation is

$$\text{standard deviation} = \sqrt{\frac{SS}{N}}$$

There is one final bit of notation before we work completely through an example computing *SS*, variance, and standard deviation. Like the mean (μ), variance and standard deviation are parameters of a population and will be identified by Greek letters. To identify the standard deviation, we use the Greek letter sigma (the Greek letter *s*, standing for standard deviation). The capital letter sigma (Σ) has been used already, so we now use the lowercase sigma, σ:

$$\text{population standard deviation} = \sigma = \sqrt{\frac{SS}{N}} \tag{4.3}$$

The symbol for population variance should help you remember the relationship between standard deviation and variance. If you square the standard deviation, you will get the variance. The symbol for variance is sigma squared, σ^2:

$$\text{population variance} = \sigma^2 = \frac{SS}{N} \tag{4.4}$$

SUMMARY OF COMPUTATION FOR VARIANCE AND STANDARD DEVIATION

The following example will demonstrate the complete process of calculating the variance and the standard deviation for a population of scores. Before we begin the example, however, we will briefly review the basic steps in the calculation.

STEP 1 Find the distance from the mean for each individual.

STEP 2 Square each distance.

STEP 3 Find the sum of the squared distances. This value is called *SS* or *sum of squares*. (*Note: SS* can also be obtained using the computational formula instead of steps 1–3.)

STEP 4 Find the mean of the squared distances. This value is called *variance* and measures the average squared distance from the mean.

STEP 5 Take the square root of the variance. This value is called *standard deviation* and provides a measure of the standard distance from the mean.

EXAMPLE 4.5 The following population of scores will be used to demonstrate the calculation of *SS*, variance, and standard deviation:

$$1, \quad 9, \quad 5, \quad 8, \quad 7$$

These five scores add up to $\Sigma X = 30$, so the mean is $\frac{30}{5} = 6$. Before we do any other calculations, remember that the purpose of variability is to determine how spread out the scores are. Standard deviation accomplishes this by providing a measurement of the standard distance from the mean. The scores we are working with have been placed in a frequency distribution histogram in Figure 4.3 so you can see the variability more easily. Note that the score closest to the mean is $X = 5$ or $X = 7$, both of which are only 1 point away. The score farthest from the mean is $X = 1$, and it is 5 points away. For this distribution, the largest distance from the mean is 5 points, and the smallest distance is 1 point. The typical, or standard, distance should be somewhere between 1 and 5. By looking quickly at a distribution in this way, you should be able to make a rough estimate of the standard deviation. In this case, the standard deviation should be between 1 and 5, probably around 3 points.

FIGURE 4.3

A frequency distribution histogram for a population of $N = 5$ scores. The mean for this population is $\mu = 6$. The smallest distance from the mean is 1 point, and the largest distance is 5 points. The standard distance (or standard deviation) should be between 1 and 5 points.

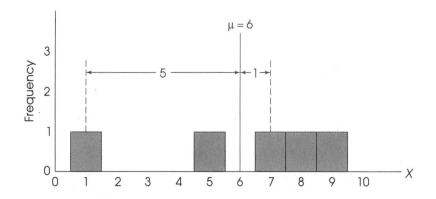

Making a preliminary judgment of standard deviation can help you avoid errors in calculation. If, for example, you worked through the formulas and ended up with a value of $\sigma = 12$, you should realize immediately that you have made an error. (If the largest deviation is only 5 points, then it is impossible for the standard deviation to be 12.)

Now we will start the calculations. The first step is to find SS for this set of scores.

Because the mean is a whole number ($\mu = 6$), we can use the definitional formula for SS:

X	$X - \mu$	$(X - \mu)^2$
1	-5	25
9	$+3$	9
5	-1	1
8	$+2$	4
7	$+1$	1

$\Sigma X = 30$
$\mu = 6$

$40 = \Sigma(X - \mu)^2 = SS$

$\sigma^2 = \dfrac{SS}{N}$

$= \dfrac{40}{5} = 8$

$\sigma = \sqrt{8} = 2.83$

For this set of scores, the variance is $\sigma^2 = 8$, and the standard deviation is $\sigma = \sqrt{8} = 2.83$. Note that the value for the standard deviation is in excellent agreement with our preliminary estimate of the standard distance from the mean.

LEARNING CHECK

1. Write brief definitions of variance and standard deviation.

2. Find SS, variance, and standard deviation for the following population of scores: 10, 10, 10, 10, 10. (*Note:* You should be able to answer this question without doing any calculations.)

3. **a.** Sketch a frequency distribution histogram for the following population of scores: 1, 3, 3, 9. Using this histogram, make an estimate of the standard deviation (i.e., the standard distance from the mean).

   b. Calculate SS, variance, and standard deviation for these scores. How well does your estimate from part a compare with the real standard deviation?

ANSWERS

1. Variance is the mean squared distance from the mean. Standard deviation is the square root of variance and provides a measure of the standard distance from the mean.

2. Because there is no variability in the population, SS, variance, and standard deviation are all equal to zero.

3. **a.** Your sketch should show a mean of $\mu = 4$. The score closest to the mean is $X = 3$, and the farthest score is $X = 9$. The standard deviation should be somewhere between 1 point and 5 points.

   b. For this population, $SS = 36$; the variance is $\frac{36}{4} = 9$; the standard deviation is $\sqrt{9} = 3$.

FIGURE 4.4

The graphic representation of a population with a mean of $\mu = 40$ and a standard deviation of $\mu = 4$.

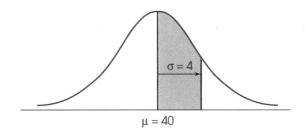

$\sigma = 4$

$\mu = 40$

GRAPHIC REPRESENTATION OF THE MEAN AND STANDARD DEVIATION

In frequency distribution graphs, we will identify the position of the mean by drawing a vertical line and labeling it with μ or $\overline{X}$ (see Figure 4.4). Because the standard deviation measures distance from the mean, it will be represented by a line drawn from the mean outward for a distance equal to the standard deviation (see Figure 4.4). For rough sketches, you can identify the mean with a vertical line in the middle of the distribution. The standard deviation line should extend approximately halfway from the mean to the most extreme score.

4.5 STANDARD DEVIATION AND VARIANCE FOR SAMPLES

The goal of inferential statistics is to use the limited information from samples to draw general conclusions about populations. The basic assumption of this process is that samples should be representative of the populations from which they come. This assumption poses a special problem for variability because samples consistently tend to be less variable than their populations. An example of this general tendency is shown in Figure 4.5. The fact that a sample tends to be less variable than its population means that sample variability gives a *biased* estimate of population

A sample statistic is said to be *biased* if, on the average, it does not provide an accurate estimate of the corresponding population parameter.

FIGURE 4.5

The population of adult heights forms a normal distribution. If you select a sample from this population, you are most likely to obtain individuals who are near average in height. As a result, the scores in the sample will be less variable (spread out) than the scores in the population.

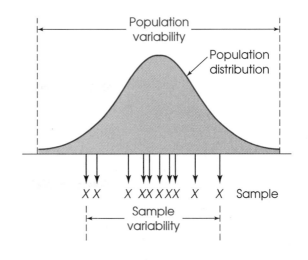

Population variability

Population distribution

$X\ X \quad X\ XX\ XX \quad X \quad X$ Sample

Sample variability

variability. This bias is in the direction of underestimating the population value, rather than being right on the mark. To correct for this bias, it is necessary to make a simple adjustment in the calculation of variability when you are working with sample data. The intent of the adjustment is to make the resulting value for sample variability a more accurate estimate of the population variability.

The calculation of variance and standard deviation for a sample follows the same general process that was used to find population variance and standard deviation. Except for minor changes in notation, the first three steps in this process are exactly the same for a sample as they are for a population. That is, the calculation of SS is the same for a sample as it is for a population. The changes in notation involve using $\overline{X}$ for the sample mean instead of μ and using n instead of N for the number of scores. Thus, SS for a sample is found as follows:

STEP 1   Find the deviation for each score:

$$\text{sample deviation score} = X - \overline{X} \tag{4.5}$$

STEP 2   Square each deviation:

$$\text{squared deviation} = (X - \overline{X})^2$$

STEP 3   Sum the squared deviations:

$$SS = \Sigma(X - \overline{X})^2$$

These three steps can be summarized in a definitional formula for SS:

$$\text{Definitional formula:} \quad SS = \Sigma(X - \overline{X})^2 \tag{4.6}$$

The value of SS can also be obtained using the computational formula. Using sample notation, this formula is

$$\text{Computational formula:} \quad SS = \Sigma X^2 - \frac{(\Sigma X)^2}{n} \tag{4.7}$$

After you compute SS, however, it becomes critical to differentiate between samples and populations. To correct for the bias in sample variability, it is necessary to make an adjustment in the formulas for sample variance and standard deviation. With this in mind, sample variance (identified by the symbol s^2) is defined as

$$\text{sample variance} = s^2 = \frac{SS}{n - 1} \tag{4.8}$$

Sample standard deviation (identified by the symbol s) is simply the square root of the variance.

$$\text{sample standard deviation} = s = \sqrt{\frac{SS}{n - 1}} \tag{4.9}$$

Remember, sample variability tends to underestimate population variability unless some correction is made.

Notice that these sample formulas use $n - 1$ instead of n. This is the adjustment that is necessary to correct for the bias in sample variability. The effect of the adjustment is to increase the value you will obtain. Dividing by a smaller number ($n - 1$ in-

FIGURE 4.6

The frequency distribution histogram for a sample of $n = 7$ scores. The sample mean is $\bar{X} = 5$. The smallest distance from the mean is 1 point, and the largest distance from the mean is 4 points. The standard distance (standard deviation) should be between 1 and 4 points.

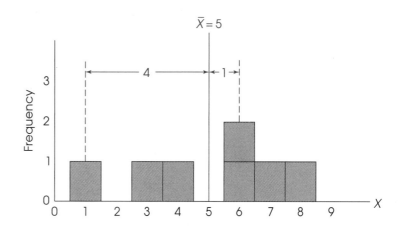

stead of n) produces a larger result and makes sample variance an accurate, or unbiased, estimator of population variance.*

The following example demonstrates the calculation of variance and standard deviation for a sample.

EXAMPLE 4.6

We have selected a sample of $n = 7$ scores. The scores are 1, 6, 4, 3, 8, 7, 6. The frequency distribution histogram for this sample is shown in Figure 4.6. Before we begin any calculations, you should be able to look at the sample distribution and make a preliminary estimate of the outcome. Remember that standard deviation measures the standard distance from the mean. For this sample, the mean is $\bar{X} = 5$ $\left(\frac{35}{7} = 5\right)$. The scores closest to the mean are $X = 4$ and $X = 6$, both of which are exactly 1 point away. The score farthest from the mean is $X = 1$, which is 4 points away. With the smallest distance from the mean equal to 1 and the largest distance equal to 4, we should obtain a standard distance somewhere around 2.5 (between 1 and 4).

Now let's begin the calculations. First, we will find SS for this sample. Because there are only a few scores and the mean is a whole number, the definitional formula will be easy to use. You should try this formula for practice. Meanwhile, we will work with the computational formula.

Caution: The calculation of SS is the same for a sample and a population. Do not use $n - 1$ in the formula for SS.

X	X^2
1	1
6	36
4	16
3	9
8	64
7	49
6	36

$\Sigma X = 35$

$\Sigma X^2 = 211$

$$SS = \Sigma X^2 - \frac{(\Sigma X)^2}{n}$$

$$= 211 - \frac{(35)^2}{7}$$

$$= 211 - \frac{1225}{7}$$

$$= 211 - 175$$

$$= 36$$

Technical note: Using $n - 1$ in the formulas for sample variance and standard deviation is intended to ensure that the sample values are unbiased estimators of the population values. Occasionally, the variance and the standard deviation are computed for a sample when there is no intention of generalizing to a larger population. In these circumstances, it is acceptable to use n (instead of $n - 1$) when computing variance and standard deviation. When the "sample" is the entire set of interest, then it is effectively a "population," and it is appropriate to use the population formulas.

SS for this sample is 36. You should obtain exactly the same answer using the definitional formula. Continuing the calculations,

$$\text{sample variance} = s^2 = \frac{SS}{n-1} = \frac{36}{7-1} = 6$$

Finally, the standard deviation is

$$s = \sqrt{s^2} = \sqrt{6} = 2.45$$

Note that the value we obtained is in excellent agreement with our preliminary prediction.

Remember that the formulas for sample variance and standard deviation were constructed so that the sample variability would provide a good estimate of population variability. For this reason, the sample variance is often called *estimated population variance,* and the sample standard deviation is called *estimated population standard deviation.* When you have only a sample to work with, the sample variance and standard deviation provide the best possible estimates of the population variability.

LEARNING CHECK

1. **a.** Sketch a frequency distribution histogram for the following sample of scores: 1, 1, 9, 1. Using your histogram, make an estimate of the standard deviation for this sample.
   b. Calculate *SS*, variance, and standard deviation for this sample. How well does your estimate from part a compare with the real standard deviation?

2. If the scores in the previous exercise were for a population, what value would you obtain for *SS*?

3. Explain why the formulas for sample variance and standard deviation use $n-1$ instead of n.

ANSWERS

1. **a.** Your graph should show a sample mean of $\overline{X} = 3$. The score farthest from the mean is $X = 9$, and the closest score is $X = 1$. You should estimate the standard deviation to be between 2 points and 6 points.
   b. For this sample, $SS = 48$; the sample variance is $\frac{48}{3} = 16$; the sample standard deviation is $\sqrt{16} = 4$.

2. $SS = 48$ whether the data are from a sample or a population.

3. Sample variability is biased because the scores in a sample tend to be less variable than the scores in the population. To correct for this bias, the formulas divide by $n-1$ instead of n.

SAMPLE VARIABILITY AND DEGREES OF FREEDOM

Although the concept of a deviation score and the calculation of *SS* are almost exactly the same for samples and populations, the minor differences in notation are really very important. When you have only a sample to work with, you must use the sample mean as the reference point for measuring deviations. Using $\overline{X}$ in place of μ

places a restriction on the amount of variability in the sample. The restriction on variability comes from the fact that you must know the value of $\overline{X}$ before you can begin to compute deviations or SS. Notice that if you know the value of $\overline{X}$, then you also must know the value of ΣX. For example, if you have a sample of $n = 3$ scores and you know that $\overline{X} = 10$, then you also know that ΣX must be equal to 30 ($\overline{X} = \Sigma X/n$).

The fact that you must know $\overline{X}$ and ΣX before you can compute variability implies that not all of the scores in the sample are free to vary. Suppose, for example, you are taking a sample of $n = 3$ scores and you know that $\Sigma X = 30$ ($\overline{X} = 10$). Once you have identified the first two scores in the sample, the value of the third score is restricted. If the first scores were $X = 0$ and $X = 5$, then the last score would have to be $X = 25$ in order for the total to be $\Sigma X = 30$. Note that the first two scores in this sample could have any values but that the last score is restricted. As a result, the sample is said to have $n - 1$ degrees of freedom; that is, only $n - 1$ of the scores are free to vary.

DEFINITION

Degrees of freedom, or *df*, for a sample are defined as

$$df = n - 1$$

where n is the number of scores in the sample.

The $n - 1$ degrees of freedom for a sample is the same $n - 1$ that is used in the formulas for sample variance and standard deviation. Remember that variance is defined as the mean squared deviation. As always, this mean is computed by finding the sum and dividing by the number of scores:

$$\text{mean} = \frac{\text{sum}}{\text{number}}$$

To calculate sample variance (mean squared deviation), we find the sum of the squared deviations (SS) and divide by the number of scores that are free to vary. This number is $n - 1 = df$.

$$s^2 = \frac{\text{sum of squared deviations}}{\text{number of scores free to vary}} = \frac{SS}{df} = \frac{SS}{n - 1}$$

Later in this book, we will use the concept of degrees of freedom in other situations. For now, you should remember that knowing the sample mean places a restriction on sample variability. Only $n - 1$ of the scores are free to vary; $df = n - 1$.

4.6 MORE ABOUT VARIANCE AND STANDARD DEVIATION

Variance and standard deviation are two of the most important elements in all of statistics. These two statistical measures are used extensively in both descriptive methods and inferential methods. Although variance and standard deviation can both be defined by mathematical formulas, it is much better to understand the concept

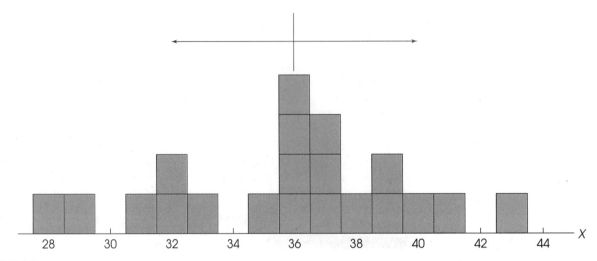

FIGURE 4.7

Image of a sample distribution with a mean of $\overline{X} = 36$ and a standard deviation of $s = 4$.

behind these two measures and to understand how they are used. In the following sections, we will examine some of the uses and characteristics of the variance and the standard deviation.

STANDARD DEVIATION AND DESCRIPTIVE STATISTICS

The general goal of descriptive statistics is to summarize and describe a set of data. The mean and the standard deviation are probably the most commonly used descriptive statistics. A research report, for example, typically will not list all of the individual scores but rather will summarize the data by reporting the mean and the standard deviation. When you are given these two descriptive statistics, you should be able to visualize the entire set of data. For example, consider a sample with a mean of $\overline{X} = 36$ and a standard deviation of $s = 4$. Although there are several different ways to picture the data, one simple technique is to imagine (or sketch) a histogram where each score is represented by a box in the graph. For this sample, the data can be pictured as a pile of boxes (scores), with the center of the pile located at a value of $\overline{X} = 36$. The individual scores or boxes are scattered on both sides of the mean, with some of the boxes relatively close to the mean and some farther away. In this example, the standard distance from the mean is $s = 4$ points. Your image of the distribution should have most of the boxes within 4 points of the mean and only a few boxes located more than 4 points away from $\overline{X} = 36$. The resulting image is shown in Figure 4.7.

Notice that Figure 4.7 not only shows the mean and the standard deviation, but also uses these two values to reconstruct the underlying scale of measurement (the X values along the horizontal line). The scale of measurement helps complete the picture of the entire distribution and helps relate each individual score to the rest of the group. In this example, you should realize that a score of $X = 34$ is located near the center of the distribution, only slightly below the mean. On the other hand, a score of $X = 44$ is an extremely high score, located far out in the right-hand tail of the distribution.

4.2   AN ANALOGY FOR THE MEAN AND THE STANDARD DEVIATION

ALTHOUGH THE basic concepts of the mean and the standard deviation are not overly complex, the following analogy often helps students gain a more complete understanding of these two statistical measures.

In our local community, the site for the new high school was selected because it provides a central location. An alternative site on the western edge of the community was considered, but this site was rejected because it would require extensive busing of students living on the east side of town. The location of the high school is analogous to the concept of the mean; that is, the mean is located in the center of the distribution of scores.

For each student in the community, it is possible to measure the distance between home and the new high school. Some students live only a few blocks from the new school, and others live as much as 3 miles away. The average distance that a student must travel to school was calculated to be 0.80 mile. The average distance from the school is analogous to the concept of the standard deviation; that is, the standard deviation measures the standard distance from an individual score to the mean.

The general point of this discussion is that the mean and the standard deviation are not simply abstract concepts or mathematical equations. Instead, these two values should be concrete and meaningful, especially in the context of a set of scores. Specifically, you should be able to see the mean and the standard deviation as if they were superimposed onto a graph showing the distribution of scores. The mean and the standard deviation are central concepts for most of the statistics that will be presented in the following chapters. A good understanding of these two statistics will help you with the more complex procedures that follow (see Box 4.2).

VARIANCE AND INFERENTIAL STATISTICS

In very general terms, the goal of inferential statistics is to detect meaningful and significant patterns in research results. The basic question is whether the sample data reflect patterns that really exist in the population or simply show random fluctuations that occur by chance. Variability plays an important role in the inferential process because the variability in the data affects how easy it is to see patterns. In general, low variability means that existing patterns can be seen clearly, whereas high variability tends to obscure any patterns that might exist. Consider the following example.

EXAMPLE 4.7   For this example, we ask you to look at a sample of $n = 4$ scores and estimate the sample mean. The first sample consists of the following scores:

X
34
35
36
35

After a few seconds examining the sample, you should realize that the sample mean is $\overline{X} = 35$, and you should be very confident that this is an accurate estimate. Now consider another sample of $n = 4$ scores.

X
26
10
57
37

This time the task of estimating the mean is much more difficult. Once again, the sample mean is $\overline{X} = 35$, but this time it is not easy to see the mean. If you did try to estimate the mean, you probably had very little confidence that your estimate was accurate.

The difference between the two samples is variability. In the first case, the scores are all clustered close together, and variability is small. In this situation, it is easy to see the sample mean. However, the scores in the second sample are spread out over a wide range, and variability is large. With high variability, it is not easy to identify the location of the mean.

The preceding example demonstrates how variability can affect the ability to identify the mean for a single sample. In most research studies, however, the goal is to compare means for two (or more) sets of data. For example:

Is the mean level of depression lower after therapy than it was before therapy?

Is the mean attitude score for men different than the mean score for women?

Is the mean reading achievement score higher for students in a special program than for students in regular classrooms?

In each case, the goal is to find a clear difference between two means that would demonstrate a significant, meaningful pattern in the results. Once again, variability plays an important role in determining whether or not a clear pattern exists. Consider the following example.

EXAMPLE 4.8   A researcher is evaluating a new therapy for treating depression. A group of depressed patients is identified, and each person is given a clinical depression test before the therapy begins. After four weeks of therapy, each patient is given the depression test again. For this example, we will assume that *everyone* who receives the therapy shows a 10-point decrease in depression. Based on this result, would you conclude that the therapy has an effect on depression, and if so, how much effect?

Example 4.8 represents an unrealistic situation where there is no variability. Everyone had exactly the same result, a 10-point decrease in depression. In this situation, it is easy to see that the therapy does have an effect and the magnitude of the effect is exactly 10 points. In the real world, however, things are usually not this neat. In an actual experiment, you might obtain a *mean* decrease of 10 points, but the scores for individual patients would not all be the same. Some patients would show more than a 10-point decrease in depression, some would show less than a 10-point decrease, and some might even show an increase in depression. In other words, the data would have *variability*.

As we saw in Example 4.7, the amount of variability will influence whether or not a clear pattern emerges from the data. Table 4.1 shows two possible outcomes for the depression therapy research study. In experiment 1, the variance is low. All patients show a decrease in depression, and the decrease is close to 10 points for each individual. In this case, it appears that the therapy has a fairly consistent effect; that is, the therapy lowers depression by about 10 points. In experiment 2, however, there is no clear pattern in the data. In this case, the variance is high, with scores ranging from -52 to $+13$. Although the mean for experiment 2 is still $\overline{X} = -10$, the mean does not represent a clear pattern. The data show no consistency or predictability for the effect of the therapy. Although some patients show a large drop in depression, others show an increase in depression, and others show no change at all. In experiment 2, the results do not provide clear evidence that the therapy has any reliable effect.

In the context of inferential statistics, the variance that exists in a set of sample data is often classified as *error variance*. This term is used to indicate that the sample variance represents unexplained and uncontrolled differences between scores. As the error variance increases, it becomes more difficult to see any systematic differences or patterns that might exist in the data. An analogy is to think of variance as the static that exists on a radio or the "snow" on a television screen. In general, variance makes it difficult to get a clear signal from the data. High variance can make it difficult or impossible to see the mean for a set of scores, or to see the mean difference between two sets of data, or to see any meaningful patterns in the results of a research study.

TABLE 4.1

Hypothetical data from two experiments evaluating a new therapy for treating depression (Individual scores represent the amount of change in depression after therapy: A negative value indicates a decrease in depression after therapy; a positive value indicates an increase in depression.)

Experiment 1	Experiment 2
-11	-34
-8	$+13$
-10	$+1$
-11	-8
-9	0
-8	$+2$
-12	-52
-10	-18
-11	$+6$
$\overline{X} = -10$	$\overline{X} = -10$

TRANSFORMATIONS OF SCALE Occasionally, it is convenient to transform a set of scores by adding a constant to each score or by multiplying each score by a constant value. This is done, for example, when you want to "curve" a set of exam scores by adding a fixed amount to each individual's grade or when you want to change the scale of measurement (to convert from minutes to seconds, multiply each X by 60). What happens to the standard deviation when the scores are transformed in this manner?

The easiest way to determine the effect of a transformation is to remember that the standard deviation is a measure of distance. If you select any two scores and see what happens to the distance between them, you also will find out what happens to the standard deviation.

1. Adding a constant to each score does not change the standard deviation. If you begin with a distribution that has $\mu = 40$ and $\sigma = 10$, what happens to σ if you add 5 points to every score? Consider any two scores in this distribution: Suppose, for example, that these are exam scores and that you had $X = 41$ and your friend had $X = 43$. The distance between these two scores is $43 - 41 = 2$ points. After adding the constant, 5 points, to each score, your score would be $X = 46$, and your friend would have $X = 48$. The distance between scores is still 2 points. Adding a constant to every score will not affect any of the distances and therefore will not change the standard deviation. This fact can be seen clearly if you imagine a frequency distribution graph. If, for example, you add 10 points to each score, then every score in the graph would be moved 10 points to the right. The result is that the entire distribution is shifted to a new position 10 points up the scale. Note that the mean moves along with the scores and is increased by 10 points. However, the variability does not change because each of the deviation scores $(X - \mu)$ does not change.

2. Multiplying each score by a constant causes the standard deviation to be multiplied by the same constant. Consider the same distribution of exam scores we looked at earlier. If $\mu = 40$ and $\sigma = 10$, what would happen to σ if each score were multiplied by 2? Again, we will look at two scores, $X = 41$ and $X = 43$, with a distance between them equal to 2 points. After all the scores have been multiplied by 2, these scores become $X = 82$ and $X = 86$. Now the distance between scores is 4 points, twice the original distance. Multiplying each score causes each distance to be multiplied, so the standard deviation also is multiplied by the same amount.

RELATIONSHIP WITH OTHER STATISTICAL MEASURES You should notice that variance and standard deviation have a direct relationship to the mean—namely, they are based on deviations from the mean. Therefore, when data are described, the mean and the standard deviation tend to be reported together. Because the mean is the most commonly reported measure of central tendency, the standard deviation will be the most common measure of variability.

Because the median and the semi-interquartile range are both based on quartiles (remember, median = $Q2$), they share a common foundation and tend to be associated. Whenever the median is used to report central tendency, the semi-interquartile range is commonly used to report variability.

The range, however, has no direct relationship to any other statistical measure. For this reason, it is rarely used in conjunction with other statistical techniques.

IN THE LITERATURE:
REPORTING THE STANDARD DEVIATION

In reporting the results of a study, the researcher often provides descriptive information for both central tendency and variability. The dependent variables in psychology research frequently involve measures taken on interval or ratio scales. Thus, the mean (central tendency) and the standard deviation (variability) are commonly reported together. In many journals, especially those following APA style, the symbol SD is used for the sample standard deviation. For example, the results might state:

> Children who viewed the violent cartoon displayed more aggressive responses ($M = 12.45$, $SD = 3.7$) than those who viewed the control cartoon ($M = 4.22$, $SD = 1.04$).

When reporting the descriptive measures for several groups, the findings may be summarized in a table. Table 4.2 illustrates the results of hypothetical data.

TABLE 4.2

The number of aggressive responses in male and female children after viewing cartoons

	Type of cartoon	
	Violent	Control
Males	$M = 15.72$	$M = 6.94$
	$SD = 4.43$	$SD = 2.26$
Females	$M = 3.47$	$M = 2.61$
	$SD = 1.12$	$SD = 0.98$

Sometimes the table will also indicate the sample size, n, for each group. You should remember that the purpose of the table is to present the data in an organized, concise, and accurate manner. ❏

SUMMARY

1. The purpose of variability is to determine how spread out the scores are in a distribution. There are four basic measures of variability: the range, the semi-interquartile range, the variance, and the standard deviation.

The range is the distance between the upper real limit of the largest X and the lower real limit of the smallest X in the distribution. The semi-interquartile range is one-half the distance between the first quartile and the third quartile. Variance is defined as the mean squared deviation. Standard deviation is the square root of the variance.

2. Standard deviation and variance are the most commonly used measures of variability. Both of these measures are based on the observation that each individual score can be described in terms of its deviation or distance from the mean. The standard deviation is a measure of the standard distance from the mean. The variance is equal to the mean of the squared deviations, that is, the mean squared distance from the mean.

3. To calculate variance or standard deviation, you first need to find the sum of the squared deviations, SS. There are two methods for calculating SS:

I. By definition, you can find SS using the following steps:

a. Find the deviation $(X - \mu)$ for each score.

b. Square each deviation.

c. Sum the squared deviations.

This process can be summarized in a formula as follows:

Definitional formula: $SS = \Sigma(X - \mu)^2$

II. The sum of the squared deviations can also be found using a computational formula, which is especially useful when the mean is not a whole number:

Computational formula: $SS = \Sigma X^2 - \dfrac{(\Sigma X)^2}{N}$

4. Variance is the mean squared deviation and is obtained by finding the sum of the squared deviations and then dividing by the number. For a population, variance is

$$\sigma^2 = \frac{SS}{N}$$

For a sample, only $n - 1$ of the scores are free to vary (degrees of freedom or $df = n - 1$), so sample variance is

$$s^2 = \frac{SS}{n-1} = \frac{SS}{df}$$

5. Standard deviation is the square root of the variance. For a population, this is

$$\sigma = \sqrt{\frac{SS}{N}}$$

Sample standard deviation is

$$s = \sqrt{\frac{SS}{n-1}} = \sqrt{\frac{SS}{df}}$$

Using $n - 1$ in the sample formulas makes the sample variance an accurate and unbiased estimate of the population variance.

6. Adding a constant value to every score in a distribution will not change the standard deviation. Multiplying every score by a constant, however, will cause the standard deviation to be multiplied by the same constant.

KEY TERMS

variability	semi-interquartile range	variance	sum of squares (*SS*)
range	deviation score	standard deviation	degrees of freedom (*df*)
interquartile range			

—— FOCUS ON PROBLEM SOLVING ——

1. The purpose of variability is to provide a measure of how spread out the scores are in a distribution. Usually this is described by the standard deviation. Because the calculations are relatively complicated, it is wise to make a preliminary estimate of the standard deviation before you begin. Remember, standard deviation provides a measure of the typical, or standard, distance from the mean. Therefore, the standard deviation must have a value somewhere between the largest and the smallest deviation scores. As a rule of thumb, the standard deviation should be about one-fourth of the range.

2. Rather than trying to memorize all the formulas for *SS*, variance, and standard deviation, you should focus on the definitions of these values and the logic that relates them to each other:

   SS is the sum of squared deviations.

   Variance is the mean squared deviation.

   Standard deviation is the square root of variance.

   The only formula you should need to memorize is the computational formula for *SS*.

3. If you heed the warnings in the following list, you may avoid some of the more common mistakes in solving variability problems.

   a. Because the calculation of standard deviation requires several steps of calculation, students often get lost in the arithmetic and forget what they are trying to compute. It helps to examine the data before you begin and make a rough estimate of the mean and the standard deviation.

b. The standard deviation formulas for populations and samples are slightly different. Be sure that you know whether the data come from a sample or a population before you begin calculations.

c. A common error is to use $n - 1$ in the computational formula for SS when you have scores from a sample. Remember, the SS formula always uses n (or N). After you compute SS for a sample, you must correct for the sample bias by using $n - 1$ in the formulas for variance and standard deviation.

DEMONSTRATION 4.1

COMPUTING MEASURES OF VARIABILITY

For the following sample data, compute the variance and the standard deviation. Scores:

$$10, \quad 7, \quad 6, \quad 10, \quad 6, \quad 15$$

Compute the sum of squares. For SS, we will use the definitional formula:

$$SS = \Sigma(X - \overline{X})^2$$

STEP 1 Calculate the sample mean for these data.

$$\overline{X} = \Sigma X/n = 54/6 = 9$$

STEP 2 Compute the deviation score, $(X - \overline{X})$, for each X value. This is facilitated by making a table listing the scores in one column and the deviation scores in another column.

X	$X - \overline{X}$
10	$10 - 9 = +1$
7	$7 - 9 = -2$
6	$6 - 9 = -3$
10	$10 - 9 = +1$
6	$6 - 9 = -3$
15	$15 - 9 = +6$

STEP 3 Square the deviation scores. This is shown in a new column labeled $(X - \overline{X})^2$.

X	$X - \overline{X}$	$(X - \overline{X})^2$
10	$+1$	1
7	-2	4
6	-3	9
10	$+1$	1
6	-3	9
15	$+6$	36

STEP 4   Sum the squared deviation scores to obtain the value for *SS*.

$$SS = \Sigma(X - \overline{X})^2 = 1 + 4 + 9 + 1 + 9 + 36 = 60$$

Compute the sample variance. For sample variance, we divide *SS* by $n - 1$ (also known as degrees of freedom).

STEP 1   Compute degrees of freedom, $n - 1$.

$$\text{degrees of freedom} = df = n - 1 = 6 - 1 = 5$$

STEP 2   Divide *SS* by *df*.

$$s^2 = \frac{SS}{n - 1} = \frac{SS}{df} = \frac{60}{5} = 12$$

Compute the sample standard deviation. The sample standard deviation is simply the square root of the sample variance.

$$s = \sqrt{\frac{SS}{n - 1}} = \sqrt{\frac{SS}{df}} = \sqrt{\frac{60}{5}} = \sqrt{12} = 3.46$$

PROBLEMS

1. In words, explain what is measured by each of the following:
   a. *SS*
   b. Variance
   c. Standard deviation

2. A population has $\mu = 100$ and $\sigma = 20$. If you select a single score from this population, on the average, how close would it be to the population mean? Explain your answer.

3. Can *SS* ever have a value less than zero? Explain your answer.

4. In general, what does it mean for a sample to have a standard deviation of zero? Describe the scores in such a sample.

5. A professor administers a 5-point quiz to a class of 40 students. The professor calculates a mean of 4.2 for the quiz with a standard deviation of 11.4. Although the mean is a reasonable value, you should realize that the standard deviation is not. Explain why it appears that the professor made a mistake computing the standard deviation.

6. A population of $N = 15$ scores has a standard deviation of 4.0. What is the variance for this population?

7. A sample of $n = 35$ scores has a variance of 100. What is the standard deviation for this sample?

8. Sketch a normal distribution with $\mu = 50$ and $\sigma = 20$.
   a. Locate each of the following scores in your sketch, and indicate whether you consider each score to be an extreme value (high or low) or a central value:

   65,   55,   40,   47

   b. Make another sketch showing a distribution with $\mu = 50$ but this time with $\sigma = 2$. Now locate each of the four scores in the new distribution, and indicate whether they are extreme or central.

   (*Note:* The value of the standard deviation can have a dramatic effect on the relative location of a score within a distribution.)

9. On an exam with $\mu = 75$, you obtain a score of $X = 80$.
   a. Would you prefer that the exam distribution had $\sigma = 2$ or $\sigma = 10$? (Sketch each distribution, and locate the position of $X = 80$ in each one.)
   b. If your score is $X = 70$, would you prefer $\sigma = 2$ or $\sigma = 10$? (Again, sketch each distribution to determine how the value of σ affects your position relative to the rest of the class.)

10. For the following scores:

   8,   4,   7,   1

a. Calculate the mean. (Note that the value of the mean does not depend on whether the set of scores is considered to be a sample or a population.)

b. Find the deviation for each score, and check that the deviations sum to zero.

c. Square each deviation, and compute SS. (Again, note that the value of SS is independent of whether the set of scores is a sample or a population.)

11. Describe the circumstances where it is easier to compute SS

a. Using the definition (or definitional formula).

b. Using the computational formula.

12. Calculate SS, variance, and standard deviation for the following sample:

0,   3,   0,   3

(*Note:* The computational formula for SS works best with these scores.)

13. Calculate SS, variance, and standard deviation for the following sample:

2,   0,   0,   0,   0,   2,   0,   2,   0

(*Note:* The computational formula for SS works best with these scores.)

14. The standard deviation measures the standard, or typical, distance from the mean. For each of the following two populations, you should be able to use this definition to determine the standard deviation without doing any serious calculations. (*Hint:* Find the mean for each population, and then look at the distances between the individual scores and the mean.)

a. Population 1:   5, 5, 5, 5

b. Population 2:   4, 6, 4, 6

15. Two samples are as follows:

Sample A: 7,   9,   10,   8,   9,   12

Sample B: 13,   5,   9,   1,   17,   9

a. Just by looking at these data, which sample has more variability? Explain your answer.

b. Compute the mean and the standard deviation for each sample.

c. In which sample is the mean more representative (more "typical") of its scores? How does the standard deviation affect the interpretation of the mean?

16. Calculate SS, variance, and standard deviation for the following population:

5,   0,   9,   3,   8,   5

17. Calculate SS, variance, and standard deviation for the following sample:

4,   7,   3,   1,   5

18. For the following population of $N = 12$ scores:

10,   14,   7,   6,   10,   12,   15,   5,   10,   8,   13,   10

a. Sketch a histogram showing the distribution.

b. Find the range, the interquartile range, and the standard deviation for the population. (*Note:* The range and the interquartile range should be easy to see in your sketch. Also, you can use your sketch to determine deviations for each score.)

c. If the two extreme scores, $X = 5$ and $X = 15$, were each moved 3 points farther from the mean, what would happen to each of the three measures of variability (increase, decrease, or stay the same)? (*Note:* In your sketch, move $X = 5$ down to $X = 2$ and move $X = 15$ up to $X = 18$.)

19. For the following sample:

2,   8,   5,   9,   1,   6,   6,   3,   6,   10,   4,   12

a. Calculate the range, the semi-interquartile range, and the standard deviation.

b. Add 2 points to every score, and then compute the range, the semi-interquartile range, and the standard deviation. How is variability affected when a constant is added to every score?

20. For the following sample of $n = 5$ scores:

10,   0,   6,   2,   2

a. Sketch a histogram showing the sample distribution.

b. Locate the value of the sample mean in your sketch, and make an estimate of the sample standard deviation.

c. Compute SS, variance, and standard deviation for the sample. (How well does your estimate compare with the actual value for *s*?)

21. For the following sample of $n = 5$ scores:

3,   11,   14,   7,   15

a. Sketch a histogram showing the sample distribution.

b. Locate the value of the sample mean in your sketch, and make an estimate of the sample standard deviation.

c. Compute SS, variance, and standard deviation for the sample. (How well does your estimate compare with the actual value for *s*?)

22. For the following population of $N = 5$ scores:

   11,   2,   0,   8,   4

   a. Sketch a histogram showing the population distribution.
   b. Locate the value of the population mean in your sketch, and make an estimate of the standard deviation.
   c. Compute SS, variance, and standard deviation for the population. (How well does your estimate compare with the actual value for σ?)

23. For the following population of $N = 6$ scores:

   8,   10,   4,   8,   5,   13

   a. Sketch a histogram showing the population distribution.
   b. Locate the value of the population mean in your sketch, and make an estimate of the standard deviation.
   c. Compute SS, variance, and standard deviation for the population. (How well does your estimate compare with the actual value for σ?)

CHAPTER 5

z-SCORES: LOCATION OF SCORES AND STANDARDIZED DISTRIBUTIONS

TOOLS YOU WILL NEED

The following items are considered essential background material for this chapter. If you doubt your knowledge of any of these items, you should review the appropriate chapter and section before proceeding.

- The mean (Chapter 3)
- The standard deviation (Chapter 4)
- Basic algebra (math review, Appendix A)

CONTENTS

5.1   OVERVIEW

At this point, we have finished the basic elements of *descriptive statistics*. You should recall that descriptive statistics are techniques that attempt to describe and summarize a set of data. The primary descriptive techniques are the following:

1. Construct a frequency distribution that displays the entire set of data.

2. Compute one or two specific values, such as the mean and the standard deviation, that summarize the distribution.

A good understanding of these descriptive statistics will be a great help as we begin to introduce inferential statistics. In particular, whenever we talk about a set of scores, either a sample or a population, you should think of a frequency distribution graph with the scores centered around the mean and the standard deviation describing the typical distance from the mean.

In this chapter and in the next two chapters, we will develop the concepts and skills that form the foundation for inferential statistics. In general, these three chapters will establish formal, quantitative relationships between samples and populations. We will begin with the situation where the sample consists of a single score and then expand to samples of any size. After we have developed the relationships between samples and populations, we can use these relationships to begin inferential statistics. That is, we can begin to use sample data as the basis for drawing conclusions about populations. A brief summary of the next three chapters is as follows:

Chapter 5: We will present a method for describing the exact location of an individual score relative to the other scores in a distribution. This will enable us, for example, to determine precisely whether an individual IQ score is close to average, slightly above average, far below average, and so on.

Chapter 6: In this chapter, we will determine probability values associated with different locations in a distribution of scores. Thus, we will be able to identify a specific score as a central value and say that it is a high probability value. For example, we could determine that a specific IQ score is close to average and is the kind of value that would be expected over 80% of the time.

Chapter 7: In this chapter, we will take the basic skills from Chapters 5 and 6 and apply them to sample means instead of individual scores. Thus, we will be able to describe how any specific sample mean is related to other sample means, and we will be able to determine a probability for each sample mean. For example, we will be able to determine that an average IQ score above 120 is extremely high for a sample of $n = 25$ people and that a sample mean this extreme has a probability that is less than 1 in 1000.

5.2   INTRODUCTION TO z-SCORES

In the previous two chapters, we introduced the concepts of the mean and the standard deviation as methods for describing an entire distribution of scores. Now we

(a)   (b)

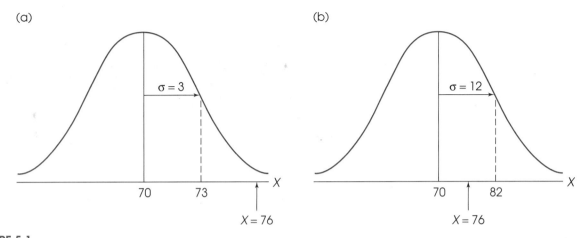

FIGURE 5.1

Two distributions of exam scores. For both distributions, $\mu = 70$, but for one distribution, $\sigma = 3$, and for the other, $\sigma = 12$. The position of $X = 76$ is very different for these two distributions.

will shift attention to the individual scores within a distribution. In this chapter, we introduce a statistical technique that uses the mean and the standard deviation to transform each score (X value) into a *z-score* or a *standard score*. The purpose of z-scores, or standard scores, is to identify and describe the exact location of every score in a distribution.

The following example demonstrates why z-scores are useful and introduces the general concept of transforming X values into z-scores.

EXAMPLE 5.1   Suppose you received a score of $X = 76$ on a statistics exam. How did you do? It should be clear that you need more information to predict your grade. Your score of $X = 76$ could be one of the best scores in the class, or it might be the lowest score in the distribution. To find the location of your score, you must have information about the other scores in the distribution. It would be useful, for example, to know the mean for the class. If the mean were $\mu = 70$, you would be in a much better position than if the mean were $\mu = 85$. Obviously, your position relative to the rest of the class depends on the mean. However, the mean by itself is not sufficient to tell you the exact location of your score. Suppose you know that the mean for the statistics exam is $\mu = 70$ and your score is $X = 76$. At this point, you know that your score is above the mean, but you still do not know exactly where it is located. You may have the highest score in the class, or you may be only slightly above average. Figure 5.1 shows two possible distributions of exam scores. Both distributions have $\mu = 70$, but for one distribution, $\sigma = 3$, and for the other, $\sigma = 12$. Notice that the relative location of $X = 76$ is very different for these two distributions. When the standard deviation is $\sigma = 3$, your score of $X = 76$ is in the extreme right-hand tail, one of the highest scores in the distribution. However, in the other distribution, where $\sigma = 12$, your score is only slightly above average. Thus, your relative location within the distribution depends on the mean and the standard deviation as well as your actual score.

The purpose of the preceding example is to demonstrate that a score *by itself* does not necessarily provide much information about its position within a distribution. These original, unchanged scores that are the direct result of measurement are often called *raw scores.* To make raw scores more meaningful, they are often transformed into new values that contain more information. This transformation is one purpose for *z*-scores. In particular, we will transform *X* values into *z*-scores so that the resulting *z*-scores tell exactly where the original scores are located.

A second purpose for *z*-scores is to *standardize* an entire distribution. A common example of a standardized distribution is the distribution of IQ scores. Although there are several different tests for measuring IQ, all of the tests are standardized so that they have a mean of 100 and a standard deviation of 15. Because all the different tests are standardized, it is possible to understand and compare IQ scores even though they come from different tests. For example, we all understand that an IQ score of 95 is a little below average, *no matter which IQ test was used.* Similarly, an IQ of 145 is extremely high, *no matter which IQ test was used.* In general terms, the process of standardizing takes different distributions and makes them equivalent. The advantage of this process is that it is possible to compare distributions even though they may have been quite different before standardization.

In summary, the process of transforming *X* values into *z*-scores serves two useful purposes:

1. Each *z*-score will tell the exact location of the original *X* value within the distribution.

2. The *z*-scores will form a standardized distribution that can be directly compared to other distributions that also have been transformed into *z*-scores.

Each of these purposes will be discussed in the following sections.

5.3 *z*-SCORES AND LOCATION IN A DISTRIBUTION

One of the primary purposes of a *z*-score is to describe the exact location of a score within a distribution. The *z*-score accomplishes this goal by transforming each *X* value into a signed number (+ or −) so that

1. The *sign* tells whether the score is located above (+) or below (−) the mean, and

2. The *number* tells the distance between the score and the mean in terms of the number of standard deviations.

Thus, in a distribution of standardized IQ scores with $\mu = 100$ and $\sigma = 15$, a score of $X = 130$ would be transformed into $z = +2.00$. The *z* value indicates that the score is located above the mean (+) by a distance of 2 standard deviations (30 points).

DEFINITION

A *z-score* specifies the precise location of each *X* value within a distribution. The sign of the *z*-score (+ or −) signifies whether the score is above the mean (positive) or below the mean (negative). The numerical value of

FIGURE 5.2

The relationship between z-score values and locations in a population distribution.

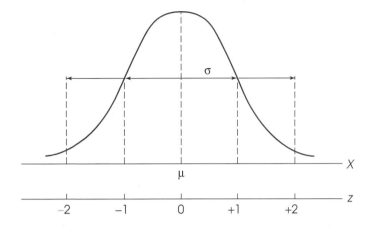

the z-score specifies the distance from the mean by counting the number of standard deviations between X and μ.

Notice that a z-score always consists of two parts: a sign ($+$ or $-$) and a magnitude. Both parts are necessary to describe completely where a raw score is located within a distribution.

Whenever you are working with z-scores, you should imagine or draw a picture similar to Figure 5.2. Although you should realize that not all distributions are normal, we will use the normal shape as an example when showing z-scores.

Figure 5.2 shows a population distribution with various positions identified by their z-score values. Notice that all z-scores above the mean are positive and all z-scores below the mean are negative. The sign of a z-score tells you immediately whether the score is located above or below the mean. Also, note that a z-score of $z = +1.00$ corresponds to a position exactly 1 standard deviation above the mean. A z-score of $z = +2.00$ is always located exactly 2 standard deviations above the mean. The numerical value of the z-score tells you the number of standard deviations from the mean. Finally, you should notice that Figure 5.2 does not give any specific values for the population mean or the standard deviation. The locations identified by z-scores are the same for *all distributions,* no matter what mean or standard deviation the distributions may have.

Now we can return to the two distributions shown in Figure 5.1 and use a z-score to describe the position of $X = 76$ within each distribution as follows:

In Figure 5.1(a), the score $X = 76$ corresponds to a z-score of $z = +2.00$. That is, the score is located *above* the mean by exactly 2 standard deviations.

In Figure 5.1(b), the score $X = 76$ corresponds to a z-score of $z = +0.50$. In this distribution, the score is located exactly $\frac{1}{2}$ standard deviation *above* the mean.

The definition of a z-score indicates that each X value has a corresponding z-score. The following examples demonstrate the relationship between X values and z-scores within a distribution.

EXAMPLE 5.2 A distribution of exam scores has a mean (μ) of 50 and a standard deviation (σ) of 8.

a. For this distribution, what is the z-score corresponding to $X = 58$? Because 58 is *above* the mean, the z-score has a positive sign. The score is

8 points greater than the mean. This distance is exactly 1 standard deviation (because $\sigma = 8$), so the z-score is

$$z = +1$$

This z-score indicates that the raw score is located 1 standard deviation above the mean.

b. What is the z-score corresponding to $X = 46$? The z-score will be negative because 46 is *below* the mean. The X value is 4 points away from the mean. This distance is exactly $\frac{1}{2}$ standard deviation; therefore, the z-score is

$$z = -\tfrac{1}{2}$$

This z-score tells us that the X value is $\frac{1}{2}$ standard deviation below the mean.

c. For this distribution, what raw score corresponds to a z-score of $+2$? This z-score indicates that the X value is 2 standard deviations above the mean. One standard deviation is 8 points, so 2 standard deviations is 16 points. Therefore, the score we are looking for is 16 points above the mean. The mean for the distribution is 50, so the X value is

$$X = 50 + 16 = 66$$

THE z-SCORE FORMULA

The relationship between X values and z-scores can be expressed symbolically in a formula. The formula for transforming raw scores into z-scores is

$$z = \frac{X - \mu}{\sigma} \tag{5.1}$$

The numerator of the equation, $X - \mu$, is a *deviation score* (Chapter 4, page 88); it measures the distance in points between X and μ and indicates whether X is located above or below the mean. We divide this difference by σ because we want the z-score to measure distance in terms of standard deviation units. Remember, the purpose of a z-score is to specify an exact location in a distribution. The z-score formula provides a standard procedure for determining a score's location by calculating the direction and distance from the mean.

EXAMPLE 5.3

A distribution of general psychology test scores has a mean of $\mu = 60$ and a standard deviation of $\sigma = 4$. What is the z-score for a student who received a 66?

Looking at a sketch of the distribution (Figure 5.3), we see that the raw score is above the mean by at least 1 standard deviation but not quite by 2. Judging from the graph, 66 appears to be $1\frac{1}{2}$ standard deviations from the mean. The computation of the z-score with the formula confirms our estimate:

$$z = \frac{X - \mu}{\sigma} = \frac{66 - 60}{4} = \frac{+6}{4} = +1.5$$

FIGURE 5.3

For the population of general psychology test scores, $\mu = 60$ and $\sigma = 4$. A student whose score is 66 is 1.5σ above the mean or has a z-score of $+1.5$.

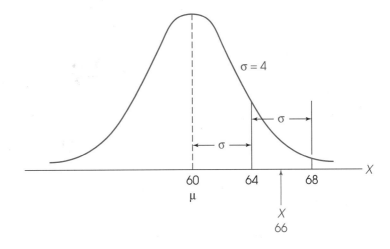

EXAMPLE 5.4   The distribution of SAT verbal scores for high school seniors has a mean of $\mu = 500$ and a standard deviation of $\sigma = 100$. Joe took the SAT and scored 430 on the verbal subtest. Locate his score in the distribution by using a z-score.

Joe's score is 70 points below the mean, so the z-score will be negative. Because 70 points is less than 1 standard deviation, the z-score should have a magnitude that is less than 1. Using the formula, his z-score is

$$z = \frac{X - \mu}{\sigma} = \frac{430 - 500}{100} = \frac{-70}{100} = -0.70$$

DETERMINING A RAW SCORE FROM A z-SCORE

There may be situations in which you have an individual's z-score and would like to determine the corresponding raw score. When you start with a z-score, you can compute the X value by using a different version of the z-score formula. Before we introduce the new formula, let's look at the logic behind converting a z-score back to a raw score.

EXAMPLE 5.5   A distribution has a mean of $\mu = 40$ and a standard deviation of $\sigma = 6$.

What raw score corresponds to $z = +1.5$? The z-score indicates that the X value is located 1.5 standard deviations *above* the mean. Because 1 standard deviation is 6 points, 1.5 standard deviations equal 9 points. Therefore, the raw score is 9 points above the mean, or $X = 49$.

In Example 5.5, we used the z-score and the standard deviation to determine the deviation for an X value; that is, how much distance lies between the raw score and the mean. The deviation score was then added to or subtracted from the mean (depending on the sign of z) to yield the X value. These steps can be incorporated into a formula so that the X value can be computed directly. This formula is obtained by solving the z-score formula for X:

$$z = \frac{X - \mu}{\sigma}$$

$z\sigma = X - \mu$ (Multiply both sides by σ.)

$X - \mu = z\sigma$ (Transpose the equation.)

$X = \mu + z\sigma$ (Add μ to both sides.)   (5.2)

Notice that the third equation in this derivation contains the expression $X - \mu$, the definition for a deviation score (Chapter 4, page 88). Therefore, the deviation score for any raw score can also be found by multiplying the z-score by the standard deviation ($z\sigma$). Essentially, this is the method we used in Example 5.5. If $z\sigma$ provides a deviation score, then we may rewrite formula 5.2 as

raw score = mean + deviation score

In using formula 5.2, always remember that the sign of the z-score ($+$ or $-$) will determine whether the deviation score is added to or subtracted from the mean.

EXAMPLE 5.6   A distribution has a mean of $\mu = 60$ and a standard deviation of $\sigma = 12$.

a. What raw score has $z = +0.25$?

$X = \mu + z\sigma$
$= 60 + 0.25(12)$
$= 60 + 3$
$= 63$

b. What X value corresponds to $z = -1.2$?

$X = \mu + z\sigma$
$= 60 + (-1.2)(12)$
$= 60 - 14.4$
$= 45.6$

LEARNING CHECK   **1.** A positive z-score always indicates a location *above* the population mean. (True or false?)

2. A population has $\mu = 80$ and $\sigma = 8$. Find the z-score for each of the following X values: 84, 72, 90, 78.

3. A population has $\mu = 100$ and $\sigma = 20$. Find the z-score and the X value for each of the following:

a. A location 3 standard deviations below the mean.

b. A location above the mean by $\frac{1}{2}$ standard deviation.

c. A location above the mean by 1 standard deviation.

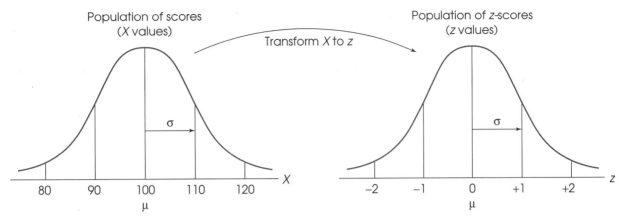

FIGURE 5.4

An entire population of scores is transformed into z-scores. The transformation does not change the shape of the population but the mean is transformed into a value of 0 and the standard deviation is transformed to a value of 1.

ANSWERS **1.** True

2. $X = 84$ corresponds to $z = +0.50$
$X = 72$ corresponds to $z = -1.00$
$X = 90$ corresponds to $z = +1.25$
$X = 78$ corresponds to $z = -0.25$

3. a. $z = -3.00$, $X = 40$
  b. $z = +0.50$, $X = 110$
  c. $z = +1.00$, $X = 120$

5.4 USING z-SCORES TO STANDARDIZE A DISTRIBUTION

It is possible to transform every X value in a distribution into a corresponding z-score. The result of this process is that the entire distribution of X scores is transformed into a distribution of z-scores (see Figure 5.4). The new distribution of z-scores has characteristics that make the *z-score transformation* a very useful tool. Specifically, if every X value is transformed into a z-score, then the distribution of z-scores will have the following properties:

1. Shape. The shape of the z-score distribution will be the same as the original distribution of raw scores. If the original distribution is negatively skewed, for example, then the z-score distribution will also be negatively skewed. If the original distribution is normal, the distribution of z-scores will also be normal. Transforming raw scores into z-scores does not change anyone's position in the distribution. For example, any raw score that is above the mean by 1 standard deviation will be transformed to a z-score of +1.00, which is still above the mean by 1 standard deviation. Transforming a distribution from X values to z values does not move scores from one position to another; the procedure simply relabels each score (see Figure 5.4). Because each individual score stays in its same position within the distribution, the overall shape of the distribution does not change.

2. The Mean. The z-score distribution will *always* have a mean of zero. In Figure 5.4, the original distribution of X values has a mean of $\mu = 100$. When this value, $X = 100$, is transformed into a z-score, the result is

$$z = \frac{X - \mu}{\sigma} = \frac{100 - 100}{10} = 0$$

Thus, the original population mean is transformed into a value of zero in the z-score distribution. The fact that the z-score distribution has a mean of zero makes it easy to identify locations. You should recall from the definition of z-scores that all positive values are above the mean and all negative values are below the mean. A value of zero makes the mean a convenient reference point.

3. The Standard Deviation. The distribution of z-scores will always have a standard deviation of 1. In Figure 5.4, the original distribution of X values has $\mu = 100$ and $\sigma = 10$. In this distribution, a value of $X = 110$ is above the mean by exactly 10 points or 1 standard deviation. When $X = 110$ is transformed, it becomes $z = +1.00$, which is above the mean by exactly 1 point in the z-score distribution. Thus, the standard deviation corresponds to a 10-point distance in the X distribution and is transformed into a 1-point distance in the z-score distribution. The advantage of having a standard deviation of 1 is that the numerical value of a z-score is exactly the same as the number of standard deviations from the mean. For example, a z-score value of 2 is exactly 2 standard deviations from the mean.

In Figure 5.4, we showed the z-score transformation as a process that changed a distribution of X values into a new distribution of z-scores. In fact, there is no need to create a whole new distribution. Instead, you can think of the z-score transformation as simply *relabeling* the values along the X-axis. That is, after a z-score transformation, you still have the same distribution, but now each individual is labeled with a z-score instead of an X value. Figure 5.5 demonstrates this concept with a single distribution that has two sets of labels: the X values along one line and the corresponding z-scores along another line. Notice that the mean for the distribution of z-scores is zero and the standard deviation is 1. Because every z-score distribution has the same mean ($\mu = 0$) and the same standard deviation ($\sigma = 1$), the z-score distribution is called a *standardized distribution.*

DEFINITION

A *standardized distribution* is composed of scores that have been transformed to create predetermined values for μ and σ. Standardized distributions are used to make dissimilar distributions comparable.

A z-score distribution is an example of a standardized distribution with $\mu = 0$ and $\sigma = 1$. That is, when any distribution (with any mean or standard deviation) is transformed into z-scores, the transformed distribution will always have $\mu = 0$ and $\sigma = 1$. The advantage of standardizing is that it makes it possible to compare different scores or different individuals even though they may come from completely different distributions. The procedure for making comparisons is discussed in the following section.

FIGURE 5.5

Following a z-score transformation, the X-axis is relabeled in z-score units. The distance that is equivalent to 1 standard deviation on the X-axis ($\sigma = 10$ points in this example) corresponds to 1 point on the z-score scale.

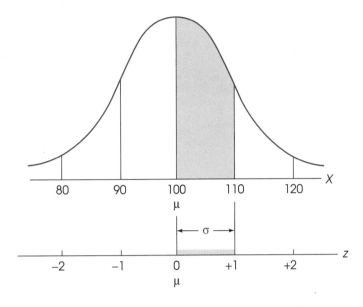

LEARNING CHECK

1. What information does a z-score provide?

2. A population of scores has $\mu = 45$ and $\sigma = 5$. Find the z-scores for the following raw scores:

   a. $X = 47$ **c.** $X = 40$ **e.** $X = 52$ **g.** $X = 45$
   b. $X = 48$ **d.** $X = 44$ **f.** $X = 39$ **h.** $X = 56$

3. For the same population, determine the raw scores that correspond to the following z-scores:

   a. $z = +1.3$ **c.** $z = -3.0$ **e.** $z = +2.8$
   b. $z = -0.4$ **d.** $z = -1.5$ **f.** $z = 0$

4. What is the advantage of having $\mu = 0$ for a distribution of z-scores?

ANSWERS

1. A z-score identifies a precise location in a distribution. The sign indicates whether the location is above or below the mean, and the magnitude of z indicates the number of standard deviations from the mean.

2. **a.** $+0.4$ **c.** -1.0 **e.** $+1.4$ **g.** 0
   b. $+0.6$ **d.** -0.2 **f.** -1.2 **h.** $+2.2$

3. **a.** 51.5   **b.** 43   **c.** 30   **d.** 37.5   **e.** 59   **f.** 45

4. With $\mu = 0$, you know immediately that any positive value is above the mean and any negative value is below the mean.

USING z-SCORES FOR MAKING COMPARISONS

When two scores come from different distributions, it is impossible to make any direct comparison between them. Suppose, for example, Bob received a score of $X = 60$ on a psychology exam and a score of $X = 56$ on a biology test. For which course should Bob expect the better grade?

Because the scores come from two different distributions, you cannot make any direct comparison. Without additional information, it is even impossible to determine whether Bob is above or below the mean in either distribution. Before you can begin to make comparisons, you must know the values for the mean and standard deviation for each distribution. Suppose the biology scores had $\mu = 48$ and $\sigma = 4$, and the psychology scores had $\mu = 50$ and $\sigma = 10$. With this new information, you could sketch the two distributions, locate Bob's score in each distribution, and compare the two locations.

An alternative procedure is to standardize the two distributions. If the biology distribution and the psychology distribution are both transformed to z-scores, then the two distributions will both have $\mu = 0$ and $\sigma = 1$. In the standardized distributions, we can compare Bob's z-score for biology with his z-score for psychology because the z-scores are coming from equivalent, standardized distributions.

In practice, it is not necessary to transform every score in a distribution to make comparisons between two scores. We need to transform only the two scores in question. In Bob's case, we must find the z-scores for his psychology and biology scores. For psychology, Bob's z-score is

Be sure to use the μ and σ values for the distribution to which X belongs.

$$z = \frac{X - \mu}{\sigma} = \frac{60 - 50}{10} = \frac{10}{10} = +1.0$$

For biology, Bob's z-score is

$$z = \frac{56 - 48}{4} = \frac{8}{4} = +2.0$$

Note that Bob's z-score for biology is $+2.0$, which means that his test score is 2 standard deviations above the class mean. On the other hand, his z-score is $+1.0$ for psychology, or 1 standard deviation above the mean. In terms of relative class standing, Bob is doing much better in the biology class. Notice that it is meaningful to make a direct comparison of the two z-scores. A z-score of $+2.00$ *always* indicates a higher position than a z-score of $+1.00$ because all z-score values are based on a standardized distribution with a $\mu = 0$ and $\sigma = 1$.

LEARNING CHECK

1. A normal-shaped distribution with $\mu = 50$ and $\sigma = 8$ is transformed into z-scores. Describe the shape, mean, and standard deviation for the z-score distribution.

2. Why is it possible to compare scores from different distributions after the distributions have been transformed into z-scores?

3. A set of mathematics exam scores has $\mu = 70$ and $\sigma = 8$. A set of English exam scores has $\mu = 74$ and $\sigma = 16$. For which exam would a score of $X = 78$ have a higher standing?

ANSWERS

1. The z-score distribution would be normal (same shape) with $\mu = 0$ and $\sigma = 1$.

2. Comparison is possible because the two distributions will have the same mean ($\mu = 0$) and the same standard deviation ($\sigma = 1$) after transformation.

3. $z = 1.00$ for the mathematics exam, which is a higher standing than $z = 0.25$ for the English exam.

5.5 OTHER STANDARDIZED DISTRIBUTIONS BASED ON z-SCORES

TRANSFORMING z-SCORES TO A DISTRIBUTION WITH A PREDETERMINED μ AND σ

Although z-score distributions have distinct advantages, many people find them cumbersome because they contain negative values and decimals. For these reasons, it is common to standardize a distribution by transforming z-scores to a distribution with a predetermined mean and standard deviation that are whole round numbers. The goal is to create a new (standardized) distribution that has "simple" values for the mean and standard deviation but does not change any individual's location within the distribution. Standardized scores of this type are frequently used in psychological or educational testing. For example, raw scores of the Scholastic Aptitude Test (SAT) are transformed to a standardized distribution that has $\mu = 500$ and $\sigma = 100$. For intelligence tests, raw scores are frequently converted to standard scores that have a mean of 100 and a standard deviation of 15. If the same standardized scale is used for several types of intelligence tests, then the exam scores on different tests can be more readily compared because the distributions will have the same mean and standard deviation. Basically, two steps are involved in standardizing a distribution so that it has a prespecified μ and σ: (1) Each of the raw scores is transformed into a z-score, and (2) each of the z-scores is then converted into a new X value so that a particular μ and σ are achieved. This process assures that each individual has exactly the same z-score (location) in the new distribution as in the original distribution.

EXAMPLE 5.7

An instructor gives an exam to a psychology class. For this exam, the distribution of raw scores has a mean of $\mu = 57$ with $\sigma = 14$. The instructor would like to simplify the distribution by transforming all scores into a new, standardized distribution with $\mu = 50$ and $\sigma = 10$. To demonstrate this process, we will consider what happens to two specific students: Joe, who has a raw score of $X = 64$ in the original distribution; and Maria, whose original raw score is $X = 43$.

STEP 1

Transform each of the original raw scores into z-scores. For Joe, $X = 64$, so his z-score is

$$z = \frac{X - \mu}{\sigma} = \frac{64 - 57}{14} = +0.5$$

Remember, the values of μ and σ are for the distribution from which X was taken.

For Maria, $X = 43$, and her z-score is

$$z = \frac{X - \mu}{\sigma} = \frac{43 - 57}{14} = -1.0$$

STEP 2

Change the z-scores to the new, standardized scores. The instructor wants to create a standardized distribution with $\mu = 50$ and $\sigma = 10$. Joe's z-score, $z = +0.50$, indicates that he is above the mean by exactly $\frac{1}{2}$ standard deviation. In the new distribution, this position would be above the mean by 5 points ($\frac{1}{2}$ of 10), so his standardized score would be 55. Maria's score is located 1 standard deviation below the mean ($z = -1.00$). In the new, standard-

TABLE 5.1

	Joe	Maria
Raw score	$X = 64$	43
Step 1: compute z-score	$z = +0.5$	-1.0
Step 2: standardized score	55	40

ized distribution, Maria is located 10 points ($\sigma = 10$) below the mean ($\mu = 50$), so her new score would be $X = 40$.

The results of this two-step transformation process are summarized in Table 5.1. Notice that Joe, for example, has exactly the same z-score ($z = +0.50$) in both the original, raw score distribution and the new, standardized distribution. This means that Joe's position relative to the other students in the class has not been changed. Similarly, *all* the students stay in the same positions relative to the rest of the class. Thus, standardizing a distribution does not change the shape of the overall distribution, and it does not move individuals around within the distribution—the process simply changes the mean and the standard deviation.

A FORMULA FOR FINDING THE STANDARDIZED SCORE

Earlier in the chapter, we derived a formula (formula 5.2) to find the raw score that corresponds to a particular z-score:

$$X = \mu + z\sigma$$

For purposes of computing the new, standardized score, we can rewrite the equation:

$$\text{standard score} = \mu_{new} + z\sigma_{new} \tag{5.3}$$

The standard score equals the mean of the new, standardized distribution plus its z-score times the standard deviation of the new, standardized distribution. The z-score in the formula is the one computed for the original raw score (step 1). Notice that $z\sigma$ is the deviation score of the standard score. If the raw score is below the mean, then its z-score and $z\sigma$ will be negative. For scores above the mean, $z\sigma$ is positive.

EXAMPLE 5.8

A psychologist has developed a new intelligence test. For years, the test has been given to a large number of people; for this population, $\mu = 65$ and $\sigma = 10$. The psychologist would like to make the scores of his subjects comparable to scores on other IQ tests, which have $\mu = 100$ and $\sigma = 15$. If the test is standardized so that it is comparable (has the same μ and σ) to other tests, what would be the standardized scores for the following individuals?

Person	X
1	75
2	45
3	67

TABLE 5.2

	Computations	
	Step 1: $z = \dfrac{X - \mu}{\sigma}$	Step 2: $X = \mu + z\sigma$
Person 1	$z = \dfrac{75 - 65}{10} = +1.0$	$X = 100 + 1(15)$ $= 100 + 15 = 115$
Person 2	$z = \dfrac{45 - 65}{10} = -2.0$	$X = 100 - 2(15)$ $= 100 - 30 = 70$
Person 3	$z = \dfrac{67 - 65}{10} = +0.2$	$X = 100 + 0.2(15)$ $= 100 + 3 = 103$

STEP 1 Compute the z-score for each individual. Remember, the original distribution has $\mu = 65$ and $\sigma = 10$.

STEP 2 Compute the standardized score for each person. Remember that the standardized distribution will have $\mu = 100$ and $\sigma = 15$. Table 5.2 summarizes the computations for these steps and the results.

	Summary		
Person	X	z	Standardized score
1	75	+1.00	115
2	45	−2.00	70
3	67	+0.20	103

LEARNING CHECK

1. A population has $\mu = 37$ and $\sigma = 2$. If this distribution is transformed into a new distribution with $\mu = 100$ and $\sigma = 20$, what new values will be obtained for each of the following scores: 35, 36, 37, 38, 39?

2. For the following population, $\mu = 7$ and $\sigma = 4$. Scores: 2, 4, 6, 10, 13.

   a. Transform this distribution so $\mu = 50$ and $\sigma = 20$.

   b. Compute μ and σ for the new population. (You should obtain $\mu = 50$ and $\sigma = 20$.)

ANSWERS

1. The five scores, 35, 36, 37, 38, and 39, are transformed into 80, 90, 100, 110, and 120, respectively.

2. **a.** The original scores, 2, 4, 6, 10, and 13, are transformed into 25, 35, 45, 65, and 80, respectively.

   b. The new scores add up to $\Sigma X = 250$, so the mean is $\frac{250}{5} = 50$. The SS for the transformed scores is 2000, the variance is 400, and the new standard deviation is 20.

SUMMARY

1. Each X value can be transformed into a z-score that specifies the exact location of X within the distribution. The sign of the z-score indicates whether the location is above (positive) or below (negative) the mean. The numerical value of the z-score specifies the number of standard deviations between X and μ.

2. The z-score formula is used to transform X values into z-scores:

$$z = \frac{X - \mu}{\sigma}$$

3. To transform z-scores back into X values, solve the z-score equation for X:

$$X = \mu + z\sigma$$

4. When an entire distribution of X values is transformed into z-scores, the result is a distribution of z-scores. The z-score distribution will have the same shape as the distribution of raw scores, and it always will have a mean of 0 and a standard deviation of 1.

5. When comparing raw scores from different distributions, it is necessary to standardize the distributions with a z-score transformation. The distributions will then be comparable because they will have the same parameters ($\mu = 0$, $\sigma = 1$). In practice, it is necessary to transform only those raw scores that are being compared.

6. In certain situations, such as in psychological testing, the z-scores are converted into standardized distributions that have a particular mean and standard deviation.

KEY TERMS

raw score   deviation score   standardized distribution   standard score

z-score z-score transformation

FOCUS ON PROBLEM SOLVING

1. When you are converting an X value to a z-score (or vice versa), do not rely entirely on the formula. You can avoid careless mistakes if you use the definition of a z-score (sign and numerical value) to make a preliminary estimate of the answer before you begin computations. For example, a z-score of $z = -0.85$ identifies a score located *below* the mean by almost 1 standard deviation. When computing the X value for this z-score, be sure that your answer is smaller than the mean, and check that the distance between X and μ is slightly less than the standard deviation.

   A common mistake when computing z-scores is to forget to include the sign of the z-score. The sign is determined by the deviation score ($X - \mu$) and should be carried through all steps of the computation. If, for example, the correct z-score is $z = -2.0$, then an answer of $z = 2.0$ is wrong. In the first case, the raw score is 2 standard deviations *below* the mean. But the second (and incorrect) answer indicates that the X value is 2 standard deviations *above* the mean. These are clearly different answers, and only one can be correct. What is the best advice to avoid careless errors? Sketch the distribution, showing the mean and the raw score (or z-score) in question. This way you will have a concrete frame of reference for each problem.

2. When comparing scores from distributions that have different standard deviations, it is important to be sure that you use the correct value for σ in the z-score formula. Use the σ value for the distribution from which the raw score in question was taken.

3. Remember, a z-score specifies a relative position within the context of a specific distribution. A z-score is a relative value, not an absolute value. For example, a z-score

of $z = -2.0$ does not necessarily suggest a very low raw score—it simply means that the raw score is among the lowest within that specific group.

DEMONSTRATION 5.1

TRANSFORMING X VALUES INTO z-SCORES

A distribution of scores has a mean of $\mu = 60$ with $\sigma = 12$. Find the z-score for $X = 75$.

STEP 1 Determine the sign of the z-score.

First, determine whether X is above or below the mean. This will determine the sign of the z-score. For this demonstration, X is larger than (above) μ, so the z-score will be positive.

STEP 2 Find the distance between X and μ.

The distance is obtained by computing a deviation score.

$$\text{deviation score} = X - \mu = 75 - 60 = 15$$

Thus, the score, $X = 75$, is 15 points above μ.

STEP 3 Convert the distance to standard deviation units.

Converting the distance from step 2 to σ units is accomplished by dividing the distance by σ. For this demonstration,

$$\frac{15}{12} = 1.25$$

Thus, $X = 75$ is 1.25 standard deviations from the mean.

STEP 4 Combine the sign from step 1 with the number from step 2.

The raw score is above the mean, so the z-score must be positive (step 1). For these data,

$$z = +1.25$$

In using the z-score formula, the sign of the z-score will be determined by the sign of the deviation score, $X - \mu$. If X is larger than μ, then the deviation score will be positive. However, if X is smaller than μ, then the deviation score will be negative. For this demonstration, formula 5.1 is used as follows:

$$z = \frac{X - \mu}{\sigma} = \frac{75 - 60}{12} = \frac{+15}{12} = +1.25$$

DEMONSTRATION 5.2

CONVERTING z-SCORES TO X VALUES

For a population with $\mu = 60$ and $\sigma = 12$, what is the X value corresponding to $z = -0.50$?

Notice that in this situation we know the z-score and must find X.

S T E P   1   Locate X in relation to the mean.

The sign of the z-score is negative. This tells us that the X value we are looking for is below μ.

S T E P   2   Determine the distance from the mean (deviation score).

The magnitude of the z-score tells us how many standard deviations there are between X and μ. In this case, X is $\frac{1}{2}$ standard deviation from the mean. In this distribution, 1 standard deviation is 12 points ($\sigma = 12$). Therefore, X is one-half of 12 points from the mean, or

$$(0.5)(12) = 6 \text{ points}$$

S T E P   3   Find the X value.

Starting with the value of the mean, use the direction (step 1) and the distance (step 2) to determine the X value. For this demonstration, we want to find the score that is 6 points below $\mu = 60$. Therefore,

$$X = 60 - 6 = 54.$$

Formula 5.2 is used to convert a z-score to an X value. For this demonstration, we obtain the following, using the formula:

$$X = \mu + z\sigma$$
$$= 60 + (-0.50)(12)$$
$$= 60 + (-6) = 60 - 6$$
$$= 54$$

Notice that the sign of the z-score determines whether the deviation score is added to or subtracted from the mean.

PROBLEMS

1. Describe exactly what information is provided by a z-score.

2. A positively skewed distribution has $\mu = 80$ and $\sigma = 12$. If this entire distribution is transformed into z-scores, describe the shape, mean, and standard deviation for the resulting distribution of z-scores.

3. A distribution of scores has a mean of $\mu = 200$. In this distribution, a score of $X = 250$ is located 50 points above the mean.
   a. Assume that the standard deviation is $\sigma = 25$. Sketch the distribution, and locate the position of $X = 250$. What is the z-score corresponding to $X = 250$ in this distribution?
   b. Now assume that the standard deviation is $\sigma = 100$. Sketch the distribution, and locate $X = 250$. What is the z-score for $X = 250$ now?

4. A distribution of scores has $\mu = 70$. The z-score for $X = 65$ is computed, and a value of $z = +1.00$ is obtained. Regardless of the value for the standard deviation, why must this z-score be incorrect?

5. For a population with $\mu = 50$ and $\sigma = 8$,
   a. Find the z-score that corresponds to each of the following X values:

   $X = 58$ $X = 34$ $X = 70$
   $X = 46$ $X = 62$ $X = 44$

   b. Find the raw score (X) for each of the following z-scores:

   $z = 2.50$ $z = -0.50$ $z = -1.50$
   $z = 0.25$ $z = -1.00$ $z = 0.75$

6. For a population with $\mu = 100$ and $\sigma = 10$,
   a. Find the z-score that corresponds to each of the following X values:

   $X = 106$ $X = 125$ $X = 93$
   $X = 90$ $X = 87$ $X = 118$

b. Find the raw score (X) for each of the following
z-scores:

$$z = 1.20 \quad z = 2.30 \quad z = -0.80$$
$$z = -0.60 \quad z = 0.40 \quad z = -3.00$$

7. A population of scores has $\mu = 60$ and $\sigma = 12$. Find the
z-score corresponding to each of the following X values
from this population:

$$X = 63 \quad X = 48 \quad X = 66$$
$$X = 54 \quad X = 72 \quad X = 84$$

8. A population of scores has $\mu = 85$ and $\sigma = 20$. Find the
raw score (X value) corresponding to each of the following z-scores in this population:

$$z = 1.25 \quad z = -2.30 \quad z = -1.50$$
$$z = 0.60 \quad z = -0.40 \quad z = 2.10$$

9. For a population with $\mu = 80$, a raw score of $X = 98$
corresponds to $z = +2.00$. What is the standard deviation
for this distribution?

10. For a population with $\mu = 50$, a raw score of $X = 44$
corresponds to $z = -1.50$. What is the standard deviation
for this distribution?

11. For a population with $\sigma = 8$, a raw score of $X = 43$ corresponds to $z = -0.50$. What is the mean for this distribution?

12. For a population with $\sigma = 20$, a raw score of $X = 110$
corresponds to $z = +1.50$. What is the mean for this distribution?

13. A list of exam scores shows that a raw score of $X = 50$
corresponds to $z = -2.00$ and a raw score of $X = 62$ corresponds to $z = +1.00$. Find the mean and the standard
deviation for the distribution of raw scores. (*Hint:* Sketch
the distribution, and locate the position of each score.
How much distance is there between the two scores?)

14. On a statistics exam, you have a score of $X = 73$. If the
mean for this exam is $\mu = 65$, would you prefer a standard deviation of $\sigma = 8$ or $\sigma = 16$?

15. Answer the question in problem 14, but this time assume
that the mean for the exam is $\mu = 81$.

16. On a psychology exam with $\mu = 72$ and $\sigma = 12$, you get
a score of $X = 78$. The same day, on an English exam
with $\mu = 56$ and $\sigma = 5$, you get a score of $X = 66$. For
which of the two exams would you expect to receive the
better grade? Explain your answer.

17. The State College requires applicants to submit test
scores from either the College Placement Exam (CPE) or
the College Board Test (CBT). Scores on the CPE have a
mean of $\mu = 200$ with $\sigma = 50$, and scores on the CBT
average $\mu = 500$ with $\sigma = 100$. Tom's application includes a CPE score of $X = 235$, and Bill's application
reports a CBT score of $X = 540$. Based on these scores,
which student is more likely to be admitted? Explain
your answer.

18. New freshmen are required to take a series of tests measuring basic skills. Scores on the Verbal Skills test have
a mean of $\mu = 80$ with $\sigma = 10$. The Math test has
$\mu = 140$ with $\sigma = 30$, and the Logical Reasoning test has
$\mu = 35$ with $\sigma = 4$. Tom's scores are 85, 130, and 45 for
Verbal Skills, Math, and Logical Reasoning, respectively.
For each skill, would you describe Tom as average, well
above average, or well below average? Explain your answers.

19. A distribution of exam scores has $\mu = 90$ and $\sigma = 10$. In
this distribution, Sharon's score is 9 points above the
mean, Jill has a z-score of $+1.20$, Steve's score is $\frac{1}{2}$ standard deviation above the mean, and Ramon has a score
of $X = 110$. List the four students in order from highest
to lowest score.

20. The Wechsler Adult Intelligence Scale is composed of a
number of subtests. Each subtest is standardized to create
a distribution with μ and $\sigma = 6$. For one subtest, the raw
scores have $\mu = 35$ and $\sigma = 6$. Following are some raw
scores from this subtest. What will these scores be when
they are standardized?

$$41, \quad 32, \quad 38, \quad 44, \quad 45, \quad 36, \quad 27$$

21. A distribution of scores has $\mu = 43$ and $\sigma = 4$. If this distribution is standardized to create a new distribution with
$\mu = 60$ and $\sigma = 20$, find the standardized value for each
of the following scores from the original distribution:

$$41, \quad 51, \quad 43, \quad 37, \quad 44, \quad 46$$

22. A distribution with $\mu = 74$ and $\sigma = 8$ is being standardized to produce a new mean of $\mu = 100$ and a new standard deviation of $\sigma = 20$. Find the standardized value for
each of the following scores from the original distribution:

$$84, \quad 78, \quad 80, \quad 66, \quad 62, \quad 72$$

23. A population consists of the following scores:

$$12, \quad 1, \quad 10, \quad 3, \quad 7, \quad 3$$

a. Compute μ and σ for the population.
b. Find the z-score for each raw score in the population.

24. A population consists of the following $N = 5$ scores:

    8, 6, 2, 4, 5

  a. Compute μ and σ for the population.
  b. Find the z-score for each raw score in the population.
  c. Transform the original population into a new population of $N = 5$ scores with $\mu = 100$ and $\sigma = 20$.

25. A population consists of the following $N = 6$ scores:

    14, 3, 12, 5, 9, 5

  a. Compute μ and σ for the population.
  b. Find the z-score for each raw score in the population.
  c. Transform the original population into a new population of $N = 5$ scores with $\mu = 60$ and $\sigma = 8$.

CHAPTER 6   PROBABILITY

TOOLS YOU WILL NEED

The following items are considered essential background material for this chapter. If you doubt your knowledge of any of these items, you should review the appropriate chapter or section before proceeding.

- Proportions (math review, Appendix A)
  - Fractions
  - Decimals
  - Percentages
- Basic algebra (math review, Appendix A)
- z-Scores (Chapter 5)

CONTENTS

6.1 INTRODUCTION TO PROBABILITY

Previously, we have seen that the observations we make during data collection are characterized by variability (Chapter 4). The fact that variability exists in data allows one to consider questions about probability. Consider two extremes: horse races with no variability versus horse races with lots of variability. In the first case, we have a horse named Cigar. Cigar will run in 10 races, but no other horses will be entered. Every time the outcome is the same: Cigar always wins, and if you are betting, you win, too. In the second case, good old Cigar runs 10 races, but each race has 14 other equally fast horses in it. From race to race, the outcomes may vary. The chances you will bet on a winning horse are slimmer. So generally speaking, there is a connection between variability and probability. When there is variability, there is also uncertainty. Probability helps us assess the amount of uncertainty.

In Chapter 1, we noted that samples may vary from their populations (sampling error). In light of this variability, probability plays an important role in inferential statistics. It can be used to describe relationships between samples and populations. Suppose, for example, you are selecting a sample of 1 marble from a jar that contains 50 black and 50 white marbles. Although you cannot guarantee the exact outcome of your sample, it is possible to talk about the potential outcomes in terms of probabilities. In this case, you have a fifty-fifty chance of getting either color. Now consider another jar (population) that has 90 black and only 10 white marbles. Again, you cannot specify the exact outcome of a sample, but now you know that the sample probably will be a black marble. By knowing the makeup of a population, we can determine the probability of obtaining specific samples. In this way, probability gives us a connection between populations and samples.

You may have noticed that the preceding examples begin with a population and then use probability to describe the samples that could be obtained. This is exactly backward from what we want to do with inferential statistics. Remember, the goal of inferential statistics is to begin with a sample and then answer general questions about the population. We will reach this goal in a two-stage process. In the first stage, we develop probability as a bridge from population to samples. This stage involves identifying the types of samples that probably would be obtained from a specific population. Once this bridge is established, we simply reverse the probability rules to allow us to move from samples to populations (see Figure 6.1). The process of reversing the probability relationship can be demonstrated by considering again the two jars of marbles we looked at earlier. (Jar 1 has 50 black and 50 white marbles; jar 2 has 90 black and only 10 white marbles.) This time, suppose that you are blindfolded when the sample is selected and that your task is to use the sample to help you to decide which jar was used. If you select a sample of $n = 4$ marbles and all are black, where did the sample come from? It should be clear that it would be relatively unlikely (low probability) to obtain this sample from jar 1; in four draws, you almost certainly would get at least 1 white marble. On the other hand, this sample would have a high probability of coming from jar 2, where nearly all the marbles are black. Your decision therefore is that the sample probably came from jar 2. Notice that you now are using the sample to make an inference about the population.

FIGURE 6.1

The role of probability in inferential statistics. The goal of inferential statistics is to use the limited information from samples to draw general conclusions about populations. The relationship between samples and populations usually is defined in terms of probability. Probability allows you to start with a population and predict what kind of sample is likely to be obtained. This forms a bridge between populations and samples. Inferential statistics use the *probability bridge* as a basis for making conclusions about populations when you have only sample data.

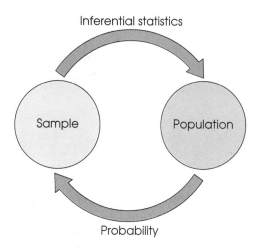

PROBABILITY DEFINITION

Probability is a huge topic that extends far beyond the limits of introductory statistics, and we will not attempt to examine it all here. Instead, we will concentrate on the few concepts and definitions that are needed for an introduction to inferential statistics. We begin with a relatively simple definition of probability.

DEFINITION

In a situation where several different outcomes are possible, we define the *probability* for any particular outcome as a fraction or proportion. If the possible outcomes are identified as *A*, *B*, *C*, *D*, etc., then

$$\text{probability of } A = \frac{\text{number of outcomes classified as } A}{\text{total number of possible outcomes}}$$

For example, when you toss a balanced coin, the outcome will be either heads or tails. Because heads is one of two possible outcomes, the probability of heads is $p = \frac{1}{2}$.

If you are selecting 1 card from a complete deck, there are 52 possible outcomes. The probability of selecting the king of hearts is $p = \frac{1}{52}$. The probability of selecting an ace is $p = \frac{4}{52}$ because there are 4 aces in the deck.

To simplify the discussion of probability, we will use a notation system that eliminates a lot of the words. The probability of a specific outcome will be expressed with a p (for probability) followed by the specific outcome in parentheses. For example, the probability of selecting a king from a deck of cards will be written as p(king). The probability of obtaining heads for a coin toss will be written as p(heads).

You should note that probability is defined as a proportion. This definition makes it possible to restate any probability problem as a proportion problem. For example, the probability problem "What is the probability of obtaining a king from a deck of cards?" can be restated as "Out of the whole deck, what proportion are kings?" In each case, the answer is $\frac{4}{52}$, or "4 out of 52." This translation from probability to proportion may seem trivial now, but it will be a great aid when the probability problems become more complex. In most situations, we are concerned with the

6.1   ZERO PROBABILITY

AN EVENT that never occurs has a probability of zero. However, the opposite of this statement is not always true: A probability of zero does not mean that the event is guaranteed never to occur. Whenever there is an extremely large number of possible events, the probability of any specific event is assigned the value zero. This is done because the probability value tends toward zero as the number of possible events gets large. Consider, for example, the series

$$\frac{1}{10} \quad \frac{1}{100} \quad \frac{1}{1000} \quad \frac{1}{10,000} \quad \frac{1}{100,000}$$

Note that the value of the fraction is getting smaller and smaller, headed toward zero. At the far extreme, when the number of possible events is so large that it cannot be specified, the probability of a single, specific event is said to be zero.

$$\frac{1}{\text{infinite number}} = 0$$

Consider, for example, the fish in the ocean. If there were only 10 fish, then the probability of selecting any particular one would be $p = \frac{1}{10}$. Note that if you add up the probabilities for all 10 fish, you get a total of 1.00. In reality, the number of fish in the ocean is essentially infinite, and, for all practical purposes, the probability of catching any specific fish would be $p = 0$. However, this does not mean that you are doomed to fail whenever you go fishing. The zero probability simply means that you cannot predict in advance which fish you will catch. Note that each individual fish has a probability near or, practically speaking, equal to zero. However, there are so many fish that when you add up all the "zeros," you still get a total of 1.00. In probability for large populations, a value of zero does not mean never. But, practically speaking, it does mean very, very close to never.

probability of obtaining a particular sample from a population. The terminology of *sample* and *population* will not change the basic definition of probability. For example, the whole deck of cards can be considered as a population, and the single card we select is the sample.

The definition we are using identifies probability as a fraction or a proportion. If you work directly from this definition, the probability values you obtain will be expressed as fractions. For example, if you are selecting a card,

$$p(\text{spade}) = \tfrac{13}{52} = \tfrac{1}{4}$$

Or if you are tossing a coin,

$$p(\text{heads}) = \tfrac{1}{2}$$

You should be aware that these fractions can be expressed equally well as either decimals or percentages:

If you are unsure how to convert from fractions to decimals or percentages, you should review the section on proportions in the math review, Appendix A.

$$p = \tfrac{1}{4} = 0.25 = 25\%$$
$$p = \tfrac{1}{2} = 0.50 = 50\%$$

By convention, probability values most often are expressed as decimal values. But you should realize that any of these three forms is acceptable.

You also should note that all the possible probability values are contained in a limited range. At one extreme, when an event never occurs, the probability is zero or 0% (see Box 6.1). At the other extreme, when an event always occurs, the probability is 1, or 100%. For example, suppose you have a jar containing 10 white marbles. The probability of randomly selecting a black marble would be

$$p(\text{black}) = \frac{0}{10} = 0$$

The probability of selecting a white marble would be

$$p(\text{white}) = \frac{10}{10} = 1$$

RANDOM SAMPLING For the preceding definition of probability to be accurate, it is necessary that the outcomes be obtained by a process called random sampling.

DEFINITION A *random sample* must satisfy two requirements:

1. Each individual in the population must have an *equal chance* of being selected.

2. If more than one individual is to be selected for the sample, there must be *constant probability* for each and every selection.

Each of the two requirements for random sampling has some interesting consequences. The first assures that there is no bias in the selection process. For a population with N individuals, each individual must have the same probability, $p = 1/N$, of being selected. Note that the "equal chance" requirement applies to the *individuals* in the population. Earlier we considered an example using a jar of 90 black marbles and 10 white marbles. For a random sample from this jar, each of the 100 marbles must have an equal chance of being selected. In this case, each marble has a probability of $p = 1/100$. The individual probabilities are then combined to find the probability for specific *groups* or *categories* in the population; for example, the probability of obtaining a black marble is $p = 90/100$.

The "equal chance" requirement also means that you must consider the definition of the population before you begin to select a sample. For example, if your population consists of "people in your city," you would not get a random sample by selecting names from the yacht club membership list. A random sample requires that *every* individual have an equal chance of being selected, not just those in the yacht club.

Finally, you should realize that the "equal chance" requirement prohibits you from using the definition of probability in situations where the *individual* outcomes are not equally likely. Consider, for example, the question of whether or not there is life on Mars. There are only two possible alternatives:

1. There is life on Mars.

2. There is no life on Mars.

With only two individual outcomes in the population of alternatives, you might conclude that the probability of life on Mars is $p = 1/2$. However, you should realize that these two individual outcomes are not equally likely, and, therefore, our definition of probability does not apply.

The second requirement also is more interesting than may be apparent at first glance. Consider, for example, the selection of $n = 2$ cards from a complete deck. For the first draw, what is the probability of obtaining the jack of diamonds?

$$p(\text{jack of diamonds}) = \frac{1}{52}$$

Now, for the second draw, what is the probability of obtaining the jack of diamonds? Assuming you still are holding the first card, there are two possibilities:

$$p(\text{jack of diamonds}) = \tfrac{1}{51} \text{ if the first card was not the jack of diamonds}$$

or

$$p(\text{jack of diamonds}) = 0 \text{ if the first card was the jack of diamonds}$$

In either case, the probability is different from its value for the first draw. This contradicts the requirement for random sampling that states the probability must stay constant. To keep the probabilities from changing from one selection to the next, it is necessary to replace each sample before you make the next selection. This solution is called *sampling with replacement*. The second requirement for random samples (constant probability) demands that you sample with replacement. You should note that the requirement for replacement becomes relatively unimportant with very large populations. With large populations, the probability values change so very little without using replacement. Consider a population with $N = 10,000,001$ individuals. Without replacement, the probability from the first to second selection changes only from $p = 1/10,000,001$ to $p = 1/10,000,000$. The type of random sampling we have been describing is known as a *simple random sample* or an *independent random sample*. We should note that there are other types of random samples, although these are beyond the scope of this book.

There are, of course, nonrandom samples, too. In the social and behavioral sciences, truly random samples are not always used. For example, a researcher doing a study on self-esteem in college students might provide a questionnaire to the students in her statistics class. Note that the sample was not randomly selected from all college students based on the requirements of a simple random sample. The students were sampled because they happened to be there congregated in a group—a statistics section. Samples obtained from existing groups are known as *convenience samples*. The problem with a convenience sample is that it is difficult to determine if it is representative of the population.

Finally, there is often confusion between the terms *random sampling* and *random assignment*. Random sampling refers to a specific method used to *select* individuals from a population for purposes of study. Random assignment is a technique for placing individuals into different treatment groups for comparison. Random assignment is used in the experimental method (Chapter 1, p. 11) to ensure that the different treatment groups are more or less equivalent at the start of the study. In random assignment, the probability of assignment to any group is the same for all individuals.

PROBABILITY AND FREQUENCY DISTRIBUTIONS

The situations where we are concerned with probability usually will involve a population of scores that can be displayed in a frequency distribution graph. If you think of the graph as representing the entire population, then different portions of the graph will represent different portions of the population. Because probability and proportion are equivalent, a particular proportion of the graph corresponds to a particular probability in the population. Thus, whenever a population is presented in a frequency distribution graph, it will be possible to represent probabilities as proportions of the graph. The relationship between graphs and probabilities is demonstrated in the following example.

FIGURE 6.2

A frequency distribution histogram for a population that consists of $N = 10$ scores. The shaded part of the figure indicates the portion of the whole population that corresponds to scores greater than $X = 4$. The shaded portion is two-tenths ($p = \frac{2}{10}$) of the whole distribution.

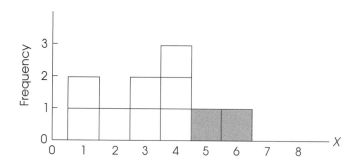

EXAMPLE 6.1   We will use a very simple population that contains only $N = 10$ scores with values 1, 1, 2, 3, 3, 4, 4, 4, 5, 6. This population is shown in the frequency distribution graph in Figure 6.2. If you are taking a random sample of $n = 1$ score from this population, what is the probability of obtaining a score greater than 4? In probability notation,

$$p(X > 4) = ?$$

Using the definition of probability, there are 2 scores that meet this criterion out of the total group of $N = 10$ scores, so the answer would be $p = \frac{2}{10}$. This answer can be obtained directly from the frequency distribution graph if you recall that probability and proportion measure the same thing. Looking at the graph (Figure 6.2), what proportion of the population consists of scores greater than 4? The answer is the shaded part of the distribution, that is, 2 squares out of the total of 10 squares in the distribution. Notice that we now are defining probability as proportion of *area* in the frequency distribution graph. This provides a very concrete and graphic way of representing probability.

Using the same population once again, what is the probability of selecting a score less than 5? In symbols,

$$p(X < 5) = ?$$

Going directly to the distribution in Figure 6.2, we now want to know what part of the graph is not shaded. The unshaded portion consists of 8 out of the 10 blocks ($\frac{8}{10}$ of the area of the graph), so the answer is $p = \frac{8}{10}$.

LEARNING CHECK   1. The animal colony in the psychology department contains 20 male rats and 30 female rats. Of the 20 males, 15 are white and 5 spotted. Of the 30 females, 15 are white and 15 spotted. Suppose you randomly select 1 rat from this colony.

   a. What is the probability of obtaining a female?

   b. What is the probability of obtaining a white male?

   c. Which selection is more likely, a spotted male or a spotted female?

2. What is the purpose of sampling with replacement?

FIGURE 6.3

The normal distribution. The exact shape of the normal distribution is specified by an equation relating each X value (score) with each Y value (frequency). The equation is

$$Y = \frac{1}{\sqrt{2\pi\sigma^2}}\, e^{-(X-\mu)^2/2\sigma^2}$$

(π and e are mathematical constants.) In simpler terms, the normal distribution is symmetrical, with a single mode in the middle. The frequency tapers off as you move farther from the middle in either direction.

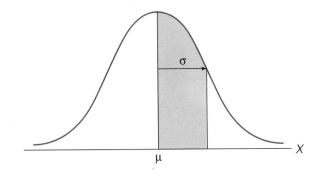

3. Suppose you are going to select a random sample of $n = 1$ score from the distribution in Figure 6.2. Find the following probabilities:

   a. $p(X > 2)$

   b. $p(X > 5)$

   c. $p(X < 3)$

ANSWERS   1. **a.** $p = 30/50 = 0.60$ **b.** $p = 15/50 = 0.30$

   c. A spotted female ($p = 0.30$) is more likely than a spotted male ($p = 0.10$).

   2. Sampling with replacement is necessary to maintain constant probabilities for each and every selection.

   3. **a.** $p = 7/10 = 0.70$ **b.** $p = 1/10 = 0.10$ **c.** $p = 3/10 = 0.30$

6.2   PROBABILITY AND THE NORMAL DISTRIBUTION

The normal distribution was first introduced in Chapter 2 as an example of a commonly occurring shape for population distributions. An example of a normal distribution is shown in Figure 6.3. Although the exact shape for the normal distribution is precisely defined by an equation, we can easily describe its general characteristics:

   1. It is symmetrical. The left side is a mirror image of the right side, and the mean, median, and mode are equal.

   2. Fifty percent of the scores are below the mean, and 50 percent above it (mean = median).

   3. Most of the scores pile up around the mean (mean = mode), and extreme scores (high or low) are relatively rare (low frequencies).

This shape describes many common variables, such as adult heights, manufacturing tolerances, intelligence scores, and personality scores.

FIGURE 6.4

The normal distribution following a
z-score transformation.

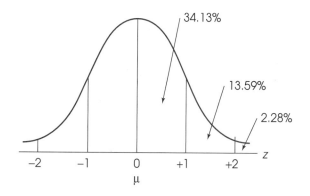

The normal shape also can be described by the proportions of area contained in each section of the distribution. Statisticians often represent these sections by using a normal distribution that has undergone a z-score transformation. Figure 6.4 displays a normal distribution. The sections are marked in z-score units (-2, -1, 0, $+1$, $+2$). Each z-score value indicates the distance from the mean in terms of standard deviations. (Remember, $z = +1$ is one standard deviation above the mean; $z = +2$ is two standard deviations above the mean, and so on.) The graph shows the percentage of scores that fall in each of these sections. For example, the section between the mean ($z = 0$) and the point that is 1 standard deviation above the mean ($z = +1$) makes up 34.13 percent of the scores. Similarly, 13.59 percent of the scores fall between 1 and 2 standard deviations from the mean. The proportions for these sections will be the same for all normal distributions (Figure 6.4). Note that the corresponding sections on the left side of the distribution have the same proportions as the right side because this is a symmetrical distribution. Thus, by this definition, a distribution is normal if and only if it has all the right proportions.

Because the normal distribution is a good model for many naturally occurring distributions and because this shape is guaranteed in some circumstances (as you will see in Chapter 7), we will devote considerable attention to this particular distribution.

The process of answering probability questions about a normal distribution is introduced in the following example.

EXAMPLE 6.2   Adult heights form a normal distribution with a mean of 68 inches and a standard deviation of 6 inches. Given this information about the population and the known proportions under the graph (Figure 6.4), we can determine the probability associated with specific samples. For example, what is the probability of randomly selecting an individual from this population who is taller than 6 feet 8 inches ($X = 80$ inches)?

Restating this question in probability notation, we get

$$p(X > 80) = ?$$

We will follow a step-by-step process to find the answer to this question.

1. First, the probability question is translated into a proportion question: Out of all possible adult heights, what proportion is greater than 80?

FIGURE 6.5

The distribution for Example 6.2.

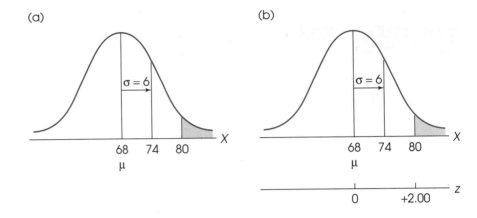

2. We know that "all possible adult heights" is simply the population distribution. This population is shown in Figure 6.5(a). The mean is $\mu = 68$, so the score $X = 80$ is to the right of the mean. Because we are interested in all heights greater than 80, we shade in the area to the right of 80. This area represents the proportion we are trying to determine.

3. Statisticians identify sections of the normal distribution in z-score units. Remember, any distribution that undergoes a z-score transformation is standardized with $\mu = 0$ and $\sigma = 1$, and it retains the original shape (Chapter 5). Thus, *any normal* distribution, regardless of its mean and standard deviation, will look like the graph in Figure 6.4. To get the proportion for the shaded area, we locate the X value using a z-score. For this example,

$$z = \frac{X - \mu}{\sigma} = \frac{80 - 68}{6} = \frac{12}{6} = 2$$

That is, a height of 80 inches is 2 standard deviations above the mean.

4. The proportion we are trying to determine may be expressed in terms of its z-score:

$$p(z > 2)$$

This area is shown in Figure 6.5(b). The shaded area consists of scores beyond 2 standard deviations above the mean. According to Figure 6.4, all normal distributions, regardless of the values for μ and σ, will have 2.28% of the scores in the tail above $z = 2$. Thus, for this population

$$p(X > 80) = 2.28\%$$

THE UNIT NORMAL TABLE   Before we answer any more probability questions, we need to examine a more useful tool than Figure 6.4. Obviously, the graph of the normal distribution in Figure 6.4 is not sufficient for many situations because it only shows proportions in sections marked by whole standard deviation units. That is, if the height for the previous example had been 1.5 standard deviations (or any fraction of a standard

FIGURE 6.6

A portion of the unit normal table. This table lists proportions of the normal distribution corresponding to each z-score value. Column A of the table lists z-scores. Column B lists the proportion in the body of the normal distribution up to the z-score value. Column C lists the proportion of the normal distribution that is located in the tail of the distribution beyond the z-score value.

(A) z	(B) Proportion in body	(C) Proportion in tail
0.00	0.5000	0.5000
0.01	0.5040	0.4960
0.02	0.5080	0.4920
0.03	0.5120	0.4880
0.21	0.5832	0.4168
0.22	0.5871	0.4129
0.23	0.5910	0.4090
0.24	0.5948	0.4052
0.25	0.5987	0.4013
0.26	0.6026	0.3974
0.27	0.6064	0.3936
0.28	0.6103	0.3897
0.29	0.6141	0.3859
0.30	0.6179	0.3821
0.31	0.6217	0.3783
0.32	0.6255	0.3745
0.33	0.6293	0.3707
0.34	0.6331	0.3669

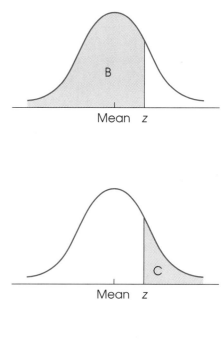

deviation) above the mean, the information in Figure 6.4 would not be adequate for answering probability questions. For this reason, the *unit normal table* was developed. It lists areas, or proportions, for a great many sections of the normal distribution.

To use the unit normal table, you need to transform the distribution of raw scores into z-scores, as was done in Example 6.2 (step 3). The unit normal table basically shows you proportions for sections marked off by a variety of z-score values, more than could be placed in the confined space of a graph.

The complete unit normal table is provided in Appendix B on page A-24; a portion of the table is reproduced in Figure 6.6. The table lists z-score values in column A. Columns B and C are for proportions. Column C represents the proportion of area in the tail beyond the z-score value. Column B contains the proportion for the remainder or "body" of the distribution. Note that for any specific z-score, column B + column C = 1.000 or 100%. Because the normal distribution is symmetrical, you could imagine negative signs in front of every z-score in the table. Again, column C represents the proportion of the distribution in the tail (this time on the left side) beyond z, and column C provides the remainder. To use the unit normal table for determining probabilities, you should keep two things in mind:

If the area you have shaded is *more* than one-half (50%) of the distribution, then use column B.

If the area is *less* than one-half of the distribution, then use column C.

We will use the phrase "body of the distribution" when referring to a portion that is greater than 50% of the total area and "tail" for the remainder (less than 50%).

(a)

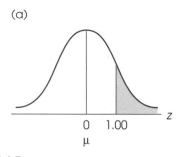

(b)

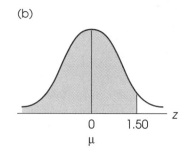

(c)

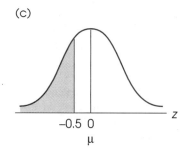

FIGURE 6.7

The distributions for Examples 6.3A–6.3C.

Using the unit normal table   The following examples demonstrate several ways the unit normal table can be used to find proportions or probabilities. *Note:* On a graph, *greater than* consists of the area to the *right* of a value, and *less than* refers to the area to the *left* of that value.

EXAMPLE 6.3A   What proportion of the normal distribution corresponds to z-score values greater than $z = 1.00$? First, you should sketch the distribution and shade in the area you are trying to determine. This is shown in Figure 6.7(a). In this case, the shaded portion is the tail of the distribution beyond $z = 1.00$. To find this shaded area, you simply look up $z = 1.00$ in column A of the unit normal table. Then you read column C (tail) for the proportion. The answer is $p(z > 1.00) = 0.1587$ (or 15.87%). Notice that the answer we obtained is consistent with Figure 6.7(a), which shows this problem. Specifically, the shaded portion of the figure appears to be approximately 15% to 20% of the distribution which agrees with our answer of $p = 0.1587$.

EXAMPLE 6.3B   What proportion of the normal distribution consists of values less than $z = 1.50$? The distribution is sketched placing $z = 1.50$ to the right of the mean. Then everything to the left of (below) $z = 1.50$ is shaded [Figure 6.7(b)]. This shaded area corresponds to column B of the unit normal table. The z-score of 1.50 is found in column A, and the proportion is taken from column B. The answer is $p(z < 1.50) = 0.9332$ (or 93.32%). Once again it is useful to compare the answer with the picture of the problem in Figure 6.7(b). In the figure it appears that approximately 95% of the distribution is shaded, which agrees with our answer of $p = 0.9332$. In general, drawing a sketch of the distribution can help you avoid errors when finding probabilities.

EXAMPLE 6.3C

Moving to the left on the X-axis results in smaller X values and smaller z-scores. Thus, a z-score of -3.00 reflects a smaller value than a z-score of -1.

Many problems will require that you find proportions for negative z-scores. For example, what proportion of the normal distribution corresponds to the tail beyond $z = -0.50$? That is, $p(z < -0.50)$. This portion has been shaded in Figure 6.7(c). To answer questions with negative z-scores, simply remember that the normal distribution is symmetrical with a z-score of zero at the mean,

FIGURE 6.8

The distributions for Examples 6.3D and 6.3E.

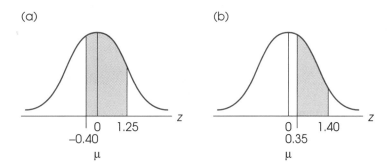

positive values to the right, and negative values to the left. The proportion in the left tail beyond $z = -0.50$ is identical to the proportion in the right tail beyond $z = +0.50$. To find this proportion, look up $z = 0.50$ in column A, and find the proportion in column C (tail). You should get an answer of 0.3085 (30.85%).

EXAMPLE 6.3D Some problems require that you find the proportion between two z-score values. For example, what proportion of the scores fall between $z = -0.40$ and $z = +1.25$? That is, $p(-0.40 < z < +1.25)$. Figure 6.8(a) shows the normal distribution, with the area between these z-scores shaded. There is no column in the unit normal table that directly gives us the answer. However, note that the unshaded portion of the graph is the two tails beyond $z = -0.40$ and $z = +1.25$. These two tails comprise exactly the area we do *not* want. The solution is to subtract both of these proportions from 1.00 (100%). Thus, for $z = -0.40$, the column C (tail) entry is 0.3446, and for $z = +1.25$, it is 0.1056. Subtracting these two tails from the whole distribution will give us the proportion left over in the shaded area:

$$p(-0.40 < z < +1.25) = 1.0000 - 0.3446 - 0.1056$$
$$= 0.5498 \text{ (or } 54.98\%)$$

EXAMPLE 6.3E If the proportion we need to find is between two z-scores that are on the same side of the mean, then the solution is a little different. For example, what proportion of the scores fall between $z = +0.35$ and $z = +1.40$? Figure 6.8(b) depicts the distribution, with the area between the two z-values shaded. Notice that for $z = +0.35$, column C gives us the proportion in the tail beyond this z-score. However, this is not exactly what we want. The area in the tail above $z = +1.40$ should be excluded. The solution is to subtract this area from the proportion of scores above $z = +0.35$. Thus, for $z = +0.35$, the proportion from column C is 0.3632, and for $z = +1.40$, it is 0.0808. Subtracting the smaller tail from the larger one will give us the proportion of scores in the shaded area:

$$p(0.35 < z < 1.40) = 0.3632 - 0.0808 = 0.2824 \text{ (or } 28.24\%)$$

You may have noticed that we have sketched distributions for each of the preceding problems. As a general rule, you should always sketch a distribution, locate the mean with a vertical line, and shade in the portion you are trying to determine. Look at your sketch. It will indicate which columns to use in the unit normal table. If you make a habit of drawing sketches, you will avoid careless errors in using the table.

LEARNING CHECK

1. Find the proportion of the normal distribution that is associated with the following sections of a graph:

   a. $z < +1.00$

   b. $z > +0.80$

   c. $z < -2.00$

   d. $z > -0.33$

   e. $z > -0.50$

   f. $z > -1.50$

   g. $z < +0.67$

   h. $z > +2.00$

2. Find the proportion of the normal distribution that is between the following sets of z-scores:

   a. $p(-0.67 < z < +0.67)$

   b. $p(1.00 < z < 2.00)$

   c. $p(-1.50 < z < +0.50)$

ANSWERS   1. a. 0.8413   b. 0.2119   c. 0.0228   d. 0.6293   e. 0.6915   f. 0.9332

   g. 0.7486   h. 0.0228

   2. a. 0.4972   b. 0.1359   c. 0.6247

ANSWERING PROBABILITY QUESTIONS WITH THE UNIT NORMAL TABLE

The unit normal table provides a listing of proportions (or probability values) corresponding to a great many z-scores in a normal distribution. To use this table to answer probability questions, you must first transform the X values into z-scores (standardize the distribution) and then use the table to look up the probability value. This process is discussed in Box 6.2. Finding probabilities with the unit normal table includes these steps:

1. Sketch the distribution, showing the mean and standard deviation.

2. On the sketch, locate the specific score identified in the problem, and draw a vertical line through the distribution at this location. An approximate location is good enough for the sketch, but make sure you place the score and the vertical line on the correct side of the mean and at roughly the correct distance.

3. Read the problem again to determine whether it asks for values *greater than* (to the right side of) or *less than* (to the left of) the specific score. Shade the appropriate section of the distribution. You should be able to make a rough preliminary estimate based on the appearance of your sketch. Does the shaded area look like 10% or 40% or 60%, and so on, of the distribution?

FIGURE 6.9

The distribution for Example 6.4.

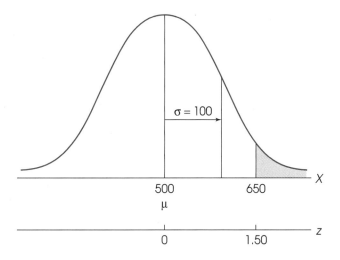

4. Transform the X value from the problem into a z-score. The z-score identifies the precise location in the distribution and will be used to find the proportion in the unit normal table.

5. Look at the shaded area in your sketch to determine which column (B or C) corresponds to the proportion you are trying to find. Ignore the sign of the z-score, and look it up in the unit normal table, taking the proportion from column B or C (whichever is appropriate).

E X A M P L E   6 . 4   Scores on the Scholastic Aptitude Test (SAT) form a normal distribution with $\mu = 500$ and $\sigma = 100$. What is the probability of selecting an individual from the population who scores above 650?

$$p(X > 650) = ?$$

Restated as a proportion question, we want to find out what proportion of the whole distribution has values greater than 650. This distribution is drawn in Figure 6.9, and the part we want has been shaded.

The next step is to change the X values to z-scores. Specifically, the score of $X = 650$ is changed to

You cannot go directly from a score to the unit normal table. You always must go by way of z-scores. (See Box 6.2.)

$$z = \frac{X - \mu}{\sigma} = \frac{650 - 500}{100} = \frac{150}{100} = 1.50$$

Next, you look up this z-score value in the unit normal table. Because we want the proportion of the distribution in the tail beyond 650 (Figure 6.9), the answer will be found in column C. A z-score of 1.5 corresponds to a proportion of 0.0668.

The probability of randomly selecting an individual who scored above 650 is 0.0668:

$$p(X > 650) = p(z > 1.5) = 0.0668 \text{ (or 6.68\%)}$$

FIGURE 6.10

The distribution for Example 6.5.

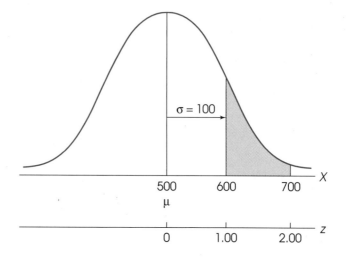

EXAMPLE   6.5   This example demonstrates the process of determining the probability associated with a specified range of scores in a normal distribution. We will use the
distribution of SAT scores, which is normal with $\mu = 500$ and $\sigma = 100$. For
this distribution, what is the probability of randomly selecting an individual
with a score between $X = 600$ and $X = 700$? In probability notation, the problem is to find

$$p(600 < X < 700) = ?$$

The distribution of SAT scores is shown in Figure 6.10, with the appropriate
area shaded. Remember, finding the probability is the same thing as finding
the proportion of the distribution located between 600 and 700. The first step
is to transform each X value into a z-score:

$$\text{For } X = 600: z = \frac{X - \mu}{\sigma} = \frac{600 - 500}{100} = \frac{100}{100} = 1.00$$

$$\text{For } X = 700: z = \frac{X - \mu}{\sigma} = \frac{700 - 500}{100} = \frac{200}{100} = 2.00$$

The problem now is to find the proportion of the distribution that is located
between $z = +1.00$ and $z = +2.00$. The solution uses the same method as
Example 6.3E. That is, you look up the proportions in the tails (column C) for
both z-score values and subtract the proportion beyond $z = 2.00$. Using column C of the table, we find that the area beyond $z = 1.00$ is 0.1587. This includes the shaded portion we want but also includes the extra area beyond
$z = 2.00$. This extra area beyond $z = 2.00$ is 0.0228 (column C) of the entire
distribution. Subtracting out this extra portion provides the final answer of

$$p(600 < X < 700) = 0.1587 - 0.0228 = 0.1359 \text{ (or } 13.59\%)$$

FINDING PROBABILITIES FROM A NORMAL DISTRIBUTION

WORKING WITH probabilities for a normal distribution involves two steps: (1) using a z-score formula and (2) using the unit normal table. However, the order of the steps may vary, depending on the type of probability question you are trying to answer.

In one instance, you may start with a known X value and have to find a probability that is associated with it (as in Example 6.4). First, you must convert the X value to a z-score using equation 5.1 (p. 116). Then you consult the unit normal table to get the probability associated with the particular area of the graph. *Note:* You cannot go directly from the X value to the unit normal table. You must find the z-score first.

However, suppose you begin with a known probability value and want to find the X value associated with it (as in Example 6.6). In this case, you use the unit normal table first to find the z-score that corresponds with the probability value. Then you convert the z-score into an X value using equation 5.2 (p. 118).

Figure 6.11 illustrates the steps you must take when moving from an X value to a probability or from a probability back to an X value. This chart, much like a map, guides you through the essential steps as you "travel" between X values and probabilities.

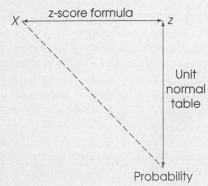

FIGURE 6.11

On this map, the solid lines are the "roads" that you must travel to find a probability value that corresponds to any specific score or to find the score that corresponds to any specific probability value. In taking these routes, you must pass through the intermediate step of finding a z-score value. *Note:* You may *not* travel directly between X values and probability (that is, along the dashed line).

In the previous two examples, we started with an X value and used the unit normal table to find a corresponding probability value. Looking at the map in Box 6.2, we started at X, moved to z, and then moved on to p. Like most maps, this one can be used to guide travel in either direction; that is, it is possible to start at p and move to X. To move in this direction, you start with a specific probability value (proportion), find its corresponding z-score, and then determine the X value. The following example demonstrates this type of problem.

EXAMPLE 6.6   Scores on the SAT form a normal distribution with $\mu = 500$ and $\sigma = 100$. What is the minimum score necessary to be in the top 15% of the SAT distribution? This problem is shown graphically in Figure 6.12.

In this problem, we begin with a proportion (15% = 0.15), and we are looking for a score. According to the map in Box 6.2, we can move from p (proportion) to X (score) by going via z-scores. The first step is to use the unit normal table to find the z-score that corresponds to a proportion of 0.15. Because the proportion is located beyond z in the tail of the distribution, we will look in column C for a proportion of 0.1500. Note that you may not find 0.1500 exactly, but locate the closest value possible. In this case, the closest value in the table is 0.1492, and the z-score that corresponds to this proportion is $z = 1.04$.

FIGURE 6.12

The distribution of SAT scores. The problem is to locate the score that separates the top 15% from the rest of the distribution. A line is drawn to divide the distribution roughly into 15% and 85% sections.

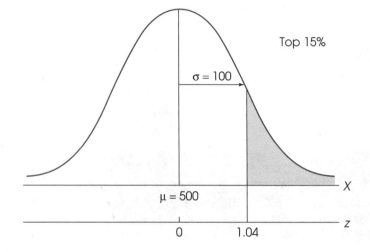

The next step is to determine whether the z-score is positive or negative. Remember, the table does not specify the sign of the z-score. Looking at the graph in Figure 6.12, you should realize that the score we want is above the mean, so the z-score is positive: $z = +1.04$.

Now you are ready for the last stage of the solution, that is, changing the z-score into an X value. Using z-score equation 5.2 (p. 118) and the known values of μ, σ, and z, we obtain

$$X = \mu + z\sigma$$
$$= 500 + 1.04(100)$$
$$= 500 + 104$$
$$= 604$$

The conclusion for this example is that you must have an SAT score of at least 604 to be in the top 15% of the distribution.

LEARNING CHECK

1. For a normal distribution with a mean of 80 and a standard deviation of 10, find each probability value requested.
   a. $p(X > 85) = ?$ c. $p(X > 70) = ?$
   b. $p(X < 95) = ?$ d. $p(75 < X < 100) = ?$

2. For a normal distribution with a mean of 100 and a standard deviation of 20, find each value requested.
   a. What score separates the top 40% from the bottom 60% of the distribution?
   b. What is the minimum score needed to be in the top 5% of this distribution?
   c. What scores form the boundaries for the middle 60% of this distribution?

3. What is the probability of selecting a score greater than 45 from a positively skewed distribution with $\mu = 40$ and $\sigma = 10$?

ANSWERS **1. a.** $p = 0.3085$ (30.85%)   **c.** $p = 0.8413$ (84.13%)
b. $p = 0.9332$ (93.32%)   **d.** $p = 0.6687$ (66.87%)

2. a. $z = +0.25$; $X = 105$ **b.** $z = +1.64$; $X = 132.8$
c. $z = \pm0.84$; boundaries are 83.2 and 116.8

3. You cannot obtain the answer. The unit normal table cannot be used to answer this question because the distribution is not normal.

6.3 PERCENTILES AND PERCENTILE RANKS

Another useful aspect of the normal distribution is that we can determine the relative standing of an individual score by using *percentile ranks* and *percentiles.* The percentile rank of a particular score is defined as the percentage of scores in the distribution that are at or below that particular score. The particular score associated with a percentile rank is called a percentile or percentile point. Suppose, for example, you have a score of $X = 43$ on an exam and you know that exactly 60% of the class had scores of 43 or lower. Then your score $X = 43$ has a percentile rank of 60%, and the score of 43 is called the 60th percentile. Remember, a percentile rank refers to a percentage of the distribution, and a percentile refers to a score.

DEFINITIONS

The *percentile rank* of a particular score is expressed as the percentage of individuals in the distribution with scores at or below that particular score.

The particular score associated with a percentile rank is called a *percentile.*

FINDING PERCENTILE RANKS

Finding percentile ranks for normal distributions is straightforward if you sketch the distribution. Because a percentile rank is the percentage of the individuals that fall *below* a particular score, we will need to find the proportion of the distribution to the *left* of the score. When finding percentile ranks, we will always be concerned with the percentage on the left-hand side of some X value or, in terms of symbols, $p(X < \text{some value})$. Therefore, if the X value is above the mean, the proportion to the left will be found in column B (the body of the distribution) of the unit normal table. Alternatively, if the X value is below the mean, everything to its left is reflected in column C (the tail). The following examples illustrate these points.

EXAMPLE 6.7A

A population is normally distributed with $\mu = 100$ and $\sigma = 10$. What is the percentile rank for $X = 114$?

Because a percentile rank indicates one's standing relative to all lower scores, we must focus on the area of the distribution to the *left* of $X = 114$. The distribution is shown in Figure 6.13, and the area of the curve containing all scores below $X = 114$ is shaded. The proportion for this shaded area will give us the percentile rank.

FIGURE 6.13

The distribution for Example 6.7A. The proportion for the shaded area provides the percentile rank for $X = 114$.

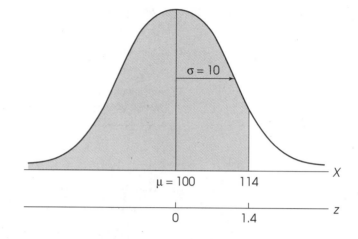

Because the distribution is normal, we can use the unit normal table to find this proportion. The first step is to compute the z-score for the X value we are considering.

$$z = \frac{X - \mu}{\sigma} = \frac{114 - 100}{10} = \frac{14}{10} = 1.40$$

The next step is to consult the unit normal table. Note that the shaded area in Figure 6.13 makes up the large body of the graph. The proportion for this area is presented in column B. For $z = 1.40$, column B indicates a proportion of 0.9192. The percentile rank for $X = 114$ is 91.92%.

E X A M P L E   6 . 7 B   For the distribution in Example 6.7A, what is the percentile rank for $X = 92$?

This example is diagrammed in Figure 6.14. The score $X = 92$ is placed in the left side of the distribution because it is below the mean. Again, percentile

FIGURE 6.14

The distribution for Example 6.7B. The proportion for the shaded area provides the percentile rank for $X = 92$.

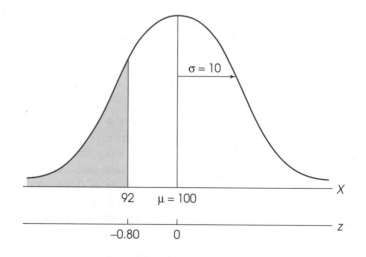

FIGURE 6.15

The distribution for Example 6.8.

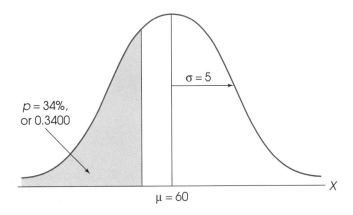

ranks deal with the area of the distribution below the score in question. Therefore, we have shaded the area to the left of $X = 92$.

First, the X value is transformed to a z-score:

$$z = \frac{X - \mu}{\sigma} = \frac{92 - 100}{10} = \frac{-8}{10} = -0.80$$

Remember, the normal distribution is symmetrical. Therefore, the proportion in the right-hand tail beyond $z = 0.80$ is identical to that in the left-hand tail beyond $z = -0.80$.

Now the unit normal table can be consulted. The proportion in the left-hand tail beyond $z = -0.80$ can be found in column C. According to the unit normal table, for $z = 0.80$, the proportion in the tail is $p = 0.2119$. This also is the area beyond $z = -0.80$. Thus, the percentile rank for $X = 92$ is 21.19%. That is, a score of 92 is greater than 21.19% of the scores in the distribution.

FINDING PERCENTILES   The process of finding a particular percentile is very similar to the process used in Example 6.6. You are given a percentage (this time a percentile rank), and you must find the corresponding X value (the percentile). You should recall that finding an X value from a percentage requires the intermediate step of determining the z-score for that proportion of the distribution. The following example demonstrates this process for percentiles.

EXAMPLE 6.8   A population is normally distributed with $\mu = 60$ and $\sigma = 5$. For this population, what is the 34th percentile?

In this example, we are looking for an X value (percentile) that has 34% (or $p = 0.3400$) of the distribution below it. This problem is illustrated in Figure 6.15. Note that 34% is roughly equal to one-third of the distribution, so the corresponding shaded area in Figure 6.15 is located entirely on the left-hand side of the mean. In this problem, we begin with a proportion (34% = 0.3400), and we are looking for a score (the percentile). The first step in moving from a proportion to a score is to find the z-score (Box 6.2). You must look at the unit normal table to find the z-score that corresponds to a proportion of 0.3400. Because the proportion is in the tail beyond z, you must look in column C for a proportion of 0.3400. However, there is no entry in the table for the proportion of exactly 0.3400. Instead, we use the closest

Be careful: The table does not differentiate positive and negative z-scores. You must look at your sketch of the distribution to determine the sign.

FIGURE 6.16

The *z*-scores corresponding to the first, second, and third quartiles in a normal distribution.

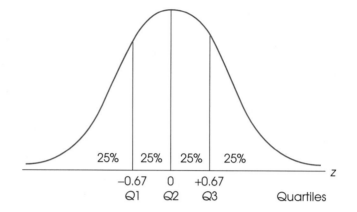

value, which you will find to be $p = 0.3409$. The *z*-score corresponding to this value is $z = -0.41$. Note that it is a negative *z*-score because it is below the mean (Figure 6.15). Thus, the *X* value for which we are looking has a *z* of -0.41.

The next step is to convert the *z*-score to an *X* value. Using the *z*-score formula solved for *X*, we obtain

$$X = \mu + z\sigma$$
$$= 60 + (-0.41)(5)$$
$$= 60 - 2.05$$
$$= 57.95$$

The 34th percentile for this distribution is $X = 57.95$. This answer makes sense because the 50th percentile for this example is 60 (the mean and median). Therefore, the 34th percentile has to be a value less than 60.

QUARTILES Percentiles divide the distribution into 100 equal parts, each corresponding to 1% of the distribution. The area in a distribution can also be divided into four equal parts called quartiles, each corresponding to 25%. We first looked at quartiles in Chapter 4 (p. 85) in considering the semi-interquartile range. The first quartile ($Q1$) is the score that separates the lowest 25% of the distribution from the rest. Thus, the first quartile is the same as the 25th percentile. Similarly, the second quartile ($Q2$) is the score that has 50% (two quarters) of the distribution below it. You should recognize that $Q2$ is the median or 50th percentile of the distribution. Finally, the third quartile ($Q3$) is the *X* value that has 75% (three quarters) of the distribution below it. The $Q3$ for a distribution is also the 75th percentile.

For a normal distribution, the first quartile always corresponds to $z = -0.67$, the second quartile corresponds to $z = 0$ (the mean), and the third quartile corresponds to $z = +0.67$ (Figure 6.16). These values can be found by consulting the unit normal table and are true of any normal distribution. This makes finding quartiles and the semi-interquartile range straightforward for normal distributions. The following example demonstrates the use of quartiles.

EXAMPLE 6.9   A population is normally distributed and has a mean of $\mu = 50$ with a standard deviation of $\sigma = 10$. Find the first, second, and third quartiles, and compute the semi-interquartile range.

The first quartile, $Q1$, is the same as the 25th percentile. The 25th percentile has a corresponding z-score of $z = -0.67$. With $\mu = 50$ and $\sigma = 10$, we can determine the X value of $Q1$.

$$X = \mu + z\sigma$$
$$= 50 + (-0.67)(10)$$
$$= 50 - 6.7$$
$$= 43.3$$

The second quartile, $Q2$, is also the 50th percentile, or median. For a normal distribution, the median equals the mean, so $Q2$ is 50. By the formula, with a z-score of 0, we obtain

$$X = \mu + z\sigma$$
$$= 50 + 0(10)$$
$$= 50$$

The third quartile, $Q3$, is also the 75th percentile. It has a corresponding z-score of $z = +0.67$. Using the z-score formula solved for X, we obtain

$$X = \mu + z\sigma$$
$$= 50 + 0.67(10)$$
$$= 50 + 6.7$$
$$= 56.7$$

In Chapter 4 (p. 86), the semi-interquartile range was defined as one-half the distance between the first and third quartiles, or

$$\text{semi-interquartile range} = \frac{(Q3 - Q1)}{2}$$

For this example, the semi-interquartile range is

$$\text{semi-interquartile range} = \frac{(Q3 - Q1)}{2}$$
$$= \frac{(56.7 - 43.3)}{2}$$
$$= \frac{13.4}{2}$$
$$= 6.7$$

Notice that $Q1$ and $Q3$ are the same distance from the mean (6.7 points in the previous example). $Q1$ and $Q3$ will always be equidistant from the mean for normal

Remember, $z\sigma$ is a deviation score, or distance from the mean (Chapter 5).

distributions because normal distributions are symmetrical. Therefore, one-half of the distance between $Q1$ and $Q3$ (the semi-interquartile range by definition) will also equal the distance of $Q3$ from mean (see Figure 6.16). The distance of $Q3$ from the mean can be obtained simply by multiplying the z-score for $Q3$ ($z = 0.67$) times the standard deviation. Using this shortcut greatly simplifies the computation. For a normal distribution,

$$\text{semi-interquartile range} = 0.67\sigma \tag{6.1}$$

Remember, this simplified formula is used *only* for normal distributions.

LEARNING CHECK

1. A population is normally distributed and has a mean of $\mu = 90$ with $\sigma = 8$. Find the following values:

   a. The percentile rank for $X = 100$
   b. The percentile rank for $X = 88$
   c. The 85th percentile
   d. The 10th percentile

2. For the population in exercise 1, find $Q1$, $Q3$, and the semi-interquartile range.

ANSWERS

1. **a.** 89.44%   **b.** 40.13%   **c.** $X = 98.32$ **d.** $X = 79.76$

2. $Q1 = 84.64$, $Q3 = 95.36$, semi-interquartile range $= 5.36$

SUMMARY

1. The probability of a particular event A is defined as a fraction or proportion:

$$p(A) = \frac{\text{number of outcomes classified as } A}{\text{total number of possible outcomes}}$$

2. This definition is accurate only for a random sample. There are two requirements that must be satisfied for a random sample:

   a. Every individual in the population has an equal chance of being selected.

   b. When more than one individual is being selected, the probabilities must stay constant. This means there must be sampling with replacement.

3. All probability problems can be restated as proportion problems. The "probability of selecting a king from a deck of cards" is equivalent to the "proportion of the deck that consists of kings." For frequency distributions, probability questions can be answered by determining proportions of area. The "probability of selecting an individual with an IQ greater than 108" is equivalent to the "proportion of the whole population that has IQs above 108."

4. For normal distributions, these probabilities (proportions) can be found in the unit normal table. This table provides a listing of the proportions of a normal distribution that correspond to each z-score value. With this table, it is possible to move between X values and probabilities using a two-step procedure:

   a. The z-score formula (Chapter 5) allows you to transform X to z or to change z back to X.

   b. The unit normal table allows you to look up the probability (proportion) corresponding to each z-score or the z-score corresponding to each probability.

5. A percentile rank measures the relative standing of a score in a distribution. Expressed as a percentage, it indicates the proportion of individuals with scores at or below a particular X value. For a normal distribution, you must determine the proportion (percentage) of the distribution that falls to the left of the score in question. This percentage of the distribution is the percentile rank of that score.

6. A percentile is the X value that is associated with a percentile rank. For example, if 80% of the scores in a distribution are less than 250, then $X = 250$ is the 80th percen-

tile. Note that the 50th percentile is the median of the distribution.

7. Quartiles are the scores that divide the distribution into four areas that make up one-quarter of the distribution in each area. Thus, the first quartile ($Q1$) is equivalent to the 25th percentile. The second quartile ($Q2$) is also the 50th percentile (the median), and the third quartile ($Q3$) is the same as the 75th percentile. For a normal distribution, the semi-interquartile range can be obtained by a special formula:

$$\text{semi-interquartile range} = 0.67\sigma$$

KEY TERMS

probability	sampling with replacement	percentile rank	percentile
simple random sample	unit normal table	quartile	

FOCUS ON PROBLEM SOLVING

1. We have defined probability as being equivalent to a proportion, which means that you can restate every probability problem as a proportion problem. This definition is particularly useful when you are working with frequency distribution graphs where the population is represented by the whole graph and probabilities (proportions) are represented by portions of the graph. When working problems with the normal distribution, you always should start with a sketch of the distribution. You should shade the portion of the graph that reflects the proportion you are looking for.

2. When using the unit normal table, you must remember that the proportions in the table (columns B and C) correspond to specific portions of the distribution. This is important because you will often need to translate the proportions from a problem into specific proportions provided in the table. Also remember that the table allows you to move back and forth between z-scores and proportions. You can look up a given z-score to find a proportion, or you can look up a given proportion to find the corresponding z-score. However, you cannot go directly from an X value to a probability in the unit normal table. You must first compute the z-score for X. Likewise, you cannot go directly from a probability value to a raw score. First, you have to find the z-score associated with the probability (see Box 6.2).

3. Remember, the unit normal table shows only positive z-scores in column A. However, since the normal distribution is symmetrical, the probability values in columns B and C also apply to the half of the distribution that is below the mean. To be certain that you have the correct sign for z, it helps to sketch the distribution showing μ and X.

4. A common error for students is to use negative values for proportions on the left-hand side of the normal distribution. Proportions (or probabilities) are always positive: 10% is 10% whether it is in the left or the right tail of the distribution.

5. The proportions in the unit normal table are accurate only for normal distributions. If a distribution is not normal, you cannot use the table.

6. When determining percentile ranks, it helps to sketch the distribution first. Remember, shade in the area to the left of the particular X value. Its percentile rank will be the proportion of the distribution in the shaded area. You will need to compute the z-score for the X value to find the proportion in the unit normal table. Percentile ranks are always expressed as a percentage (%) of the distribution.

(a)   (b)   (c)

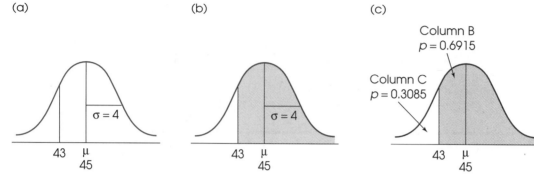

FIGURE 6.17

Sketches of the distribution for Demonstration 6.1.

---

7. Percentiles are *X* values. When asked to find the 90th percentile, you are looking for the *X* value that has 90% (or 0.9000) of the distribution below (to the left of) it. The procedure of finding a percentile requires that you go from the proportion to the *z*-score and then from the *z*-score to the *X* value (Box 6.2).

── DEMONSTRATION 6.1 ──

FINDING PROBABILITY FROM THE UNIT NORMAL TABLE

A population is normally distributed with a mean of $\mu = 45$ and a standard deviation of $\sigma = 4$. What is the probability of randomly selecting a score that is greater than 43? In other words, what proportion of the distribution consists of scores greater than 43?

STEP 1   Sketch the distribution.

You should always start by sketching the distribution, identifying the mean and the standard deviation ($\mu = 45$ and $\sigma = 4$ in this example). You should also find the approximate location of the specified score and draw a vertical line through the distribution at that score. The score of $X = 43$ is lower than the mean, and, therefore, it should be placed somewhere to the left of the mean. Figure 6.17(a) shows the preliminary sketch.

STEP 2   Shade in the distribution.

Read the problem again to determine whether you want the proportion greater than the score (to the right of your vertical line) or less than the score (to the left of the line). Then shade in the appropriate portion of the distribution. In this demonstration, we are considering scores greater than 43. Thus, we shade in the distribution to the right of this score [Figure 6.17(b)]. Notice that if the shaded area covers more than one-half of the distribution, then the probability should be greater than 0.5000.

STEP 3   Transform the *X* value into a *z*-score.

Remember, to get a probability from the unit normal table, we must first convert the *X* value to a *z*-score.

$$z = \frac{X - \mu}{\sigma} = \frac{43 - 45}{4} = \frac{-2}{4} = -0.5$$

STEP 4 Consult the unit normal table.

Look up the z-score (ignoring the sign) in the unit normal table. Find the two proportions in the table that are associated with the z-score, and write the two proportions in the appropriate regions on the figure. Remember, column B gives the proportion of area in the body of the distribution (more than 50%), and column C gives the area in the tail beyond z. Figure 6.17(c) shows the proportions for different regions of the normal distribution.

STEP 5 Determine the probability.

If one of the proportions in step 4 corresponds exactly to the entire shaded area, then that proportion is your answer. Otherwise, you will need to do some additional arithmetic. In this example, the proportion in the shaded area is given in column B:

$$p(X > 43) = 0.6915$$

PROBLEMS

1. In a psychology class of 90 students, there are 30 males and 60 females. Of the 30 men, only 5 are freshmen. Of the 60 women, 10 are freshmen. If you randomly sample an individual from this class,
   a. What is the probability of obtaining a female?
   b. What is the probability of obtaining a freshman?
   c. What is the probability of obtaining a male freshman?

2. A jar contains 10 black marbles and 40 white marbles.
   a. If you randomly select a marble from the jar, what is the probability that you will get a white marble?
   b. If you are selecting a random sample of $n = 3$ marbles and the first 2 marbles are both white, what is the probability that the third marble will be black?

3. Drawing a vertical line through a normal distribution will separate the distribution into two parts: the section to the right of the line and the section to the left. The size of each section depends on where you draw the line. Sketch a normal distribution, and draw a vertical line for each of the following locations (z-scores). Then find the proportion of the distribution on each side of the line.
   a. $z = 1.00$ **c.** $z = 0.25$
   b. $z = -1.50$ **d.** $z = -0.50$

4. For each of the following z-score values, sketch a normal distribution, and draw a vertical line through the distribution at the location specified by the z-score. Then find the proportion of the distribution in the tail of the distribution beyond the line.
   a. $z = 0.50$ **c.** $z = -1.50$
   b. $z = 1.75$ **d.** $z = -0.25$

5. Find the proportion of the normal distribution that lies in the tail beyond each of the following z-scores:
   a. $z = 0.43$ **c.** $z = -1.35$
   b. $z = 1.68$ **d.** $z = -0.29$

6. What is sampling with replacement, and why is it used?

7. For a normal distribution, find the z-score location of a vertical line that would separate the distribution into two parts as specified in each of the following:
   a. Where do you draw a line that separates the top 40% (right side) from the bottom 60% (left side)?
   b. Where do you draw a line that separates the top 10% (right side) from the bottom 90% (left side)?
   c. Where do you draw a line that separates the top 80% (right side) from the bottom 20% (left side)?

8. A normal distribution has $\mu = 50$ and $\sigma = 10$. For each of the following locations (X values), draw a sketch of the distribution, and draw a vertical line through the distribution at the location specified by the score. Then find the proportion of the distribution on each side of the line.
   a. $X = 55$ **c.** $X = 48$
   b. $X = 50$ **d.** $X = 40$

9. For a normal distribution with $\mu = 500$ and $\sigma = 100$, find the X value (location) of a vertical line that would separate the distribution into two parts as specified in each of the following:
   a. Where do you draw a line that separates the top 30% (right side) from the bottom 70% (left side)?
   b. Where do you draw a line that separates the top 25% (right side) from the bottom 75% (left side)?
   c. Where do you draw a line that separates the top 90% (right side) from the bottom 10% (left side)?

10. A normal distribution has $\mu = 225$ with $\sigma = 15$. Find the following probabilities:
   a. $p(X > 227)$ **c.** $p(X < 230)$
   b. $p(X > 212)$ **d.** $p(X < 220)$

11. A population forms a normal distribution with $\mu = 68$ and $\sigma = 6$. Find the following probabilities:
   a. The probability of selecting a score with a value less than 60

b. The probability of selecting a score with a value greater than 58

c. The probability of selecting a score with a value greater than 74

12. For a normal distribution with $\mu = 100$ and $\sigma = 16$,

    a. What is the probability of randomly selecting a score greater than 104?

    b. What is the probability of randomly selecting a score greater than 96?

    c. What is the probability of randomly selecting a score less than 108?

    d. What is the probability of randomly selecting a score less than 92?

13. Find the proportion of the normal distribution that is located between the following z-score boundaries:

    a. Between $z = 0.25$ and $z = 0.75$

    b. Between $z = -1.00$ and $z = +1.00$

    c. Between $z = 0$ and $z = 1.50$

    d. Between $z = -0.75$ and $z = 2.00$

14. For a normal distribution with $\mu = 80$ and $\sigma = 12$,

    a. What is the probability of randomly selecting a score greater than 83?

    b. What is the probability of randomly selecting a score greater than 74?

    c. What is the probability of randomly selecting a score less than 92?

    d. What is the probability of randomly selecting a score less than 62?

15. One question on a multiple-choice test asked for the probability of selecting a score greater than $X = 50$ from a normal population with $\mu = 60$ and $\sigma = 20$. The answer choices were

    a. 0.1915   **b.** 0.3085   **c.** 0.6915

    Sketch a distribution showing this problem, and without looking at the unit normal table, explain why answers a and b cannot be correct.

16. A normal distribution has a mean of 120 and a standard deviation of 20. For this distribution,

    a. What score separates the top 40% (highest scores) from the rest?

    b. What score corresponds to the 90th percentile?

    c. What range of scores would form the middle 60% of this distribution?

17. A normal distribution has $\mu = 75$ with $\sigma = 9$. Find the following probabilities:

    a. $p(X < 86) = ?$ **c.** $p(X > 80) = ?$

    b. $p(X > 60) = ?$ **d.** $p(X > 94) = ?$

    e. $p(63 < X < 88) = ?$

    f. The probability of randomly selecting a score within 3 points of the mean.

18. The scores on a psychology exam form a normal distribution with $\mu = 80$ and $\sigma = 8$. On this exam, Tom has a score of $X = 84$. Mary's score is located at the 60th percentile. John's score corresponds to a z-score of 0.75. If these three students are listed from highest score to lowest score, what is the correct ordering?

19. A normal distribution has a mean of $\mu = 120$ with $\sigma = 15$. Find the following values:

    a. The 15th percentile

    b. The 88th percentile

    c. The percentile rank for $X = 142$

    d. The percentile rank for $X = 102$

    e. The percentile rank for $X = 120$

    f. The semi-interquartile range

20. For a normal distribution with $\mu = 500$ and $\sigma = 100$,

    a. What is the percentile rank for $X = 550$?

    b. What is the percentile rank for $X = 650$?

    c. What is the percentile rank for $X = 450$?

    d. What is the percentile rank for $X = 350$?

21. A normal distribution has a mean of 60 and a standard deviation of 10.

    a. Find the semi-interquartile range for this distribution.

    b. If the standard deviation were 20, what would be the value for the semi-interquartile range?

    c. In general, what is the relationship between the standard deviation and the semi-interquartile range for a normal distribution?

22. A normal distribution has a mean of 80 and a standard deviation of 10. For this distribution, find each of the following probability values:

    a. $p(X > 75) = ?$ **d.** $p(65 < X < 95) = ?$

    b. $p(X < 65) = ?$ **e.** $p(84 < X = 90)?$

    c. $p(X < 100) = ?$

23. A positively skewed distribution has a mean of 100 and a standard deviation of 12. What is the probability of randomly selecting a score greater than 106 from this distribution? (Be careful; this is a trick problem.)

24. For a normal distribution with $\mu = 200$ and $\sigma = 50$,

    a. Find the 90th percentile.

    b. Find the 30th percentile.

    c. Find the 75th percentile.

25. For a normal distribution with $\mu = 100$ and $\sigma = 30$,

    a. Find $Q1$, $Q2$, and $Q3$.

    b. Find the semi-interquartile range.

26. A normal distribution has $\mu = 200$ and $\sigma = 20$. Find the following:

    a. The first quartile ($Q1$).

    b. The second quartile ($Q2$).

    c. The third quartile ($Q3$).

    d. The semi-interquartile range.

CHAPTER 7

PROBABILITY AND SAMPLES: THE DISTRIBUTION OF SAMPLE MEANS

TOOLS YOU WILL NEED

The following items are considered essential background material for this chapter. If you doubt your knowledge of any of these items, you should review the appropriate chapter and section before proceeding.

- Random sampling (Chapter 6)

- Probability and the normal distribution (Chapter 6)

- z-Scores (Chapter 5)

CONTENTS

7.1 SAMPLES AND SAMPLING ERROR

The preceding two chapters presented the topics of z-scores and probability. Whenever a score is selected from a population, you should be able to compute a z-score that describes exactly where the score is located in the distribution. And, if the population is normal, you should be able to determine the probability value for each score. In a normal distribution, for example, a z-score of $+2.00$ corresponds to an extreme score out in the tail of the distribution, and a score this large has a probability of only $p = .0228$.

However, the z-scores and probabilities that we have considered so far are limited to situations where the sample consists of a single score. Most research studies use much larger samples, such as $n = 25$ preschool children or $n = 100$ laboratory rats. In this chapter, we will extend the concepts of z-scores and probability to cover situations with larger samples. Thus, a researcher will be able to compute a z-score that describes an entire sample. A z value near zero indicates a central, representative sample; a z value beyond $+2.00$ or -2.00 indicates an extreme sample. Also, it will be possible to determine exact probabilities for samples, no matter how many scores the sample contains.

The difficulty of working with samples is that samples generally are not identical to the populations from which they come. More precisely, the statistics calculated for a sample will differ from the corresponding parameters for the population. For example, the sample mean may differ from the population mean. This difference, or *error*, is referred to as *sampling error*.

DEFINITION   *Sampling error* is the discrepancy, or amount of error, between a sample statistic and its corresponding population parameter.

Furthermore, samples are variable; they are not all the same. If you take two separate samples from the same population, the samples will be different. They will contain different individuals, they will have different scores, and they will have different sample means. How can you tell which sample is giving the best description of the population? Can you even predict how well a sample will describe its population? What is the probability of selecting a sample that has a certain sample mean? These questions can be answered once we establish the set of rules that relate samples to populations.

7.2 THE DISTRIBUTION OF SAMPLE MEANS

As noted, two separate samples probably will be different even though they are taken from the same population. The samples will have different individuals, different scores, different means, and the like. In most cases, it is possible to obtain thousands of different samples from one population. With all these different samples coming from the same population, it may seem hopeless to try to establish some simple rules for the relationships between samples and populations. Fortunately, the huge set of possible samples forms a relatively simple, orderly, and predictable pat-

FIGURE 7.1

Frequency distribution histogram for a population of four scores: 2, 4, 6, 8.

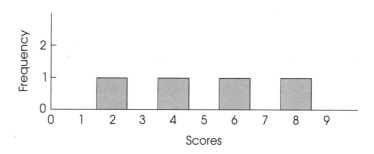

tern that makes it possible to predict the characteristics of a sample with some accuracy. The ability to predict sample characteristics is based on the *distribution of sample means*.

DEFINITION

The *distribution of sample means* is the collection of sample means for all the possible random samples of a particular size (*n*) that can be obtained from a population.

You should notice that the distribution of sample means contains *all the possible samples*. It is necessary to have all the possible values in order to compute probabilities. For example, if the entire set contains exactly 100 samples, then the probability of obtaining any specific sample would be 1 out of 100; $p = 1/100$. You should also notice that the distribution of sample means is different from distributions we have considered before. Until now we always have discussed distributions of scores; now the values in the distribution are not scores, but statistics (sample means). Because statistics are obtained from samples, a distribution of statistics is referred to as a sampling distribution.

DEFINITION

A *sampling distribution* is a distribution of statistics obtained by selecting all the possible samples of a specific size from a population.

Thus, the distribution of sample means is an example of a sampling distribution. In fact, it often is called the sampling distribution of $\overline{X}$.

Before we consider the general rules concerning this distribution, we will look at a simple example that provides an opportunity to examine the distribution in detail.

EXAMPLE 7.1

Consider a population that consists of only four scores: 2, 4, 6, 8. This population is pictured in the frequency distribution histogram in Figure 7.1.

We are going to use this population as the basis for constructing the distribution of sample means for $n = 2$. Remember, this distribution is the collection of sample means from all the possible random samples of $n = 2$ from this population. We begin by looking at all the possible samples. Each of the 16 different samples is listed in Table 7.1.

Next, we compute the mean, $\overline{X}$, for each of the 16 samples (see the last column of Table 7.1). The 16 sample means form the distribution of sample means. These 16 values are organized in a frequency distribution histogram in Figure 7.2.

Remember, random sampling requires sampling with replacement.

FIGURE 7.2

The distribution of sample means for $n = 2$. This distribution shows the 16 sample means from Table 7.1.

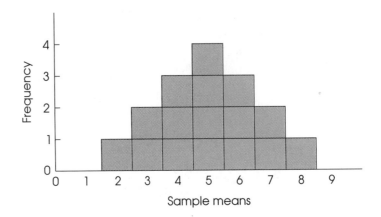

Notice that the distribution of sample means has some predictable and some very useful characteristics:

1. The sample means tend to pile up around the population mean. For this example, the population mean is $\mu = 5$, and the sample means are clustered around a value of 5. It should not surprise you that the sample means tend to approximate the population mean. After all, samples are supposed to be representative of the population.

2. The distribution of sample means is approximately normal in shape. This is a characteristic that will be discussed in detail later and will be extremely useful because we already know a great deal about probabilities and the normal distribution (Chapter 6).

Remember, our goal in this chapter is to answer probability questions about samples with $n > 1$.

3. Finally, you should notice that we can use the distribution of sample means to answer probability questions about sample means. For ex-

TABLE 7.1

All the possible samples of $n = 2$ scores that can be obtained from the population presented in Figure 7.1

Notice that the table lists *random samples*. This requires sampling with replacement so it is possible to select the same score twice. Also note that samples are listed systematically. The first four samples are all the possible samples that have $X = 2$ as the first score, the next four samples all have $\bar{X} = 4$ as the first score, and so on. This way we are sure to have all the possible samples listed, although the samples probably would not be selected in this order.

Sample	First score	Second score	Sample mean ($\bar{X}$)
1	2	2	2
2	2	4	3
3	2	6	4
4	2	8	5
5	4	2	3
6	4	4	4
7	4	6	5
8	4	8	6
9	6	2	4
10	6	4	5
11	6	6	6
12	6	8	7
13	8	2	5
14	8	4	6
15	8	6	7
16	8	8	8

ample, if you take a sample of $n = 2$ scores from the original population, what is the probability of obtaining a sample mean greater than 7? In symbols, $p(\overline{X} > 7) = ?$.

Because probability is equivalent to proportion, the probability question can be restated as follows: Of all the possible sample means, what proportion has values greater than 7? In this form, the question is easily answered by looking at the distribution of sample means. All the possible sample means are pictured (Figure 7.2), and only 1 out of the 16 means has a value greater than 7. The answer therefore is 1 out of 16, or $p = \frac{1}{16}$.

THE CENTRAL LIMIT THEOREM

Example 7.1 demonstrates the construction of the distribution of sample means for a relatively simple, specific situation. In most cases, however, it will not be possible to list all the samples and compute all the possible sample means. As the size (n) of the samples increases, the number of possible samples increases rapidly. Therefore, it is necessary to develop the general characteristics of the distribution of sample means that can be applied in any situation. Fortunately, these characteristics are specified in a mathematical proposition known as the *central limit theorem*. This important and useful theorem serves as a cornerstone for much of inferential statistics. Following is the essence of the theorem.

DEFINITION

Central Limit Theorem: For any population with mean of μ and standard deviation of σ, the distribution of sample means for sample size n will approach a normal distribution with a mean of μ and a standard deviation of $\sigma/\sqrt{n}$ as n approaches infinity.

The value of this theorem comes from two simple facts. First, it describes the distribution of sample means for *any population*, no matter what shape, or mean, or standard deviation. Second, the distribution of sample means "approaches" a normal distribution very rapidly. By the time the sample size reaches $n = 30$, the distribution is almost perfectly normal.

Notice that the central limit theorem describes the distribution of sample means by identifying the three basic characteristics that describe any distribution: shape, central tendency, and variability. Each of these will be examined.

THE SHAPE OF THE DISTRIBUTION OF SAMPLE MEANS

It has been observed that the distribution of sample means tends to be a normal distribution. In fact, this distribution will be almost perfectly normal if either one of the following two conditions is satisfied.

1. The population from which the samples are selected is a normal distribution.

2. The number of scores (n) in each sample is relatively large, around 30 or more.

(As n gets larger, the distribution of sample means will closely approximate a normal distribution. In most situations when $n > 30$, the distribution is almost perfectly normal regardless of the shape of the original population.)

The fact that the distribution of sample means tends to be normal should not be surprising. Whenever you take a sample from a population, you expect the sample

7.1   A CAUTION ABOUT SUBSCRIPTS

SUBSCRIPTS ARE used to clarify the identity of a mathematical symbol. Often they are used to describe the set of values or the variable to which a symbol applies. Thus, the subscript $\overline{X}$ in the notation for standard error ($\sigma_{\overline{X}}$) simply indicates that we are measuring the standard deviation (σ) for a set of sample means. Students often become confused and try to include the subscript in mathematical computations. A typical error would be to take the notation $\sigma_{\overline{X}}$ and multiply the standard deviation (σ) by the sample mean ($\overline{X}$), or symbolically (σ)($\overline{X}$). *This error is caused by a misinterpretation of subscripts.* If you simply consider the subscript a name for a symbol, then you can think of standard error as follows:

$$\sigma_{\overline{X}} = \sigma_{\text{of sample means}}$$

mean to be near to the population mean. When you take lots of different samples, you expect the sample means to "pile up" around μ, resulting in a normal-shaped distribution.

THE MEAN OF THE DISTRIBUTION OF SAMPLE MEANS: THE EXPECTED VALUE OF $\overline{X}$

You probably noticed in Example 7.1 that the distribution of sample means is centered around the mean of the population from which the samples were obtained. In fact, the average value of all the sample means is exactly equal to the value of the population mean. This fact should be intuitively reasonable; the sample means are expected to be close to the population mean, and they do tend to pile up around μ. The formal statement of this phenomenon is that the mean of the distribution of sample means always will be identical to the population mean. This mean value is called the *expected value of $\overline{X}$*.

DEFINITION

The mean of the distribution of sample means will be equal to μ (the population mean) and is called the *expected value of $\overline{X}$*.

The expected value of $\overline{X}$ is often identified by the symbol $\mu_{\overline{X}}$, signifying the "mean of the sample means." However, $\mu_{\overline{X}}$ is always equal to μ, so we will continue to use the symbol μ to refer to the mean for the population of scores and the mean for the distribution of sample means.

In commonsense terms, a sample mean is "expected" to be near its population mean. When all of the possible sample means are obtained, the average value will be identical to μ.

THE STANDARD ERROR OF $\overline{X}$

So far we have considered the shape and the central tendency of the distribution of sample means. To completely describe this distribution, we need one more characteristic, variability. The measure we will be working with is the standard deviation for the distribution of sample means, and it is called the *standard error of $\overline{X}$*. Like any measure of standard deviation, the standard error defines the standard, or typical, distance from the mean. In this case, we are measuring the standard distance between a single sample mean, $\overline{X}$, and the population mean, μ. Remember, a sample is not expected to provide a perfectly accurate reflection of its population. Although a sample mean should be representative of the population mean, there typically will be some error between the sample and the population. The standard error measures exactly how much difference should be expected, on average, between $\overline{X}$ and μ.

7.2 THE LAW OF LARGE NUMBERS

CONSIDER THE following problem:

Imagine an urn filled with balls. Two-thirds of the balls are one color, and the remaining one-third are a second color. One individual selects 5 balls from the urn and finds that 4 are red and 1 is white. Another individual selects 20 balls and finds that 12 are red and 8 are white. Which of these two individuals should feel more confident that the urn contains two-thirds red balls and one-third white balls rather than the opposite?*

When Tversky and Kahneman presented this problem to a group of experimental subjects, they found that most people felt that the first sample (4 out of 5) provided much stronger evidence and therefore should give more confidence. At first glance, it may appear that this is the correct decision. After all, the first sample contained $4/5 = 80\%$ red balls, and the second sample contained only $12/20 = 60\%$ red balls.

However, you should also notice that the two samples differ in another important aspect: the sample size. One sample contains only $n = 5$ scores, and the other sample contains $n = 20$. The correct answer to the problem is that the larger sample (12 out of 20) gives a much stronger justification for concluding that the balls in the urn are predominantly red. It appears that most people tend to focus on the sample proportion and pay very little attention to the sample size.

The importance of sample size may be easier to appreciate if you approach the urn problem from a different perspective. Suppose you are the individual assigned the responsibility for selecting a sample and then deciding which color is in the majority. Before you select your sample, you are offered a choice of selecting a sample of $n = 5$ balls or a sample of $n = 20$ balls. Which would you prefer? It should be clear that the larger sample would be better. The larger sample is much more likely to provide an accurate representation of the population. This is an example of the law of large numbers, which states that large samples will be representative of the population from which they are selected.

*Reprinted with permission from "Judgments Under Uncertainty: Heuristics and Biases," by A. Tversky, and D. Kahneman, 1974, *Science, 185,* 1124–1131. Copyright © 1974 American Association for the Advancement of Science.

DEFINITION   The standard deviation of the distribution of sample means is called the *standard error of $\overline{X}$*. The standard error measures the standard amount of difference one should expect between $\overline{X}$ and μ simply due to chance.

$$\text{standard error} = \sigma_{\overline{X}} = \text{standard distance between } \overline{X} \text{ and } \mu$$

The notation that is used to identify the standard error is $\sigma_{\overline{X}}$. The σ indicates that we are measuring a standard deviation or a standard distance from the mean. The subscript $\overline{X}$ indicates that we are measuring the standard deviation for a distribution of sample means. (See Box 7.1 for a word of caution.)

The magnitude of the standard error is determined by two factors: (1) the size of the sample and (2) the standard deviation of the population from which the sample is selected. We will examine each of these factors.

The sample size   It should be intuitively reasonable that the size of a sample should influence how accurately the sample represents its population. Specifically, a large sample should be more accurate than a small sample. In general, as the sample size increases, the error between the sample mean and the population mean should decrease. This rule is also known as the *law of large numbers* (see Box 7.2).

The population standard deviation   As we noted earlier, when the sample size decreases, the standard error increases. At the extreme, the smallest possible sample and the largest possible error occur when the sample consists of $n = 1$ score. At this extreme, the sample is a single score, X, and the sample mean also equals X. In this case, the standard error measures the standard distance between the score, X, and the population mean, μ. However, we already know that the standard distance between X and μ is the standard deviation. By definition, σ is the standard distance between X and μ. Thus, when $n = 1$, the standard error and the standard deviation are identical.

When $n = 1$,

$$\text{standard error} = \sigma_{\bar{X}} = \sigma = \text{standard deviation}$$

You can think of the standard deviation as the "starting point" for standard error. When $n = 1$, σ and $\sigma_{\bar{X}}$ are the same. As the sample size increases, the standard error decreases in relation to n.

The formula for standard error incorporates both the standard deviation and the sample size.

$$\text{standard error} = \sigma_{\bar{X}} = \frac{\sigma}{\sqrt{n}} \tag{7.1}$$

Note that the formula satisfies all the requirements for the concept of standard error. Specifically, as the sample size increases, the error decreases. As n decreases, the error increases. At the extreme, when $n = 1$, the standard error is equal to the standard deviation. In mathematical terms, the standard error is *directly* related to the population standard deviation and it is *inversely* related to the sample size. That is, when the standard deviation increases, the standard error also increases. But when the sample size increases, the standard error decreases.

In equation 7.1 and in most of the preceding discussion, we have defined standard error in terms of the population standard deviation. However, the population standard deviation (σ) and the population variance (σ^2) are directly related, and it is easy to substitute variance into the equation for standard error. Using the simple equality $\sigma = \sqrt{\sigma^2}$, the equation for standard error can be rewritten as follows:

$$\text{standard error} = \sigma_{\bar{X}} = \frac{\sqrt{\sigma^2}}{\sqrt{n}} = \sqrt{\frac{\sigma^2}{n}} \tag{7.2}$$

Throughout the rest of this chapter (and in Chapter 8), we will continue to define standard error in terms of the standard deviation (equation 7.1). However, in later chapters (starting in Chapter 9), the formula based on variance (equation 7.2) will become more useful.

LEARNING CHECK

1. A population of scores is normal with $\mu = 80$ and $\sigma = 20$. Describe the distribution of sample means for samples of size $n = 16$ selected from this population. (Describe shape, central tendency, and variability for this distribution.)

2. As sample size increases, the value of the standard error also increases. (True or false?)

3. Under what circumstances is the distribution of sample means guaranteed to be normal?

ANSWERS
1. The distribution of sample means will be normal with an expected value of $\mu = 80$ and a standard error of $\sigma_{\overline{X}} = 20/\sqrt{16} = 5$.

2. False. Standard error decreases as n increases.

3. The distribution of sample means will be normal if the sample size is relatively large (around $n = 30$ or more) or if the population of scores is normal.

7.3 PROBABILITY AND THE DISTRIBUTION OF SAMPLE MEANS

The primary use of the distribution of sample means is to find the probability associated with any specific sample. You should recall that probability is equivalent to proportion. Because the distribution of sample means presents the entire set of all possible $\overline{X}$'s, we can use proportions of this distribution to determine probabilities. The following example demonstrates this process.

EXAMPLE 7.2   The population of scores on the SAT forms a normal distribution with $\mu = 500$ and $\sigma = 100$. If you take a random sample of $n = 25$ students, what is the probability that the sample mean would be greater than $\overline{X} = 540$?

First, you can restate this probability question as a proportion question: Out of all the possible sample means, what proportion has values greater than 540? You know about "all the possible sample means"; this is simply the distribution of sample means. The problem is to find a specific portion of this distribution. The parameters of this distribution are the following:

Caution: Whenever you have a probability question about a sample mean, you must use the distribution of sample means.

a. The distribution is normal because the population of SAT scores is normal.

b. The distribution has a mean of 500 because the population mean is $\mu = 500$.

c. The distribution has a standard error of $\sigma_{\overline{X}} = 20$:

$$\sigma_{\overline{X}} = \frac{\sigma}{\sqrt{n}} = \frac{100}{\sqrt{25}} = \frac{100}{5} = 20$$

This distribution of sample means is shown in Figure 7.3.

We are interested in sample means greater than 540 (the shaded area in Figure 7.3), so the next step is to use a z-score to locate the exact position of $\overline{X} = 540$ in the distribution. The value 540 is located above the mean by 40 points, which is exactly 2 standard deviations (in this case, exactly 2 standard errors). Thus, the z-score for $\overline{X} = 540$ is $z = +2.00$.

Because this distribution of sample means is normal, you can use the unit normal table to find the probability associated with $z = +2.00$. The table indicates that 0.0228 of the distribution is located in the tail of the distribution beyond $z = +2.00$. Our conclusion is that it is very unlikely, $p = 0.0228$ (2.28%), to obtain a random sample of $n = 25$ students with an average SAT score greater than 540.

FIGURE 7.3

The distribution of sample means for $n = 25$. Samples were selected from a normal population with $\mu = 500$ and $\sigma = 100$.

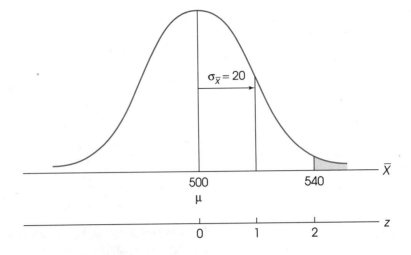

As demonstrated in Example 7.2, it is possible to use a z-score to describe the position of any specific sample within the distribution of sample means. The z-score tells exactly where a specific sample is located in relation to all the other possible samples that could have been obtained. A z-score of $z = +2.00$, for example, indicates that the sample mean is much larger than usually would be expected: It is greater than the expected value of $\overline{X}$ by twice the standard distance. The z-score for each sample mean can be computed by using the standard z-score formula with a few minor changes. First, the value we are locating is a sample mean, rather than a score, so the formula uses $\overline{X}$ in place of X. Second, the standard deviation for this distribution is measured by the standard error, so the formula uses $\sigma_{\overline{X}}$ in place of σ (see Box 7.3). The resulting formula, giving the z-score value corresponding to any sample mean, is

$$z = \frac{\overline{X} - \mu}{\sigma_{\overline{X}}} \qquad (7.3)$$

Every sample mean has a z-score that describes its position in the distribution of sample means. Using z-scores and the unit normal table, it is possible to find the probability associated with any specific sample mean (as in Example 7.2). The following example demonstrates that it also is possible to make quantitative predictions about the kinds of samples that should be obtained from any population.

EXAMPLE 7.3 Suppose you simply wanted to predict the kind of value that would be expected for the mean SAT score for a random sample of $n = 25$ students. For example, what range of values would be expected for the sample mean 80% of the time? The simplest way of answering this question is to look at the distribution of sample means. Remember, this distribution is the collection of all the possible sample means, and it will show which samples are likely to be obtained and which are not.

As demonstrated in Example 7.2, the distribution of sample means for $n = 25$ will be normal, will have an expected value of $\mu = 500$, and will have a standard error of $\sigma_{\overline{X}} = 20$. Looking at this distribution, shown again in Fig-

7.3   THE DIFFERENCE BETWEEN STANDARD DEVIATION AND STANDARD ERROR

A CONSTANT source of confusion for many students is the difference between standard deviation and standard error. You should remember that standard deviation measures the standard distance between a *score* and the population mean, $X - \mu$. Whenever you are working with a distribution of scores, the standard deviation is the appropriate measure of variability. Standard error, on the other hand, measures the standard distance between a *sample mean* and the population mean, $\overline{X} - \mu$. Whenever you have a question concerning a sample, the standard error is the appropriate measure of variability.

If you still find the distinction confusing, there is a simple solution. Namely, if you always use standard error, you always will be right. Consider the formula for standard error:

$$\text{standard error} = \sigma_{\overline{X}} = \frac{\sigma}{\sqrt{n}}$$

If you are working with a single score, then $n = 1$, and the standard error becomes

$$\text{standard error} = \sigma_{\overline{X}} = \frac{\sigma}{\sqrt{n}} = \frac{\sigma}{\sqrt{1}}$$

$$= \sigma = \text{standard deviation}$$

Thus, standard error always measures the standard distance from the population mean, whether you have a sample of $n = 1$ or $n = 100$.

Remember, when answering probability questions, it always is helpful to sketch a distribution and shade in the portion you are trying to find.

ure 7.4, it is clear that the most likely value to expect for a sample mean is around 500. To be more precise, we can identify the range of values that would be expected 80% of the time by locating the middle 80% of the distribution. Because the distribution is normal, we can use the unit normal table. To find the middle 80%, we need to separate the extreme 20% in the tails. This corresponds to 10% (or .1000) in each of the two tails. Looking up a proportion of .1000 in the unit normal table (column C) gives a z-score of 1.28. Thus, the middle 80% is bounded by z-scores of $z = \pm 1.28$. By definition, $z = 1.28$ indicates a location exactly 1.28 standard error units from the

FIGURE 7.4

The middle 80% of the distribution of sample means for $n = 25$. Samples were selected from a normal population with $\mu = 500$ and $\sigma = 100$.

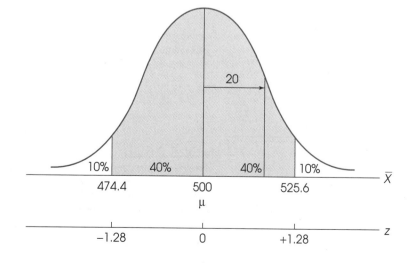

mean. This distance is $1.28 \times 20 = 25.6$ points. The mean is 500, so 25.6 points in either direction would give a range from 474.4 to 525.6. This is the middle 80% of all the possible sample means, so you can expect any particular sample mean to be in this range 80% of the time.

LEARNING CHECK

1. A population has $\mu = 80$ and $\sigma = 12$. Find the z-score corresponding to each of the following sample means:
   a. $\overline{X} = 84$ for a sample of $n = 9$ scores
   b. $\overline{X} = 74$ for a sample of $n = 16$ scores
   c. $\overline{X} = 81$ for a sample of $n = 36$ scores

2. A random sample of $n = 25$ scores is selected from a normal population with $\mu = 90$ and $\sigma = 10$.
   a. What is the probability that the sample mean will have a value greater than 94?
   b. What is the probability that the sample mean will have a value less than 91?

3. A positively skewed distribution has $\mu = 60$ and $\sigma = 8$.
   a. What is the probability of obtaining a sample mean greater than $\overline{X} = 62$ for a sample of $n = 4$ scores? (Be careful. This is a trick question.)
   b. What is the probability of obtaining a sample mean greater than $\overline{X} = 62$ for a sample of $n = 64$ scores?

ANSWERS

1. a. The standard error is $12/\sqrt{9} = 4$. The z-score is $z = +1.00$.
   b. The standard error is $12/\sqrt{16} = 3$. The z-score is $z = -2.00$.
   c. The standard error is $12/\sqrt{36} = 2$. The z-score is $z = +0.50$.

2. a. The standard error is $10/\sqrt{25} = 2$. A mean of $\overline{X} = 94$ corresponds to $z = +2.00$, and $p = 0.0228$.
   b. A mean of $\overline{X} = 91$ corresponds to $z = +0.50$, and $p = 0.6915$.

3. a. The distribution of sample means does not satisfy either of the criteria for being normal. Therefore, you cannot use the unit normal table, and it is impossible to find the probability.
   b. With $n = 64$, the distribution of sample means is normal. The standard error is $8/\sqrt{64} = 1$, the z-score is $+2.00$, and the probability is 0.0228.

7.4 MORE ABOUT STANDARD ERROR

At the beginning of this chapter, we introduced the idea that it is possible to obtain thousands of different samples from a single population. Each sample will have its own individuals, its own scores, and its own sample mean. The distribution of sample means provides a method for organizing all of the different sample means

FIGURE 7.5

An example of a typical distribution of sample means. Each of the small boxes represents the mean obtained for one sample. The normal curve is superimposed on the frequency distribution.

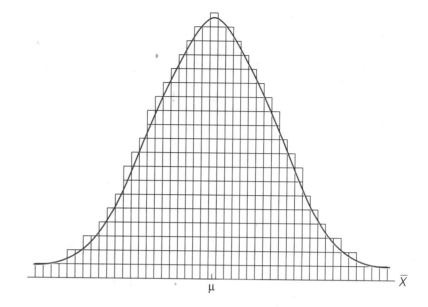

into a single picture that shows how the sample means are related to each other and how they are related to the overall population mean. Figure 7.5 shows a prototypical distribution of sample means. To emphasize the fact that the distribution contains many different samples, we have constructed this figure so that the distribution is made up of hundreds of small boxes, each box representing a single sample mean. Also notice that the sample means tend to pile up around the population mean (μ), forming a normal-shaped distribution, as predicted by the central limit theorem.

The distribution shown in Figure 7.5 provides a concrete example for reviewing the general concepts of *sampling error* and *standard error*. Although the following points may seem obvious, they are intended to provide you with a better understanding of these two statistical concepts.

1. Sampling Error: The general concept of sampling error is that a sample typically will not provide a perfectly accurate representation of its population. More specifically, there typically will be some discrepancy (or error) between a statistic computed for a sample and the corresponding parameter for the population. As you look at Figure 7.5, notice that the individual sample means tend to *underestimate* or *overestimate* the population mean. In fact, 50% of the samples will have means that are smaller than μ (the entire left-hand side of the distribution). Similarly, 50% of the samples will produce means that overestimate the true population mean. In general, there will be some discrepancy, or *sampling error*, between the mean for a sample and the mean for the population from which the sample was obtained.

2. Standard Error: Again, looking at Figure 7.5, notice that most of the sample means are relatively close to the population mean (those in the center of the distribution). These samples provide a fairly accurate representation of the population. On the other hand, some samples will produce means that are out in the tails of the distribution, relatively far away from the population mean. These extreme sample means do not accurately represent the population. For each individual sample, you can measure the error (or distance) between the sample mean and the population mean. For some samples, the error will be relatively small, but for other samples, the error will be relatively large. The *standard error* provides a way to measure the "average" or standard distance between a sample mean and the population mean.

Thus, the standard error provides a method for defining and measuring sampling error. The advantage of knowing the standard error is that it gives researchers a good indication of how accurately their sample data represent the populations they are studying. In most research situations, for example, the population mean is unknown, and the researcher selects a sample to help obtain information about the unknown population. Specifically, the sample mean provides information about the value of the unknown population mean. The sample mean is not expected to give a perfectly accurate representation of the population mean; there will be some error, and the standard error tells *exactly how much error*, on average, should exist between the sample mean and the unknown population mean.

The following two examples demonstrate some important aspects of the standard error and provide additional information about the relation between standard error and standard deviation.

EXAMPLE 7.4   Consider a population consisting of $N = 3$ scores: 1, 8, 9. The mean for this population is $\mu = 6$. Using this population, try to find a random sample of $n = 2$ scores with a sample mean ($\overline{X}$) exactly equal to the population mean. (Try selecting a few samples—write down the scores and the sample mean for each of your samples.)

You may have guessed that we constructed this example so that it is impossible to obtain a sample mean that is identical to μ. The point of the example is to emphasize the notion of sampling error. Samples are not identical to their populations, and a sample mean generally will not provide a perfect estimate of the population mean. The purpose of standard error is to provide a quantitative measure of the difference between sample means and the population mean. Standard error is the standard distance between $\overline{X}$ and μ.

EXAMPLE 7.5   We will begin with a population that is normally distributed with a mean of $\mu = 80$ and a standard deviation of $\sigma = 20$. Next, we will take a sample from this population and examine how accurately the sample mean represents the population mean. More specifically, we will examine how sample size affects accuracy by considering three different samples: one with $n = 1$ score, one with $n = 4$ scores, and one with $n = 100$ scores.

Figure 7.6 shows the distributions of sample means based on samples of $n = 1$, $n = 4$, and $n = 100$. Each distribution shows the collection of all possible sample means that could be obtained for that particular sample size. Notice that all three sampling distributions are normal (because the original population is normal), and all three have the same mean, $\mu = 80$, which is the expected value of $\overline{X}$. However, the three distributions differ greatly with respect to variability. We will consider each one separately.

The smallest sample size is $n = 1$. When a sample consists of a single score, the mean for the sample will equal the value of that score, $\overline{X} = X$.

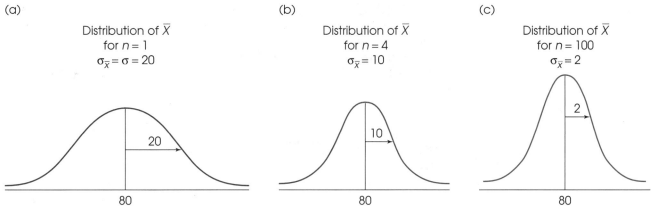

FIGURE 7.6

The distribution of sample means for random samples of size (a) $n = 1$, (b) $n = 4$, and (c) $n = 100$ obtained from a normal population with $\mu = 80$ and $\sigma = 20$. Notice that the size of the standard error decreases as the sample size increases.

Thus, when $n = 1$, the distribution of sample means is identical to the original population of scores. In this case, the standard error for the distribution of sample means is equal to the standard deviation for the original population. Equation 7.1 confirms this observation.

$$\sigma_{\overline{X}} = \frac{\sigma}{\sqrt{n}} = \frac{20}{\sqrt{1}} = 20$$

In one sense, the population standard deviation is the "starting point" for the standard error. With the smallest possible sample, $n = 1$, the standard error is equal to the standard deviation [see Figure 7.6(a)].

As the sample size increases, however, the standard error gets smaller. For $n = 4$, the standard error is

$$\sigma_{\overline{X}} = \frac{\sigma}{\sqrt{n}} = \frac{20}{\sqrt{4}} = \frac{20}{2} = 10$$

That is, the typical (or standard) distance between $\overline{X}$ and μ is 10 points. Figure 7.6(b) illustrates this distribution. Notice that the sample means in this distribution approximate the population mean more closely than in the previous distribution, where $n = 1$.

With samples of $n = 100$, the standard error is still smaller.

$$\sigma_{\overline{X}} = \frac{\sigma}{\sqrt{n}} = \frac{20}{\sqrt{100}} = \frac{20}{10} = 2$$

A sample of $n = 100$ scores should provide a sample mean that is a much better estimate of μ than you would obtain with a sample of $n = 4$ or $n = 1$. As shown in Figure 7.6(c), there is very little error between $\overline{X}$ and μ (you would expect only a 2-point error, on average). The sample means pile up very close to μ.

In summary, this example illustrates that with the smallest possible sample ($n = 1$), the standard error and the population standard deviation are the same. When sample size is increased, the standard error gets smaller, and the sample means tend to approximate μ more closely. Thus, standard error defines the relation between sample size and the accuracy with which $\overline{X}$ represents μ.

STANDARD ERROR AND STATISTICAL INFERENCE

Inferential statistics are methods that use sample data as the basis for drawing general conclusions about populations. However, we have noted that a sample is not expected to give a perfectly accurate reflection of its population. In particular, there will be some error or discrepancy between a sample statistic and the corresponding population parameter. In this chapter, we have observed that a sample mean will not be exactly equal to the population mean. The standard error of $\overline{X}$ specifies how much difference should be expected between the mean for a sample and the mean for the population.

The natural differences that exist between samples and populations introduce a certain amount of uncertainty and error into all inferential processes. For example, there is always a margin of error that must be considered whenever a researcher uses a sample mean as the basis for drawing a conclusion about a population mean. Remember, the sample mean is not perfect. In the next five chapters, we will introduce a variety of statistical methods that all use sample means to draw inferences about population means. In each case, the standard error will be a critical element in the inferential process. Before we begin this series of chapters, we will pause briefly to reexamine the concept of the standard error and to consider how this concept will influence the inferential statistics that follow.

Standard error as a measure of chance   Most inferential statistics are used in the context of a research study. Typically, the researcher begins with a general question about how a treatment will affect the individuals in a population. For example,

Will the drug affect blood pressure?

Will the hormone affect growth?

Will the special training affect students' reading scores?

Because populations are usually too large to observe every individual, most research is conducted with samples. The research process typically involves selecting a sample from the population, administering the treatment to the sample, and then comparing the treated sample with the original (untreated) population. If the sample that received the treatment is noticeably different, then the researcher can conclude that the treatment has an effect. This process is shown in Figure 7.7.

Notice that the final step in the research process is a comparison between the treated sample and the original, untreated population. The difference between the sample and the population is viewed as evidence that the treatment has an effect. In Figure 7.7, for example, the original population has a mean of $\mu = 30$, and the treated sample has a mean of $\overline{X} = 34$, resulting in a 4-point difference between the sample and the population. The question for the researcher is how to interpret the 4-point difference. Specifically, there are two possible explanations:

1. The treatment may have caused the scores in the sample to be 4 points higher.

FIGURE 7.7

The structure of a research study to investigate whether or not a treatment has an effect on the population.

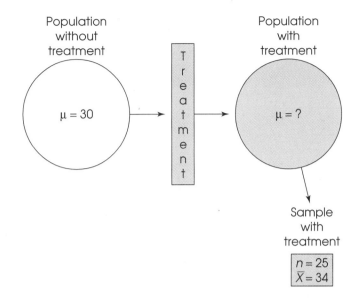

2. The 4-point difference may be sampling error. Remember, a sample mean is not expected to be exactly the same as the population mean. Perhaps the treatment has no effect at all, and the 4-point difference has occurred just by chance.

The standard error can help the researcher decide between these two alternatives. In particular, the standard error tells exactly how much difference is reasonable to expect just by chance. For example, if the standard error is only 1 point, then the researcher could conclude that the observed difference (4 points) is much larger than would be expected by chance. In this case, it would be reasonable to conclude that the treatment has caused the difference. On the other hand, if the standard error is 4 points (or more), then the researcher should conclude that results could have occurred just by chance. In this case, the data do not provide convincing evidence that the treatment has any effect. Thus, the magnitude of the standard error is an important element in interpreting the results of a research study.

Standard error as a measure of reliability   In most research studies, the researcher must rely on a *single* sample to provide an accurate representation of the population being investigated. As we have noted, however, if you take two different samples from the same population, you will get different individuals, with different scores, and different sample means. Thus, every researcher must face the nagging question, "If I had taken a different sample, would I have obtained different results?"

The importance of this question is directly related to the degree of similarity among all the different samples. For example, if there is a high level of consistency from one sample to another, then a researcher can be reasonably confident that the specific sample being studied will provide a good measurement of the population; that is, when all the samples are similar, then it does not matter which one you have selected. On the other hand, if there are big differences from one sample to another, then the researcher is left with some doubts about the accuracy of his/her specific sample. In this case, a different sample could have produced vastly different results.

In this context, the standard error can be viewed as a measure of the *reliability* of a sample mean. The term *reliability* refers to the consistency of different measurements of the same thing. More specifically, a measurement procedure is said to be reliable if you make two different measurements of the same thing and obtain identical (or nearly identical) values. If you view a sample as a "measurement" of a population, then a sample mean is a "measurement" of a population mean. If the means from different samples are all nearly identical, then the sample mean provides a reliable measure of the population. On the other hand, if there are big differences from one sample to another, then the sample mean is an unreliable measure of the population mean. Figure 7.6 demonstrates how the standard error can be used as a measure of reliability for a sample mean.

Figure 7.6(b) shows a distribution of sample means based on $n = 4$ scores. In this distribution, the standard error is $\sigma_{\bar{X}} = 10$, and it is relatively easy to find two samples with means that differ by 10 or 20 points. In this situation, a researcher cannot be confident that the mean for any individual sample is reliable; that is, a different sample could provide a very different mean.

Now consider what happens when the sample size is increased. Figure 7.6(c) shows the distribution of sample means based on $n = 100$ scores. For this distribution, the standard error is $\sigma_{\bar{X}} = 2$. With a larger sample and a smaller standard error, all of the sample means are clustered close together. With $n = 100$, a researcher can be reasonably confident that the mean for any individual sample is reliable; that is, if a second sample were taken, it would probably produce a sample mean that is essentially equivalent to the mean for the first sample.

In summary, standard error plays a critical role in inferential statistics. First, the magnitude of the standard error provides researchers with a benchmark for evaluating research results. The standard error measures how much difference is reasonable to expect just by chance and therefore allows researchers to determine whether their research results are greater than chance. Second, the magnitude of the standard error helps researchers evaluate the credibility of their results. A small standard error indicates that there are small differences between samples, so a researcher does not have to worry that a different sample might have produced different results. In the series of chapters that follow, we will introduce several different inferential techniques. In each case, however, you will find that the standard error is a central element in the inferential process.

IN THE LITERATURE:
REPORTING STANDARD ERROR

As we will see later, the standard error plays a very important role in inferential statistics. Because of its crucial role, the standard error for a sample mean, rather than the sample standard deviation, is often reported in scientific papers. Scien-

TABLE 7.2

The mean self-consciousness scores for subjects who were working in front of a video camera and those who were not (controls)

	n	Mean	*SE*
Control	17	32.23	2.31
Camera	15	45.17	2.78

FIGURE 7.8

The mean ($\pm SE$) score for treatment groups A and B.

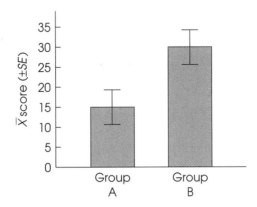

tific journals vary in how they refer to the standard error, but frequently the symbols *SE* and *SEM* (for standard error of the mean) are used. The standard error is reported in two ways. Much like the standard deviation, it may be reported in a table along with sample means (see Table 7.2). Alternatively, the standard error may be reported in graphs.

Figure 7.8 illustrates the use of a bar graph to display information about the sample mean and the standard error. In this experiment, two samples (groups A and B) are given different treatments, and then the subjects' scores on a dependent variable are recorded. The mean for group A is $\overline{X} = 15$, and for group B, it is $\overline{X} = 30$. For both samples, the standard error of $\overline{X}$ is $\sigma_{\overline{X}} = 4$. Note that the mean is represented by the height of the bar and the standard error is depicted on the graph by brackets at the top of each bar. Each bracket extends 1 standard error above and 1 standard error below the sample mean. Thus, the graph illustrates the mean for each group plus or minus 1 standard error ($\overline{X} \pm SE$). When you glance at Figure 7.8, not only do you get a "picture" of the sample means, but also you get an idea of how much error you should expect for those means.

Figure 7.9 shows how sample means and standard error are displayed on a line graph. In this study, two samples (groups A and B) receive different treatments and then are tested on a task for four trials. The number of mistakes committed on each trial is recorded for all subjects. The graph shows the mean ($\overline{X}$) number of mistakes committed for each group on each trial. The brackets show

FIGURE 7.9

The mean ($\pm SE$) number of mistakes made for groups A and B on each trial.

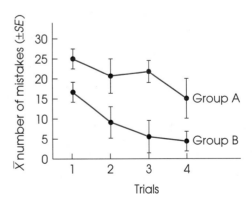

the size of the standard error for each sample mean. Again, the brackets extend 1 standard error above and below the value of the mean.   ❏

LEARNING CHECK

1. A population has a standard deviation of $\sigma = 20$.

   a. If a single score is selected from this population, how close, on average, would you expect the score to be to the population mean?

   b. If a sample of $n = 4$ scores is selected from the population, how close, on average, would you expect the sample mean to be to the population mean?

   c. If a sample of $n = 25$ scores is selected from the population, how close, on average, would you expect the sample mean to be to the population mean?

2. Can the magnitude of the standard error ever be larger than the population standard deviation? Explain your answer.

3. A researcher plans to select a random sample of $n = 36$ individuals from a population with a standard deviation of $\sigma = 18$. On average, how much error should the researcher expect between the sample mean and the population mean?

4. A researcher plans to select a random sample from a population with a standard deviation of $\sigma = 20$.

   a. How large a sample is needed to have a standard error of 10 points or less?

   b. How large a sample is needed if the researcher wants the standard error to be 2 points or less?

ANSWERS

1. a. The standard deviation, $\sigma = 20$, measures the standard distance between a score and the mean.

   b. The standard error is $20/\sqrt{4} = 10$ points.

   c. The standard error is $20/\sqrt{25} = 4$ points.

2. No. The standard error is always less than or equal to the standard deviation. The two values are equal only when $n = 1$.

3. The standard error is $18/\sqrt{36} = 3$ points.

4. a. A sample of $n = 4$ or larger.

   b. A sample of $n = 100$ or larger.

SUMMARY

1. The distribution of sample means is defined as the set of $\overline{X}$'s for all the possible random samples for a specific sample size (n) that can be obtained from a given population. According to the *central limit theorem,* the parameters of the distribution of sample means are as follows:

   a. *Shape.* The distribution of sample means will be normal if either one of the following two conditions is satisfied:

      (1) The population from which the samples are selected is normal.

      (2) The size of the samples is relatively large (around $n = 30$ or more).

   b. *Central Tendency.* The mean of the distribution of sample means will be identical to the mean of the population from which the samples are selected. The mean of the distribution of sample means is called the expected value of $\overline{X}$.

   c. *Variability.* The standard deviation of the distribution of sample means is called the standard error of $\overline{X}$ and is defined by the formula

$$\sigma_{\overline{X}} = \frac{\sigma}{\sqrt{n}}$$

Standard error measures the standard distance between a sample mean ($\overline{X}$) and the population mean (μ).

2. One of the most important concepts in this chapter is the standard error. The standard error is the standard deviation of the distribution of sample means. It measures the standard distance between a sample mean ($\overline{X}$) and the population mean (μ). The standard error tells how much error to expect if you are using a sample mean to estimate a population mean.

3. The location of each $\overline{X}$ in the distribution of sample means can be specified by a z-score:

$$z = \frac{\overline{X} - \mu}{\sigma_{\overline{X}}}$$

Because the distribution of sample means tends to be normal, we can use these z-scores and the unit normal table to find probabilities for specific sample means. In particular, we can identify which sample means are likely and which are very unlikely to be obtained from any given population. This ability to find probabilities for samples is the basis for the inferential statistics in the chapters ahead.

4. In general terms, the standard error measures how much discrepancy you should expect, due to chance, between a sample statistic and a population parameter. Statistical inference involves using sample statistics to make a general conclusion about a population parameter. Thus, standard error plays a crucial role in inferential statistics.

KEY TERMS

sampling error	sampling distribution	expected value of $\overline{X}$	law of large numbers
distribution of sample means	central limit theorem	standard error of $\overline{X}$	

—— FOCUS ON PROBLEM SOLVING ——

1. Whenever you are working probability questions about sample means, you must use the distribution of sample means. Remember, every probability question can be restated as a proportion question. Probabilities for sample means are equivalent to proportions of the distribution of sample means.

2. When computing probabilities for sample means, the most common error is to use standard deviation (σ) instead of standard error ($\sigma_{\overline{X}}$) in the z-score formula. Standard deviation measures the typical deviation (or "error") for a single score. Standard error measures the typical deviation (or error) for a sample. Remember, the larger the sample is, the more accurately the sample represents the population—that is, the larger the sample, the smaller the error.

$$\text{standard error} = \sigma_{\overline{X}} = \frac{\sigma}{\sqrt{n}}$$

3. Although the distribution of sample means is often normal, it is not always a normal distribution. Check the criteria to be certain the distribution is normal before you use the unit normal table to find probabilities (see item 1a of the Summary). Remember, all probability problems with the normal distribution are easier if you sketch the distribution and shade in the area of interest.

—— DEMONSTRATION 7.1 ——

PROBABILITY AND THE DISTRIBUTION OF SAMPLE MEANS

For a normally distributed population with $\mu = 60$ and $\sigma = 12$, what is the probability of selecting a random sample of $n = 36$ scores with a sample mean greater than 64?

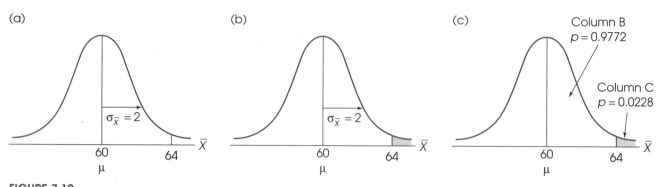

FIGURE 7.10

Sketches of the distribution for Demonstration 7.1.

In symbols, for $n = 36$, $p(\overline{X} > 64) = ?$

Notice that we may rephrase the probability question as a proportion question. Out of all the possible sample means for $n = 36$, what proportion have values greater than 64?

STEP 1   Sketch the distribution.

We are looking for a specific proportion of *all possible sample means.* Therefore, we will have to sketch the distribution of sample means. We should include the expected value, the standard error, and the specified sample mean.

The expected value for this demonstration is $\mu = 60$. The standard error is

$$\sigma_{\overline{X}} = \frac{\sigma}{\sqrt{n}} = \frac{12}{\sqrt{36}} = \frac{12}{6} = 2$$

Remember, we must use the standard error, *not* the standard deviation, because we are dealing with the distribution of sample means.

Find the approximate location of the sample mean, and place a vertical line through the distribution. In this demonstration, the sample mean is 64. It is larger than the expected value of $\mu = 60$ and therefore is placed on the right side of the distribution. Figure 7.10(a) depicts the preliminary sketch.

STEP 2   Shade the appropriate area of the distribution.

Determine whether the problem asks for a proportion greater than ($>$) or less than ($<$) the specified sample mean. Then shade the appropriate area of the distribution. In this demonstration, we are looking for the area greater than $\overline{X} = 64$, so we shade the area on the right-hand side of the vertical line [Figure 7.10(b)].

STEP 3   Compute the z-score for the sample mean.

Use the z-score formula for sample means. Remember that it uses standard error in the denominator.

$$z = \frac{\overline{X} - \mu}{\sigma_{\overline{X}}} = \frac{64 - 60}{2} = \frac{4}{2} = 2.00$$

STEP 4   Consult the unit normal table.

Look up the z-score in the unit normal table, and note the two proportions in columns B and C. Jot them down in the appropriate areas of the distribution [Figure

7.10(c)]. For this demonstration, the value in column C (the tail beyond z) corresponds exactly to the proportion we want (shaded area). Thus, for $n = 36$,

$$p(\overline{X} > 64) = p(z > +2.00) = 0.0228$$

PROBLEMS

1. Briefly define each of the following:
   a. Distribution of sample means
   b. Expected value of $\overline{X}$
   c. Standard error of $\overline{X}$

2. Two samples are randomly selected from a population. One sample has $n = 5$ scores, and the second has $n = 30$. Which sample should have a mean $(\overline{X})$ that is closer to μ? Explain your answer.

3. You have a population with $\mu = 100$ and $\sigma = 30$.
   a. If you randomly select a single score from this population, then, on average, how close would you expect the score to be to the population mean?
   b. If you randomly select a sample of $n = 100$ scores, then, on average, how close would you expect the sample mean to be to the population mean?

4. The distribution of SAT scores is normal with $\mu = 500$ and $\sigma = 100$.
   a. If you selected a random sample of $n = 4$ scores from this population, how much error would you expect between the sample mean and the population mean?
   b. If you selected a random sample of $n = 25$ scores, how much error would you expect between the sample mean and the population mean?
   c. How much error would you expect for a sample of $n = 100$ scores?

5. Standard error measures the standard distance between a sample mean and the population mean. For a population with $\sigma = 30$,
   a. How large a sample would be needed to have a standard error of 10 points or less?
   b. How large a sample would be needed to have a standard error of 5 points or less?
   c. Can the standard error ever be larger than the population standard deviation? Explain your answer.

6. IQ scores form normal distributions with $\sigma = 15$. However, the mean IQ varies from one population to another. For example, the mean IQ for registered voters is different from the mean for nonregistered voters. A researcher would like to use a sample to obtain information about the mean IQ for the population of licensed clinical psychologists in the state of California. Which sample mean should provide a more accurate representation of the population: a sample of $n = 9$ or a sample of $n = 25$? In

each case, compute exactly how much error the researcher should expect between the sample mean and the population mean.

7. A distribution of scores has $\sigma = 6$, but the value of the mean is unknown. A researcher plans to select a sample from the population in order to learn more about the unknown mean.
   a. If the sample consists of a single score $(n = 1)$, how accurately should the score represent the population mean? That is, how much error, on average, should the researcher expect between the score and the population mean?
   b. If the sample consists of $n = 9$ scores, how accurately should the sample mean represent the population mean?
   c. If the sample consists of $n = 36$ scores, how much error, on average, would be expected between the sample mean and the population mean?

8. A population has $\mu = 60$ and $\sigma = 10$. Find the z-score corresponding to each of the following sample means:
   a. A sample of $n = 4$ with $\overline{X} = 55$
   b. A sample of $n = 25$ with $\overline{X} = 64$
   c. A sample of $n = 100$ with $\overline{X} = 62$

9. If you are taking a random sample from a normal population with $\mu = 100$ and $\sigma = 16$, which of the following outcomes is more likely? Explain your answer. (*Hint:* Calculate the z-score for each sample mean.)
   a. A sample mean greater than 106 for a sample of $n = 4$
   b. A sample mean greater than 103 for a sample of $n = 36$

10. A population consists of exactly three scores: $X = 0$, $X = 2$, and $X = 4$.
    a. List all the possible random samples of $n = 2$ scores from this population. (You should obtain nine different samples—same procedure as in Example 7.1.)
    b. Compute the sample mean for each of the nine samples, and draw a histogram showing the distribution of sample means.

11. A normal population has $\mu = 100$ and $\sigma = 20$.
    a. Sketch the distribution of sample means for random samples of $n = 25$.
    b. Using z-scores, find the boundaries that separate the middle 95% of the sample means from the extreme 5% in the tails of the distribution.

c. A sample mean of $\overline{X} = 106$ is computed for a sample of $n = 25$ scores. Is this sample mean in the extreme 5%?

12. A normal population has $\mu = 80$ and $\sigma = 12$.
   a. Using this population, sketch the distribution of sample means for $n = 4$. Mark the location of the mean (expected value), and show the standard error in your sketch.
   b. Using your sketch from part a, what proportion of samples of $n = 4$ will have sample means greater than 86? (Locate the position of $\overline{X} = 86$, find the corresponding z-score, and use the unit normal table to determine the proportion.)
   c. Sketch the distribution of sample means based on $n = 16$.
   d. Using your sketch from part c, what proportion of samples of $n = 16$ will have sample means greater than 86?

13. A normal population has $\mu = 55$ and $\sigma = 8$.
   a. Sketch the distribution of sample means for samples of size $n = 16$ selected from this population.
   b. What proportion of samples based on $n = 16$ will have sample means greater than 59?
   c. What proportion of samples with $n = 16$ will have sample means less than 56?
   d. What proportion of samples with $n = 16$ will have sample means between 53 and 57?

14. A normal population has $\mu = 70$ and $\sigma = 12$.
   a. Sketch the population distribution. What proportion of the scores have values greater than $X = 73$?
   b. Sketch the distribution of sample means for samples of size $n = 16$. What proportion of the sample means have values greater than 73?

15. A normal population has $\mu = 50$ and $\sigma = 10$.
   a. Sketch the population distribution. What proportion of the scores have values within 5 points of the population mean? That is, find $p(45 < \overline{X} < 55)$.
   b. Sketch the distribution of sample means for samples of size $n = 25$. What proportion of the sample means will have values within 5 points of the population mean? That is, find $p(45 < \overline{X} < 55)$ for $n = 25$.

16. For a normal population with $\mu = 70$ and $\sigma = 20$, what is the probability of obtaining a sample mean greater than 75
   a. For a random sample of $n = 4$ scores?
   b. For a random sample of $n = 16$ scores?
   c. For a random sample of $n = 100$ scores?

17. For a negatively skewed population with $\mu = 85$ and $\sigma = 18$, what is the probability of obtaining a sample mean greater than 88
   a. For a random sample of $n = 4$ scores? (*Caution:* This is a trick question.)
   b. For a random sample of $n = 36$ scores?

18. A population of scores forms a normal distribution with $\mu = 75$ and $\sigma = 12$.
   a. What is the probability of obtaining a random sample of $n = 4$ scores with a sample mean that is within 5 points of the population mean? That is, find $p(70 < \overline{X} < 80)$.
   b. For a sample of $n = 16$ scores, what is the probability of obtaining a sample mean that is within 5 points of the population mean?

19. A sample is selected from a population with $\sigma = 10$.
   a. If the standard error for the sample mean is $\sigma_{\overline{X}} = 2$, then how many scores are in the sample?
   b. If the standard error for the sample mean is $\sigma_{\overline{X}} = 1$, then how many scores are in the sample?

20. A population has $\mu = 300$ and $\sigma = 80$.
   a. If a random sample of $n = 25$ scores is selected from this population, how much error would be expected between the sample mean and the population mean?
   b. If the sample size were 4 times larger, $n = 100$, how much error would be expected?
   c. If the sample size were increased again by a factor of 4 to $n = 400$, how much error would be expected?
   d. Comparing your answers to parts a, b, and c, what is the relationship between sample size and standard error? (Note that the sample size was increased by a factor of 4 each time.)

21. The local hardware store sells screws in 1-pound bags. Because the screws are not identical, the number of screws varies from bag to bag, with $\mu = 140$ and $\sigma = 10$. A carpenter needs a total of 600 screws for a project. What is the probability that the carpenter will get enough screws if only four bags are purchased? Assume that the distribution is normal. (*Note:* In order to obtain a total of 600 screws, the 4 bags must average at least $\overline{X} = 150$.)

22. A large grocery store chain in upstate New York has received a shipment of 1000 cases of oranges. The shipper claims that the cases average $\mu = 40$ oranges with a standard deviation of 2. To check this claim, the store manager randomly selects four cases and counts the number of oranges in each case. For these four cases, the average number of oranges is $\overline{X} = 38$.
   a. Assuming that the shipper's claim is true, what is the probability of obtaining a sample mean at least this small?
   b. Based on your answer for part a, does the grocery store manager have reason to suspect that he has been cheated? Explain your answer.

23. A researcher evaluated the effectiveness of relaxation training in reducing anxiety. One sample of anxiety-ridden people received relaxation training, while a second sample did not. Then anxiety scores were measured for all subjects, using a standardized test. Use the information that is summarized in the following table to complete this exercise.

The effect of relaxation training on anxiety scores		
Group	Mean anxiety score	*SE*
Control group	36	7
Relaxation training	18	5

a. Construct a bar graph that incorporates all of the information in the table.

b. Looking at your graph, do you think the relaxation training really worked? Explain your answer.

24. Suppose a researcher did the same study described in problem 23. This time the results of the experiment were as follows:

The effect of relaxation training on anxiety scores		
Group	Mean anxiety score	*SE*
Control group	36	12
Relaxation training	18	10

a. Construct a bar graph that incorporates all of the information in the table.

b. Looking at your graph, do you think the relaxation training really worked? Explain your answer.

25. A researcher assessed the effects of a new drug on migraine headaches. One sample of migraine sufferers received a placebo pill (0 milligrams of the drug) every day for a month. A second sample received a 10-mg dose of the drug daily for a month, and a third sample received daily doses of 20 mg. The number of headaches each person had during the month was recorded. The results are summarized in the following table:

The mean number of migraines during drug treatment		
Dose of drug (mg)	Mean number of migraines	*SE*
0	28	5
10	12	4
20	5	2

Draw a line graph (polygon) that illustrates the information provided by the table.

CHAPTER 8

INTRODUCTION TO HYPOTHESIS TESTING

TOOLS YOU WILL NEED

The following items are considered essential background material for this chapter. If you doubt your knowledge of any of these items, you should review the appropriate chapter or section before proceeding.

- z-Scores (Chapter 5)
- Distribution of sample means (Chapter 7)
  - Expected value
  - Standard error
  - Probability of sample means

CONTENTS

8.1   THE LOGIC OF HYPOTHESIS TESTING

It usually is impossible or impractical for a researcher to observe every individual in a population. Therefore, researchers usually collect data from a sample and then use the sample data to help answer questions about the population. Hypothesis testing is a statistical procedure that allows researchers to use sample data to draw inferences about the population of interest.

Hypothesis testing is one of the most commonly used inferential procedures. In fact, most of the remainder of this book will examine hypothesis testing in a variety of different situations and applications. Although the details of a hypothesis test will change from one situation to another, the general process will remain constant. In this chapter, we will introduce the general procedure for a hypothesis test. You should notice that we will use the statistical techniques that have been developed in the preceding three chapters: That is, we will combine the concepts of z-scores, probability, and the distribution of sample means to create a new statistical procedure known as a *hypothesis test*.

DEFINITION   A *hypothesis test* is a statistical method that uses sample data to evaluate a hypothesis about a population parameter.

In very simple terms, the logic underlying the hypothesis-testing procedure is as follows:

1. First, we state a hypothesis about a population. Usually the hypothesis concerns the value of a population parameter. For example, we might hypothesize that the mean IQ for registered voters in the United States is $\mu = 110$.
2. Next, we obtain a random sample from the population. For example, we might select a random sample of $n = 200$ registered voters.
3. Finally, we compare the sample data with the hypothesis. If the data are consistent with the hypothesis, we will conclude that the hypothesis is reasonable. But if there is a big discrepancy between the data and the hypothesis, we will decide that the hypothesis is wrong.

A hypothesis test is typically used in the context of a research study. That is, a researcher completes a research study and then uses a hypothesis test to evaluate the results. Depending on the type of research and the type of data, the details of the hypothesis test will change from one research situation to another. In later chapters, we will examine different versions of hypothesis testing that are used for different kinds of research. For now, however, we will focus on the basic elements that are common to all hypothesis tests. To accomplish this general goal, we will examine a hypothesis test as it applies to the simplest possible research study. In particular, we will look at the situation where a researcher is using one sample to examine one unknown population.

Figure 8.1 shows the general research situation that we will use to introduce the process of hypothesis testing. Notice that the researcher begins with a known population. This is the set of individuals as they exist *before treatment*. For this example, we are assuming that the original set of scores forms a normal distribution with $\mu = 26$ and $\sigma = 4$. The purpose of the research is to determine the effect of the

FIGURE 8.1

The basic experimental situation for hypothesis testing. It is assumed that the parameter μ is known for the population before treatment. The purpose of the experiment is to determine whether or not the treatment has an effect on the population mean.

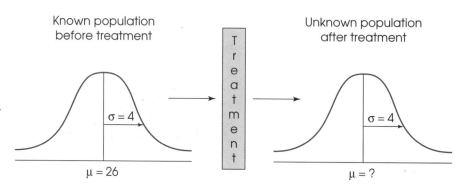

Known population before treatment

$\sigma = 4$

$\mu = 26$

Treatment

Unknown population after treatment

$\sigma = 4$

$\mu = ?$

treatment on the individuals in the population. That is, the goal is to determine what happens to the population *after the treatment is administered.*

To simplify the hypothesis-testing situation, one basic assumption is made about the effect of the treatment: If the treatment has any effect, it is simply to add a constant amount to (or subtract a constant amount from) each individual's score. You should recall from Chapters 3 and 4 that adding (or subtracting) a constant will not change the shape of the population, nor will it change the standard deviation. Thus, we will assume that the population after treatment has the same shape as the original population and the same standard deviation as the original population. This assumption is incorporated into the situation shown in Figure 8.1.

A hypothesis test is a formalized procedure that follows a standard series of operations. In this way, researchers have a standardized method for evaluating the results of their research studies. Other researchers will recognize and understand exactly how the data were evaluated and how conclusions were reached. To emphasize the formal structure of a hypothesis test, we will present hypothesis tests as a four-step process that will be used throughout the rest of the book. The following example will be used to provide a concrete foundation for introducing the hypothesis-testing procedure.

EXAMPLE 8.1   Psychologists have noted that stimulation during infancy can have profound effects on the development of infant rats. It has been demonstrated that stimulation (for example, increased handling) results in eyes opening sooner, more rapid brain maturation, faster growth, and larger body weight (see Levine, 1960). Based on the data, one might theorize that increased stimulation early in life can be beneficial. Suppose a researcher who is interested in this developmental theory would like to determine whether or not stimulation during infancy has an effect on human development.

It is known from national health statistics that the mean weight for 2-year-old children is $\mu = 26$ pounds. The distribution of weights is normal with $\sigma = 4$ pounds. The researcher's plan is to obtain a sample of $n = 16$ newborn infants and give their parents detailed instructions for giving their children increased handling and stimulation. At age 2, each of the 16 infants will be weighed, and the mean weight for the sample will be computed. If the mean weight for the sample is substantially more (or less) than the general population mean, then the researcher can conclude that the increased handling did have an effect on development. On the other hand, if the mean weight for the sample is around 26 pounds (the same as the general population mean), then the researcher must conclude that increased handling does not appear to have any effect.

Figure 8.1 depicts the research situation that was described in the preceding example. Notice that the population after treatment is unknown. Specifically, we do not know what will happen to the mean weight for 2-year-old children if the whole population receives special handling. However, we do have a sample of $n = 16$ children who have received special handling, and we can use this sample to draw inferences about the unknown population. The following four steps outline the hypothesis-testing procedure that allows us to use the sample data to answer questions about the unknown population.

STEP 1: STATE THE HYPOTHESES

The goal of inferential statistics is to make general statements about the population by using sample data. Therefore, when testing hypotheses, we make our predictions about the population parameters.

As the name implies, the process of hypothesis testing begins by stating a hypothesis about the unknown population. Actually, we state two opposing hypotheses. Notice that both hypotheses are stated in terms of population parameters.

The first and most important of the two hypotheses is called the *null hypothesis* and is identified by the symbol H_0. The null hypothesis states that the treatment has no effect. In general, the null hypothesis states that there is no change, no effect, no difference—nothing happened, hence the name *null*. For the study in Example 8.1, the null hypothesis states that increased handling during infancy has *no effect* on body weight for the population of infants. In symbols, this hypothesis would be

$$H_0:\ \mu_{\text{infants handled}} = 26 \text{ pounds} \quad \text{(Even with extra handling, the mean weight at 2 years is still 26 pounds.)}$$

DEFINITION

The *null hypothesis* (H_0) states that in the general population there is no change, no difference, or no relationship. In the context of an experiment, H_0 predicts that the independent variable (treatment) will have *no effect* on the dependent variable for the population.

The second hypothesis is simply the opposite of the null hypothesis, and it is called the *scientific* or *alternative hypothesis* (H_1). This hypothesis states that the treatment has an effect on the dependent variable.

DEFINITION

The *alternative hypothesis* (H_1) states that there is a change, a difference, or a relationship for the general population. In the context of an experiment, H_1 predicts that the independent variable (treatment) *will have an effect* on the dependent variable.

The null hypothesis and the alternative hypothesis are mutually exclusive and exhaustive. They cannot both be true. The data will determine which one should be rejected.

For this example, the alternative hypothesis predicts that handling does alter growth for the population. In symbols, H_1 is represented as

$$H_1:\ \mu_{\text{infants handled}} \neq 26 \quad \text{(With handling, the mean will be different from 26 pounds.)}$$

Notice that the alternative hypothesis simply states that there will be some type of change. It does not specify whether the effect will be increased or decreased growth.

In some circumstances, it is appropriate to specify the direction of the effect in H_1. For example, the researcher might hypothesize that increased handling will increase growth ($\mu > 26$ pounds). This type of hypothesis results in a directional hypothesis test, which will be examined in detail later in this chapter. For now we will concentrate on nondirectional tests, where the hypotheses simply state that the treatment has some effect (H_1) or has no effect (H_0). A nondirectional hypothesis test is always appropriate, even when a researcher has a definite prediction that the treatment will increase (or decrease) scores. For this example, we are examining whether handling in infancy does alter growth in some way (H_1) or has no effect (H_0). You should also note that both hypotheses refer to a population whose mean is unknown—namely, the population of infants who receive extra handling early in life.

STEP 2: SET THE CRITERIA FOR A DECISION

The researcher will eventually use the data from the sample to evaluate the credibility of the null hypothesis. The data will either provide support for the null hypothesis or tend to refute the null hypothesis. In particular, if there is a big discrepancy between the data and the hypothesis, we will conclude that the hypothesis is wrong.

To formalize the decision process, we will use the null hypothesis to determine exactly what kind of data would be expected if the treatment has no effect. In particular, we will examine all the possible sample means that could be obtained if the null hypothesis is true. For our example, this is the distribution of sample means for $n = 16$, centered at a value of $\mu = 26$ pounds. The distribution is then divided into two sections:

1. Sample means that are likely to be obtained if H_0 is true; that is, sample means that are close to the null hypothesis.
2. Sample means that are very unlikely to be obtained if H_0 is true; that is, sample means that are very different from the null hypothesis.

The distribution of sample means divided into these two sections is shown in Figure 8.2. Notice that the high-probability samples are located in the center of the distribution and have sample means close to the value specified in the null hypothesis. On the other hand, the low-probability samples are located in the extreme tails of the distribution. After the distribution has been divided in this way, we can compare our sample data with the values in the distribution. Specifically, we can determine whether our sample mean is consistent with the null hypothesis (like the values in the center of the distribution) or whether our sample mean is very different from the null hypothesis (like the values in the extreme tails).

To find the boundaries that separate the high-probability samples from the low-probability samples, we must define exactly what is meant by "low" probability and "high" probability. This is accomplished by selecting a specific probability value, which is known as the *level of significance* or the *alpha level* for the hypothesis test. The alpha (α) value is a small probability that is used to identify the low-probability samples. By convention, commonly used alpha levels are $\alpha = .05$ (5%), $\alpha = .01$ (1%), and $\alpha = .001$ (0.1%). For example, with $\alpha = .05$, we will separate the most unlikely 5% of the sample means (the extreme values) from the most likely 95% of the sample means (the central values).

The extremely unlikely values, as defined by the alpha level, make up what is called the *critical region*. These extreme values in the tails of the distribution define outcomes that are inconsistent with the null hypothesis; that is, they are very un-

FIGURE 8.2

The set of potential samples is divided into those that are likely to be obtained and those that are very unlikely if the null hypothesis is true.

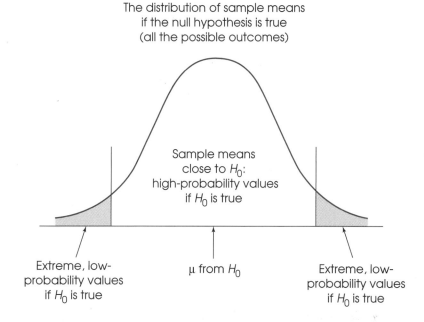

The distribution of sample means if the null hypothesis is true (all the possible outcomes)

Sample means close to H_0: high-probability values if H_0 is true

μ from H_0

Extreme, low-probability values if H_0 is true

Extreme, low-probability values if H_0 is true

likely to occur if the null hypothesis is true. Whenever the data from a research study produce a sample mean that is located in the critical region, we will conclude that the data are inconsistent with the null hypothesis, and we will reject the null hypothesis.

DEFINITIONS

The *alpha level* or the *level of significance* is a probability value that is used to define the very unlikely sample outcomes if the null hypothesis is true.

The *critical region* is composed of extreme sample values that are very unlikely to be obtained if the null hypothesis is true. The boundaries for the critical region are determined by the alpha level. If sample data fall in the critical region, the null hypothesis is rejected.

To determine the exact location for the boundaries that define the critical region, we will use the alpha-level probability and the unit normal table. In most cases, the distribution of sample means is normal, and the unit normal table will provide the precise z-score location for the critical region boundaries. With $\alpha = .05$, for example, the boundaries separate the extreme 5% from the middle 95%. Because the extreme 5% is split between two tails of the distribution, there is exactly 2.5% (or 0.0250) in each tail. In the unit normal table, you can look up a proportion of 0.0250 in column C (the tail) and find that the z-score boundary is $z = 1.96$. Thus, for any normal distribution, the extreme 5% is in the tails of the distribution beyond $z = 1.96$ and $z = -1.96$. These values are shown in Figure 8.3.

Similarly, an alpha level of $\alpha = .01$ means that 1% or .0100 is split between the two tails. In this case, the proportion in each tail is .0050, and the corresponding z-score boundaries are $z = \pm 2.58$ (± 2.57 is equally good). For $\alpha = .001$, the boundaries are located at $z = \pm 3.30$. You should verify these values in the unit normal table and be sure that you understand exactly how they are obtained.

FIGURE 8.3

The critical region (very unlikely outcomes) for $\alpha = .05$.

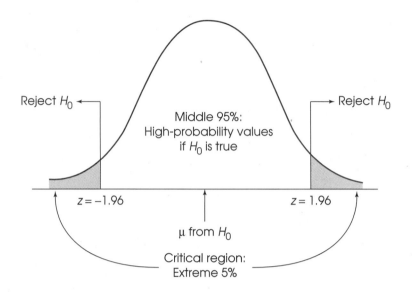

Reject H_0

Middle 95%:
High-probability values
if H_0 is true

Reject H_0

$z = -1.96$

$z = 1.96$

μ from H_0

Critical region:
Extreme 5%

**STEP 3: COLLECT DATA AND
COMPUTE SAMPLE STATISTICS**

The next step in hypothesis testing is to obtain the sample data. At this time, a random sample of infants would be selected, and their parents would be trained to provide the additional daily handling that constitutes the treatment for this study. Then the body weight for each infant would be measured at 2 years of age. Notice that the data are collected *after* the researcher has stated the hypotheses and established the criteria for a decision. This sequence of events helps ensure that a researcher makes an honest, objective evaluation of the data and does not tamper with the decision criteria after the experimental outcome is known.

Next, the raw data from the sample are summarized with the appropriate statistics: For this example, the researcher would compute the sample mean. Now it is possible for the researcher to compare the sample mean (the data) with the null hypothesis. This is the heart of the hypothesis test: comparing the data with the hypothesis.

The comparison is accomplished by computing a z-score that describes exactly where the sample mean is located relative to the hypothesized population mean from H_0. In step 2, we constructed the distribution of sample means that would be expected if the null hypothesis were true, that is, the entire set of sample means that could be obtained if the treatment has no effect (see Figure 8.3). Now we will calculate a z-score that identifies where our sample mean is located in this hypothesized distribution. The z-score formula for a sample mean is

$$z = \frac{\overline{X} - \mu}{\sigma_{\overline{X}}}$$

In the formula, the value of the sample mean ($\overline{X}$) is obtained from the sample data, and the values for μ and $\sigma_{\overline{X}}$ come from the hypothesized distribution in step 2. In particular, you should note that the value of μ is obtained from the null hypothesis. Thus, the z-score formula can be expressed in words as follows:

$$z = \frac{\text{sample mean} - \text{hypothesized population mean}}{\text{standard error between } \overline{X} \text{ and } \mu}$$

8.1   REJECTING THE NULL HYPOTHESIS VERSUS PROVING THE ALTERNATIVE HYPOTHESIS

IT MAY seem awkward to pay so much attention to the null hypothesis. After all, the purpose of most experiments is to show that a treatment does have an effect, and the null hypothesis states that there is no effect. The reason for focusing on the null hypothesis, rather than the alternative hypothesis, comes from the limitations of inferential logic. Remember, we want to use the sample data to draw conclusions, or inferences, about a population. Logically, it is much easier to demonstrate that a universal (population) hypothesis is false than to demonstrate that it is true. This principle is shown more clearly in a simple example. Suppose you make the universal statement "all dogs have four legs" and you intend to test this hypothesis by using a sample of one dog. If the dog in your sample does have four legs, have you

proved the statement? It should be clear that one four-legged dog does not prove the general statement to be true. On the other hand, suppose the dog in your sample has only three legs. In this case, you have proved the statement to be false. Again, it is much easier to show that something is false than to prove that it is true.

Hypothesis testing uses this logical principle to achieve its goals. It would be difficult to state "the treatment has an effect" as the hypothesis and then try to prove that this is true. Therefore, we state the null hypothesis, "the treatment has no effect," and try to show that it is false. The end result still is to demonstrate that the treatment does have an effect. That is, we find support for the alternative hypothesis by disproving (rejecting) the null hypothesis.

Notice that the top of the z-score formula measures how much difference there is between the data and the hypothesis. The bottom of the formula measures the standard distance that ought to exist between the sample mean and the population mean.

STEP 4: MAKE A DECISION   In the final step, the researcher uses the z-score value obtained in step 3 to make a decision about the null hypothesis according to the criteria established in step 2. There are two possible decisions, and both are stated in terms of the null hypothesis (see Box 8.1).

One possible decision is to *reject the null hypothesis.* This decision is made whenever sample data fall in the critical region. By definition, a sample value in the critical region indicates that there is a big discrepancy between the sample and the null hypothesis (the sample is in the extreme tail of the distribution). This kind of outcome is very unlikely to occur if the null hypothesis is true, so our conclusion is to reject H_0. In this case, the data provide convincing evidence that the null hypothesis is wrong, and we conclude that the treatment really did have an effect on the individuals in the sample. For the example we have been considering, suppose the sample of $n = 16$ infants produced a sample mean of $\overline{X} = 30$ pounds at age 2. With $n = 16$ and $\sigma = 4$, the standard error for the sample mean would be

$$\sigma_{\overline{X}} = \frac{\sigma}{\sqrt{n}} = \frac{4}{\sqrt{16}} = \frac{4}{4} = 1$$

The z-score for the sample is

$$z = \frac{\overline{X} - \mu}{\sigma_{\overline{X}}} = \frac{30 - 26}{1} = \frac{4}{1} = 4.00$$

With an alpha level of $\alpha = .05$, this z-score is far beyond the boundary of 1.96. Because the sample z-score is in the critical region, we would reject the null hypothesis and conclude that the special handling did have an effect on the infants' weights.

The second possibility occurs when the sample data are not in the critical region. In this case, the data are reasonably close to the null hypothesis (in the center of the distribution). Because the data do not provide strong evidence that the null hypothesis is wrong, our conclusion is to *fail to reject the null hypothesis.* This conclusion means that the treatment does not appear to have an effect. For the research study examining extra handling of infants, suppose our sample of $n = 16$ produced a mean weight of $\overline{X} = 27$ pounds. As before, the standard error for the sample mean is $\sigma_{\overline{X}} = 1$, and the null hypothesis states that $\mu = 26$. These values produce a z-score of

$$z = \frac{\overline{X} - \mu}{\sigma_{\overline{X}}} = \frac{27 - 26}{1} = \frac{1}{1} = 1.00$$

The z-score of 1.00 is not in the critical region. Therefore, we would fail to reject the null hypothesis and conclude that the special handling does not appear to have an effect on weight.

The two possible decisions may be easier to understand if you think of a research study as an attempt to gather evidence to prove that a treatment works. From this perspective, the research study has two possible outcomes:

1. You gather enough evidence to convincingly demonstrate that the treatment really works. That is, you reject the null hypothesis and conclude that the treatment does have an effect.

2. The evidence you gather from the research study is not convincing. In this case, all you can do is conclude that there is not enough evidence. The research study has failed to demonstrate that the treatment has an effect, and the statistical decision is to fail to reject the null hypothesis.

A CLOSER LOOK AT THE z-SCORE STATISTIC   The z-score statistic that is used in the hypothesis test is the first specific example of what is called a *test statistic.* The term *test statistic* simply indicates that the sample data are converted into a single, specific statistic that is used to test the hypotheses. In the chapters that follow, we will introduce several other test statistics that are used in a variety of different research situations. However, each of the new test statistics will have the same basic structure and will serve the same purpose as the z-score. We have already described the z-score equation as a formal method for comparing the sample data and the population hypothesis. In this section, we will discuss the z-score from two other perspectives that may give you a better understanding of hypothesis testing and the role that z-scores play in this inferential technique. In each case, keep in mind that the z-score will serve as a general model for other test statistics that will come in future chapters.

The z-score formula as a recipe   The z-score formula, like any formula, can be viewed as a recipe. If you follow instructions and use all the right ingredients, the formula will produce a z-score. In the hypothesis-testing situation, however, you do not have all the necessary ingredients. Specifically, you do not know the value for the population mean (μ), which is one component or ingredient in the formula.

This situation is similar to trying to follow a cake recipe where one of the ingredients is not clearly listed. For example, there may be a grease stain that obscures one part of your recipe. In this situation, what do you do? One possibility is to make a hypothesis about the missing ingredient. You might, for example, hypothesize that the missing ingredient is 2 cups of flour. To test the hypothesis, you simply add the rest of the ingredients along with the hypothesized flour and see what happens. If the cake turns out to be good, then you can be reasonably confident that your hypothesis was correct. But if the cake is terrible, then you conclude that your hypothesis was wrong.

In a hypothesis test with z-scores, you do essentially the same thing. You do not know the value for the population mean, μ. Therefore, you make a hypothesis (the null hypothesis) about the unknown value μ and plug it into the formula along with the rest of the ingredients. To evaluate your hypothesis, you simply look at the result. If the formula produces a z-score near zero (which is where z-scores are supposed to be), then you conclude that your hypothesis was correct. On the other hand, if the formula produces an unlikely result (an extremely large or small value), then you conclude that your hypothesis was wrong.

The z-score formula as a ratio   In the context of a hypothesis test, the z-score formula has the following structure:

$$z = \frac{\overline{X} - \mu}{\sigma_{\overline{X}}} = \frac{\text{sample mean} - \text{hypothesized population mean}}{\text{standard error between } \overline{X} \text{ and } \mu}$$

Notice that the numerator of the formula involves a direct comparison between the sample data and the null hypothesis. In particular, the numerator measures the obtained difference between the sample mean and the hypothesized population mean. The standard error in the denominator of the formula measures how much difference ought to exist between the sample mean and the population mean just by chance. Thus, the z-score formula (and most other test statistics) forms a ratio

$$z = \frac{\text{obtained difference}}{\text{difference due to chance}}$$

In general, the purpose of a test statistic is to determine whether the result of the research study (the obtained difference) is more than would be expected by chance alone. Whenever the test statistic has a value greater than 1.00, it means that the obtained result (numerator) is greater than chance (denominator). However, you should realize that most researchers are not satisfied with results that are simply more than chance. Instead, conventional research standards require results that are substantially more than chance. Specifically, most hypothesis tests require that the test statistic ratio have a value of at least 2 or 3; that is, the obtained difference must be 2 or 3 times bigger than chance before the researcher will reject the null hypothesis.

LEARNING CHECK

1. What does the null hypothesis predict about a population or a treatment effect?

2. Define the *critical region* for a hypothesis test.

3. As the alpha level gets smaller, the size of the critical region also gets smaller. (True or false?)

4. A small value (near zero) for the z-score statistic is evidence that the null hypothesis should be rejected. (True or false?)

5. A decision to reject the null hypothesis means that you have demonstrated that the treatment has no effect. (True or false?)

ANSWERS **1.** The null hypothesis predicts that the treatment has no effect and the population is unchanged.

2. The critical region consists of sample outcomes that are very unlikely to be obtained if the null hypothesis is true.

3. True. As alpha gets smaller, the critical region is moved farther out into the tails of the distribution.

4. False. A z-score near zero indicates that the data support the null hypothesis.

5. False. Rejecting the null hypothesis means that you have concluded that the treatment does have an effect.

8.2 UNCERTAINTY AND ERRORS IN HYPOTHESIS TESTING

Hypothesis testing is an *inferential process,* which means that it uses limited information as the basis for reaching a general conclusion. Specifically, a sample provides only limited or incomplete information about the whole population, and yet a hypothesis test uses a sample to draw a conclusion about the population. In this situation, there is always the possibility that an incorrect conclusion will be made. Although sample data are usually representative of the population, there is always a chance that the sample is misleading and will cause a researcher to make the wrong decision about the research results. In a hypothesis test, there are two different kinds of errors that can be made.

TYPE I ERRORS It is possible that the data will lead you to reject the null hypothesis when in fact the treatment has no effect. Remember, samples are not expected to be identical to their populations, and some extreme samples can be very different from the populations they are supposed to represent. If a researcher selects one of these extreme samples, just by chance, then the data from the sample may give the appearance of a strong treatment effect, even though there is no real effect. In the previous section, for example, we discussed a research study examining how increased handling during infancy affects body weight. Suppose the researcher selected a sample of $n = 16$ infants who were genetically destined to be bigger than average even without any special treatment. When these infants are measured at 2 years of age, they are going to look bigger than average even though the special handling (the treatment) may have had no effect. In this case, the researcher is likely to conclude that the treatment has had an effect, when in fact it really did not. This is an example of what is called a *Type I error.*

DEFINITION

A *Type I error* occurs when a researcher rejects a null hypothesis that is actually true. In a typical research situation, a Type I error means that the researcher concludes that a treatment does have an effect when in fact the treatment has no effect.

You should realize that a Type I error is not a stupid mistake in the sense that a researcher is overlooking something that should be perfectly obvious. On the contrary, the researcher is looking at sample data that appear to show a clear treatment effect. The researcher then makes a careful decision based on the available information. The problem is that the information from the sample is misleading. In general, a researcher never knows for certain whether a hypothesis is true or false. Instead, we must rely on the data to tell us whether the hypothesis *appears* to be true or false.

In most research situations, the consequences of a Type I error can be very serious. Because the researcher has rejected the null hypothesis and believes that the treatment has a real effect, it is likely that the researcher will report or even publish the research results. A Type I error, however, means that this is a false report. Thus, Type I errors lead to false reports in the scientific literature. Other researchers may try to build theories or develop other experiments based on the false results. A lot of precious time and resources may be wasted.

Fortunately, the hypothesis test is structured to control and minimize the risk of committing a Type I error. Although extreme and misleading samples are possible, they are relatively unlikely. In order for an extreme sample to produce a Type I error, the sample mean must be in the critical region. However, the critical region is structured so that it is *very unlikely* to obtain a sample mean in the critical region when H_0 is true. Specifically, the alpha level determines the probability of obtaining a sample mean in the critical region when H_0 is true. Thus, the alpha level defines the probability or risk of a Type I error. This fact leads to an alternative definition of the alpha level.

DEFINITION

The *alpha level* for a hypothesis test is the probability that the test will lead to a Type I error. That is, the alpha level determines the probability of obtaining sample data in the critical region even though the null hypothesis is true.

In summary, whenever the sample data are in the critical region, the appropriate decision for a hypothesis test is to reject the null hypothesis. Normally this is the correct decision because the treatment has caused the sample to be different from the original population; that is, the treatment effect has literally pushed the sample mean into the critical region. In this case, the hypothesis test has correctly identified a real treatment effect. Occasionally, however, sample data will be in the critical region just by chance, without any treatment effect. When this occurs, the researcher will make a Type I error; that is, the researcher will conclude that a treatment effect exists, when in fact it does not. Fortunately, the risk of a Type I error is small and is under the control of the researcher. Specifically, the probability of a Type I error is equal to the alpha level.

TABLE 8.1

Possible outcomes of a statistical decision

		Actual situation	
		No effect, H_0 true	Effect exists, H_0 false
Experimenter's decision	Reject H_0	Type I error	Decision correct
	Retain H_0	Decision correct	Type II error

TYPE II ERRORS   Whenever a researcher rejects the null hypothesis, there is a risk of a Type I error. Similarly, whenever a researcher fails to reject the null hypothesis, there is a risk of a *Type II error*. By definition, a Type II error is failing to reject a false null hypothesis. In more straightforward English, a Type II error means that a treatment effect really exists, but the hypothesis test fails to detect it.

DEFINITION   A *Type II error* occurs when a researcher fails to reject a null hypothesis that is really false. In a typical research situation, a Type II error means that the hypothesis test has failed to detect a real treatment effect.

A Type II error occurs when the sample mean is not in the critical region even though the treatment has had an effect on the sample. Often this happens when the effect of the treatment is relatively small. In this case, the treatment does influence the sample, but the magnitude of the effect is not big enough to move the sample mean into the critical region. Because the sample is not substantially different from the original population (it is not in the critical region), the statistical decision is to fail to reject the null hypothesis and to conclude that there is not enough evidence to say there is a treatment effect.

The consequences of a Type II error are usually not very serious. A Type II error means that the research data do not show the results that the researcher had hoped to obtain. The researcher can accept this outcome and conclude that the treatment either has no effect or has only a small effect that is not worth pursuing. Or the researcher can repeat the experiment (usually with some improvements) and try to demonstrate that the treatment really does work.

Unlike a Type I error, it is impossible to determine a single, exact probability value for a Type II error. Instead, the probability of a Type II error depends on a variety of factors and therefore is a function, rather than a specific number. Nonetheless, the probability of a Type II error is represented by the symbol β, the Greek letter beta.

In summary, a hypothesis test always leads to one of two decisions: Either you reject the null hypothesis and conclude that the treatment has an effect, or you fail to reject the null hypothesis and conclude that the treatment does not appear to have an effect. In either case, there is a chance that the data are misleading and the decision is wrong. The complete set of decisions and outcomes is shown in Table 8.1. The risk of an error is especially important in the case of a Type I error, which can lead to a false report. Fortunately, the probability of a Type I error is determined by the alpha level, which is completely under the control of the researcher. At the beginning of a hypothesis test, the researcher states the hypotheses and selects the alpha level, which immediately determines the risk that a Type I error will be made.

SELECTING AN ALPHA LEVEL

As you have seen, the alpha level for a hypothesis test serves two very important functions. First, alpha helps determine the boundaries for the critical region by defining the concept of "very unlikely" outcomes. More importantly, alpha determines the probability of a Type I error. When you select a value for alpha at the beginning of a hypothesis, your decision influences both of these functions.

The primary concern when selecting an alpha level is to minimize the risk of a Type I error. Thus, alpha levels tend to be very small probability values. By convention, the largest permissible value is $\alpha = .05$. An alpha level of .05 means that there is a 5% risk, or a 1-in-20 probability, of committing a Type I error. For many research situations, the .05 level of significance is viewed as an unacceptably large risk. Consider, for example, a scientific journal containing 20 research articles, each evaluated with an alpha level of .05. With a 5% risk of a Type I error, 1 out of the 20 articles is probably a false report. For this reason, most scientific publications require a more stringent alpha level, such as .01 or .001. (For more information on the origins of the .05 level of significance, see the excellent short article by Cowles and Davis, 1982.)

At this point, it may appear that the best strategy for selecting an alpha level is to choose the smallest possible value in order to minimize the risk of a Type I error. However, there is a different kind of risk that develops as the alpha level is lowered. Specifically, a lower alpha level means less risk of a Type I error, but it also means that the hypothesis test demands more evidence from the research results.

The trade-off between the risk of a Type I error and the demands of the test is controlled by the boundaries of the critical region. In order for the hypothesis test to conclude that the treatment does have an effect, the sample data must be in the critical region. If the treatment really has an effect, it should cause the sample to be different from the original population; essentially, the treatment should push the sample into the critical region. However, as the alpha level is lowered, the boundaries for the critical region move farther out and become more difficult to reach. Figure 8.4 shows how the boundaries for the critical region move farther away as the alpha level decreases. Notice that lower alpha levels produce more extreme boundaries for the critical region. Thus, an extremely small alpha level would mean almost no risk of a Type I error but would push the critical region so far out that it would become essentially impossible to ever reject the null hypothesis; that is, it would require an enormous treatment effect before the sample data would reach the critical boundaries.

In general, researchers try to maintain a balance between the risk of a Type I error and the demands of the hypothesis test. Alpha levels of .05, .01, and .001 are considered reasonably good values because they provide a relatively low risk of error without placing excessive demands on the research results.

LEARNING CHECK

1. What is a Type I error?

2. Why is the consequence of a Type I error considered serious?

3. If the alpha level is changed from $\alpha = .05$ to $\alpha = .01$, the probability of a Type I error increases. (True or false?)

FIGURE 8.4

The locations of the critical region boundaries for three different levels of significance: $\alpha = .05$, $\alpha = .01$, and $\alpha = .001$.

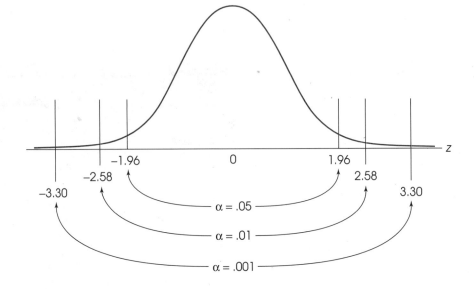

---

4. Define a Type II error.

5. What research situation is likely to lead to a Type II error?

ANSWERS   **1.** A Type I error is rejecting a true null hypothesis, that is, saying that the treatment has an effect when in fact it does not.

2. A Type I error often results in a false report. A researcher reports or publishes a treatment effect that does not exist.

3. False. The probability of a Type I error is α.

4. A Type II error is failing to reject a false null hypothesis. In terms of a research study, a Type II error occurs when a study fails to detect a treatment effect that really exists.

5. A Type II error is likely to occur when the treatment effect is very small. In this case, a research study is more likely to overlook or fail to detect the effect.

8.3   AN EXAMPLE OF A HYPOTHESIS TEST

At this time, we have introduced all the elements of a hypothesis test. In this section, we will present a complete example of the hypothesis-testing process and discuss how the results from a hypothesis test are presented in a research report. For purposes of demonstration, the following scenario will be used to provide a concrete background for the hypothesis-testing process.

EXAMPLE 8.2   Alcohol appears to be involved in a variety of birth defects, including low birth weight and retarded growth. A researcher would like to investigate the effect of prenatal alcohol on birth weight. A random sample of $n = 16$ preg-

FIGURE 8.5

The structure of a research study to determine whether prenatal alcohol affects birth weight.

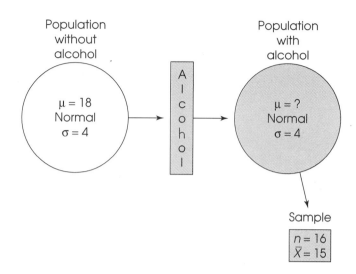

nant rats is obtained. The mother rats are given daily doses of alcohol. At birth, one pup is selected from each litter to produce a sample of $n = 16$ newborn rats. The average weight for the sample is $\overline{X} = 15$ grams. The researcher would like to compare the sample with the general population of rats. It is known that regular newborn rats (not exposed to alcohol) have an average weight of $\mu = 18$ grams. The distribution of weights is normal with $\sigma = 4$. Figure 8.5 shows the overall research situation. Notice that the researcher's question concerns the unknown population that is exposed to alcohol and the data consist of a sample from this unknown population. The following steps outline the hypothesis test that evaluates the effect of alcohol exposure on birth weight.

STEP 1 *State the hypotheses, and select the alpha level.* Both hypotheses concern the population that is exposed to alcohol where the population mean is unknown (the population on the right-hand side of Figure 8.5). The null hypothesis states that exposure to alcohol has no effect on birth weight. Thus, the population of rats with alcohol exposure should have the same mean birth weight as the regular, unexposed rats. In symbols,

$$H_0: \mu_{\text{alcohol exposure}} = 18 \quad \text{(even with alcohol exposure, the rats still average 18 grams at birth)}$$

The alternative hypothesis states that alcohol exposure does affect birth weight, so the exposed population should be different from the regular rats. In symbols,

$$H_1: \mu_{\text{alcohol exposure}} \neq 18 \quad \text{(alcohol exposure will change birth weight)}$$

Notice that both hypotheses concern the unknown population. For this test, we will use an alpha level of $\alpha = .05$. That is, we are taking a 5% risk of committing a Type I error.

STEP 2 *Set the decision criteria by locating the critical region.* By definition, the critical region consists of outcomes that are very unlikely if the null hypothesis is true. With $\alpha = .05$, the critical region consists of the most unlikely 5% of the

FIGURE 8.6

Decision criteria for the hypothesis test in Example 8.2.

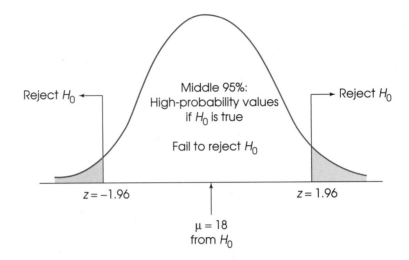

possible outcomes. To locate the critical region, first sketch the distribution of sample means that could be obtained if H_0 is true. This is the set of all possible sample means using the value of n from the research study. For this example, the distribution of sample means is normal with a standard error of

$$\sigma_{\overline{X}} = \frac{\sigma}{\sqrt{n}} = \frac{4}{\sqrt{16}} = \frac{4}{4} = 1$$

Finally, the distribution is centered at a value of $\mu = 18$ if the null hypothesis is true. This distribution is shown in Figure 8.6.

The next step is to find the boundaries that separate the extreme 5% (with $\alpha = .05$) from the rest of the distribution. As we saw earlier, for any normal distribution, z-scores of $z = \pm 1.96$ separate the middle 95% from the extreme 5% (a proportion of 0.0250 in each tail). Thus, if the null hypothesis is true, the most unlikely outcomes correspond to z-score values beyond ± 1.96. This is the critical region for our test.

STEP 3   *Collect the data, and compute the test statistic.* At this point, we would select our sample of $n = 16$ pups whose mothers had received alcohol during pregnancy. The birth weight would be recorded for each pup and the sample mean computed. As noted, we obtained a sample mean of $\overline{X} = 15$ grams. The sample mean is then converted to a z-score, which is our test statistic.

$$z = \frac{\overline{X} - \mu}{\sigma_{\overline{X}}} = \frac{15 - 18}{1} = \frac{-3}{1} = -3.00$$

STEP 4   *Make a decision.* The z-score computed in step 3 has a value of -3.00, which is beyond the boundary of -1.96. Therefore, the sample mean is located in the critical region. This is a very unlikely outcome if the null hypothesis is true, so our decision is to reject the null hypothesis. In addition to this statistical decision concerning the null hypothesis, it is customary to state a conclusion about the results of the research study. For this example, we conclude that prenatal exposure to alcohol does have a significant effect on birth weight.

IN THE LITERATURE:
REPORTING THE RESULTS OF THE STATISTICAL TEST

A special jargon and notational system are used in published reports of hypothesis tests. When you are reading a scientific journal, for example, you typically will not be told explicitly that the researcher evaluated the data using a z-score as a test statistic with an alpha level of .05. Nor will you be told that "the null hypothesis is rejected." Instead, you will see a statement such as

> The treatment with medication had a significant effect on people's depression scores, $z = 3.85$, $p < .05$.

Let us examine this statement piece by piece. First, what is meant by the term *significant?* In statistical tests, this word indicates that the result is different from what would be expected due to chance. A significant result means that the null hypothesis has been rejected. (Note that the null hypothesis for this study would state that the medication *has no effect* on depression scores for the population.) That is, the data are significant because the sample mean falls in the critical region and is not what we would have expected to obtain if H_0 were true.

DEFINITION

Findings are said to be *statistically significant* when the null hypothesis has been rejected. Thus, if results achieve statistical significance, the researcher concludes that a treatment effect occurred.

Next, what is the meaning of $z = 3.85$? The z indicates that a z-score was used as the test statistic to evaluate the sample data and that its value is 3.85. Finally, what is meant by $p < .05$? This part of the statement is a conventional way of specifying the alpha level that was used for the hypothesis test. More specifically, we are being told that the result of the experiment would occur by chance with a probability (p) that is less than .05 (alpha).

The APA style does not use a leading zero in a probability value that refers to a level of significance.

In circumstances where the statistical decision is to *fail to reject H_0*, the report might state that

> There was no evidence that the medication had an effect on depression scores, $z = 1.30$, $p > .05$.

In this case, we are saying that the obtained result, $z = 1.30$, is not unusual (not in the critical region) and is relatively likely to occur by chance (the probability is greater than .05). Thus, H_0 was not rejected.

Sometimes students become confused when trying to determine whether $p > .05$ or $p < .05$. Note that if H_0 were really true, then it would be extremely unlikely to obtain a test statistic (z) that falls in the critical region. Therefore, we reject H_0. Now we are defining *extremely unlikely* as meaning "having a probability less than alpha." With this idea in mind, whenever we reject H_0, we are saying that the results are unlikely, or $p < \alpha$ that we got these results just by chance. On the other hand, when we fail to reject H_0, we are stating that the results are not unusual; they are what we would expect by chance alone and $p > \alpha$. Thus, when the results are in the critical region, H_0 is rejected, and $p < \alpha$. When the data do not fall in the critical region, H_0 is not rejected, and $p > \alpha$. Figure 8.7 illustrates this point.

Finally, you should notice that in scientific reports, the researcher does not actually state that "the null hypothesis was rejected." Instead, the researcher

FIGURE 8.7

Sample means that fall in the critical region (shaded areas) have a probability
less than alpha ($p < \alpha$). H_0 should be
rejected. Sample means that do not fall in
the critical region have a probability
greater than alpha ($p > \alpha$).

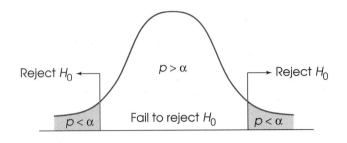

reports that the effect of the treatment was statistically significant. Likewise,
when H_0 is not rejected, one simply states that the treatment effect was not statistically significant or that there was no evidence for a treatment effect. In fact,
when you read scientific reports, you will note that the terms *null hypothesis* and
alternative hypothesis are rarely mentioned. Nevertheless, H_0 and H_1 are part of
the logic of hypothesis testing even if they are not formally stated in a scientific
report. Because of their central role in the process of hypothesis testing, you
should be able to identify and state these hypotheses.   □

**ASSUMPTIONS FOR HYPOTHESIS
TESTS WITH z-SCORES**

It will become evident in later chapters that certain conditions must be present
for each type of hypothesis test to be an appropriate and accurate procedure. These
conditions are assumed to be satisfied when the results of a hypothesis test are interpreted. However, the decisions based on the test statistic (that is, rejecting or not
rejecting H_0) may be compromised if these assumptions are not satisfied. In practice, researchers are not overly concerned with the conditions of a statistical test unless they have strong suspicions that the assumptions have been violated. Nevertheless, it is crucial to keep in mind the fundamental conditions that are associated with
each type of statistical test to ensure that it is being used appropriately. The assumptions for hypothesis tests with z-scores that involve one sample are summarized
below.

Random sampling   It is assumed that the subjects used to obtain the sample data
were selected randomly. Remember, we wish to generalize our findings from the
sample to the population. This task is accomplished when we use sample data to test
a hypothesis about the population. Therefore, the sample must be representative of
the population from which it has been drawn. Random sampling helps to ensure that
it is representative.

Independent observations   The values in the sample must consist of *independent*
observations. In everyday terms, two observations are independent if there is no
consistent, predictable relationship between the first observation and the second.
More precisely, two events (or observations) are independent if the occurrence of
the first event has no effect on the probability of the second event. Specific examples of independence and nonindependence are examined in Box 8.2. Usually
this assumption is satisfied by using a *random* sample, which also helps ensure that
the sample is representative of the population and that the results can be generalized
to the population.

The value of σ is unchanged by the treatment   The general purpose of hypothesis testing is to determine whether or not a treatment (independent variable) pro-

8.2   INDEPENDENT OBSERVATIONS

INDEPENDENT OBSERVATIONS are a basic requirement for nearly all hypothesis tests. The critical concern is that the observation or score obtained for one individual is not influenced by the observation or score obtained for another individual. The concept of independent observations is best demonstrated by counterexample. The following two situations demonstrate circumstances where observations are *not* independent.

1. A researcher is interested in examining television preferences for children. To obtain a sample of $n = 20$ children, the researcher selects 4 children from family A, 3 children from family B, 5 children from family C, 2 children from family D, and 6 children from family E.

   It should be obvious that the researcher does *not* have 20 independent observations. Within each family, the children probably share television preference (at least they watch the same shows). Thus, the response for each child is likely to be related to the responses of his or her siblings.

2. A researcher is interested in people's ability to judge distances. A sample of 20 people is obtained. All 20 subjects are gathered together in the same room. The researcher asks the first subject to estimate the distance between New York and Miami. The second subject is asked the same question, then the third subject, the fourth subject, and so on.

   Again, the observations that the researcher obtains are not independent: The response of each subject is probably influenced by the responses of the previous subjects. For example, consistently low estimates by the first few subjects could create pressure to conform and thereby increase the probability that the following subjects would also produce low estimates.

duces a change in the population mean. The null hypothesis, the critical region, and the z-score statistic all are concerned with the treated population. In the z-score formula, we use $\overline{X}$ from the treated sample, a hypothesized value of μ for the treated population, and a standard error that indicates how close $\overline{X}$ should be to μ. However, you may have noticed that when we compute the standard error, we use the standard deviation from the untreated population. Thus, the z-score appears to be using values from two different populations: $\overline{X}$ and μ from the treated population and σ from the untreated population. To justify this apparent contradiction, we must make an assumption. Specifically, we must assume that the value of σ is the same after treatment as it was before treatment.

Actually, this assumption is the consequence of a more general assumption that is part of many statistical procedures. This general assumption states that the effect of the treatment is to add a constant amount to (or subtract a constant amount from) every score in the population. You should recall that adding (or subtracting) a constant will change the mean but will have no effect on the standard deviation. You also should note that this assumption is a theoretical ideal. In actual experiments, a treatment generally will not show a perfect and consistent additive effect.

Normal sampling distribution   To evaluate hypotheses with z-scores, we have used the unit normal table to identify the critical region. This table can be used only if the distribution of sample means is normal.

CONCERNS ABOUT HYPOTHESIS TESTING

Although hypothesis testing is the most commonly used technique for evaluating and interpreting research data, a number of scientists have expressed a variety of concerns about the hypothesis-testing procedure (for example, see Loftus, 1996 and

Hunter, 1997). The following are some of the questions and criticisms that have been presented:

1. One common criticism is that a hypothesis test produces an absolute, all-or-none decision: Either you reject the null hypothesis, or you don't. Suppose, for example, a hypothesis test for one treatment produces a z-score of $z = 2.00$ and a test for a second treatment produces a z-score of 1.90. Assuming both tests are done with $\alpha = .05$, then both tests have the same critical boundary of $z = 1.96$. According to this standard, the first test would reject H_0 and conclude that the treatment has a significant effect. The second test would fail to reject H_0 and conclude that the treatment does not have a significant effect. Although the hypothesis test makes completely different decisions about the two treatments, it could be argued that they are really almost identical: How much difference is there between $z = 1.90$ and $z = 2.00$? The counterargument is that you have to draw a line somewhere. If you decided to reject H_0 with $z = 1.90$ just because it is close to the criterion, what about $z = 1.80$, or 1.70, and so on? Although the .05 level of significance may be arbitrary, some criterion is essential.

2. A second criticism is that the whole idea of a null hypothesis is artificial. Remember, the null hypothesis states that the treatment has no effect, meaning absolutely none. The criticism is based on the real-world observation that *every* treatment must have *some* effect. Anytime a researcher administers any treatment, it has to have some effect, however small, on the subjects. If this is true, it means that the null hypothesis is always wrong. If you follow this line of reasoning, it means that you can never make a Type I error: If H_0 cannot be true, you can never reject a true H_0. The counterargument to this criticism is that the null hypothesis, even if it is flawed, is a logical necessity. Because we are using an *inferential* process, it is impossible to prove that a hypothesis is true; however, it is possible to show that a hypothesis is false (or at least very unlikely). See Box 8.1 for more discussion of this issue.

3. A third criticism is that people routinely misinterpret statistical decisions. In this case, it is not the hypothesis test that is being questioned, but rather the fact that many people do not understand what a hypothesis test does.

   For example, when a test rejects the null hypothesis and concludes that the treatment has a statistically significant effect, many people automatically assume that the treatment has a "substantial" effect. This is wrong. In fact, rejecting the null hypothesis simply means that the hypothesis test has ruled out *chance* as a reasonable explanation for the data. More precisely, the hypothesis test has demonstrated that it is very unlikely (alpha level) that the result has occurred just by chance. You should recall that "chance" is measured by the standard error. If the standard error is very small, then it is possible for a treatment effect to be very small and still be greater than chance. In general, a significant treatment effect does not necessarily mean a big treatment effect.

   On the other hand, people often misunderstand what it means to fail to reject the null hypothesis. In simple terms, failing to reject H_0 means that the data are inconclusive; you cannot say whether the treatment has an effect or not. Nonetheless, many people assume that failing to reject H_0 is the same as proving that the null hypothesis is true. Again, this is wrong. You cannot prove that a treatment has no effect. All you can say is that the data do not provide convincing evidence of an effect.

Our purpose for presenting these arguments is not to criticize or to defend the hypothesis-testing process. Instead, our hope is that presenting these issues will give you a better understanding of the hypothesis-testing process, including some of its limitations.

LEARNING CHECK

1. A researcher would like to know whether room temperature will affect eating behavior. The lab temperature is usually kept at 72 degrees, and under these conditions, the rats eat an average of $\mu = 10$ grams of food each day. The amount of food varies from one rat to another, forming a normal distribution with $\sigma = 2$. The researcher selects a sample of $n = 4$ rats and places them in a room where the temperature is kept at 65 degrees. The daily food consumption for these rats averaged $\overline{X} = 13$ grams. Do these data indicate that temperature has a significant effect on eating? Test at the .05 level of significance.

2. In a research report, the term *significant effect* is used when the null hypothesis is rejected. (True or false?)

3. In a research report, the results of a hypothesis test include the phrase "$p < .01$." This means that the test failed to reject the null hypothesis. (True or false?)

ANSWERS

1. The null hypothesis states that temperature has no effect, or $\mu = 10$ grams even at the lower temperature. With $\alpha = .05$, the critical region consists of z-scores beyond $z = \pm 1.96$. The sample data produce $z = 3.00$. The decision is to reject H_0 and conclude that temperature has a significant effect on eating behavior.

2. true

3. False. The probability is *less than* .01, which means it is very unlikely that the result occurred by chance. In this case, the data are in the critical region, and H_0 is rejected.

8.4 DIRECTIONAL (ONE-TAILED) HYPOTHESIS TESTS

The hypothesis-testing procedure presented in Section 8.2 was the standard, or *two-tailed,* test format. The *two-tailed* comes from the fact that the critical region is located in both tails of the distribution. This format is by far the most widely accepted procedure for hypothesis testing. Nonetheless, there is an alternative that will be discussed in this section.

Usually a researcher begins an experiment with a specific prediction about the direction of the treatment effect. For example, a special training program is expected to *increase* student performance, or alcohol consumption is expected to *slow* reaction times. In these situations, it is possible to state the statistical hypotheses in a manner that incorporates the directional prediction into the statement of H_0 and H_1. The result is a directional test, or what commonly is called a *one-tailed test.*

FIGURE 8.8

The distribution of sample means for $n = 16$ if H_0 is true. The null hypothesis states that the diet pill has no effect, so the population mean will be 10 or larger.

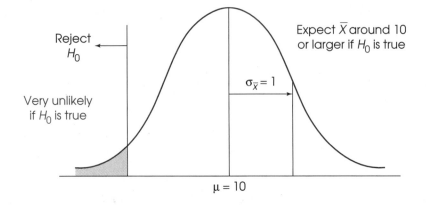

Reject H_0

Very unlikely if H_0 is true

$\sigma_{\bar{X}} = 1$

Expect $\bar{X}$ around 10 or larger if H_0 is true

$\mu = 10$

DEFINITION

In a *directional hypothesis test*, or a *one-tailed test*, the statistical hypotheses (H_0 and H_1) specify either an increase or a decrease in the population mean score. That is, they make a statement about the direction of the effect.

Suppose, for example, a researcher is using a sample of $n = 16$ laboratory rats to examine the effect of a new diet drug. It is known that under regular circumstances these rats eat an average of 10 grams of food each day. The distribution of food consumption is normal with $\sigma = 4$. The expected effect of the drug is to reduce food consumption. The purpose of the experiment is to determine whether or not the drug really works.

THE HYPOTHESES FOR A DIRECTIONAL TEST

Because a specific direction is expected for the treatment effect, it is possible for the researcher to perform a directional test. The first step (and the most critical step) is to state the statistical hypotheses. Remember that the null hypothesis states that there is no treatment effect and that the alternative hypothesis says that there is an effect. For directional tests, it is easier to begin with the alternative hypothesis. In words, this hypothesis says that with the drug the mean food consumption is *less than* 10 grams per day; that is, the drug does reduce food consumption. In symbols, H_1 would say the following:

$$H_1: \mu_{\text{with drug}} < 10 \quad \text{(mean food consumption is reduced)}$$

The null hypothesis states that the treatment did not work (the opposite of H_1). In this case, H_0 states that the drug does not reduce food consumption; that is, even with the drug, the rats still will eat at least 10 grams per day. In symbols, H_0 would say the following:

$$H_0: \mu_{\text{with drug}} \geq 10 \quad \text{(the mean is at least 10 grams per day)}$$

THE CRITICAL REGION FOR DIRECTIONAL TESTS

The critical region is determined by sample values that are very unlikely if the null hypothesis is true, that is, sample values that refute H_0 and provide evidence that the treatment really does work. In this example, the treatment is intended to reduce food consumption. Therefore, only sample values that are substantially less than $\mu = 10$ would indicate that the treatment worked and thereby lead to rejecting H_0. Thus, the critical region is located entirely in one tail of the distribution (see Figure 8.8). This is why the directional test commonly is called *one-tailed*.

A complete example of a one-tailed test is presented next. Once again, the experiment examining the effect of extra handling on the physical growth of infants will be used to demonstrate the hypothesis-testing procedure.

EXAMPLE 8.3   It is known that under regular circumstances the population of 2-year-old children has an average weight of $\mu = 26$ pounds. The distribution of weights is normal with $\sigma = 4$. The researcher selects a random sample of $n = 4$ newborn infants, instructs the parents to provide each child with extra handling, and then records the weight of each child at age 2. The average weight for the sample is $\overline{X} = 29.5$ pounds.

STEP 1   *State the hypotheses.* Because the researcher is predicting that extra handling will produce an increase in weight, it is possible to do a directional test. It usually is easier to begin with H_1, the alternative hypothesis, which states that the treatment does have an effect. In symbols,

$H_1: \mu > 26$ (there is an increase in weight)

Simply because you can do a directional test does not mean that you must use a directional test. A two-tailed test is always acceptable and generally preferred.

The null hypothesis states that the treatment does not have an effect. In symbols,

$H_0: \mu \leq 26$ (there is no increase)

STEP 2   *Locate the critical region.* To find the critical region, we look at all the possible sample means for $n = 4$ that could be obtained if H_0 were true. This is the distribution of sample means. It will be normal (because the population is normal), it will have a standard error of $\sigma_{\overline{X}} = 4/\sqrt{4} = 2$, and it will have a mean of $\mu = 26$ if the null hypothesis is true. The distribution is shown in Figure 8.9.

If H_0 is true and extra handling does not increase weight, we would expect the sample to average around 26 pounds or less. Only large-sample means would provide evidence that H_0 is wrong and that extra handling actually does increase weight. Therefore, it is only large values that comprise the critical region. With $\alpha = .05$, the most likely 95% of the distribution is separated from the most unlikely 5% by a z-score of $z = +1.65$ (see Figure 8.9).

FIGURE 8.9

Critical region for Example 8.3.

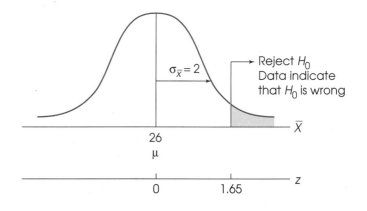

STEP 3   *Obtain the sample data.* The mean for the sample is $\overline{X} = 29.5$. This value corresponds to a z-score of

$$z = \frac{\overline{X} - \mu}{\sigma_{\overline{X}}} = \frac{29.5 - 26}{2} = \frac{3.5}{2} = 1.75$$

STEP 4   *Make a statistical decision.* A z-score of $z = +1.75$ indicates that our sample mean is in the critical region. This is a very unlikely outcome if H_0 is true, so the statistical decision is to reject H_0. The conclusion is that extra handling does result in increased growth for infants.

COMPARISON OF ONE-TAILED VERSUS TWO-TAILED TESTS

The general goal of hypothesis testing is to determine whether or not a particular treatment has any effect on a population. The test is performed by selecting a sample, administering the treatment to the sample, and then comparing the result with the original population. If the treated sample is noticeably different from the original population, then we conclude that the treatment has an effect, and we reject H_0. On the other hand, if the treated sample is still similar to the original population, then we conclude that there is no evidence for a treatment effect, and we fail to reject H_0. The critical factor in this decision is the *size of the difference* between the treated sample and the original population. A large difference is evidence that the treatment worked; a small difference is not sufficient to say that the treatment has any effect.

The major distinction between one-tailed and two-tailed tests is in the criteria they use for rejecting H_0. A one-tailed test allows you to reject the null hypothesis when the difference between the sample and the population is relatively small, provided the difference is in the specified direction. A two-tailed test, on the other hand, requires a relatively large difference independent of direction. This point is illustrated in the following example.

EXAMPLE 8.4   Consider again the experiment examining extra handling and infant growth (see Example 8.3). If we had used a standard two-tailed test, then the hypotheses would have been

H_0: $\mu = 26$ pounds   (no treatment effect)

H_1: $\mu \neq 26$ pounds   (handling does affect growth)

With $\alpha = .05$, the critical region would consist of any z-score beyond the $z = \pm 1.96$ boundaries.

If we obtained the sample data, $\overline{X} = 29.5$ pounds, which corresponds to a z-score of $z = 1.75$, our statistical decision would be "fail to reject H_0."

Notice that with the two-tailed test (Example 8.4), the difference between the data ($\overline{X} = 29.5$) and the hypothesis ($\mu = 26$) is not big enough to conclude that the hypothesis is wrong. In this case, we are saying that the data do not provide sufficient evidence to justify rejecting H_0. However, with the one-tailed test (Example 8.3), the same data led us to reject H_0.

All researchers agree that one-tailed tests are different from two-tailed tests. However, there are several ways to interpret the difference. One group of researchers (the present authors included) contends that one-tailed tests make it too easy to reject H_0 and therefore too easy to make a Type I error. According to this position, a two-tailed test requires strong evidence to reject H_0 and thus provides a convincing demonstration that a treatment effect has occurred. A one-tailed test, on the other hand, can result in rejecting the null hypothesis even when the evidence is relatively weak. For this reason, you usually will find two-tailed tests used for research that is published in journal articles for scrutiny by the scientific community. In this type of research, it is important that the results be convincing, and it is important to avoid a Type I error (publishing a false report). The two-tailed (nondirectional) test satisfies these criteria.

> Remember, you risk a Type I error every time H_0 is rejected.

Another group of researchers focuses on the advantages of one-tailed tests. As we have noted, one-tailed tests can lead to rejecting H_0 when the evidence is relatively weak. The statement can be rephrased by saying that one-tailed tests are more sensitive in detecting a treatment effect. If you think of a hypothesis test as a detection device that is used to seek out a treatment effect, then the advantages of a more sensitive test should be obvious. Although a more sensitive test will generate more false alarms (Type I errors), it also is more likely to find a significant treatment effect. Thus, one-tailed tests can be very useful in situations where a researcher does not want to overlook any possible significant outcome and where a Type I error is not very damaging. A good example of this situation is in exploratory research. The intent of exploratory research is to investigate a new area or to try an approach that is new and different. Instead of producing publishable results, exploratory research is directed toward generating new research possibilities. In this atmosphere, a researcher is willing to tolerate a few false alarms (Type I errors) in exchange for the increased sensitivity of a one-tailed test.

For the reasons we have outlined, most researchers do not use one-tailed tests except in very limited situations. Even though most experiments are designed with an expectation that the treatment effect will be in a specific direction, the directional prediction is not included in the statement of the statistical hypothesis. Nevertheless, you will probably encounter one-tailed tests, and you may find occasion to use them, so you should understand the rationale and the procedure for conducting directional tests.

LEARNING CHECK

1. A researcher predicts that a treatment will lower scores. If this researcher uses a one-tailed test, will the critical region be in the right- or left-hand tail of the distribution?

2. A psychologist is examining the effects of early sensory deprivation on the development of perceptual discrimination. A sample of $n = 9$ newborn kittens is obtained. These kittens are raised in a completely dark environment for four weeks, after which they receive normal visual stimulation. At age 6 months, the kittens are tested on a visual discrimination task. The average score for this sample is $\overline{X} = 32$. It is known that under normal circumstances cats score an average of $\mu = 40$ on this task. The distribution of scores is normal with $\sigma = 12$. The researcher is predicting that the early sensory deprivation will reduce the kittens' performance on the discrimination task. Use a one-tailed test with $\alpha = .01$ to test this hypothesis.

1. the left-hand tail

2. The hypotheses are H_0: $\mu \geq 40$ and H_1: $\mu < 40$. The critical region is determined by z-scores less than -2.33. The z-score for these sample data is $z = -2.00$. Fail to reject H_0.

8.5 THE GENERAL ELEMENTS OF HYPOTHESIS TESTING: A REVIEW

The z-score hypothesis test presented in this chapter is one specific example of the general process of hypothesis testing. In later chapters, we will examine many other hypothesis tests. Although the details of the hypothesis test will vary from one situation to another, all of the different tests use the same basic logic and consist of the same elements. In this section, we will present a generic hypothesis test that outlines the basic elements and logic common to all tests. The better you understand the general process of a hypothesis test, the better you will understand inferential statistics.

Hypothesis tests consist of five basic elements organized in a logical pattern. The elements and their relationships to each other are as follows:

1. Hypothesized Population Parameter. The first step in a hypothesis test is to state a null hypothesis. You will notice that the null hypothesis typically provides a specific value for an unknown population parameter. The specific value predicted by the null hypothesis is the first element of hypothesis testing.

2. Sample Statistic. The sample data are used to calculate a sample statistic that corresponds to the hypothesized population parameter. Thus far we have considered only situations where the null hypothesis specifies a value for the population mean, and the appropriate sample statistic is the sample mean. Similarly, if the hypothesis specified a value for the population variance, then the appropriate statistic would be the sample variance.

3. Estimate of Error. To evaluate our findings, we must know what types of sample data are likely to occur simply due to chance. Typically, a measure of standard error is used to provide this information. The general purpose of standard error is to provide a measure of how much difference is expected between a sample statistic and the corresponding population parameter. You should remember that samples (statistics) are not expected to provide a perfectly accurate picture of populations (parameters). There always will be some error or discrepancy between a statistic and a parameter; this is the concept of sampling error. The standard error tells how much error is expected just by chance.

In Chapter 7, we introduced the standard error of $\overline{X}$. This is the first specific example of the general concept of standard error, and it provides a measure of how much error is expected between a sample mean (statistic) and the corresponding population mean (parameter). In later chapters, you will encounter other examples of standard error, each of which measures the standard distance (expected by chance) between a specific statistic and its corresponding parameter. Although the exact formula for standard error will change from one situation to another, you should recall that standard error is basically determined by two factors:

a. The variability of the scores. Standard error is directly related to the variability of the scores: The larger the variability, the larger the standard error.

b. The size of the sample. Standard error is inversely related to sample size: The larger the sample, the smaller the standard error.

4. The Test Statistic. In this chapter, the test statistic is a z-score:

$$z = \frac{\overline{X} - \mu}{\sigma_{\overline{X}}}$$

This z-score statistic provides a model for other test statistics that will follow. In particular, a test statistic typically forms a *ratio*. The numerator of the ratio is simply the obtained difference between the sample statistic and the hypothesized parameter. The denominator of the ratio is the standard error measuring how much difference is expected by chance. Thus, the generic form of a test statistic is

$$\text{test statistic} = \frac{\text{obtained difference}}{\text{difference expected by chance}}$$

In general, the purpose of a test statistic is to determine whether or not the result of an experiment (the obtained difference) is more than expected by chance alone. Thus, a test statistic that has a value greater than 1.00 indicates that the obtained result (numerator) is greater than chance (denominator). In addition, the test statistic provides a measure of *how much* greater than chance. A value of 3.00, for example, indicates that the obtained difference is three times greater than would be expected by chance.

5. The Alpha Level. The final element in a hypothesis test is the alpha level, or level of significance. The alpha level provides a criterion for interpreting the test statistic. As we noted, a test statistic greater than 1.00 indicates that the obtained result is greater than expected by chance. However, researchers demand that the likelihood of a result be not simply "more than chance," but *significantly more than chance.* The alpha level provides a criterion for "significance."

Earlier in this chapter (page 201), we noted that when a significant result is reported in the literature, it always is accompanied by a p value; for example, $p < .05$ means that there is less than a 5% probability that the result occurred by chance. In general, a *significant* result means that the researcher is very confident that the result is not simply due to chance. A result that is significant at the .05 (5%) level, for example, means that the researcher is 95% confident that the obtained difference is greater than what one would expect by chance alone. Similarly, the .01 (1%) level indicates 99% confidence, and the .001 (.1%) level indicates 99.9% confidence.

To gain more confidence (rejecting H_0 with a smaller alpha), the results must demonstrate a larger difference (treatment effect). As you know, the alpha level is used to obtain a critical value for the test statistic. With the z-score test, for example, $\alpha = .05$ produces a critical value of ± 1.96. In general, the alpha level determines how big the test statistic ratio must be before you can claim a *significant* result. The following values are *not* intended to be precise, but rather they should provide you with a *rough idea* of how the alpha level is related to the test statistic:

a. For $\alpha = .05$, the criterion for significance is a test statistic ratio of at least 2.00. That is, the obtained difference (numerator) must be at least *twice as big* as chance (denominator). *Note:* The exact value for the z-score test is 1.96.

b. For $\alpha = .01$, the criterion for significance is a test statistic ratio of at least 2.50. That is, the obtained difference must be at least *two and one-half times bigger* than chance. *Note:* The exact value for the *z*-score test is 2.58.

c. For $\alpha = .001$, the criterion for significance is a test statistic ratio of at least 3.00. That is, the obtained difference must be at least *three times bigger* than chance. *Note:* The exact value for the *z*-score test is 3.30.

Again, these values are *approximate* and provide a rough "rule of thumb." You always will need to consult a statistical table (for example, the unit normal table) to determine the exact critical values for your hypothesis test. Finally, we remind you to watch for the five basic elements described in this section as you encounter different examples of hypothesis testing throughout the rest of the book.

SUMMARY

1. Hypothesis testing is an inferential procedure that uses the data from a sample to draw a general conclusion about a population. The procedure begins with a hypothesis about an unknown population. Then a sample is selected, and the sample data provide evidence that either supports or refutes the hypothesis.

2. In this chapter, we introduced hypothesis testing using the simple situation where a sample mean is used to test a hypothesis about an unknown population mean. We begin with an unknown population, generally a population that has received a treatment. The question is to determine whether or not the treatment has had an effect on the population mean (see Figure 8.1).

3. Hypothesis testing is structured as a four-step process that will be used throughout the remainder of the book.
   a. State the null hypothesis (H_0), and select an alpha level. The null hypothesis states that there is no effect or no change. In this case, H_0 states that the mean for the treated population is the same as the mean before treatment. The alpha level, usually $\alpha = .05$ or $\alpha = .01$, provides a definition of the term *very unlikely* and determines the risk of a Type I error. An alternative hypothesis (H_1), which is the exact opposite of the null hypothesis, is also stated.
   b. Locate the critical region. The critical region is defined as sample outcomes that would be very unlikely to occur if the null hypothesis is true. The alpha level defines "very unlikely." For example, with $\alpha = .05$, the critical region is defined as sample means in the extreme 5% of the distribution of sample means. When the distribution is normal, the extreme 5% corresponds to z-scores beyond $z = \pm 1.96$.
   c. Collect the data, and compute the test statistic. The sample mean is transformed into a z-score by the formula

$$z = \frac{\overline{X} - \mu}{\sigma_{\overline{X}}}$$

The value of μ is obtained from the null hypothesis. The z-score test statistic identifies the location of the sample mean in the distribution of sample means. Expressed in words, the z-score formula is

$$z = \frac{\text{sample mean} - \text{hypothesized population mean}}{\text{standard error}}$$

   d. Make a decision. If the obtained z-score is in the critical region, we reject H_0 because it is very unlikely that these data would be obtained if H_0 were true. In this case, we conclude that the treatment has changed the population mean. If the z-score is not in the critical region, we fail to reject H_0 because the data are not significantly different from the null hypothesis. In this case, the data do not provide sufficient evidence to indicate that the treatment has had an effect.

4. Whatever decision is reached in a hypothesis test, there is always a risk of making the incorrect decision. There are two types of errors that can be committed.

   A Type I error is defined as rejecting a true H_0. This is a serious error because it results in falsely reporting a treatment effect. The risk of a Type I error is determined by the alpha level and therefore is under the experimenter's control.

   A Type II error is defined as failing to reject a false H_0. In this case, the experiment fails to report an effect that actually occurred. The probability of a Type II error cannot be specified as a single value and depends in part on the size of the treatment effect. It is identified by the symbol β (beta).

5. When a researcher predicts that a treatment effect will be in a particular direction (increase or decrease), it is possible to do a directional or one-tailed test. The first step in this procedure is to state the alternative hypothesis (H_1). This hypothesis states that the treatment works and, for directional tests, specifies the direction of the predicted

treatment effect. The null hypothesis is the opposite of H_1. To locate the critical region, you must identify the kind of experimental outcome that refutes the null hypothesis and demonstrates that the treatment works. These outcomes will be located entirely in one tail of the distribution. The entire critical region (5% or 1%, depending on α) will be in one tail.

6. Directional tests should be used with caution because they may allow the rejection of H_0 when the experimental evidence is relatively weak. Even though a researcher may have a specific directional prediction for an experiment, it is generally safer and never inappropriate to use a nondirectional (two-tailed) test.

KEY TERMS

hypothesis testing

null hypothesis

alternative hypothesis

Type I error

Type II error

level of significance

alpha level

critical region

test statistic

beta

statistically significant

directional test

one-tailed test

FOCUS ON PROBLEM SOLVING

1. Hypothesis testing involves a set of logical procedures and rules that enable us to make general statements about a population when all we have are sample data. This logic is reflected in the four steps that have been used throughout this chapter. Hypothesis-testing problems will become easier to tackle when you learn to follow the steps.

STEP 1 State the hypotheses, and set the alpha level.

STEP 2 Locate the critical region.

STEP 3 Compute the z-score for the sample data.

STEP 4 Make a decision about H_0 based on the result of step 3.

A nice benefit of mastering these steps is that all hypothesis tests that will follow use the same basic logic outlined in this chapter.

2. Students often ask, "What alpha level should I use?" Or a student may ask, "Why is an alpha of .05 used?" as opposed to something else. There is no single correct answer to either of these questions. Keep in mind the idea of setting an alpha level in the first place: *to reduce the risk of committing a Type I error.* Therefore, you would not want to set alpha to something like .20. In that case, you would be taking a 20% risk of committing a Type I error—reporting an effect when one actually does not exist. Most researchers would find this level of risk unacceptable. Instead, researchers generally agree to the convention that $\alpha = .05$ is the greatest risk one should take of making a Type I error. Thus, the .05 level of significance is frequently used and has become the "standard" alpha level. However, some researchers prefer to take even less risk and use alpha levels of .01 and smaller.

3. Take time to consider the implications of your decision about the null hypothesis. The null hypothesis states that there is no effect. Therefore, if your decision is to reject H_0, you should conclude that the sample data provide evidence for a treatment effect. However, it is an entirely different matter if your decision is to fail to reject H_0. Remember that when you fail to reject the null hypothesis, the results are inconclusive. It is impossible to *prove* that H_0 is correct; therefore, you cannot state with certainty that "there is no effect" when H_0 is not rejected. At best, all you can state is that "there is insufficient evidence for an effect" (see Box 8.1).

4. It is very important that you understand the structure of the z-score formula (page 190). It will help you understand many of the other hypothesis tests that will be covered later.

5. When you are doing a directional hypothesis test, read the problem carefully, and watch for key words (such as increase or decrease, raise or lower, and more or less) that tell you which direction the researcher is predicting. The predicted direction will determine the alternative hypothesis (H_1) and the critical region. For example, if a treatment is expected to *increase* scores, H_1 would contain a *greater than* symbol, and the critical region would be in the tail associated with high scores.

DEMONSTRATION 8.1

HYPOTHESIS TEST WITH z

A researcher begins with a known population—in this case, scores on a standardized test that are normally distributed with $\mu = 65$ and $\sigma = 15$. The researcher suspects that special training in reading skills will produce a change in the scores for the individuals in the population. Because it is not feasible to administer the treatment (the special training) to everyone in the population, a sample of $n = 25$ individuals is selected, and the treatment is given to this sample. Following treatment, the average score for this sample is $\overline{X} = 70$. Is there evidence that the training has an effect on test scores?

STEP 1   State the hypothesis, and select an alpha level.

Remember, the goal of hypothesis testing is to use sample data to make general conclusions about a population. The hypothesis always concerns an unknown population. For this demonstration, the researcher does not know what would happen if the entire population were given the treatment. Nevertheless, it is possible to make hypotheses about the treated population.

Specifically, the null hypothesis says that the treatment has no effect. According to H_0, the unknown population (after treatment) is identical to the original population (before treatment). In symbols,

$$H_0: \mu = 65 \quad \text{(After special training, the mean is still 65.)}$$

The alternative hypothesis states that the treatment does have an effect that causes a change in the population mean. In symbols,

$$H_1: \mu \neq 65 \quad \text{(After special training, the mean is different from 65.)}$$

At this time, you also select the alpha level. Traditionally, alpha is set at .05 or .01. If there is particular concern about a Type I error or if a researcher desires to present overwhelming evidence for error treatment effect, a smaller alpha level can be used (such as $\alpha = .001$). For this demonstration, we will set alpha to .05. Thus, we are taking a 5% risk of committing a Type I error.

FIGURE 8.10

The critical region for Demonstration 8.1 consists of the extreme tails with boundaries of $z = -1.96$ and $z = +1.96$.

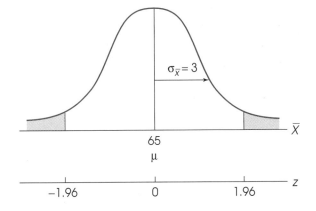

STEP 2   Locate the critical region.

You should recall that the critical region is defined as the set of outcomes that is very unlikely to be obtained if the null hypothesis is true. Therefore, obtaining sample data from this region would lead us to reject the null hypothesis and conclude there is an effect. Remember, the critical region is a region of rejection.

We begin by looking at all possible outcomes that could be obtained and then use the alpha level to determine the outcomes that are unlikely. For this demonstration, we look at the distribution of sample means for samples of $n = 25$, that is, all possible sample means that could be obtained if H_0 were true. The distribution of sample means will be normal because the original population is normal. It will have an expected value of $\mu = 65$ and a standard error of

$$\sigma_{\overline{X}} = \frac{\sigma}{\sqrt{n}} = \frac{15}{\sqrt{25}} = \frac{15}{5} = 3$$

With $\alpha = .05$, we want to identify the most unlikely 5% of this distribution. The most unlikely part of a normal distribution is in the tails. Therefore, we divide our alpha level evenly between the two tails, 2.5% or $p = .0250$ per tail. In column C of the unit normal table, find $p = .0250$. Then find its corresponding z-score in column A. The entry is $z = 1.96$. The boundaries for the critical region are -1.96 (on the left side) and $+1.96$ (on the right side). The distribution with its critical region is shown in Figure 8.10.

STEP 3   Obtain the sample data, and compute the test statistic.

For this demonstration, the researcher obtained a sample mean of $\overline{X} = 70$. This sample mean corresponds to a z-score of

$$z = \frac{\overline{X} - \mu}{\sigma_{\overline{X}}} = \frac{70 - 65}{3} = \frac{5}{3} = +1.67$$

STEP 4   Make a decision about H_0, and state the conclusion.

The z-score we obtained is not in the critical region. This indicates that our sample mean of $\overline{X} = 70$ is not an extreme or unusual value to be obtained from a population with $\mu = 65$. Therefore, our statistical decision is to *fail to reject* H_0. Our conclusion for the study is that the data do not provide sufficient evidence that the special training changes test scores.

PROBLEMS

1. In the z-score formula as it is used in a hypothesis test
   a. Explain what is measured by $\overline{X} - \mu$ in the numerator.
   b. Explain what is measured by the standard error in the denominator.

2. Discuss the errors that can be made in hypothesis testing.
   a. What is a Type I error? Why might it occur?
   b. What is a Type II error? How does it happen?

3. Briefly explain the advantage of using an alpha level of .01 versus a level of .05. What is the advantage of using a smaller value for alpha?

4. The term *error* is used two different ways in the context of a hypothesis test. First, there is the concept of standard error, and, second, there is the concept of a Type I error.
   a. What factor can a researcher control that will reduce the risk of a Type I error?
   b. What factor can a researcher control that will reduce the standard error?

5. A researcher did a one-tailed hypothesis test using an alpha level of .01. For this test, H_0 was rejected. A colleague analyzed the same data but used a two-tailed test with $\alpha = .05$. In this test, H_0 was *not* rejected. Can both analyses be correct? Explain your answer.

6. A sample of $n = 4$ individuals is selected from a normal population with $\mu = 70$ and $\sigma = 10$. A treatment is administered to the individuals in the sample, and after the treatment, the sample mean is found to be $\overline{X} = 75$.
   a. On the basis of the sample data, can you conclude that the treatment has a significant effect? Use a two-tailed test with $\alpha = .05$.
   b. Suppose that the sample consisted of $n = 25$ individuals and produced a mean of $\overline{X} = 75$. Repeat the hypothesis test at the .05 level of significance.
   c. Compare the results from part a and part b. How does the sample size influence the outcome of a hypothesis test?

7. A mood questionnaire has been standardized so that the scores form a normal distribution with $\mu = 50$ and $\sigma = 15$. A psychologist would like to use this test to examine how the environment affects mood. A sample of $n = 25$ individuals is obtained, and the individuals are given the mood test in a darkly painted, dimly lit room with plain metal desks and no windows. The average score for the sample is $\overline{X} = 43$.
   a. Do the sample data provide sufficient evidence to conclude that the environment has a significant effect on mood? Test at the .05 level of significance.
   b. Repeat the hypothesis test using the .01 level of significance.

   c. Compare the results from part a and part b. How does the level of significance influence the outcome of a hypothesis test?

8. A normal population has a mean of $\mu = 100$. A sample of $n = 36$ is selected from the population, and a treatment is administered to the sample. After treatment, the sample mean is computed to be $\overline{X} = 106$.
   a. Assuming that the population standard deviation is $\sigma = 12$, use the data to test whether or not the treatment has a significant effect. Test with $\alpha = .05$.
   b. Repeat the hypothesis test, but this time assume that the population standard deviation is $\sigma = 30$.
   c. Compare the results from part a and part b. How does the population standard deviation influence the outcome of a hypothesis test?

9. A researcher would like to test the effectiveness of a newly developed growth hormone. The researcher knows that under normal circumstances laboratory rats reach an average weight of $\mu = 950$ grams at 10 weeks of age. The distribution of weights is normal with $\sigma = 30$. A random sample of $n = 25$ newborn rats is obtained, and the hormone is given to each rat. When the rats in the sample reached 10 weeks old, each rat was weighed. The mean weight for this sample was $\overline{X} = 974$.
   a. Identify the independent and the dependent variables for this study.
   b. State the null hypothesis in a sentence that includes the independent variable and the dependent variable.
   c. Using symbols, state the hypotheses (H_0 and H_1) that the researcher is testing.
   d. Sketch the appropriate distribution, and locate the critical region for $\alpha = .05$.
   e. Calculate the test statistic (z-score) for the sample.
   f. What decision should be made about the null hypothesis, and what decision should be made about the effect of the hormone?

10. Scores on the Scholastic Aptitude Test (SAT) form a normal distribution with $\mu = 500$ and $\sigma = 100$. A high school counselor has developed a special course designed to boost SAT scores. A random sample of $n = 16$ students is selected to take the course and then take the SAT. The sample had a mean score of $\overline{X} = 554$. Does the course have an effect on SAT scores?
   a. Set $\alpha = .05$, and perform a two-tailed hypothesis test using the four-step procedure outlined in the text.
   b. Repeat the hypothesis test using $\alpha = .01$.
   c. Explain why the change in α from .05 to .01 (part a versus part b) caused a change in the conclusion.

11. A psychologist has developed a standardized test for measuring the vocabulary skills of 4-year-old children. The scores on the test form a normal distribution with $\mu = 60$ and $\sigma = 10$. A researcher would like to use this test to investigate the hypothesis that children who grow up as an only child develop vocabulary skills at a different rate than children in large families. A sample of $n = 25$ only children is obtained, and the mean test score for this sample is $\overline{X} = 63$.

    a. On the basis of this sample, can the researcher conclude that vocabulary skills for only children are significantly different from those of the general population? Test at the .05 level of significance.

    b. Perform the same test assuming that the researcher had used a sample of $n = 100$ only children and obtained the same sample mean, $\overline{X} = 63$.

    c. You should find that the larger sample (part b) produces a different conclusion than the smaller sample (part a). Explain how the sample size influences the outcome of the hypothesis test.

12. In 1975, a nationwide survey revealed that U.S. grade-school children spent an average of $\mu = 8.4$ hours per week doing homework. The distribution of homework times was normal with $\sigma = 3.2$. Last year, a sample of $n = 100$ grade-school students was given the same survey. For this sample, the mean number of homework hours was $\overline{X} = 7.1$. Has there been a significant change in the homework habits of grade-school children? Test with $\alpha = .05$.

13. A psychologist examined the effect of chronic alcohol abuse on memory. In this experiment, a standardized memory test was used. Scores on this test for the general population form a normal distribution with $\mu = 50$ and $\sigma = 6$. A sample of $n = 22$ alcohol abusers has a mean score of $\overline{X} = 47$. Is there evidence for memory impairment among alcoholics? Use $\alpha = .01$ for a one-tailed test.

14. Over a 20-year period, first-year students at a state university had a mean grade-point average (GPA) of $\mu = 2.27$ with $\sigma = 0.42$. Recently the university has tried to increase admissions standards. A sample of $n = 30$ first-year students taken this year had a mean GPA of $\overline{X} = 2.63$. Have the new admissions standards resulted in a significant change in GPA of first-year students? Use the .01 level of significance for two tails.

15. A psychologist has developed a new test to measure depression. Using a very large group of "normal" individuals, the mean score on this test is found to be $\mu = 55$ with $\sigma = 12$, and the scores form a normal distribution. A researcher would like to use this test to evaluate a theory predicting that increased depression is a common part of the aging process. To test the theory, a sample of $n = 36$ elderly subjects is obtained. The depression test is administered to each subject, and the mean score for the sample is $\overline{X} = 59$. Do these data support the theory?

    a. Use a one-tailed test with $\alpha = .05$.

    b. Repeat the hypothesis test using a one-tailed test with $\alpha = .01$.

    c. Explain why the change in alpha changes the conclusion for the hypothesis test.

16. Performance scores on a motor skills task form a normal distribution with $\mu = 20$ and $\sigma = 4$. A psychologist is using this task to determine the extent to which increased self-awareness affects performance. The prediction for this experiment is that increased self-awareness will reduce a subject's concentration and result in lower performance scores. A sample of $n = 16$ subjects is obtained, and each subject is tested on the motor skills task while seated in front of a large mirror. The purpose of the mirror is to make the subjects more self-aware. The mean score for this sample is $\overline{X} = 15.5$. Use a one-tailed test with $\alpha = .05$ to test the psychologist's prediction.

17. A sample of $n = 9$ scores is obtained from a normal population with $\sigma = 12$. The sample mean is $\overline{X} = 60$.

    a. Use the sample data to test the hypothesis that the population mean is $\mu = 65$. Test at the .05 level of significance.

    b. Use the sample data to test the hypothesis that the population mean is $\mu = 55$. Test at the .05 level of significance.

    c. In parts a and b of this problem, you should find that $\mu = 65$ and $\mu = 55$ are both acceptable hypotheses. Explain how two different values can both be acceptable.

18. Suppose that during impersonal social interactions (that is, with business or casual acquaintances) people in the United States maintain a mean social distance of $\mu = 7$ feet from the other individual. This distribution is normal and has a standard deviation of $\sigma = 1.5$. A researcher examines whether or not this is true for other cultures. A random sample of $n = 8$ individuals of Middle Eastern culture is observed in an impersonal interaction. For this sample, the mean distance is $\overline{X} = 4.5$ feet. Is there a cultural difference in social distance? Use an alpha of .05 for two tails.

19. A psychological theory predicts that individuals who grow up as an only child will have above-average IQs. A sample of $n = 64$ people from single-child families is obtained. The average IQ for this sample is $\overline{X} = 104.9$. In the general population, IQs form a normal distribution with $\mu = 100$ and $\sigma = 15$. Use a one-tailed test with $\alpha = .01$ to evaluate the theory.

20. Researchers have often noted increases in violent crimes when it is very hot. In fact, Reifman, Larrick, and Fein (1991) noted that this relationship even extends to baseball. That is, there is a much greater chance of a batter being hit by a pitch when the temperature increases. Consider the following hypothetical data. Suppose over the past 30 years, during any given week of the major league season, an average of $\mu = 12$ players are hit by wild pitches. Assume the distribution is nearly normal with $\sigma = 3$. For a sample of $n = 4$ weeks in which the daily temperature was extremely hot, the weekly average of hit-by-pitch players was $\bar{X} = 15.5$. Are players more likely to get hit by pitches during hot weeks? Set alpha to .05 for two tails.

21. A psychologist develops a new inventory to measure depression. Using a very large standardization group of "normal" individuals, the mean score on this test is $\mu = 55$ with $\sigma = 12$, and the scores are normally distributed. To determine if the test is sensitive in detecting those individuals that are severely depressed, a random sample of patients who are described as depressed by a therapist is selected and given the test. Presumably, the higher the score on the inventory is, the more depressed the patient is. The data are as follows: 59, 60, 60, 67, 65, 90, 89, 73, 74, 81, 71, 71, 83, 83, 88, 83, 84, 86, 85, 78, 79. Do patients score significantly differently on the test? Test with the .01 level of significance for two tails.

CHAPTER 9

INTRODUCTION TO THE *t* STATISTIC

TOOLS YOU WILL NEED

The following items are considered essential background material for this chapter. If you doubt your knowledge of any of these items, you should review the appropriate chapter or section before proceeding.

- Sample standard deviation (Chapter 4)

- Degrees of freedom (Chapter 4)

- Hypothesis testing (Chapter 8)

CONTENTS

9.1 INTRODUCTION

In the previous chapter, we presented the statistical procedures that permit researchers to use a sample mean to test hypotheses about a population. These statistical procedures were based on a few basic notions, which we summarize as follows:

Remember, the expected value of the distribution of sample means is μ, the population mean.

1. A sample mean ($\overline{X}$) is expected more or less to approximate its population mean (μ). This permits us to use the sample mean to test a hypothesis about the population mean.

2. The standard error provides a measure of how well a sample mean approximates the population mean.

$$\sigma_{\overline{X}} = \frac{\sigma}{\sqrt{n}}$$

3. To quantify our inferences about the population, we compare the obtained sample mean ($\overline{X}$) with the hypothesized population mean (μ) by computing a *z*-score test statistic:

$$z = \frac{\overline{X} - \mu}{\sigma_{\overline{X}}} = \frac{\text{obtained difference between data and hypothesis}}{\text{standard distance expected by chance}}$$

When the *z*-scores form a normal distribution, we are able to use the unit normal table (Appendix B) to find the critical region for the hypothesis test.

The shortcoming of using the *z*-score as an inferential statistic is that the *z*-score formula requires more information than is usually available. Specifically, *z*-scores require that we know the value of the population standard deviation, which is needed to compute the standard error. Most often the standard deviation of the population is not known, and the standard error of sample means cannot be computed. Without the standard error, we have no way of quantifying the expected amount of distance (or error) between $\overline{X}$ and μ. We have no way of making precise, quantitative inferences about the population based on *z*-scores.

THE *t* STATISTIC—A SUBSTITUTE FOR *z*

As previously noted, the limitation of *z*-scores in hypothesis testing is that the population standard deviation (or variance) must be known. More often than not, however, the variability of the scores in the population is not known. In fact, the whole reason for conducting a hypothesis test is to gain knowledge about an *unknown* population. This situation appears to create a paradox: You want to use a *z*-score to find out about an unknown population, but you must know about the population before you can compute a *z*-score. Fortunately, there is a relatively simple solution to this problem. When the variability for the population is not known, we use the sample variability in its place.

In Chapter 4, the sample variance was developed specifically to provide an unbiased estimate of the corresponding population variance. You should recall the formulas for sample variance and sample standard deviation as follows:

$$\text{sample variance} = \frac{SS}{n-1} = \frac{SS}{df}$$

$$\text{sample standard deviation} = \sqrt{\frac{SS}{n-1}} = \sqrt{\frac{SS}{df}}$$

Using the sample values, we can now *estimate* the standard error. You should recall from Chapters 7 and 8 that the value of the standard error can be computed using either standard deviation or variance:

$$\text{standard error} = \sigma_{\overline{X}} = \frac{\sigma}{\sqrt{n}} = \sqrt{\frac{\sigma^2}{n}}$$

The process of estimating the standard error simply involves substituting the sample variance or sample standard deviation in place of the unknown population value:

$$\text{estimated standard error} = s_{\overline{X}} = \frac{s}{\sqrt{n}} = \sqrt{\frac{s^2}{n}} \tag{9.1}$$

Notice that the symbol for the estimated standard error is $s_{\overline{X}}$ instead of $\sigma_{\overline{X}}$, indicating that the estimated value is computed from sample data rather than the actual population parameter.

Finally, you should realize that the standard error (actual or estimated) can be computed using either the standard deviation or the variance. In the past (Chapters 7 and 8), we have concentrated on the formula using standard deviation. However, the formula based on variance is better suited to the variety of inferential statistics that we will encounter in later chapters. Therefore, we will now shift our focus to the variance-based formula. Throughout the remainder of this chapter, and in following chapters, the *estimated standard error of* $\overline{X}$ typically will be presented and computed using

$$s_{\overline{X}} = \sqrt{\frac{s^2}{n}}$$

Additional justification for using sample variance to compute the estimated standard error is given in Box 9.1.

DEFINITION   The *estimated standard error* ($s_{\overline{X}}$) is used as an estimate of $\sigma_{\overline{X}}$ when the value of σ is unknown. It is computed from the sample standard deviation or sample variance and provides an estimate of the standard distance (expected by chance) between a sample mean ($\overline{X}$) and the population mean (μ).

Now we can substitute the estimated standard error in the denominator of the z-score formula. This new test statistic is called a *t statistic:*

$$t = \frac{\overline{X} - \mu}{s_{\overline{X}}} \tag{9.2}$$

9.1 USING VARIANCE TO COMPUTE STANDARD ERROR

IN CHAPTERS 7 and 8, the standard error was always computed using the standard deviation and the sample size:

$$\sigma_{\overline{X}} = \frac{\sigma}{\sqrt{n}}$$

Now, we are asking you to make two adjustments to this formula:

1. We have introduced a situation where the population standard deviation is unknown and you must substitute the sample standard deviation in its place. As a result, we are now computing the *estimated* standard error.

$$s_{\overline{X}} = \frac{s}{\sqrt{n}}$$

2. We are asking you to square the standard deviation and then place it under a square root so that it appears in the formula as variance.

$$s_{\overline{X}} = \frac{\sqrt{s^2}}{\sqrt{n}} = \sqrt{\frac{s^2}{n}}$$

The first of the two adjustments is a necessity: If the population standard deviation (or variance) is unknown, you must use the sample value in its place. The second adjustment, however, is not as simple and so we are offering two explanations for the shift from standard deviation to variance.

Explanation No. 1: The sample standard deviation *s* is a descriptive statistic that summarizes or describes the set of scores in the sample. On the other hand, the standard error, $s_{\overline{X}}$, is an inferential statistic that measures how well the sample mean represents the population mean. To emphasize this distinction, we will use the standard deviation exclusively as a descriptive statistic, and switch to variance when we are computing the inferential statistic.

Explanation No. 2: The sample variance, s^2, provides an unbiased estimate of the population variance. (You should remember that this is why we use $n - 1$ in the formula for sample variance.) However, the sample standard deviation *s* does not provide an unbiased estimate of the population standard deviation. The problem is that variance and standard deviation are related by a square-root relationship, so they cannot both be unbiased. For example, consider a population with $\sigma^2 = 52$ and two samples, one with $s^2 = 100$ and one with $s^2 = 4$. Notice that the mean of the two sample variances is exactly equal to the population variance. This is the definition of *unbiased;* on average the sample value equals the population value. Now consider the standard deviations. The population standard deviation is $\sigma = \sqrt{52} = 7.21$. The two sample standard deviations are $s = \sqrt{100} = 10$ and $s = \sqrt{4} = 2$. The average of the two sample values is 6.00 which *underestimates* the population value. In other words, the sample standard deviations are biased.

In the formula for estimated standard error we will use the *unbiased* sample variance instead of the *biased* standard deviation. The goal is to obtain an unbiased estimate of the real standard error.

The only difference between the *t* formula and the *z*-score formula is that the *z*-score uses the actual population variance, σ^2 (or the standard deviation), and the *t* formula uses the corresponding sample variance (or sample standard deviation) when the population value is not known.

$$z = \frac{\overline{X} - \mu}{\sigma_{\overline{X}}} = \frac{\overline{X} - \mu}{\sqrt{\frac{\sigma^2}{n}}} \qquad t = \frac{\overline{X} - \mu}{s_{\overline{X}}} = \frac{\overline{X} - \mu}{\sqrt{\frac{s^2}{n}}}$$

Structurally, these two formulas have the same form:

$$z \ or \ t = \frac{\text{sample mean} - \text{population mean}}{\text{(estimated) standard error}}$$

Because both z-score and t formulas are used for hypothesis testing, there is one rule to remember:

RULE   When you know the value of σ, use a z-score. If σ is unknown, use the t statistic.

DEFINITION   The t *statistic* is used to test hypotheses about μ when the value for σ is not known. The formula for the t statistic is similar in structure to the z-score, except that the t statistic uses estimated standard error.

DEGREES OF FREEDOM AND THE t STATISTIC

In this chapter, we have introduced the t statistic as a substitute for a z-score. The basic difference between these two is that the t statistic uses sample variance and the z-score uses the population variance. To determine how well a t statistic approximates a z-score, we must determine how well the sample variance approximates the population variance.

In Chapter 4, we introduced the concept of degrees of freedom. Reviewing briefly, you must know the sample mean before you can compute the sample variance. This places a restriction on sample variability such that only $n - 1$ scores in a sample are free to vary. The value $n - 1$ is called the *degrees of freedom* (or *df*) for the sample variance.

$$\text{degrees of freedom} = df = n - 1 \tag{9.3}$$

DEFINITION   *Degrees of freedom* describe the number of scores in a sample that are free to vary. Because the sample mean places a restriction on the value of 1 score in the sample, there are $n - 1$ degrees of freedom for the sample (see Chapter 4).

The greater the value of *df* is for a sample, the better s^2 represents σ^2, and the better the t statistic approximates the z-score. This should make sense because the larger the sample (n) is, the better the sample represents its population. Thus, the degrees of freedom associated with s^2 also describe how well t represents z.

THE t DISTRIBUTIONS

Every sample from a population can be used to compute a z-score or a t statistic. If you select all the possible samples of a particular size (n), then the entire set of resulting z-scores will form a z-score distribution. In the same way, the set of all possible t statistics will form a t *distribution*. As we saw in Chapter 7, the distribution of z-scores computed from sample means tends to be a normal distribution. For this reason, we consulted the unit normal table to find the critical region when using z-scores to test hypotheses about a population. The t distribution, on the other hand, is generally not normal. However, the t distribution will approximate a normal distribution in the same way that a t statistic approximates a z-score. How well a t distribution approximates a normal distribution is determined by degrees of freedom. In general, the greater the sample size (n) is, the larger the degrees of freedom ($n - 1$) are, and the better the t distribution approximates the normal distribution (see Figure 9.1).

FIGURE 9.1

Distributions of the *t* statistic for different values of degrees of freedom are compared to a normal *z*-score distribution. Like the normal distribution, *t* distributions are bell-shaped and symmetrical and have a mean of zero. However, *t* distributions have more variability, indicated by the flatter and more spread-out shape. The larger the value of *df* is, the more closely the *t* distribution approximates a normal distribution.

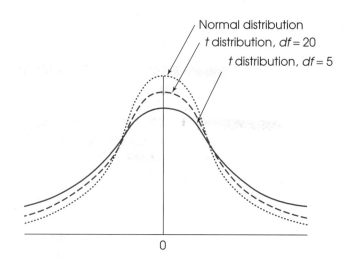

THE SHAPE OF THE *t* DISTRIBUTION

The exact shape of a *t* distribution changes with degrees of freedom. In fact, statisticians speak of a "family" of *t* distributions. That is, there is a different sampling distribution of *t* (a distribution of all possible sample *t* values) for each possible number of degrees of freedom. As *df* gets very large, the *t* distribution gets closer in shape to a normal *z*-score distribution. A quick glance at Figure 9.1 reveals that distributions of *t* are bell-shaped and symmetrical and have a mean of zero. However, the *t* distribution has more variability than a normal *z* distribution, especially when *df* values are small (Figure 9.1). The *t* distribution tends to be flatter and more spread out, whereas the normal *z* distribution has more of a central peak.

Why is the *t* distribution flatter and more variable than a normal *z* distribution? For a particular population, the top of the *z*-score formula, $\overline{X} - \mu$, can take on different values because $\overline{X}$ will vary from one sample to another. However, the value of the bottom of the *z*-score formula, $\sigma_{\overline{X}}$, is constant. The standard error will not vary from sample to sample because it is derived from the population variance. The implication is that samples that have the same value for $\overline{X}$ should also have the same *z*-score.

On the other hand, the standard error in the *t* formula is not a constant because it is estimated. That is, $s_{\overline{X}}$ is based on the sample variance, which will vary in value from sample to sample. The result is that samples can have the same value for $\overline{X}$, yet different values of *t* because the estimated error will vary from one sample to another. Therefore, a *t* distribution will have more variability than the normal *z* distribution. It will look flatter and more spread out. When the value of *df* increases, the variability in the *t* distribution decreases, and it more closely resembles the normal distribution because with greater *df*, $s_{\overline{X}}$ will more closely estimate $\sigma_{\overline{X}}$, and when *df* is very large, they are nearly the same.

DETERMINING PROPORTIONS AND PROBABILITIES FOR *t* DISTRIBUTIONS

Just as we used the unit normal table to locate proportions associated with *z*-scores, we will use a *t*-distribution table to find proportions for *t* statistics. The complete *t*-distribution table is presented in Appendix B, page A-27, and a portion of this table is reproduced in Table 9.1. The two rows at the top of the table show proportions of the *t* distribution contained in either one or two tails, depending on which

FIGURE 9.2

The *t* distribution with *df* = 3. Note that 5% of the distribution is located in the tail beyond *t* = 2.353. Also, 5% is in the tail beyond *t* = −2.353. Thus, a total proportion of 10% (0.10) is in the two tails beyond *t* = ±2.353.

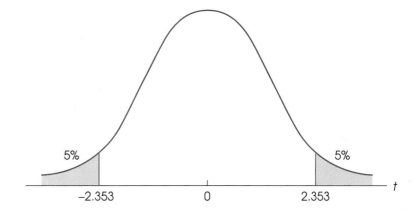

row is used. The first column of the table lists degrees of freedom for the *t* statistic. Finally, the numbers in the body of the table are the *t* values that mark the boundary between the tails and the rest of the *t* distribution.

For example, with *df* = 3, exactly 5% of the *t* distribution is located in the tail beyond *t* = 2.353 (see Figure 9.2). To find this value, you locate *df* = 3 in the first column and 0.05 (5%) in the one-tail proportion row. When you line up these two values in the table, you should find *t* = 2.353. Similarly, 5% of the *t* distribution is located in the tail beyond *t* = −2.353 (see Figure 9.2). Finally, you should notice that a total of 10% is contained in the two tails beyond *t* = ±2.353 (check the proportion value in the "two-tails" row at the top of the table).

A close inspection of the *t*-distribution table in Appendix B will demonstrate a point we made earlier: As the value for *df* increases, the *t* distribution becomes more and more similar to a normal distribution. For example, examine the column containing *t* values for a 0.05 proportion in two tails. You will find that when *df* = 1, the *t* values that separate the extreme 5% (0.05) from the rest of the distribution are *t* = ±12.706. As you read down the column, however, you should find that the critical *t* values become smaller and smaller, ultimately reaching ±1.96. You should recognize ±1.96 as the *z*-score values that separate the extreme 5% in a normal distribution. Thus, as *df* increases, the proportions in a *t* distribution become more like the proportions in a normal distribution. When the sample size (and degrees of freedom) is sufficiently large, the difference between a *t* distribution and the normal distribution becomes negligible.

TABLE 9.1

A portion of the *t*-distribution table

The numbers in the table are the values of *t* that separate the tail from the main body of the distribution. Proportions for one or two tails are listed at the top of the table, and *df* values for *t* are listed in the first column.

			Proportion in one tail			
	0.25	0.10	0.05	0.025	0.01	0.005
			Proportion in two tails			
df	0.50	0.20	0.10	0.05	0.02	0.01
1	1.000	3.078	6.314	12.706	31.821	63.657
2	0.816	1.886	2.920	4.303	6.965	9.925
3	0.765	1.638	2.353	3.182	4.541	5.841
4	0.741	1.533	2.132	2.776	3.747	4.604
5	0.727	1.476	2.015	2.571	3.365	4.032
6	0.718	1.440	1.943	2.447	3.143	3.707

Caution: The *t*-distribution table printed in this book has been abridged and does not include entries for every possible *df* value. For example, the table lists *t* values for *df* = 40 and for *df* = 60, but it does not list any entries for *df* values between 40 and 60. Occasionally, you will encounter a situation where your *t* statistic has a *df* value that is not listed in the table. In these situations, you should look up the critical *t* for both of the surrounding *df* values listed and then use the *larger* value for *t*. If, for example, you have *df* = 53 (not listed), you should look up the critical *t* value for *both df* = 40 and *df* = 60 and then use the larger *t* value. If your sample *t* statistic is greater than the larger value listed, you can be certain that the data are in the critical region, and you can confidently reject the null hypothesis.

LEARNING CHECK

1. Under what circumstances is a *t* statistic used instead of a *z*-score for a hypothesis test?

2. To conduct a hypothesis test with a *z*-score statistic, you must know the variance or the standard deviation for the population of scores. (True or false?)

3. In general, a distribution of *t* statistics is flatter and more spread out than a normal distribution. (True or false?)

4. A sample of *n* = 15 scores would produce a *t* statistic with *df* = 16. (True or false?)

5. For *df* = 15, find the value(s) of *t* associated with each of the following:
   a. The top 5% of the distribution.
   b. The middle 95% versus the extreme 5% of the distribution.
   c. The middle 99% versus the extreme 1% of the distribution.

ANSWERS

1. A *t* statistic is used instead of a *z*-score when the population standard deviation or variance is not known.

2. true

3. true

4. False. With *n* = 15, the *df* value would be 14.

5. a. $t = +1.753$
   b. $t = \pm 2.131$
   c. $t = \pm 2.947$

9.2 HYPOTHESIS TESTS WITH THE *t* STATISTIC

The *t*-statistic formula is used in exactly the same way that the *z*-score formula is used to test a hypothesis about a population mean. Once again, the *t* formula and its structure are

$$t = \frac{\overline{X} - \mu}{s_{\overline{X}}} = \frac{\text{sample mean} - \text{population mean}}{\text{estimated standard error}}$$

FIGURE 9.3

The basic experimental situation for using the *t* statistic or the *z*-score is presented. It is assumed that the parameter μ is known for the population before treatment. The purpose of the experiment is to determine whether or not the treatment has an effect. We ask, Is the population mean after treatment the same as or different from the mean before treatment? A sample is selected from the treated population to help answer this question.

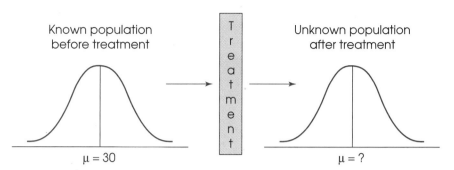

In the hypothesis-testing situation, we begin with a population with an unknown mean and an unknown variance, often a population that has received some treatment (Figure 9.3). The goal is to use a sample from the treated population (a treated sample) as the basis for determining whether or not the treatment has any effect. As always, the null hypothesis states that the treatment has no effect; specifically, H_0 states that the population mean is unchanged. Thus, the null hypothesis provides a specific value for the unknown population mean. The sample data provide a value for the sample mean. Finally, the variance and the estimated standard error are computed from the sample data. When these values are used in the *t* formula, the result becomes

$$t = \frac{\text{sample mean (from the data)} - \text{population mean (hypothesized from } H_0)}{\text{estimated standard error}}$$

As with the *z*-score formula, the *t* statistic forms a ratio. The numerator measures the actual difference between the data and the hypothesis. The denominator measures how much difference is expected just by chance (error). When the obtained difference between the data and the hypothesis (numerator) is much greater than chance (denominator), we will obtain a large value for *t* (either large positive or large negative). In this case, we conclude that the data are not consistent with the hypothesis, and our decision is to "reject H_0." On the other hand, when the difference between the data and the hypothesis is small relative to the standard error, we will obtain a *t* statistic near zero, and our decision will be "fail to reject H_0."

The basic steps of the hypothesis-testing procedure will now be reviewed.

STEPS AND PROCEDURES

For hypothesis tests with a *t* statistic, we use the same steps that we used with *z*-scores (Chapter 8). The major difference is that we are now required to estimate standard error because σ is unknown. Consequently, we compute a *t* statistic, rather than a *z*-score, and consult the *t*-distribution table, rather than the unit normal table, to find the critical region.

STEP 1

The hypotheses are stated, and the alpha level is set. The experimenter states the null hypothesis, that is, what should happen if no treatment effect exists. On the other hand, the alternative hypothesis predicts the outcome if an effect does occur. These hypotheses are always stated in terms of the population parameter, μ.

FIGURE 9.4

Apparatus used in Example 9.1.

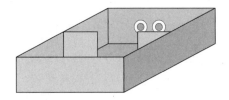

STEP 2   Locate the critical region. The exact shape of the *t* distribution and therefore the critical *t* values vary with degrees of freedom. Thus, to find a critical region in a *t* distribution, it is necessary to determine the value for *df*. Then the critical region can be located by consulting the *t*-distribution table (Appendix B).

STEP 3   The sample data are collected, and the test statistic is computed. When σ is unknown, the test statistic is a *t* statistic (formula 9.2).

STEP 4   The null hypothesis is evaluated. If the *t* statistic we obtained in step 3 falls within the critical region (exceeds the value of a critical *t*), then H_0 is rejected. It can be concluded that a treatment effect exists. However, if the obtained *t* value does not lie in the critical region, then we fail to reject H_0, and we conclude that we failed to observe evidence for an effect in our study.

HYPOTHESIS-TESTING EXAMPLE   The following research situation will be used to demonstrate the procedures of hypothesis testing with the *t* statistic.

EXAMPLE 9.1

Minitab only.

Many studies have shown that direct eye contact and even patterns that look like eyes are avoided by many animals. Some insects, such as moths, have even evolved large eye-spot patterns on their wings to help ward off predators. This research example, modeled after Scaife's (1976) study, examines how eye-spot patterns affect the behavior of moth-eating birds.

A sample of *n* = 16 insectivorous birds is selected. The animals are tested in a box that has two separate chambers (see Figure 9.4). The birds are free to roam from one chamber to another through a doorway in a partition. On the wall of one chamber, two large eye-spot patterns have been painted. The other chamber has plain walls. The birds are tested one at a time by placing them in the doorway in the center of the apparatus. Each animal is left in the box for 60 minutes, and the amount of time spent in the plain chamber is recorded. Suppose that the sample of *n* = 16 birds spent an average of $\overline{X}$ = 35 minutes in the plain side with *SS* = 1215. Can we conclude that eye-spot patterns have an effect on behavior? Note that while it is possible to predict a value for μ, we have no information about the population standard deviation.

STEP 1   State the hypotheses, and select an alpha level. If the null hypothesis were true, then the eye-spot patterns would have no effect on behavior. The animals should show no preference for either side of the box. That is, they should spend half of the 60-minute test period in the plain chamber. In symbols, the null hypothesis would state that

$$H_0: \mu_{\text{plain side}} = 30 \text{ minutes}$$

FIGURE 9.5

The critical region in the *t* distribution for $\alpha = .05$ and $df = 15$.

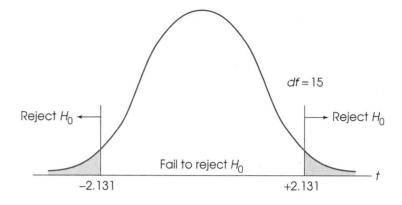

Directional hypotheses could be used and would specify whether the average time on the plain side is more or less than 30 minutes.

The alternative hypothesis would state that the eye patterns have an effect on behavior. There are two possibilities: (1) The animals may avoid staying in the chamber with the eye spots as we suspect, or (2) for some reason the animals may show a preference for the patterns painted on the wall. A nondirectional hypothesis (for a two-tailed test) would be represented in symbols as follows:

$$H_1: \mu_{\text{plain side}} \neq 30 \text{ minutes}$$

We will set the level of significance at $\alpha = .05$ for two tails.

STEP 2 Locate the critical region. The test statistic is a *t* statistic because the population standard deviation is not known. The exact shape of the *t* distribution and therefore the proportions under the *t* distribution depend on the number of degrees of freedom associated with the sample. To find the critical region, *df* must be computed:

$$df = n - 1 = 16 - 1 = 15$$

For a two-tailed test at the .05 level of significance and with 15 degrees of freedom, the critical region consists of *t* values greater than $+2.131$ or less than -2.131. Figure 9.5 depicts the critical region in this *t* distribution.

STEP 3 Calculate the test statistic. The *t* statistic typically requires much more computation than is necessary for a *z*-score. Therefore, we recommend that you divide the calculations into a three-stage process as follows.

a. First, calculate the sample variance. Remember, the population variance is unknown, and you must use the sample value in its place. (This is why we are using a *t* statistic instead of a *z*-score.)

$$s^2 = \frac{SS}{n-1} = \frac{SS}{df}$$
$$= \frac{1215}{15}$$
$$= 81$$

b. Next, use the sample variance (s^2) to compute the estimated standard error. This value will be the denominator of the *t* statistic and measures how much error is expected by chance between a sample mean and the corresponding population mean.

$$s_{\overline{X}} = \sqrt{\frac{s^2}{n}}$$
$$= \sqrt{\frac{81}{16}}$$
$$= 2.25$$

c. Finally, compute the *t* statistic for the sample data.

$$t = \frac{\overline{X} - \mu}{s_{\overline{X}}}$$
$$= \frac{35 - 30}{2.25}$$
$$= \frac{5}{2.25}$$
$$= 2.22$$

STEP 4 Make a decision regarding H_0. The obtained *t* statistic of 2.22 falls into the critical region on the right-hand side of the *t* distribution (Figure 9.5). Our statistical decision is to reject H_0 and conclude that the presence of eye-spot patterns does influence behavior. As can be seen from the sample mean, there is a tendency for animals to avoid the eyes and spend more time on the plain side of the box.

IN THE LITERATURE:
REPORTING THE RESULTS OF A *t* TEST

In Chapter 8, we noted the conventional style for reporting the results of a hypothesis test, according to APA format. First, you should recall that a scientific report typically uses the term *significant* to indicate that the null hypothesis has been rejected and the term *not significant* to indicate failure to reject H_0. Additionally, there is a prescribed format for reporting the calculated value of the test statistic, degrees of freedom, and alpha level for a *t* test. This format parallels the style introduced in Chapter 8 (page 201). For Example 9.1, we calculated a *t* statistic of +2.22 with $df = 15$ and decided to reject H_0 with alpha set at .05. In a scientific report, this information is conveyed in a concise statement, as follows:

> The subjects spent more time on the plain side of the apparatus ($M = 35$, $SD = 9$). These data demonstrate that the birds spent significantly more time in the chamber without eye-spot patterns; $t(15) = +2.22$, $p < .05$, two-tailed.

In the first statement, the mean ($M = 35$) and the standard deviation ($SD = 9$) are reported as previously noted (Chapter 4, page 105). The next statement provides the results of the statistical analysis. Note that the degrees of freedom are reported in parentheses immediately after the symbol *t*. The value for the obtained *t* statistic follows (2.22), and next is the probability of committing a Type I error (less than 5%). Finally, the type of test (one- versus two-tailed) is noted.

The statement $p < .05$ was explained in Chapter 8, page 201.

❑

DIRECTIONAL HYPOTHESES AND ONE-TAILED TESTS

As we noted in Chapter 8, the nondirectional (two-tailed) test is commonly used for research that is intended for publication in a scientific journal. On the other hand, a directional (one-tailed) test may be used in some research situations, such as exploratory investigations or pilot studies. Although one-tailed tests are used occasionally, you should remember that the two-tailed test is preferred by most researchers. Even though a researcher may have a specific directional prediction for an experiment, it is generally safer and always appropriate to use a nondirectional (two-tailed) test. The following example demonstrates a directional hypothesis test with a *t* statistic, using the same experimental situation that was presented in Example 9.1.

EXAMPLE 9.2

The research question is whether eye-spot patterns will affect the behavior of birds placed in a special testing box. The researcher is expecting the birds to avoid the eye-spot patterns. Therefore, the researcher predicts that the birds will spend most of the hour on the plain side of the box.

STEP 1

State the hypotheses, and select an alpha level. With most directional tests, it is easier to begin by stating the alternative hypothesis. Remember, H_1 states that the treatment does have an effect. For this example, the eye patterns should cause the birds to spend most of their time on the plain side. In symbols,

$$H_1: \mu_{\text{plain side}} > 30 \text{ minutes}$$

The null hypothesis states that the treatment will not have the predicted effect. In this case, H_0 says that the eye patterns will not cause the birds to spend more time on the plain side. In symbols,

$$H_0: \mu_{\text{plain side}} \leq 30 \text{ minutes} \quad \text{(not greater than 30 minutes)}$$

We will set the level of significance at $\alpha = .05$.

STEP 2

Locate the critical region. In this example, the researcher is predicting that the sample mean ($\overline{X}$) will be greater than 30. The null hypothesis states that the population mean is $\mu = 30$ (or less). If you examine the structure of the *t*-statistic formula, it should be clear that a positive *t* statistic would support the researcher's prediction and refute the null hypothesis.

$$t = \frac{\overline{X} - \mu}{s_{\overline{X}}}$$

FIGURE 9.6

The critical region in the *t* distribution for
α = .05, *df* = 15, one-tailed test.

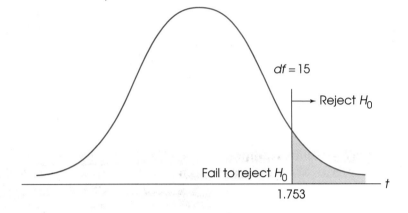

The problem is to determine how large a positive value is necessary to reject
H_0. To find the critical value, you must look in the *t*-distribution table using
the one-tail proportions. With a sample of *n* = 16, the *t* statistic will have
df = 15; using α = .05, you should find a critical *t* value of 1.753. Figure 9.6
depicts the critical region in the *t* distribution.

STEP 3   Calculate the test statistic. The computation of the *t* statistic is the same for
either a one-tailed or a two-tailed test. Earlier (in Example 9.1), we found that
the data for this experiment produce a test statistic of *t* = 2.22.

STEP 4   Make a decision. The test statistic is in the critical region, so we reject H_0. In
terms of the experimental variables, we have decided that the birds spent significantly more time on the plain side of the box than on the side with eye-
spot patterns.

ASSUMPTIONS OF THE *t* TEST   Two basic assumptions are necessary for hypothesis tests with the *t* statistic.

1. The values in the sample must consist of *independent* observations.
   In everyday terms, two observations are independent if there is no consistent, predictable relationship between the first observation and the second. More precisely, two events (or observations) are independent if the
   occurrence of the first event has no effect on the probability of the second
   event. Specific examples of independence and nonindependence were examined in Box 8.2 (page 203).

2. The population sampled must be normal.
   This assumption is a necessary part of the mathematics underlying the
   development of the *t* statistic and the *t*-distribution table. However, violating this assumption has little practical effect on the results obtained for a *t*
   statistic, especially when the sample size is relatively large. With very small
   samples, a normal population is important. With larger samples, this assumption can be violated without affecting the validity of the hypothesis
   test. If you have reason to suspect that the population is not normal, use a
   large sample to be safe.

THE VERSATILITY OF THE *t* TEST   The obvious advantage of hypothesis testing with the *t* statistic (as compared to *z*-scores) is that you do not need to know the value of the population standard deviation. This means that we still can do a hypothesis test even though we have little or no information about the population.

Both the *t*-statistic and *z*-score tests were introduced as methods for determining whether or not a treatment has any effect on a dependent variable. For both tests, we assumed that you start with a *known* population mean before treatment, and the research question (hypotheses) concerned the *unknown* population mean after treatment (see Figures 8.1 and 9.3). For example, if the population mean is known to be $\mu = 80$ before treatment, the null hypothesis would state that the population mean is still $\mu = 80$ after treatment. For the *z*-score test, a known population is essential to determine the value for σ and to determine the hypothesized mean for H_0. Although the *t* statistic can be used in this "before and after" type of experiment, the *t* test also permits hypothesis testing in situations where you do not have a "known" population mean to serve as a standard. Specifically, the *t* test does not require any prior knowledge about the population mean or the population variance. All you need to compute a *t* statistic is a sensible null hypothesis and a sample from the unknown population. Thus, a *t* test can be used in situations where the value for the null hypothesis can be obtained from a theory, a logical prediction, or just wishful thinking.

You may have noticed in the preceding examples (9.1 and 9.2) that there was no "known" value for the mean time that birds spend in each side of the box. For both examples, the null hypothesis was based on logic: If the eye-spot patterns have no effect, then the birds should show no preference between the two sides of the box. Some similar examples follow. Notice that in each case the hypothesis is not dependent on knowing the actual population mean before treatment and that the rest of the *t* statistic can be computed entirely from the obtained sample data.

1. A researcher would like to examine the accuracy of people's judgment of time when they are distracted. Individuals are placed in a waiting room where many distracting events occur for a period of 12 minutes. The researcher then asks each person to judge how much time has passed. The null hypothesis would state that distraction has no effect and that time judgments are accurate. That is,

   H_0: $\mu = 12$ minutes

2. A local fund-raising organization has set a goal of receiving $25 per contributor. After 1 week, a sample of contributions is selected to see if they deviate significantly from the goal. The null hypothesis would state that the contributions do not deviate from the goal of the organization. In symbols, it is

   H_0: $\mu = \$25.00$

3. A soft-drink company has developed a new, improved formula for its product and would like to determine how consumers respond to the new formula. A sample is obtained, and each individual is asked to taste the original soft drink and the new formula. After tasting, the subjects are required to state whether the new formula is better or worse than the original formula using a 7-point scale. A rating of 4 indicates no preference between the old and new formulas. Ratings above 4 indicate that the new formula is better (5 = slightly better, 6 = better, and 7 = much better). Similarly, ratings below 4 indicate

that the new formula tastes worse than the old formula. The null hypothesis states that there is no perceived difference between the two formulas. In symbols, the new formula would receive an average rating of

$$H_0: \mu = 4$$

LEARNING CHECK

1. A researcher would like to evaluate the effect of a new cold medication on reaction time. It is known that under regular circumstances the distribution of reaction times is normal with $\mu = 200$. A sample of $n = 4$ subjects is obtained. Each person is given the new cold medication, and one hour later reaction time is measured for each individual. The average reaction time for this sample is $\bar{X} = 215$ with $SS = 300$. On the basis of these data, can the researcher conclude that the cold medication has an effect on reaction time? Test at the .05 level of significance.

   a. State the hypotheses.

   b. Locate the critical region.

   c. Compute the t statistic.

   d. Make a decision.

2. Suppose the researcher in the previous problem predicted that the medication would increase reaction time and decided to use a one-tailed test to evaluate this prediction.

   a. State the hypotheses for the one-tailed test.

   b. Locate the one-tailed critical region for $\alpha = .05$.

ANSWERS

1. **a.** $H_0: \mu = 200$ (the medication has no effect on reaction time)
      $H_1: \mu \neq 200$ (reaction time is changed)

   b. With $df = 3$ and $\alpha = .05$, the critical region consists of t values beyond ± 3.182.

   c. The sample variance is $s^2 = 100$, the standard error is 5, and $t = 3.00$.

   d. The t statistic is not in the critical region. Fail to reject the null hypothesis, and conclude that the data do not provide enough evidence to say that the medication affects reaction time.

2. **a.** $H_0: \mu \leq 200$ (reaction time is not increased)
      $H_1: \mu > 200$ (reaction time is increased)

   b. With $df = 3$ and $\alpha = .05$, the one-tailed critical region consists of values beyond $t = +2.353$.

SUMMARY

1. When the population variance (σ^2) or the population standard deviation (σ) is unknown, the standard error cannot be computed, and a hypothesis test based on a z-score is impossible.

2. In order to test a hypothesis about μ when the population variance and standard deviation are unknown, you must first compute the sample variance or sample standard deviation as an estimate of the unknown population value.

$$s^2 = \frac{SS}{df} \qquad s = \sqrt{\frac{SS}{df}}$$

Next, the standard error is *estimated* by substituting s^2 or s in the formula for standard error. The estimated standard error is calculated in the following manner:

$$s_{\bar{X}} = \sqrt{\frac{s^2}{n}} \qquad \text{or} \qquad s_{\bar{X}} = \frac{s}{\sqrt{n}}$$

Finally, a *t* statistic is computed using the estimated standard error. The *t* statistic is used as a substitute for a *z*-score that cannot be computed when the population variance or standard deviation is unknown.

$$t = \frac{\overline{X} - \mu}{s_{\overline{X}}}$$

3. The structure of the *t* formula is similar to that of the *z*-score in that

$$z \text{ or } t = \frac{\text{sample mean} - \text{population mean}}{\text{(estimated) standard error}}$$

4. The *t* distribution is an approximation of the normal *z* distribution. To evaluate a *t* statistic that is obtained for a sample mean, the critical region must be located in a *t* distribution. There is a family of *t* distributions, with the exact shape of a particular distribution of *t* values depending on degrees of freedom ($n - 1$). Therefore, the critical *t* values will depend on the value for *df* associated with the *t* test. As *df* increases, the shape of the *t* distribution approaches a normal distribution.

KEY TERMS

estimated standard error   *t* statistic   degrees of freedom   *t* distribution

FOCUS ON PROBLEM SOLVING

1. The first problem we confront in analyzing data is determining the appropriate statistical test. Remember, you can use a *z*-score for the test statistic only when the value for σ is known. If the value for σ is not provided, then you must use the *t* statistic.

2. For a *t* test, students sometimes use the unit normal table to locate the critical region. This, of course, is a mistake. The critical region for a *t* test is obtained by consulting the *t*-distribution table. Notice that to use this table you first must compute the value for degrees of freedom (*df*).

3. For the *t* test, the sample variance is used to find the value for estimated standard error. Remember, when computing the sample variance, use $n - 1$ in the denominator (see Chapter 4). When computing estimated standard error, use $\sqrt{n}$ in the denominator.

DEMONSTRATION 9.1

A HYPOTHESIS TEST WITH THE *t* STATISTIC

A psychologist has prepared an "Optimism Test" that is administered yearly to graduating college seniors. The test measures how each graduating class feels about its future—the higher the score, the more optimistic the class. Last year's class had a mean score of $\mu = 15$. A sample of $n = 9$ seniors from this year's class was selected and tested. The scores for these seniors are as follows:

7,   12,   11,   15,   7,   8,   15,   9,   6

On the basis of this sample, can the psychologist conclude that this year's class has a different level of optimism than last year's class?

Note that this hypothesis test will use a *t* statistic because the population standard deviation (σ) is not known.

STEP 1   State the hypotheses, and select an alpha level.

The statements for the null hypothesis and the alternative hypothesis follow the same form for the *t* statistic and the *z*-score test.

$$H_0: \mu = 15 \quad \text{(there is no change)}$$

$$H_1: \mu \neq 15 \quad \text{(this year's mean is different)}$$

For this demonstration, we will use $\alpha = .05$, two tails.

STEP 2   Locate the critical region.

Remember, for hypothesis tests with the *t* statistic, we must now consult the *t*-distribution table to find the critical *t* values. With a sample of $n = 9$ students, the *t* statistic will have degrees of freedom equal to

$$df = n - 1 = 9 - 1 = 8$$

For a two-tailed test with $\alpha = .05$ and $df = 8$, the critical *t* values are $t = \pm 2.306$. These critical *t* values define the boundaries of the critical region. The obtained *t* value must be more extreme that either of these critical values to reject H_0.

STEP 3   Obtain the sample data, and compute the test statistic.

For the *t* formula, we need to determine the values for the following:

1. The sample mean, $\overline{X}$
2. The estimated standard error, $s_{\overline{X}}$

We also have to compute sum of squares (*SS*) and the variance for the sample in order to get the estimated standard error.

The sample mean. For these data, the sum of the scores is

$$\Sigma X = 7 + 12 + 11 + 15 + 7 + 8 + 15 + 9 + 6 = 90$$

Therefore, the sample mean is

$$\overline{X} = \frac{\Sigma X}{n} = \frac{90}{9} = 10$$

Sum of squares. We will use the definitional formula for sum of squares,

$$SS = \Sigma(X - \overline{X})^2$$

The following table summarizes the steps in the computation of *SS*.

X	$X - \overline{X}$	$(X - \overline{X})^2$
7	$7 - 10 = -3$	9
12	$12 - 10 = +2$	4
11	$11 - 10 = +1$	1
15	$15 - 10 = +5$	25
7	$7 - 10 = -3$	9
8	$8 - 10 = -2$	4
15	$15 - 10 = +5$	25
9	$9 - 10 = -1$	1
6	$6 - 10 = -4$	16

For this demonstration problem, the sum of squares value is

$$SS = \Sigma(X - \bar{X})^2 = 9 + 4 + 1 + 25 + 9 + 4 + 25 + 1 + 16$$
$$= 94$$

Sample variance. The variance for the sample is

$$s^2 = \frac{SS}{n-1} = \frac{94}{8} = 11.75$$

Estimated standard error. The estimated standard error for these data is

$$s_{\bar{X}} = \sqrt{\frac{s^2}{n}} = \sqrt{\frac{11.75}{9}} = 1.14$$

The t statistic. Now that we have the estimated standard error and the sample mean, we can compute the *t* statistic. For this demonstration,

$$t = \frac{\bar{X} - \mu}{s_{\bar{X}}} = \frac{10 - 15}{1.14} = \frac{-5}{1.14} = -4.39$$

STEP 4 Make a decision about H_0 and a conclusion.
   The *t* statistic we obtained ($t = -4.39$) is in the critical region. Thus, our sample data are unusual enough to reject the null hypothesis at the .05 level of significance. We can conclude that there is a significant difference in level of optimism between this year's and last year's graduating classes, $t(8) = -4.39$, $p < .05$, two-tailed.

PROBLEMS

1. Why is it necessary to *estimate* the standard error when a *t* statistic is used?

2. Briefly describe what is measured by the sample standard deviation, *s*, and what is measured by the estimated standard error, $s_{\bar{X}}$.

3. Several different factors influence the value obtained for a *t* statistic. For each of the following, describe how the value of *t* is affected. In each case, assume all other factors are held constant.
   a. What happens to the value of *t* when the variability of the scores in the sample increases?
   b. What happens to the value of *t* when the number of scores in the sample increases?
   c. What happens to the value of *t* when the difference between the sample mean and the hypothesized population mean increases?

4. What assumptions (or set of conditions) must be satisfied for a *t* test to be valid?

5. Why is a *t* distribution generally more variable than a normal distribution?

6. What is the relationship between the value for degrees of freedom and the shape of the *t* distribution? What happens to the critical value of *t* for a particular alpha level when *df* increases in value?

7. The following sample was obtained from a population with unknown parameters. Scores:

   15, 3, 15, 7

   a. Find the sample mean, the sample variance, and the sample standard deviation. (Note that these values are descriptive statistics that summarize the sample data.)
   b. How accurately does the sample mean represent the unknown population mean? Compute the estimated standard error for $\bar{X}$. (Note that this value is an inferential statistic that allows you to use the sample data to answer questions about the population.)

8. The following sample was obtained from an unknown population. Scores:

   3, 11, 3, 3

   a. Compute the sample mean, the sample variance, and the sample standard deviation. (Note that these values are descriptive statistics that summarize the sample data.)

b. How accurately does the sample mean represent the unknown population mean? Compute the estimated standard error for $\overline{X}$. (Note that this value is an inferential statistic that allows you to use the sample data to answer questions about the population.)

9. A developmental psychologist would like to know whether success in one specific area can affect a person's general self-esteem. The psychologist selects a sample of $n = 25$ 10-year-old children who all excel in athletics. These children are given a standardized self-esteem test for which the general population of 10-year-old children averages $\mu = 70$. The average score for the sample is $\overline{X} = 73$ with $SS = 2400$. On the basis of these data, can the psychologist conclude that excelling in athletics has a general effect on self-esteem? Use a two-tailed test with $\alpha = .05$.

10. A college professor has noted that this year's freshman class appears to be smarter than classes from previous years. The professor obtains a sample of $n = 36$ freshmen and computes an average IQ of $\overline{X} = 114.5$ with a variance of $s^2 = 324$ for this group. College records indicate that the mean IQ for entering freshmen from earlier years is $\mu = 110.3$. On the basis of the sample data, can the professor conclude that this year's students have IQs that are significantly different from those of previous students? Use a two-tailed test with $\alpha = .05$.

11. Educational administrators have complained for years that American high school students are dismally ignorant about world geography. To evaluate this complaint, a history teacher prepared a 40-question multiple-choice geography test. Each question had four choices for the answer, so the probability of guessing correctly is $p = 1/4$. Thus, chance performance would result in an average score of $\mu = 10$ out of 40. The test was administered to a random sample of $n = 36$ students, and the mean score for the sample was $\overline{X} = 13.5$ with $s^2 = 144$. On the basis of these data, can the teacher conclude that the students' scores are significantly different from what would be expected by chance? Use a two-tailed test with $\alpha = .05$.

12. A recent survey of college seniors evaluated how the students view the "real world" that they are about to enter. One question asked the seniors to evaluate the future job market on a 10-point scale from 1 (dismal) to 10 (excellent). The average score for the sample of $n = 100$ students was $\overline{X} = 7.3$ with $SS = 99$. Ten years ago the same survey produced an average score of $\mu = 7.9$ for this question. Do the sample data indicate a significant change in the perceived job market during the past 10 years? Test at the .01 level of significance.

13. A random sample has a mean of $\overline{X} = 81$ with $s = 10$. Use this sample to test the null hypothesis that $\mu = 85$ for each of the following situations.

a. Assume that the sample size is $n = 25$, and use $\alpha = .05$ for the hypothesis test.

b. Assume that the sample size is $n = 400$, and use $\alpha = .05$ for the hypothesis test.

c. In general, how does sample size contribute to the outcome of a hypothesis test?

14. A random sample of $n = 16$ scores has a mean of $\overline{X} = 48$. Use this sample to test the null hypothesis that $\mu = 45$ for each of the following situations.

a. Assume that the sample has $SS = 60$, and use $\alpha = .05$.

b. Assume that the sample has $SS = 6000$, and use $\alpha = .05$.

c. In general, how does the variability of the scores in the sample contribute to the outcome of a hypothesis test?

15. A population has a mean of $\mu = 50$. A sample of $n = 16$ individuals is selected from the population, and a treatment is administered to the sample.

a. After treatment, the scores in the sample are 54, 55, 60, 55, 55, 52, 54, 57, 56, 55, 57, 53, 55, 54, 52, 56. Compute the mean and the standard deviation (s) for the sample, and draw a sketch showing the sample distribution.

b. Locate the position of $\mu = 50$ on your sketch. Does it appear from your sketch that the sample is centered around $\mu = 50$, or does it appear that the scores in the sample pile up around a location significantly different from $\mu = 50$?

c. Now use a hypothesis test to determine whether these data provide sufficient evidence to conclude that the treatment has a significant effect? Test at the .05 level of significance. Are the results of the hypothesis test consistent with the appearance of your sketch?

16. The local grocery store sells potatoes in 5-pound bags. Because potatoes come in unpredictable sizes, it is almost impossible to get a bag that weighs exactly 5 pounds. Therefore, the store advertises that the bags *average* 5 pounds. To check this claim, a sample of $n = 25$ bags was randomly selected. The average weight for the sample was $\overline{X} = 5.20$ pounds with $s = 0.50$. On the basis of this sample,

a. Can you conclude that the average weight for a bag of potatoes is *significantly different* from $\mu = 5$ pounds. Use $\alpha = .05$.

b. Can you conclude that the average weight for a bag of potatoes is *significantly more* than $\mu = 5$ pounds. Use a one-tailed test with $\alpha = .05$.

17. A researcher would like to examine the effects of humidity on eating behavior. It is known that laboratory rats normally eat an average of $\mu = 21$ grams of food each day. The researcher selects a random sample of $n = 100$

rats and places them in a controlled atmosphere room where the relative humidity is maintained at 90%. The daily food consumption for the sample averages $\overline{X} = 18.7$ with $SS = 2475$. On the basis of this sample, can the researcher conclude that humidity affects eating behavior? Use a two-tailed test at the .05 level.

18. A psychologist would like to determine whether there is a relation between depression and aging. It is known that the general population averages $\mu = 40$ on a standardized depression test. The psychologist obtains a sample of $n = 36$ individuals who are all over the age of 70. The average depression score for this sample is $\overline{X} = 44.5$ with $SS = 5040$. On the basis of this sample, can the psychologist conclude that depression for elderly people is significantly different from depression in the general population? Test at the .05 level of significance.

19. A social psychologist recently developed a childhood-memories test that is intended to measure how people recall pleasant and unpleasant experiences from childhood. Scores on the test range from 1 (very unpleasant) to 100 (very pleasant). The psychologist standardized the test with a large group of college students (ages 18–25), and the mean score was $\mu = 60$. The test was then given to a sample of $n = 25$ individuals who were all between the ages of 40 and 45. The average score for this sample was $\overline{X} = 64.3$ with $SS = 600$. Does this sample provide sufficient evidence to conclude that childhood memories for older adults (ages 40–45) are significantly more pleasant than memories for college students? Use a one-tailed test with $\alpha = .05$.

20. Fifteen years ago the average weight of American men between the ages of 30 and 50 was $\mu = 166$ pounds. A researcher would like to determine whether there has been any change in this figure during the past 15 years. A sample of $n = 100$ men is obtained. The average weight for this sample is $\overline{X} = 173$ with $s = 23$. On the basis of these data, can the researcher conclude that there has been a significant change in weight? Use a two-tailed test with $\alpha = .01$.

21. A fund raiser for a charitable organization has set a goal of averaging at least $25 per donation. To see if the goal is being met, a random sample of recent donations is selected. The data for this sample are as follows: 20, 50, 30, 25, 15, 20, 40, 50, 10, 20. Use a one-tailed test with $\alpha = .05$ to determine whether the donations are averaging significantly greater than $25.

22. One of the original tests of ESP involves using Zener cards. Each card shows 1 of 5 different symbols (square, circle, star, wavy lines, cross). One person randomly picks a card and concentrates on the symbol it shows. A second person, in a different room, attempts to identify the symbol that was selected. Chance performance on this task (just guessing) should lead to correct identification of 1 out of 5 cards. A psychologist used the Zener cards to evaluate a sample of $n = 9$ subjects who claimed to have ESP. Each subject was tested on a series of 100 cards, and the number correct for each individual is as follows: 18, 23, 24, 22, 19, 28, 15, 26, 25.

Chance performance on a series of 100 cards would be $\mu = 20$ correct. Did this sample perform significantly better than chance? Use a one-tailed test at the .05 level of significance.

CHAPTER 10

HYPOTHESIS TESTS WITH TWO INDEPENDENT SAMPLES

TOOLS YOU WILL NEED

The following items are considered essential background material for this chapter. If you doubt your knowledge of any of these items, you should review the appropriate chapter or section before proceeding.

- Sample variance (Chapter 4)
- The *t* statistic (Chapter 9)
  - Distribution of *t* values
  - *df* for the *t* statistic
  - Estimated standard error

CONTENTS

10.1 INTRODUCTION

Until this point, all the inferential statistics we have considered involve using one sample as the basis for drawing conclusions about one population. Although these *single-sample* techniques are used occasionally in real research, most research studies require the comparison of two (or more) sets of data. For example, a social psychologist may want to compare men and women in terms of their attitudes toward abortion, or an educational psychologist may want to compare two methods for teaching mathematics, or a clinical psychologist may want to evaluate a therapy technique by comparing depression scores for patients before therapy with their scores after therapy. In each case, the research question concerns a mean difference between two sets of data.

There are two general research strategies that can be used to obtain the two sets of data to be compared:

1. The two sets of data could come from two completely separate samples. For example, the study could involve a sample of men compared with a sample of women. Or the study could compare one sample of students taught by method A and a second sample taught by method B.

2. The two sets of data could both come from the same sample. For example, the study could obtain one set of scores by measuring depression for a sample of patients before they begin therapy and then obtain a second set of data by measuring the same individuals after 6 weeks of therapy.

The first research strategy, using completely separate samples, is called an *independent-measures* research design or a *between-subjects* design. These terms emphasize the fact that the design involves separate and independent samples and makes a comparison between two groups of individuals. In this chapter, we will ex-

FIGURE 10.1

Do the achievement scores for children taught by method A differ from the scores for children taught by method B? In statistical terms, are the two population means the same or different? Because neither of the two population means is known, it will be necessary to take two samples, one from each population. The first sample will provide information about the mean for the first population, and the second sample will provide information about the second population.

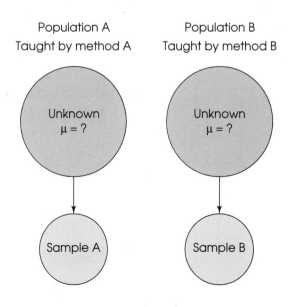

amine the statistical techniques used to evaluate the data from an independent-measures design. More precisely, we will introduce the hypothesis test that allows researchers to use the data from two separate samples to evaluate the mean difference between two populations or between two treatment conditions.

The second research strategy, where the two sets of data are obtained from the same sample, is called a *repeated-measures* research design or a *within-subjects* design. The statistical analysis for repeated measures will be introduced in Chapter 11. Also, at the end of Chapter 11, we will discuss some of the advantages and disadvantages of independent-measures versus repeated-measures research designs.

The typical structure of an independent-measures research study is shown in Figure 10.1. Notice that the research study is using two separate samples to answer a question about two populations.

DEFINITION   A research design that uses a separate sample for each treatment condition (or for each population) is called an *independent-measures* research design.

10.2 THE *t* STATISTIC FOR AN INDEPENDENT-MEASURES RESEARCH DESIGN

Because an independent-measures experiment involves two separate samples, we will need some special notation to help specify which data go with which sample. This notation involves the use of subscripts, which are small numbers written beside each sample statistic. For example, the number of scores in the first sample would be identified by n_1; for the second sample, the number of scores would be n_2. The sample means would be identified by $\overline{X}_1$ and $\overline{X}_2$. The sums of squares would be SS_1 and SS_2.

THE HYPOTHESES FOR AN INDEPENDENT-MEASURES TEST

The goal of an independent-measures research study is to evaluate the mean difference between two populations (or between two treatment conditions). Using subscripts to differentiate the two populations, the mean for the first population is μ_1, and the second population mean is μ_2. The difference between means is simply $\mu_1 - \mu_2$. As always, the null hypothesis states that there is no change, no effect, or, in this case, no difference. Thus, in symbols, the null hypothesis for the independent-measures test is

$$H_0: \mu_1 - \mu_2 = 0 \quad \text{(no difference between the population means)}$$

The null hypothesis could be stated as $\mu_1 = \mu_2$. However, the first version of H_0 (which states that the difference between means is equal to zero) gives us a specific numerical value (zero) that will be used in the calculation of the t statistic. Therefore, we prefer to phrase the null hypothesis in terms of the difference between the two population means. The alternative hypothesis states that there is a mean difference between the two populations:

$$H_1: \mu_1 - \mu_2 \neq 0 \quad \text{(there is a mean difference)}$$

Equivalently, the alternative hypothesis can simply state that the two population means are not equal: $\mu_1 \neq \mu_2$.

THE FORMULAS FOR AN INDEPENDENT-MEASURES HYPOTHESIS TEST

Although the formulas that are used to compute the independent-measures *t* statistic may appear to be a bit overpowering, they are easier to understand if you view them in relation to the single-sample *t* formulas from Chapter 9. In particular, there are two points to remember:

1. The basic structure of the *t* statistic is the same for both the independent-measures and the single-sample hypothesis tests. In both cases,

$$t = \frac{\text{sample statistic} - \text{hypothesized population parameter}}{\text{estimated standard error}}$$

2. The independent-measures *t* is basically a *two-sample t that doubles all the elements of the single-sample t formulas.*

To demonstrate the second point, we will examine the two *t* formulas piece by piece.

The overall *t* formula   The single-sample *t* uses one sample mean to test a hypothesis about one population mean. The sample mean and the population mean appear in the numerator of the *t* formula, which measures how much difference there is between the sample data and the population hypothesis.

$$t = \frac{\overline{X} - \mu}{\text{standard error}}$$

The independent-measures *t* uses two sample means to evaluate a hypothesis about two population means. The numerator of the *t* formula compares the sample mean difference with the hypothesized population mean difference.

$$t = \frac{(\overline{X}_1 - \overline{X}_2) - (\mu_1 - \mu_2)}{\text{standard error}}$$

In this formula, the value of $\overline{X}_1 - \overline{X}_2$ is obtained from the sample data, and the value for $\mu_1 - \mu_2$ is taken from the null hypothesis.

The standard error   The standard error in the denominator of the *t* formula measures how accurately the sample statistic represents the population parameter. The single-sample standard error is identified by the symbol $s_{\overline{X}}$ and measures how much error is expected between a sample mean ($\overline{X}$) and the population mean. Similarly, the two-sample standard error measures how much error is expected when you are using a sample mean difference ($\overline{X}_1 - \overline{X}_2$) to represent a population mean difference. The two-sample standard error is identified by the symbol $s_{\overline{X}_1 - \overline{X}_2}$.

Caution: Do not let the notation for standard error confuse you. In general, standard error measures how accurately a statistic represents a parameter. The symbol for standard error takes the form, $s_{\text{statistic}}$. When the statistic is a sample mean, $\overline{X}$, the symbol for standard error is $s_{\overline{X}}$. For the independent-measures test, the statistic is a sample mean difference ($\overline{X}_1 - \overline{X}_2$), and the symbol for standard error is $s_{\overline{X}_1 - \overline{X}_2}$. In each case, the standard error tells how much discrepancy is reasonable to expect just by chance between the statistic and the corresponding population parameter.

In general, the magnitude of the standard error is determined by two factors: the variance of the scores and the size of the sample. To compute the standard error, the first step is to calculate the variance for the sample data. For the single-sample t statistic, there is only one sample, and the sample variance is computed as

$$s^2 = \frac{SS}{df}$$

For the independent-measures t statistic, there are two SS values and two df values (one from each sample). The values from the two samples are combined to compute what is called the *pooled variance*. The pooled variance is identified by the symbol s_p^2 and is computed as

$$s_p^2 = \frac{SS_1 + SS_2}{df_1 + df_2} \qquad (10.1)$$

With one sample, variance is computed as SS divided by df. With two samples, the two SSs are divided by the two dfs to compute pooled variance. The pooled variance is actually an average of the two sample variances, and the value of the pooled variance will always be located between the two sample variances. A more detailed discussion of the pooled variance is presented in Box 10.1.

After variance is computed, the next step is to find the standard error that will become the denominator of the t statistic. With one sample, standard error is

$$\text{one-sample standard error} = s_{\overline{X}} = \sqrt{\frac{s^2}{n}}$$

This standard error measures how accurately the sample mean $\overline{X}$ represents the population mean μ.

For the independent-measures t, there are two samples, and the formula for the standard error simply adds the error for the first sample and the error for the second sample:

$$\text{two-sample standard error} = s_{\overline{X}_1 - \overline{X}_2} = \sqrt{\frac{s_p^2}{n_1} + \frac{s_p^2}{n_2}} \qquad (10.2)$$

Conceptually, this standard error measures how accurately the difference between two sample means represents the difference between the two population means. The formula combines the error for the first sample mean with the error for the second sample mean. Notice that the pooled variance from the two samples is used to compute the standard error for the two samples.

THE FINAL FORMULA AND DEGREES OF FREEDOM

The complete equation for the independent-measures t statistic is as follows:

$$t = \frac{(\overline{X}_1 - \overline{X}_2) - (\mu_1 - \mu_2)}{s_{\overline{X}_1 - \overline{X}_2}} = \frac{(\overline{X}_1 - \overline{X}_2) - (\mu_1 - \mu_2)}{\sqrt{\frac{s_p^2}{n_1} + \frac{s_p^2}{n_2}}} \qquad (10.3)$$

with the pooled variance, s_p^2, defined by formula 10.1.

The degrees of freedom for this t statistic are determined by the df values for the two separate samples:

10.1   THE POOLED VARIANCE

THE POOLED variance as defined by formula 10.1 is actually a *weighted mean* of the two sample variances. You may recall that we first computed a weighted mean in Chapter 3. At that time, we found the weighted mean for two samples as

$$\text{weighted mean} = \frac{\text{combined sums}}{\text{combined } ns} = \frac{\Sigma X_1 + \Sigma X_2}{n_1 + n_2}$$

Now we are using a similar procedure to compute the pooled variance for two samples:

$$\text{pooled variance} = \frac{\text{combined } SS}{\text{combined } df} = \frac{SS_1 + SS_2}{df_1 + df_2}$$

To compute the pooled variance, each of the two sample variances is weighted by its degrees of freedom, and then the average is computed. The procedure can be described as a two-step process:

1. Multiply each variance by its *df*, and then add the results together. This weights each variance.

2. Divide the total by the sum of the two *df* values.

To demonstrate this process, consider the following two samples. The first sample has $n = 4$ scores with $SS = 36$. The second sample has $n = 8$ with $SS = 70$. The two sample variances are

$$s_1^2 = \frac{SS}{df} = \frac{36}{3} = 12 \quad \text{and} \quad s_2^2 = \frac{SS}{df} = \frac{70}{7} = 10$$

To find the weighted mean, we take 3 of the first variance ($df = 3$) and 7 of the second variance ($df = 7$). This gives a total of 10 variances. To find the mean, divide the total by 10. Expressed in an equation,

$$\text{weighted mean} =$$

$$\frac{3(12) + 7(10)}{3 + 7} = \frac{36 + 70}{3 + 7} = \frac{106}{10} = 10.6$$

There are two important facts to notice about the weighted mean. First, the obtained value of 10.6 is not halfway between the two sample variances. Instead, the weighted mean is closer to $s_2^2 = 10$ than it is to $s_1^2 = 12$. This is appropriate because the second sample is larger and therefore carries more weight when computing the mean. The second point is that the calculation of the weighted mean includes the formula we used to define the pooled variance (formula 10.1). During the calculation of the weight mean, we obtained a fraction with $36 + 70$ in the numerator and $7 + 3$ in the denominator. You should recognize the numerator values as the two *SS*s and the two denominator values as the two *df*s. To demonstrate, we will use formula 10.1 to calculate the pooled variance for the two samples.

$$\text{pooled variance} =$$

$$\frac{SS_1 + SS_2}{df_1 + df_2} = \frac{36 + 70}{3 + 7} = \frac{106}{10} = 10.6$$

Remember, we pooled the two sample variances to compute the *t* statistic. Now we combine the two *df* values to obtain the overall *df* for the *t* statistic.

$$df = df \text{ for first sample} + df \text{ for second sample} \qquad (10.4)$$
$$= df_1 + df_2$$

Occasionally, you will see degrees of freedom written in terms of the number of scores in each sample:

$$df = (n_1 - 1) + (n_2 - 1) \qquad (10.5)$$
$$= n_1 + n_2 - 2$$

10.2 THE VARIABILITY OF DIFFERENCE SCORES

IT MAY seem odd that the independent-measures t statistic *adds* together the standard errors from the two samples when it *subtracts* to find the difference between the two sample means. The logic behind this apparently unusual procedure is demonstrated here.

We begin with two populations, I and II (see Figure 10.2). The scores in population I range from a high of 70 to a low of 50. The scores in population II range from 30 to 20. We will use the range as a measure of how spread out (variable) each population is:

For population I, the scores cover a range of 20 points.

For population II, the scores cover a range of 10 points

If we randomly select a score from population I and a score from population II and compute the difference between these two scores ($X_1 - X_2$), what range of values is possible for these differences? To answer this question,

we need to find the biggest possible difference and the smallest possible difference. Look at Figure 10.2; the biggest difference occurs when $X_1 = 70$ and $X_2 = 20$. This is a difference of $X_1 - X_2 = 50$ points. The smallest difference occurs when $X_1 = 50$ and $X_2 = 30$. This is a difference of $X_1 - X_2 = 20$ points. Notice that the differences go from a high of 50 to a low of 20. This is a range of 30 points:

range for population I (X_1 scores) = 20 points

range for population II (X_2 scores) = 10 points

range for the differences ($X_1 - X_2$) = 30 points

The variability for the difference scores is found by *adding* together the variabilities for the two populations.

In the independent-measures t statistic, we are computing the variability (standard error) for a sample mean difference. To compute this value, we add together the variability for each of the two sample means.

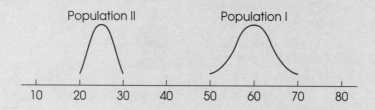

FIGURE 10.2

Two population distributions. The scores in population I vary from 50 to 70 (a 20-point spread), and the scores in population II range from 20 to 30 (a 10-point spread). If you select one score from each of these two populations, the closest two values would be $X_1 = 50$ and $X_2 = 30$. The two values that are farthest apart would be $X_1 = 70$ and $X_2 = 20$.

This t formula will be used for hypothesis testing. We will use the sample statistic $(\overline{X}_1 - \overline{X}_2)$ as the basis for testing hypotheses about the population parameter $(\mu_1 - \mu_2)$. See Box 10.2 for a final comment about the t-statistic formula.

LEARNING CHECK

1. Describe the general characteristics of an independent-measures research study.

2. Identify the two sources of error that are reflected in the standard error for the independent-measures t statistic.

3. One sample has $n = 5$ scores and $SS = 40$. A second sample has $n = 7$ scores and $SS = 36$.

   a. Calculate the variance for each of the two samples.

b. Calculate the pooled variance for the two samples. You should find that the value for the pooled variance falls between the two sample variances and is closer to the variance for the larger sample.

ANSWERS   **1.** An independent-measures study uses a separate sample to represent each of the treatment conditions or populations being compared.

2. The two sources of error come from the fact that two sample means are being used to represent two population means. Thus, $\overline{X}_1$ represents its population mean with some error, and $\overline{X}_2$ represents its population with some error. The two errors combine in the estimated standard error for the independent-measures *t* statistic.

3. a. The first sample has $s^2 = 10$, and the second sample has $s^2 = 6$.

   b. The pooled variance is $\frac{76}{10} = 7.6$.

10.3 HYPOTHESIS TESTS WITH THE INDEPENDENT-MEASURES *t* STATISTIC

A complete example of a hypothesis test with two independent samples follows. Notice that the hypothesis-testing procedure follows the same four steps that we have used before.

STEP 1   State hypotheses H_0 and H_1, and select an alpha level. For the independent-measures *t* test, the hypotheses concern the difference between two population means.

STEP 2   Locate the critical region. The critical region is defined as sample data that would be extremely unlikely ($p < \alpha$) if the null hypothesis were true. In this case, we will be locating extremely unlikely *t* values.

STEP 3   Get the data, and compute the test statistic. Here we compute the *t* value for our data using the value from H_0 in the formula.

STEP 4   Make a decision. If the *t* statistic we compute is in the critical region, we reject H_0. Otherwise, we conclude that the data do not provide sufficient evidence that the two populations are different.

EXAMPLE 10.1   In recent years, psychologists have demonstrated repeatedly that using mental images can greatly improve memory. A hypothetical experiment, designed to examine this phenomenon, is presented here.

The psychologist first prepares a list of 40 pairs of nouns (for example, dog/bicycle, grass/door, lamp/piano). Next, two groups of subjects are obtained (two separate samples). Subjects in the first group are given the list for 5 minutes and instructed to memorize the 40 noun pairs. Subjects in the second group receive the same list of words, but in addition to the regular instructions, these people are told to form a mental image for each pair of nouns (imagine a dog riding a bicycle, for example). Notice that the two samples are identical except that the second group is using mental images to help learn the list.

Remember, an independent-measures design means that there are separate samples for each treatment condition.

FIGURE 10.3

The t distribution with $df = 18$. The critical region for $\alpha = .05$ is shown.

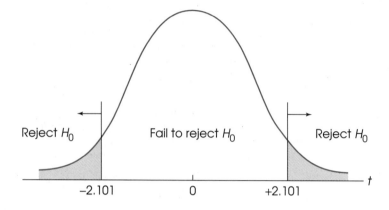

Reject H_0 Fail to reject H_0 Reject H_0

−2.101   0   +2.101 t

Later each group is given a memory test, and the psychologist records the number of words correctly recalled for each individual. The data from this experiment are as follows. On the basis of these data, can the psychologist conclude that mental images affected memory?

Data (number of words recalled)			
Group 1 (no images)		Group 2 (images)	
24	13	18	31
23	17	19	29
16	20	23	26
17	15	29	21
19	26	30	24
$n = 10$		$n = 10$	
$\overline{X} = 19$		$\overline{X} = 25$	
$SS = 160$		$SS = 200$	

STEP 1 State the hypotheses, and select the alpha level.

Directional hypotheses could be used and would specify whether imagery should increase or decrease recall scores.

$H_0: \mu_1 - \mu_2 = 0$ (no difference; imagery has no effect)

$H_1: \mu_1 - \mu_2 \neq 0$ (imagery produces a difference)

We will set $\alpha = .05$.

STEP 2 This is an independent-measures design. The t statistic for these data will have degrees of freedom determined by

$$df = df_1 + df_2$$
$$= (n_1 - 1) + (n_2 - 1)$$
$$= 9 + 9$$
$$= 18$$

The t distribution for $df = 18$ is presented in Figure 10.3. For $\alpha = .05$, the critical region consists of the extreme 5% of the distribution and has boundaries of $t = +2.101$ and $t = -2.101$.

STEP 3   Obtain the data, and compute the test statistic. The data are as given, so all that remains is to compute the *t* statistic. Because the independent-measures *t* formula is relatively complex, the calculations can be simplified by dividing the process into three parts.

First, find the pooled variance for the two samples:

Caution: The pooled variance combines the two samples to obtain a single estimate of variance. In the formula, the two samples are combined in a single fraction.

$$s_p^2 = \frac{SS_1 + SS_2}{df_1 + df_2}$$

$$= \frac{160 + 200}{9 + 9}$$

$$= \frac{360}{18}$$

$$= 20$$

Second, use the pooled variance to compute the estimated standard error:

Caution: The standard error adds the errors from two separate samples. In the formula, these two errors are added as two separate fractions. In this case, the two errors are equal because the sample sizes are the same.

$$s_{\overline{X}_1 - \overline{X}_2} = \sqrt{\frac{s_p^2}{n_1} + \frac{s_p^2}{n_2}}$$

$$= \sqrt{\frac{20}{10} + \frac{20}{10}}$$

$$= \sqrt{2 + 2}$$

$$= \sqrt{4}$$

$$= 2$$

Third, compute the *t* statistic:

$$t = \frac{(\overline{X}_1 - \overline{X}_2) - (\mu_1 - \mu_2)}{s_{\overline{X}_1 - \overline{X}_2}} = \frac{(19 - 25) - 0}{2}$$

$$= \frac{-6}{2}$$

$$= -3.00$$

STEP 4   Make a decision. The obtained value ($t = -3.00$) is in the critical region. In this example, the obtained sample mean difference is three times greater than would be expected by chance (the standard error). This result is very unlikely if H_0 is true. Therefore, we reject H_0 and conclude that using mental images produced a significant difference in memory performance. More specifically, the group using images recalled significantly more words than the group with no images.

IN THE LITERATURE:
REPORTING THE RESULTS OF AN INDEPENDENT-MEASURES *t* TEST

In Chapter 4 (page 105), we demonstrated how the mean and the standard deviation are reported in APA format. In Chapter 9 (page 230), we illustrated the

APA style for reporting the results of a *t* test. Now we will use the APA format to report the results of Example 10.2, an independent-measures *t* test. A concise statement might read as follows:

> The group using mental images recalled more words ($M = 25$, $SD = 4.22$) than the group that did not use mental images ($M = 19$, $SD = 4.71$). This difference was significant, $t(18) = -3.00$, $p < .05$, two-tailed.

You should note that standard deviation is not a step in the computations for the independent-measures *t* test. Yet it is useful when providing descriptive statistics for each treatment group. It is easily computed when doing the *t* test because you need *SS* and *df* for both groups to determine the pooled variance. Note that the format for reporting *t* is exactly the same as that described in Chapter 9.   ❏

THE *t* STATISTIC AS A RATIO

You should note that the magnitude of the *t* statistic is determined not only by the mean difference between the two samples, but also by the sample variability. The bigger the difference between the sample means (the numerator of the *t* formula), the bigger the *t* value. This relationship is reasonable because a big difference between the samples is a clear indication of a difference between the two populations. However, the sample variability is just as important. The variability of the samples, in the form of SS_1 and SS_2, is a major component of the standard error in the denominator of the *t* statistic. If the variability is large, *t* will tend to be small. The role of variability becomes clearer if you consider a simplified version of the *t* formula:

$$t = \frac{\text{sample mean difference}}{\text{variability}}$$

Notice that we have left out the population mean difference because the null hypothesis says that this is zero. Also, we have used the general term *variability* in place of standard error. In this simplified form, *t* becomes a ratio involving only the sample mean difference and the sample variability. It should be clear that large variability will make the *t* value smaller (and small variability will make *t* larger). Thus, variability plays an important role in determining whether or not a *t* statistic is significant. When variability is large, even a big difference between the two sample means may not be enough to make the *t* statistic significant. On the other hand, when variability is low, even a small difference between the two sample means may be enough to produce a significant *t* statistic.

DIRECTIONAL HYPOTHESES AND ONE-TAILED TESTS

When planning an independent-measures experiment, a researcher usually has some expectation or specific prediction for the outcome. For the memory experiment described in Example 10.1, the psychologist clearly expects the group using images to have higher memory scores than the group without images. This kind of directional prediction can be incorporated into the statement of the hypotheses, resulting in a directional, or one-tailed, test. You should recall from Chapter 8 that one-tailed tests are used only in limited situations, where a researcher wants to increase the chances of finding a significant difference. In general, a nondirectional (two-tailed) test is preferred, even in research situations where there is a definite directional prediction. The following example demonstrates the procedure for stating hypotheses and locating the critical region for a one-tailed test using the independent-measures *t* statistic.

EXAMPLE 10.2   We will use the same experimental situation that was described in Example 10.1. The researcher is using an independent-measures design to examine the effect of mental images on memory. The prediction is that the imagery group will have higher memory scores.

STEP 1   State the hypotheses, and select the alpha level. As always, the null hypothesis says that there is no effect, and the alternative hypothesis says that there is an effect. For this example, the predicted effect is that images will produce higher scores. Thus, the two hypotheses are

$$H_0: \mu_{images} \leq \mu_{no\ images} \quad \text{(imagery scores are not higher)}$$

$$H_1: \mu_{images} > \mu_{no\ images} \quad \text{(imagery scores are higher)}$$

Notice that it is usually easier to begin with the alternative hypothesis (H_1), which states that the treatment works as predicted. Then the null hypothesis is just the opposite of H_1. Also notice that the equal sign goes in the null hypothesis, indicating *no difference* between the two treatment conditions. The idea of zero difference is the essence of the null hypothesis, and the numerical value of zero will be used for $(\mu_1 - \mu_2)$ during the calculation of the *t* statistic.

STEP 2   Locate the critical region. For a directional test, the critical region will be located entirely in one tail of the distribution. Rather than trying to determine which tail, positive or negative, is the correct location, we suggest you identify the criteria for the critical region in a two-step process as follows. First, look at the data, and determine whether or not the sample mean difference is in the direction that was predicted. If the answer is no, then the data obviously do not support the predicted treatment effect, and you can stop the analysis. On the other hand, if the difference is in the predicted direction, then the second step is to determine whether the difference is large enough to be significant. To test for significance, simply find the one-tailed critical value in the *t*-distribution table. If the sample *t* statistic is more extreme (either positive or negative) than the critical value, then the difference is significant.

For this example, the imagery group scored 6 points higher, as predicted. With $df = 18$, the one-tailed critical value for $\alpha = .05$ is $t = 1.734$.

STEP 3   Collect the data, and calculate the test statistic. The details of the calculations were shown in Example 10.1. The data produce a *t* statistic of $t = -3.00$.

STEP 4   Make a decision. The *t* statistic of $t = 3.00$ (ignoring the sign) is well beyond the critical boundary of $t = 1.734$. Therefore, we reject the null hypothesis and conclude that scores in the imagery condition are significantly higher than scores in the no-imagery condition, as predicted.

LEARNING CHECK   1. An independent-measures *t* statistic is used to evaluate the mean difference between two treatments using a sample of $n = 10$ in one treatment and a separate sample of $n = 15$ in the other treatment. What is the *df* value for this *t* test?

2. A researcher reports an independent-measures t statistic with $df = 20$. What is the total number of subjects who participated in the research study?

3. A developmental psychologist would like to examine the difference in mathematical skills for 10-year-old boys versus 10-year-old girls. A sample of 10 boys and 10 girls is obtained, and each child is given a standardized mathematical abilities test. The data from this experiment are as follows:

Boys	Girls
$\overline{X} = 37$	$\overline{X} = 31$
$SS = 150$	$SS = 210$

Do these data indicate a significant difference in mathematical skills for boys versus girls? Test at the .05 level of significance.

4. Suppose the psychologist in the previous question predicted that boys will have higher mathematics scores. What value of t would be necessary to conclude that boys score significantly higher, using a one-tailed test with $\alpha = .01$?

ANSWERS   1. The t statistic has $df = df_1 + df_2 = 9 + 14 = 23$.

2. There were 22 subjects in the whole experiment. When computing df, you lose 1 for the first sample $(n_1 - 1)$, and you lose another 1 for the second sample $(n_2 - 1)$. Thus, df is 2 points less than the total number of subjects.

3. Pooled variance = 20; estimated standard error = 2; $t = 3.00$. With $df = 18$, this value is in the critical region, so the decision is to reject H_0: There is a significant difference.

4. With $df = 18$ and $\alpha = .01$, the t value must be beyond 2.552 to be significant for a one-tailed test.

10.4 ASSUMPTIONS UNDERLYING THE INDEPENDENT-MEASURES t FORMULA

There are three assumptions that should be satisfied before you use the independent-measures t formula for hypothesis testing:

1. The observations within each sample must be independent (see page 202).
2. The two populations from which the samples are selected must be normal.
3. The two populations from which the samples are selected must have equal variances.

The first two assumptions should be familiar from the single-sample t hypothesis test presented in Chapter 9. As before, the normality assumption is the less important of the two, especially with large samples. When there is reason to suspect that the populations are far from normal, you should compensate by ensuring that the samples are relatively large.

The third assumption is referred to as *homogeneity of variance* and states that the two populations being compared must have the same variance. You may recall a

Remember, adding a constant to (or subtracting a constant from) each score does not change the standard deviation.

similar assumption for both the *z*-score and the single-sample *t*. For those tests, we assumed that the effect of the treatment was to add a constant amount to (or subtract a constant amount from) each individual score. As a result, the population standard deviation after treatment was the same as it had been before treatment. We now are making essentially the same assumption but phrasing it in terms of variances.

You should recall that the pooled variance in the *t*-statistic formula is obtained by averaging together the two sample variances. It makes sense to average these two values only if they both are estimating the same population variance—that is, if the homogeneity of variance assumption is satisfied. If the two sample variances represent different population variances, then the average would be meaningless. (*Note:* There is no meaning to the value obtained by averaging two unrelated numbers. For example, what is the significance of the number obtained by averaging your shoe size and the last two digits of your social security number?)

The importance of the homogeneity assumption increases when there is a large discrepancy between the sample sizes. With equal (or nearly equal) sample sizes, this assumption is less critical but still important.

The homogeneity of variance assumption is quite important because violating this assumption can negate any meaningful interpretation of the data from an independent-measures experiment. Specifically, when you compute the *t* statistic in a hypothesis test, all the numbers in the formula come from the data except for the population mean difference, which you get from H_0. Thus, you are sure of all the numbers in the formula except for one. If you obtain an extreme result for the *t* statistic (a value in the critical region), you conclude that the hypothesized value was wrong. But consider what happens when you violate the homogeneity of variance assumption. In this case, you have two questionable values in the formula (the hypothesized population value and the meaningless average of the two variances). Now if you obtain an extreme *t* statistic, you do not know which of these two values is responsible. Specifically, you cannot reject the hypothesis because it may have been the pooled variance that produced the extreme *t* statistic. Without satisfying the homogeneity of variance requirement, you cannot accurately interpret a *t* statistic, and the hypothesis test becomes meaningless.

How do you know whether or not the homogeneity of variance assumption is satisfied? One simple test involves just looking at the two sample variances. Logically, if the two population variances are equal, then the two sample variances should be very similar. When the two sample variances are reasonably close, you can be reasonably confident that the homogeneity assumption has been satisfied and proceed with the test. However, if one sample variance is more than three or four times larger than the other, then there is reason for concern. A more objective procedure involves a statistical test to evaluate the homogeneity assumption. Although there are many different statistical methods for determining whether or not the homogeneity of variance assumption has been satisfied, Hartley's *F*-max test is one of the simplest to compute and to understand. An additional advantage is that this test can also be used to check homogeneity of variance with more than two independent samples. Later, in Chapter 13, we will examine statistical methods for comparing several different samples, and Hartley's test will be useful again.

The following example demonstrates the *F*-max test for two independent samples.

EXAMPLE 10.3   The *F*-max test is based on the principle that a sample variance provides an unbiased estimate of the population variance. Therefore, if the population variances are the same, the sample variances should be very similar. The procedure for using the *F*-max test is as follows:

1. Compute the sample variance, $s^2 = SS/df$, for each of the separate samples.

2. Select the largest and the smallest of these sample variances and compute

$$F\text{-max} = \frac{s^2(\text{largest})}{s^2(\text{smallest})}$$

A relatively large value for F-max indicates a large difference between the sample variances. In this case, the data suggest that the population variances are different and that the homogeneity assumption has been violated. On the other hand, a small value of F-max (near 1.00) indicates that the sample variances are similar and that the homogeneity assumption is reasonable.

3. The F-max value computed for the sample data is compared with the critical value found in Table B.3 (Appendix B). If the sample value is larger than the table value, then you conclude that the variances are different and that the homogeneity assumption is not valid.

To locate the critical value in the table, you need to know

a. k = number of separate samples. (For the independent-measures t test, $k = 2$.)

b. $df = n - 1$ for each sample variance. The Hartley test assumes that all samples are the same size.

c. The alpha level. The table provides critical values for $\alpha = .05$ and $\alpha = .01$. Generally a test for homogeneity would use the larger alpha level.

Example: Two independent samples each have $n = 10$. The sample variances are 12.34 and 9.15. For these data,

$$F\text{-max} = \frac{s^2(\text{largest})}{s^2(\text{smallest})} = \frac{12.34}{9.15} = 1.35$$

With $\alpha = .05$, $k = 2$, and $df = n - 1 = 9$, the critical value from the table is 4.03. Because the obtained F-max is smaller than this critical value, you conclude that the data do not provide evidence that the homogeneity of variance assumption has been violated.

SUMMARY

1. The independent-measures t statistic is used to draw inferences about the mean difference between two populations or between two treatment conditions. The term *independent* is used because this t statistic requires data from two separate (or independent) samples.

2. The formula for the independent-measures t statistic has the same structure as the original z-score or the single-sample t:

$$t = \frac{\text{sample statistic} - \text{population parameter}}{\text{estimated standard error}}$$

For the independent-measures t, the sample statistic is the difference between the two sample means $(\overline{X}_1 - \overline{X}_2)$. The

population parameter of interest is the difference between the two population means $(\mu_1 - \mu_2)$. The standard error is computed by combining the errors for the two sample means. The resulting formula is

$$t = \frac{(\overline{X}_1 - \overline{X}_2) - (\mu_1 - \mu_2)}{s_{\overline{X}_1 - \overline{X}_2}}$$

$$= \frac{(\overline{X}_1 - \overline{X}_2) - (\mu_1 - \mu_2)}{\sqrt{\dfrac{s_p^2}{n_1} + \dfrac{s_p^2}{n_2}}}$$

The pooled variance in the formula, s_p^2, is the weighted mean of the two sample variances:

$$s_p^2 = \frac{SS_1 + SS_2}{df_1 + df_2}$$

This t statistic has degrees of freedom determined by the sum of the df values for the two samples:

$$df = df_1 + df_2$$
$$= (n_1 - 1) + (n_2 - 1)$$

3. For hypothesis testing, the formula has the following structure:

$$t = \frac{\text{sample statistic} - \text{hypothesized population parameter}}{\text{estimated standard error}}$$

The null hypothesis normally states that there is no difference between the two population means:

$$H_0: \mu_1 = \mu_2 \quad \text{or} \quad \mu_1 - \mu_2 = 0$$

4. Appropriate use and interpretation of the t statistic require that the data satisfy the homogeneity of variance assumption. This assumption stipulates that the two populations have equal variances. An informal test of this assumption can be made by simply comparing the two sample variances: If the two sample variances are approximately equal, the t test is justified. Hartley's F-max test provides a statistical technique for determining whether or not the data satisfy the homogeneity assumption.

KEY TERMS

independent-measures research design	between-subjects research design	pooled variance	homogeneity of variance

FOCUS ON PROBLEM SOLVING

1. As you learn more about different statistical methods, one basic problem will be deciding which method is appropriate for a particular set of data. Fortunately, it is easy to identify situations where the independent-measures t statistic is used. First, the data will always consist of two separate samples (two ns, two $\bar{X}$s, two SSs, and so on). Second, this t statistic always is used to answer questions about a mean difference: On average, is one group different (better, faster, smarter) than the other group? If you examine the data and identify the type of question that a researcher is asking, you should be able to decide whether or not an independent-measures t is appropriate.

2. When computing an independent-measures t statistic from sample data, we suggest that you routinely divide the formula into separate stages, rather than trying to do all the calculations at once. First, find the pooled variance. Second, compute the standard error. Third, compute the t statistic.

3. One of the most common errors for students involves confusing the formulas for pooled variance and standard error. When computing pooled variance, you are "pooling" the two samples together into a single variance. This variance is computed as a *single fraction*, with two SS values in the numerator and two df values in the denominator. When computing the standard error, you are adding the error from the first sample and the error from the second sample. These two separate errors add as *two separate fractions* under the square root symbol.

DEMONSTRATION 10.1

THE INDEPENDENT-MEASURES t TEST

In a study of jury behavior, two samples of subjects were provided details about a trial in which the defendant was obviously guilty. Although group 2 received the same details as group 1, the second group was also told that some evidence had been withheld from the jury by the judge. Later the subjects were asked to recommend a jail sentence. The length of term suggested by each subject is presented here. Is there a significant difference between the two groups in their responses?

Group 1 scores: 4,   4,   3,   2,   5,   1,   1,   4

Group 2 scores: 3,   7,   8,   5,   4,   7,   6,   8

There are two separate samples in this study. Therefore, the analysis will use the independent-measures t test.

STEP 1 State the hypotheses, and select an alpha level.

H_0: $\mu_1 - \mu_2 = 0$ (For the population, knowing evidence has been withheld has no effect on the suggested sentence.)

H_1: $\mu_1 - \mu_2 \neq 0$ (For the population, knowledge of withheld evidence has an effect on the jury's response.)

We will set the level of significance to $\alpha = .05$, two tails.

STEP 2 Identify the critical region.
For the independent-measures t statistic, degrees of freedom are determined by

$$df = n_1 + n_2 - 2$$
$$= 8 + 8 - 2$$
$$= 14$$

The t-distribution table is consulted for a two-tailed test with $\alpha = .05$ and $df = 14$. The critical t values are $+2.145$ and -2.145.

STEP 3 Compute the test statistic.
We are computing an independent-measures t statistic. To do this, we will need the mean and SS for each sample, pooled variance, and estimated standard error.

Sample means and sums of squares. The means ($\overline{X}$) and sums of squares (SS) for the samples are computed as follows:

Sample 1		Sample 2	
X	X^2	X	X^2
4	16	3	9
4	16	7	49
3	9	8	64
2	4	5	25
5	25	4	16
1	1	7	49
1	1	6	36
4	16	8	64
$\Sigma X = 24$	$\Sigma X^2 = 88$	$\Sigma X = 48$	$\Sigma X^2 = 312$

$$n_1 = 8 \qquad\qquad n_2 = 8$$

$$\bar{X}_1 = \frac{\Sigma X}{n} = \frac{24}{8} = 3 \qquad \bar{X}_2 = \frac{\Sigma X}{n} = \frac{48}{8} = 6$$

$$SS_1 = \Sigma X^2 - \frac{(\Sigma X)^2}{n} \qquad SS_2 = \Sigma X^2 - \frac{(\Sigma X)^2}{n}$$

$$= 88 - \frac{(24)^2}{8} \qquad\qquad = 312 - \frac{(48)^2}{8}$$

$$= 88 - \frac{576}{8} \qquad\qquad = 312 - \frac{2304}{8}$$

$$= 88 - 72 \qquad\qquad = 312 - 288$$

$$SS_1 = 16 \qquad\qquad SS_2 = 24$$

Pooled variance. For these data, the pooled variance equals

$$s_p^2 = \frac{SS_1 + SS_2}{df_1 + df_2} = \frac{16 + 24}{7 + 7} = \frac{40}{14} = 2.86$$

Estimated standard error. Now we can calculate the estimated standard error for mean differences.

$$s_{\bar{X}_1 - \bar{X}_2} = \sqrt{\frac{s_p^2}{n_1} + \frac{s_p^2}{n_2}} = \sqrt{\frac{2.86}{8} + \frac{2.86}{8}} = \sqrt{0.358 + 0.358}$$

$$= \sqrt{0.716} = 0.85$$

The t statistic. Finally, the t statistic can be computed.

$$t = \frac{(\bar{X}_1 - \bar{X}_2) - (\mu_1 - \mu_2)}{s_{\bar{X}_1 - \bar{X}_2}} = \frac{(3 - 6) - 0}{0.85} = \frac{-3}{0.85}$$

$$= -3.53$$

STEP 4 Make a decision about H_0, and state a conclusion.
The obtained t value of -3.53 falls in the critical region of the left tail (critical $t = \pm2.145$). Therefore, the null hypothesis is rejected. The subjects that were informed about the withheld evidence gave significantly longer sentences, $t(14) = -3.53$, $p < .05$, two tails.

PROBLEMS

1. Describe the general characteristics of a research study for which an independent-measures t would be the appropriate test statistic.

2. What is measured by the estimated standard error that is used for the independent-measures t statistic?

3. What happens to the value of the independent-measures t statistic as the difference between the two sample means increases? What happens to the t value as the variability of the scores in the two samples increases?

4. One sample has $n = 8$ scores with $SS = 56$, and a second sample has $n = 10$ scores with $SS = 108$.
   a. Find the variance for each sample.
   b. Find the pooled variance for the two samples.

5. One sample has $SS = 63$, and a second sample has $SS = 45$.
   a. Assuming that $n = 10$ for both samples, find each of the sample variances, and calculate the pooled variance. You should find that the pooled variance is exactly halfway between the two sample variances.

b. Now assume that $n = 10$ for the first sample and $n = 16$ for the second. Find each of the sample variances, and calculate the pooled variance. You should find that the pooled variance is closer to the variance for the larger sample ($n = 16$).

6. One sample has $n = 15$ with $SS = 1660$, and a second sample has $n = 15$ with $SS = 1700$.
   a. Find the pooled variance for the two samples.
   b. Find the standard error for the sample mean difference.
   c. If the sample mean difference is 8 points, is this enough to reject the null hypothesis and conclude that there is a significant difference at the .05 level?
   d. If the sample mean difference is 12 points, is this enough to indicate a significant difference at the .05 level?

7. A person's gender can have a tremendous influence on his/her personality and behavior. Psychologists classify individuals as masculine, feminine, or androgynous. Androgynous individuals possess both masculine and feminine traits. Among other things, androgynous individuals appear to cope better with stress than do traditionally masculine or feminine people. In a typical study, depression scores are recorded for a sample of traditionally masculine or feminine subjects and for a sample of androgynous subjects, all of whom have recently experienced a series of strongly negative events. The average depression score for the sample of $n = 10$ androgynous subjects is $\overline{X} = 63$ with $SS = 700$. The sample of $n = 10$ traditional sex-typed subjects averaged $\overline{X} = 71$ with $SS = 740$. Do these data indicate that the traditional subjects experienced significantly more depression than the androgynous subjects? Use a one-tailed test with $\alpha = .05$.

8. A researcher would like to compare the political attitudes for college freshmen with those for college seniors. Samples of $n = 10$ freshmen and $n = 10$ seniors are obtained, and each student is given a questionnaire measuring political attitude on a scale from 0 (very conservative) to 100 (very liberal). The average score for the freshmen is $\overline{X} = 52$ with $SS = 4800$, and the seniors average $\overline{X} = 39$ with $SS = 4200$. Do these data indicate a significant difference in political attitude for freshmen versus seniors? Test at the .05 level of significance.

9. A psychologist would like to examine the effects of fatigue on mental alertness. An attention test is prepared that requires subjects to sit in front of a blank TV screen and press a response button each time a dot appears on the screen. A total of 110 dots are presented during a 90-minute period, and the psychologist records the number of errors for each subject. Two groups of subjects are selected. The first group ($n = 5$) is tested after they have been kept awake for 24 hours. The second group ($n = 10$) is tested in the morning after a full night's sleep. The data for these two samples are as follows:

Awake 24 hours	Rested
$\overline{X} = 35$	$\overline{X} = 24$
$SS = 120$	$SS = 270$

On the basis of these data, can the psychologist conclude that fatigue significantly increases errors on an attention task? Use a one-tailed test with $\alpha = .05$.

10. Anagrams are words that have had their letters scrambled into a different order. The task is to unscramble the letters and determine the correct word. One factor that influences performance on this task is the appearance of the scrambled letters. For example, the word SHORE could be presented as OSHER (a pronounceable form) or as HRSOE (an unpronounceable, random-looking sequence). Typically, the pronounceable form is more difficult, probably because it already looks "good," which makes it difficult for people to change the letters around. A researcher studying this phenomenon presented one group of $n = 8$ subjects with a series of pronounceable anagrams and a second group of $n = 8$ with unpronounceable versions of the same anagrams. The amount of time to solve the set was recorded for each subject. For the pronounceable condition, the mean time was $\overline{X} = 48$ seconds with $SS = 3100$, and for the unpronounceable condition, the mean was $\overline{X} = 27$ seconds with $SS = 2500$. Is there a significant difference between the two conditions? Test with $\alpha = .01$.

11. In a study examining the power of imagination, Cervone (1989) asked subjects to think about factors that could influence performance on solving mazelike puzzles. In one condition, subjects imagined positive factors that could make the task easier, and in another condition, subjects imagined negative factors that would make the task more difficult. Later Cervone measured task persistence: how long the subjects continued working on the task, or how quickly they gave up. Hypothetical data, similar to Cervone's, are as follows:

   Positive condition: $n = 12$, $\overline{X} = 28$ with $SS = 280$

   Negative condition: $n = 12$, $\overline{X} = 22$ with $SS = 248$

Do these data support the conclusion that imagination can have a significant effect on task persistence? Test at the .05 level of significance.

12. Extensive data indicate that first-born children develop different characteristics than later-born children. For example, first-borns tend to be more responsible, more hard-working, higher achieving, and more self-disciplined than their later-born siblings. The following data represent scores on a test measuring self-esteem and pride. Samples of $n = 10$ first-born college freshmen and

$n = 20$ later-born freshmen were each given the self-esteem test. The first-borns averaged $\overline{X} = 48$ with $SS = 670$, and the later-borns averaged $\overline{X} = 41$ with $SS = 1010$. Do these data indicate a significant difference between the two groups? Test at the .01 level of significance.

13. A researcher reports an independent-measures t statistic of $t = 2.53$ with $df = 24$.
   a. How many subjects participated in the researcher's experiment?
   b. Based on the reported t value, can the researcher conclude that there is a significant difference between the two samples with $\alpha = .05$?
   c. Based on the value of t, can the researcher conclude that the mean difference is significant at the .01 level?

14. The following data are from two separate independent-measures experiments. Without doing any calculation, which experiment is more likely to demonstrate a significant difference between treatments A and B? Explain your answer. (*Note:* You do not need to compute the t statistics; just look carefully at the data.)

Experiment I		Experiment II	
Treatment A	Treatment B	Treatment A	Treatment B
$n = 10$	$n = 10$	$n = 10$	$n = 10$
$\overline{X} = 42$	$\overline{X} = 52$	$\overline{X} = 61$	$\overline{X} = 71$
$SS = 180$	$SS = 120$	$SS = 986$	$SS = 1042$

15. A researcher would like to compare the effectiveness of video versus reading for communicating information to adolescents. A sample of $n = 10$ adolescents is given a pamphlet explaining nuclear energy and told that they will be tested on the information. A separate sample of $n = 10$ adolescents is shown a video that contains the same information presented in a fast-paced visual form and told that they will be tested. One week later all 20 subjects are given a test on nuclear energy. The average score for the pamphlet group was $\overline{X} = 49$ with $SS = 160$. The video group averaged $\overline{X} = 46$ with $SS = 200$. Do these data indicate a significant difference between the two modes of presentation? Perform the appropriate test using $\alpha = .05$.

16. Friedman and Rosenman (1983) have classified people into two categories: Type A personalities and Type B personalities. Type As are hard-driving, competitive, and ambitious. Type Bs are more relaxed, easy-going people. One factor that differentiates these two groups is the chronically high level of frustration experienced by Type As. To demonstrate this phenomenon, separate samples of Type As and Type Bs are obtained, with

$n = 8$ in each sample. The individual subjects are all given a frustration inventory measuring level of frustration. The average score for the Type As is $\overline{X} = 84$ with $SS = 740$, and the Type Bs average $\overline{X} = 71$ with $SS = 660$. Do these data indicate a significant difference between the two groups? Test at the .01 level of significance.

17. In a classic study of problem solving, Duncker (1945) asked subjects to mount a candle on a wall in an upright position so that it would burn normally. One group of subjects was given a candle, a book of matches, and a box of tacks. A second group was given the same items, except the tacks and the box were presented separately as two distinct items. The solution to this problem involves using the tacks to mount the box on the wall, which creates a shelf for the candle. Duncker reasoned that the first group of subjects would have trouble seeing a "new" function for the box (a shelf) because it was already serving a function (holding tacks). For each subject, the amount of time to solve the problem was recorded. Data similar to Duncker's are as follows:

Time to solve problem (in sec.)	
Box of tacks	Tacks & box separate
128	42
160	24
53	68
101	35
94	47

Do these data indicate a significant difference between the two conditions? Test at the .01 level of significance.

18. A psychologist studying human memory would like to examine the process of forgetting. One group of subjects is required to memorize a list of words in the evening just before going to bed. Their recall is tested 10 hours later in the morning. Subjects in the second group memorize the same list of words in the morning, and then their memories are tested 10 hours later after being awake all day. The psychologist hypothesizes that there will be less forgetting during sleep than during a busy day. The recall scores for two samples of college students are as follows:

Asleep scores				Awake scores			
15	13	14	14	15	13	14	12
16	15	16	15	14	13	11	12
16	15	17	14	13	13	12	14

a. Sketch a frequency distribution polygon for the "asleep" group. On the same graph (in a different color), sketch the distribution for the "awake" group. Just by looking at these two distributions, would you predict a significant difference between the two treatment conditions?

b. Use the independent-measures t statistic to determine whether there is a significant difference between the treatments. Conduct the test with $\alpha = .05$.

19. Siegel (1990) found that elderly people who owned dogs were less likely to pay visits to their doctors after upsetting events than were those who did not own pets. Similarly, consider the following hypothetical data. A sample of elderly dog owners is compared to a similar group (in terms of age and health) who do not own dogs. The researcher records the number of visits to the doctor during the past year for each person. For the following data, is there a significant difference in the number of doctor visits between dog owners and control subjects? Use the .05 level of significance.

Control group	Dog owners
12	8
10	5
6	9
9	4
15	6
12	
14	

20. A researcher examines the short-term effect of a single treatment with acupuncture on cigarette smoking. Subjects who smoke approximately a pack of cigarettes a day are assigned to one of two groups. One group receives the treatment with the acupuncture needles. The second group of subjects serves as a control group and is given relaxation instruction while on the examining table but no acupuncture treatment. The number of cigarettes smoked the next day is recorded for all subjects. For the following data, determine if there is a significant effect of acupuncture on cigarette consumption. Use an alpha level of .05.

Control				Acupuncture			
23	20	24	17	17	9	0	12
30	18	10	22	3	24	14	10

21. In a study examining the effects of environment on development, Krech and his colleagues (1962) divided a sample of infant rats into two groups. One group was housed in a *stimulus-rich* environment containing ladders, platforms, tunnels, and colorful decorations. The second group was housed in *stimulus-poor* conditions consisting of plain gray cages. At maturity, maze-learning performance was measured for all the rats. The following hypothetical data simulate Krech's results. The scores are the number of errors committed by each rat before it successfully solved the maze:

Rich rats: 18, 24, 27, 23, 31, 29, 20, 33, 25, 30

Poor rats: 37, 27, 26, 31, 35, 43, 40, 36, 28, 39

Do these data indicate a significant difference between the two groups? Test at the .01 level of significance.

22. A personality *trait* is defined as a long-lasting, relatively stable characteristic of an individual's personality. Traits such as honesty, optimism, sensitivity, and seriousness help define your personality. Cattell (1973) identified 16 basic traits that serve to differentiate individuals. One of these traits, relaxed/tense, has been shown to be a consistent factor differentiating creative artists from airline pilots. The following data, similar to Cattell's, represent relaxed/tense scores for two groups of subjects. Lower scores indicate a more relaxed personality.

Artists: 7, 7, 6, 9, 8, 8, 7, 9, 5, 3, 6, 8, 7, 9

Pilots: 4, 2, 2, 3, 1, 5, 4, 3, 2, 2, 6, 2, 5, 3

Do these data indicate a significant difference between the two professions? Test at the .05 level of significance.

CHAPTER 11

HYPOTHESIS TESTS WITH TWO RELATED SAMPLES

TOOLS YOU WILL NEED

The following items are considered essential background material for this chapter. If you doubt your knowledge of any of these items, you should review the appropriate chapter or section before proceeding.

- Introduction to the *t* statistic (Chapter 9)
  - Estimated standard error
  - Degrees of freedom
  - *t* distribution
  - Hypothesis tests with the *t* statistic
- Independent-measures design (Chapter 10)

CONTENTS

11.1   INTRODUCTION TO RELATED SAMPLES

In the previous chapter, we introduced the independent-measures research design as one strategy for comparing two treatment conditions or two populations. The independent-measures design is characterized by the fact that two separate samples are used to obtain the two sets of data that are to be compared. In this chapter, we will examine an alternative research strategy known as a *repeated-measures* research design. With a repeated-measures design, two sets of data are obtained from the same sample of individuals. For example, a group of patients could be measured before therapy and then measured again after therapy. Or a group of individuals could be tested in a quiet, comfortable environment and then tested again in a noisy, stressful environment to examine the effects of stress on performance. In each case, notice that the same variable is being measured twice for the same set of individuals; that is, we are literally repeating measurements on the same sample.

DEFINITION   A *repeated-measures* study is one in which a single sample of individuals is measured more than once on the same dependent variable. The same subjects are used in all of the treatment conditions.

The main advantage of a repeated-measures study is that it uses exactly the same subjects in all treatment conditions. Thus, there is no risk that the subjects in one treatment are substantially different from the subjects in another. With an independent-measures design, on the other hand, there is always a risk that the results are biased because the individuals in one sample are much different (smarter, faster, more extroverted, and so on) than the individuals in the other sample.

Occasionally, researchers will try to approximate the advantages of a repeated-measures study by using a technique known as *matched subjects*. A matched-subjects design involves two separate samples, but each individual in one sample is matched one-to-one with an individual in the second sample. The matching is based on a variable that the researcher identifies as being especially relevant to the particular study. Suppose, for example, a researcher plans to compare learning performance in two different treatment conditions. For this study, the researcher might want to match the subjects in terms of intelligence. If there is a subject with an IQ of 110 in one sample, the researcher would find a matched subject with an IQ of 110 for the second sample. Thus, the two samples are perfectly matched in terms of intelligence.

DEFINITION   In a *matched-subjects* study, each individual in one sample is matched with a subject in the other sample. The matching is done so that the two individuals are equivalent (or nearly equivalent) with respect to a specific variable that the researcher would like to control.

Of course, it is possible to match subjects on more than one variable. For example, a researcher could match pairs of subjects on age, gender, race, and IQ. In this case, for example, a 22-year-old white female with an IQ of 115 who was in one sample would be matched with another 22-year-old white female with an IQ of 115 in the second sample. The more variables that are used, however, the more difficult it is to find matching subjects. The goal of the matching process is to simulate a repeated-

TABLE 11.1

Scores on a depression inventory before and after treatment

Person	Before treatment (X_1)	After treatment (X_2)	D
A	72	64	−8
B	68	60	−8
C	60	50	−10
D	71	66	−5
E	55	56	+1

$$\Sigma D = -30$$

$$\bar{D} = \frac{\Sigma D}{n} = \frac{-30}{5} = -6$$

measures design as closely as possible. In a repeated-measures design, the matching is perfect because the same individual is used in both conditions. In a matched-subjects design, however, the best you can get is a degree of match that is limited to the variable(s) that are used for the matching process.

In a repeated-measures design, or a matched-subjects design, the data consist of two sets of scores (two samples) with the scores in one sample directly related, one-to-one, with the scores in the second sample. For this reason, the two research designs are statistically equivalent and are grouped together under the common name *related-samples* designs (or correlated-samples designs). In this chapter, we will focus our discussion on repeated-measures designs because they are overwhelmingly the more common example of related-groups designs. However, you should realize that the statistical techniques used for repeated-measures studies can be applied directly to data from a matched-subjects study.

Now we will examine the statistical techniques that allow a researcher to use the sample data from a repeated-measures study to draw inferences about the general population.

A matched-subjects study occasionally is called a *matched-samples design.* But the subjects in the samples must be matched one-to-one before you can use the statistical techniques in this chapter.

DIFFERENCE SCORES: THE DATA FOR A RELATED-SAMPLES STUDY

Table 11.1 presents hypothetical data for a drug evaluation study. The first score for each person (X_1) is the score obtained on the depression inventory before the drug treatment. The second score (X_2) was obtained after the drug treatment. Because we are interested in how much change occurs as a result of the treatment, each person's scores are summarized as a single difference score. This is accomplished by subtracting the first score (before treatment) from the second score (after treatment) for each person:

$$\text{difference score} = D = X_2 - X_1 \tag{11.1}$$

In a matched-subjects design, D is the difference between scores for two matched subjects.

The difference scores, or D values, are shown in the last column of the table. Note that the sign of each D score tells you the direction of change. For example, person A showed a decrease in depression, as indicated by the negative difference score.

THE HYPOTHESES FOR A RELATED-SAMPLES TEST

The researcher's goal is to use the sample of difference scores to answer questions about the general population. Notice that we are interested in a population of *difference scores.* More specifically, we are interested in the mean for this population of

11.1 | ANALOGIES FOR H_0 AND H_1 IN THE REPEATED-MEASURES TEST

An Analogy for H_0: Intelligence is a fairly stable characteristic; that is, you do not get noticeably smarter or dumber from one day to the next. However, if we gave you an IQ test every day for a week, we probably would get seven different numbers. The day-to-day changes in your IQ scores would probably be small and would be random. Some days your IQ score is slightly higher, and some days it is slightly lower. On average, the day-to-day changes in IQ should balance out to zero. This is the situation that is predicted by the null hypothesis for a repeated-measures test. According to H_0, any changes that occur either for an individual or for a sample are just due to chance, and in the long run, they will average out to zero.

An Analogy for H_1: On the other hand, suppose we measure the strength of the muscles in your right arm as you begin a fitness training program. We will measure your grip strength every day over a four-week period. We probably will find small differences in your scores from one day to the next, just as we did with the IQ scores. However, the day-to-day changes in grip strength will not be random. Although your grip strength may decrease occasionally, there should be a general trend toward increased strength as you go through the training program. Thus, most of the day-to-day changes should show an increase. This is the situation that is predicted by the alternative hypothesis for the repeated-measures test. According to H_1, the changes that occur are systematic and predictable and will not average out to zero.

difference scores. We will identify this population mean difference with the symbol μ_D (using the subscript letter D to indicate that we are dealing with D values, rather than X scores).

As always, the null hypothesis states that for the general population there is no effect, no change, or no difference. For a repeated-measures study, the null hypothesis states that the population mean difference is zero. In symbols,

$$H_0: \mu_D = 0$$

Again, this hypothesis refers to the mean for the entire population of difference scores. According to this hypothesis, it is possible that some individuals will show positive difference scores and some will show negative scores, but the differences are random and unsystematic and will balance out to zero.

The alternative hypothesis states that there is a treatment effect that causes the scores in one treatment condition to be systematically higher (or lower) than the scores in the other condition. In symbols,

$$H_1: \mu_D \neq 0$$

According to H_1, the difference scores for the individuals in the population tend to be consistently positive (or negative), indicating a consistent, predictable difference between the two treatments. See Box 11.1 for further discussion of H_0 and H_1.

THE *t* STATISTIC FOR RELATED SAMPLES

Figure 11.1 shows the general situation that exists for a repeated-measures hypothesis test. You may recognize that we are facing essentially the same situation that we encountered in Chapter 9. In particular, we have a population where the mean and the standard deviation are unknown, and we have a sample that will be used to test a hypothesis about the unknown population. In Chapter 9, we introduced a *t* statistic that allowed us to use the sample mean as a basis for testing hypotheses about the

FIGURE 11.1

Because populations are usually too large to test in a study, the researcher selects a random sample from the population. The sample data are used to make inferences about the population mean, μ_D. The data of interest in the repeated-measures study are difference scores (D scores) for each subject.

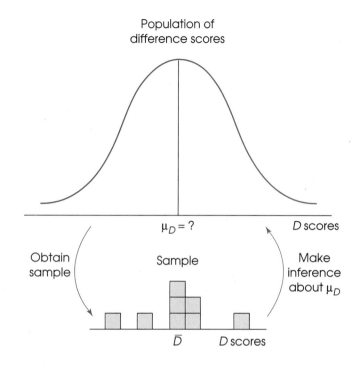

population mean. This *t*-statistic formula will be used again here to develop the repeated-measures *t* test. To refresh your memory, the single-sample *t* statistic (Chapter 9) is defined by the formula

$$t = \frac{\overline{X} - \mu}{s_{\overline{X}}}$$

In this formula, the sample mean, $\overline{X}$, is calculated from the data, and the value for the population mean, μ, is obtained from the null hypothesis. The estimated standard error, $s_{\overline{X}}$, is also calculated from the data and provides a measure of how much difference it is reasonable to expect between the sample mean and the population mean.

For the repeated-measures design, the sample data are difference scores and are identified by the letter D, rather than X. Therefore, we will substitute Ds in the formula in place of Xs to emphasize that we are dealing with difference scores instead of X values. Also, the population mean that is of interest to us is the population mean difference (the mean amount of change for the entire population), and we identify this parameter with the symbol μ_D. With these simple changes, the *t* formula for the repeated-measures design becomes

$$t = \frac{\overline{D} - \mu_D}{s_{\overline{D}}} \qquad (11.2)$$

In this formula, the estimated standard error, $s_{\overline{D}}$, is computed in exactly the same way as $s_{\overline{X}}$ is computed for the single-sample *t* statistic. The change in notation simply indicates that the sample mean is computed for a sample of D scores instead of a sample of X scores. To calculate the estimated standard error, the first step is to compute the variance (or the standard deviation) for the sample of D scores.

$$s^2 = \frac{SS}{n-1} = \frac{SS}{df} \qquad or \qquad s = \sqrt{\frac{SS}{df}}$$

The estimated standard error is then computed using the sample variance (or sample standard deviation) and the sample size, n.

$$s_{\overline{D}} = \sqrt{\frac{s^2}{n}} \qquad or \qquad s_{\overline{D}} = \frac{s}{\sqrt{n}}$$

You should notice that all of the calculations are done using the difference scores (the D scores) and that there is only one D score for each subject. With a sample of n subjects, there will be exactly n D scores, and the t statistic will have $df = n - 1$. Remember that n refers to the number of D scores, not the number of X scores in the original data.

LEARNING CHECK

1. What characteristics differentiate a repeated-measures study from an independent-measures research study?

2. How are the difference scores (D values) obtained for a repeated-measures hypothesis test?

3. Stated in words and stated in symbols, what is the null hypothesis for a repeated-measures hypothesis test?

ANSWERS

1. A repeated-measures study obtains two sets of data from one sample by measuring each individual twice. An independent-measures study uses two separate samples to obtain two sets of data.

2. The difference score for each subject is obtained by subtracting the subject's first score from the second score. In symbols, $D = X_2 - X_1$.

3. The null hypothesis states that the mean difference for the entire population is zero. That is, if every individual is measured in both treatments and a difference score is computed for each individual, then the average difference score for the entire population will be equal to zero. In symbols, $\mu_D = 0$.

11.2 HYPOTHESIS TESTS FOR THE REPEATED-MEASURES DESIGN

The hypothesis test with the repeated-measures t statistic follows the same four-step process that we have used for other tests. The complete hypothesis-testing procedure is demonstrated in Example 11.1.

EXAMPLE 11.1   A researcher in behavioral medicine believes that stress often makes asthma symptoms worse for people who suffer from this respiratory disorder. Because of

TABLE 11.2

The number of doses of medication needed for asthma attacks before and after relaxation training

Patient	Week before training	Week after training	D	D^2
A	9	4	−5	25
B	4	1	−3	9
C	5	5	0	0
D	4	0	−4	16
E	5	1	−4	16
			$\Sigma D = -16$	$\Sigma D^2 = 66$

Because the data consist of difference scores (D values) instead of X scores, the formulas for the mean and SS also use D in place of X. However, you should recognize the computational formula for SS that was introduced in Chapter 4.

$$\bar{D} = \frac{\Sigma D}{n} = \frac{-16}{5} = -3.2$$

$$SS = \Sigma D^2 - \frac{(\Sigma D)^2}{n} = 66 - \frac{(-16)^2}{5}$$

$$= 66 - 51.2 = 14.8$$

the suspected role of stress, the investigator decides to examine the effect of relaxation training on the severity of asthma symptoms. A sample of 5 patients is selected for the study. During the week before treatment, the investigator records the severity of their symptoms by measuring how many doses of medication are needed for asthma attacks. Then the patients receive relaxation training. For the week following training, the researcher once again records the number of doses required by each patient. Table 11.2 shows the data and summarizes the findings. Do these data indicate that relaxation training alters the severity of symptoms?

STEP 1   State the hypotheses, and select the alpha level.

H_0: $\mu_D = 0$ (no change in symptoms)

H_1: $\mu_D \neq 0$ (there is a change)

The level of significance is set at $\alpha = .05$ for a two-tailed test.

STEP 2   Locate the critical region. For this example, $n = 5$, so the t statistic will have $df = n - 1 = 4$. From the t-distribution table, you should find that the critical values are $+2.776$ and -2.776. These values are shown in Figure 11.2.

FIGURE 11.2

The critical regions with $\alpha = .05$ and $df = 4$ begin at $+2.776$ and -2.776 in the t distribution. Obtained values of t that are more extreme than these values will lie in a critical region. In that case, the null hypothesis would be rejected.

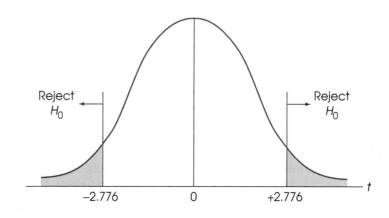

STEP 3   Calculate the t statistic. Table 11.2 shows the sample data and the calculations for $\bar{D} = -3.6$ and $SS = 14.8$. As we have done with the other t statistics, we will present the computation of the t statistic as a three-step process.

First, compute the variance for the sample. Remember, the population variance (σ^2) is unknown, and we must use the sample value in its place.

$$s^2 = \frac{SS}{n-1} = \frac{14.8}{4} = 3.7$$

Next, use the sample variance to compute the estimated standard error.

$$s_{\bar{D}} = \sqrt{\frac{s^2}{n}} = \sqrt{\frac{3.7}{5}} = \sqrt{0.74} = 0.86$$

Finally, use the sample mean ($\bar{D}$) and the hypothesized population mean (μ_D) along with the estimated standard error to compute the value for the t statistic.

$$t = \frac{\bar{D} - \mu_D}{s_{\bar{D}}} = \frac{-3.2 - 0}{.86} = -3.72$$

STEP 4   Make a decision. The t value we obtained falls in the critical region (see Figure 11.2). The investigator rejects the null hypothesis and concludes that relaxation training does affect the amount of medication needed to control the asthma symptoms.

IN THE LITERATURE:
REPORTING THE RESULTS OF A REPEATED-MEASURES t TEST

As we have seen in Chapters 9 and 10, the APA format for reporting the results of t tests consists of a concise statement that incorporates the t value, degrees of freedom, and alpha level. One typically includes values for means and standard deviations, either in a statement or a table (Chapter 4). For Example 11.1, we observed a mean difference of $\bar{D} = -3.2$ with $s = 1.92$. Also, we obtained a t statistic of $t = -3.72$ with $df = 4$, and our decision was to reject the null hypothesis at the .05 level of significance. A published report of this study might summarize the conclusion as follows:

> Relaxation training resulted in a decrease ($M = 3.2$, $SD = 1.92$) in the number of doses of medication needed to control asthma symptoms. This reduction was statistically significant, $t(4) = -3.72$, $p < .05$, two-tailed.

In reporting the mean change, we did not bother to include the negative sign because we explicitly noted that the change was a decrease.   ❏

DIRECTIONAL HYPOTHESES AND ONE-TAILED TESTS

In many repeated-measures and matched-subjects studies, the researcher has a specific prediction concerning the direction of the treatment effect. For example, in the study described in Example 11.1, the researcher expects relaxation training to reduce the severity of asthma symptoms and therefore to reduce the amount of medication needed for asthma attacks. This kind of directional prediction can be incorporated into the statement of hypotheses, resulting in a directional, or one-tailed, hypothesis test. You should recall that directional tests should be used with caution. The standard two-tailed test is usually preferred and is always appropriate, even in situations where the researcher has a specific directional prediction. The following example demonstrates how the hypotheses and critical region are determined in a directional test.

EXAMPLE 11.2

We will re-examine the experiment presented in Example 11.1. The researcher is using a repeated-measures design to investigate the effect of relaxation training on the severity of asthma symptoms. The researcher predicts that people will need less medication after training than before, which will produce negative difference scores.

$$D = X_2 - X_1 = \text{after} - \text{before}$$

STEP 1

State the hypotheses, and select the alpha level. As always, the null hypothesis states that there is no effect. For this example, the researcher predicts that relaxation training will reduce the need for medication. The null hypothesis says it will not. In symbols,

$$H_0: \mu_D \geq 0 \quad \text{(medication is not reduced after training)}$$

The alternative hypothesis says that the treatment does work. For this example, H_1 states that the training does reduce medication. In symbols,

$$H_1: \mu_D < 0 \quad \text{(medication is reduced after training)}$$

It often is easier to start by stating H_1, which says that the treatment works as predicted. Then H_0 is simply the opposite of H_1.

STEP 2

Locate the critical region. The researcher is predicting negative difference scores if the treatment works. Hence, a negative t statistic would tend to support the experimental prediction and refute H_0. With a sample of $n = 5$ subjects, the t statistic will have $df = 4$. Looking in the t-distribution table for $df = 4$ and $\alpha = .05$ for a one-tailed test, we find a critical value of 2.132. Thus, a t statistic that is more extreme than -2.132 will be sufficient to reject H_0. The t distribution with the one-tailed critical region is shown in Figure 11.3.

STEP 3

Compute the t statistic. The t statistic was calculated in Example 11.1, where we obtained $t = -3.72$.

STEP 4

Make a decision. The obtained t statistic is well beyond the critical boundary. Therefore, we reject H_0 and conclude that the relaxation training did significantly reduce the amount of medication needed for asthma attacks.

LEARNING CHECK

1. A researcher would like to examine the effect of hypnosis on cigarette smoking. A sample of smokers ($n = 4$) is selected for the study. The number of cigarettes smoked on the day prior to treatment is recorded. The subjects are

FIGURE 11.3

The one-tailed critical region for $\alpha = .05$ in the t distribution with $df = 4$.

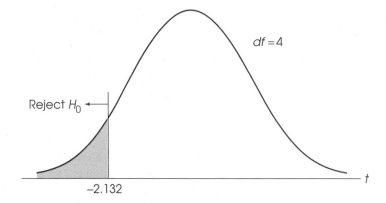

then hypnotized and given the posthypnotic suggestion that each time they light a cigarette, they will experience a horrible taste and feel nauseous. The data are as follows:

The number of cigarettes smoked before and after hypnosis		
Subject	Before treatment	After treatment
1	19	13
2	35	37
3	20	14
4	31	25

a. Find the difference scores (D values) for this sample.

b. The difference scores have $\bar{D} = -4$ and $SS = 48$. Do these data indicate that hypnosis has a significant effect on cigarette smoking? Test with $\alpha = .05$.

A N S W E R S   **1. a.** The difference scores are $-6, 2, -6, -6$.

   b. For these data, $s^2 = 16$, $s_{\bar{D}} = 2$, and $t = -2.00$. With $\alpha = .05$, we fail to reject H_0; these data do not provide sufficient evidence to conclude that hypnosis has a significant effect on cigarette smoking.

11.3 USES AND ASSUMPTIONS FOR RELATED-SAMPLES t TESTS

USES OF RELATED-SAMPLES STUDIES   The repeated-measures study differs from the independent-measures study in a fundamental way. In the latter type of study, a separate sample is used for each treatment. In the repeated-measures design, only one sample of subjects is used, and measurements are repeated for the same sample in each treatment. There are many situations where it is possible to examine the effect of a treatment by using either type of study. However, there are situations where one type of design is more desirable or appropriate than the other. For example, if a researcher would like to study a particular type of subject that is not commonly found (a rare species, people with an unusual illness, etc.), a repeated-measures study will be more economical in the sense that fewer subjects are needed. Rather than selecting several samples for the study (one sample per treatment), a single sample can be used for the entire experiment.

Another factor in determining the type of design is the specific question being asked by the researcher. Some questions are better studied with a repeated-measures design, especially those concerning changes in response across time. For example, a psychologist may wish to study the effect of practice on how well a person performs a task. To show that practice is improving a person's performance, the researcher would typically measure the person's responses very early in the study (when there is little or no practice) and repeat the measurement later when the person has had a certain amount of practice. Most studies of skill acquisition examine practice effects by using a repeated-measures design. Another situation where a repeated-measures study is useful is in developmental psychology. By repeating observations of the same individuals at various points in time, an investigator can watch behavior unfold and obtain a better understanding of developmental processes.

Finally, there are situations where a repeated-measures design cannot be used. Specifically, when you are comparing two different populations (men versus women, first-born versus second-born children, and the like), you must use separate samples from each population. Typically, these situations require an independent-measures design, but it is occasionally possible to match subjects with respect to a critical variable and conduct a matched-subjects study.

ADVANTAGES OF RELATED-SAMPLES STUDIES

Each subject enters a research study with his/her own individual characteristics. Factors such as age, gender, race, personality, height, weight, and so on differentiate one subject from another. These *individual differences* can influence the scores that are obtained for subjects and can create problems for interpreting research results. With an independent-measures design, there is always the potential that the individuals in one sample are substantially different (for example, smarter or stronger) than the individuals in the other sample. Although a researcher would like to think that the mean differences found in an independent-measures study are caused by the treatment, it is always possible that the differences are simply due to preexisting individual differences. As we noted at the beginning of the chapter, one of the advantages of a repeated-measures research design is that it eliminates this problem because the same individuals are used in every treatment. In other words, there are no individual differences between treatments with a repeated-measures design.

Individual differences can cause another statistical problem. Specifically, large differences from one individual to another tend to increase the sample variance. When the concept of variance was first introduced in Chapter 4, we noted that large variance can obscure any patterns in a set of data (p. 101). In more recent chapters, you should have noticed that variance contributes directly to the magnitude of the standard error. In general, as variance increases, it becomes increasingly difficult to demonstrate a significant treatment effect. The repeated-measures design also helps to reduce this problem. Specifically, a repeated-measures design removes the individual differences from the data and therefore tends to reduce the sample variance.

To demonstrate this fact, consider the sample data in Table 11.3. Notice that we have constructed the data so that there are large differences between individuals: Subject A consistently has scores in the 70s, subject B has scores in the 50s, and so on. Normally, large individual differences cause the scores to be spread over a wide range, so that the sample variance is large. In Table 11.3, the data show SS values around 2000 for the scores in treatment 1 and treatment 2. However, with a repeated-measures design, you do not make comparisons *between* individuals. Instead, this design makes comparisons *within* individuals; that is, each subject's score in treatment 1 is compared with the same subject's score in treatment 2. The comparison is accomplished by computing difference scores. Notice in Table 11.3 that when the difference scores are computed, the huge individual differences disappear. The SS value for the D scores is only $SS = 10$. The process of subtracting to obtain the difference scores has eliminated the absolute level of performance for each subject (70 versus 50, etc.) and thus has eliminated the individual differences from the sample variance. Reducing the sample variance also reduces the standard error and produces a more precise and sensitive test for mean differences.

CARRYOVER EFFECTS AND PROGRESSIVE ERROR

The general goal of a repeated-measures t test is to demonstrate that two different treatment conditions result in two different means. However, the outcome of a repeated-measures study may be contaminated by other factors that can cause the two means to be different when there actually is no difference between the two treatments. Two such factors, specifically associated with repeated-measures designs, are *carryover effects* and *progressive error.*

A carryover effect occurs when a subject's response in the second treatment is altered by lingering aftereffects from the first treatment. Some examples of carryover effects follow:

1. Imagine that a researcher is comparing the effectiveness of two drugs by testing both drugs, one after the other, on the same group of subjects. If the second drug is tested too soon after the first, there may be some of the first drug still in the subject's system, and this residual could exaggerate or minimize the effects of the second drug.

2. Imagine a researcher comparing performance on two tasks that vary in difficulty. If subjects are given a very difficult task first, their poor performance may cause them to lose motivation, so that performance suffers when they get to the second task.

TABLE 11.3

Hypothetical data showing large individual differences that are eliminated when the difference scores are computed

Subject	Treatment I	Treatment II	Difference scores
A	72	78	+6
B	55	59	+4
C	31	37	+6
D	14	18	+4
	$\overline{X} = 43$	$\overline{X} = 48$	$\overline{D} = +5$
	$SS = 1970$	$SS = 2042$	$SS = 10$

In each of these examples, the researcher will observe a difference between the two treatment means. However, the difference in performance is not caused by the treatments; instead, it is caused by carryover effects.

Progressive error occurs when a subject's performance or response changes consistently over time. For example, a subject's performance may decline over time as a result of fatigue. Or a subject's performance may improve over time as a result of practice. In either case, a researcher would observe a mean difference in performance between the first treatment and the second treatment, but the change in performance is not due to the treatments; instead, it is caused by progressive error.

To help distinguish between carryover effects and progressive error, remember that carryover effects are related directly to the first treatment. Progressive error, on the other hand, occurs as a function of time, independent of which treatment condition is presented first or second. Both of these factors, however, can influence the outcome of a repeated-measures study, and they can make it difficult for a researcher to interpret the results: Is the mean difference caused by the treatments, or is it caused by other factors?

One way to deal with carryover effects or progressive error is to *counterbalance* the order of presentation of treatments. That is, the subjects are randomly divided into two groups, with one group receiving treatment 1 followed by treatment 2 and the other group receiving treatment 2 followed by treatment 1. When there is reason to expect strong carryover effects or large progressive error, your best strategy is not to use a repeated-measures design. Instead, use independent measures with a separate sample for each treatment condition, or use a matched-subjects design so that each individual participates in only one treatment condition.

ASSUMPTIONS OF THE RELATED-SAMPLES *t* TEST

The related-samples *t* statistic requires two basic assumptions:

1. The observations within each treatment condition must be independent (see page 202). Notice that the assumption of independence refers to the scores *within* each treatment. Inside each treatment, the scores are obtained from different individuals and should be independent of one another.

2. The population distribution of difference scores (*D* values) must be normal.

As before, the normality assumption is not a cause for concern unless the sample size is relatively small. In the case of severe departures from normality, the validity of the *t* test may be compromised with small samples. However, with relatively large samples ($n > 30$), this assumption can be ignored.

LEARNING CHECK

1. What assumptions must be satisfied for repeated-measures *t* tests to be valid?

2. Describe some situations for which a repeated-measures design is well suited.

3. How is a matched-subjects design similar to a repeated-measures design? How do they differ?

4. The data from a research study consist of 10 scores in each of two different treatment conditions. How many individual subjects would be needed to produce these data

   a. For an independent-measures design?

   b. For a repeated-measures design?

   c. For a matched-subjects design?

A N S W E R S   1. The observations within a treatment are independent. The population distribution of D scores is assumed to be normal.

2. The repeated-measures design is suited to situations where a particular type of subject is not readily available for study. This design is helpful because it uses fewer subjects (only one sample is needed). Certain questions are addressed more adequately by a repeated-measures design—for example, anytime one would like to study changes across time in the same individuals. Also, when individual differences are large, a repeated-measures design is helpful because it reduces the amount of this type of error in the statistical analysis.

3. They are similar in that the role of individual differences in the experiment is reduced. They differ in that there are two samples in a matched-subjects design and only one in a repeated-measures study.

4. **a.** The independent-measures design would require 20 subjects (two separate samples with $n = 10$ in each).

   b. The repeated-measures design would require 10 subjects (the same 10 individuals are measured in both treatments).

   c. The matched-subjects design would require 20 subjects (10 matched pairs).

SUMMARY

1. In a related-samples research study, the individuals in one treatment condition are directly related, one-to-one, with the individuals in the other treatment condition(s). The most common related-samples study is a repeated-measures design, where the same sample of individuals is tested in all of the treatment conditions. This design literally repeats measurements on the same subjects. An alternative is a matched-subjects design, where the individuals in one sample are matched one-to-one with individuals in another sample. The matching is based on a variable relevant to the study.

2. The repeated-measures t test begins by computing a difference between the first and second measurements for each subject (or the difference for each matched pair). The difference scores, or D scores, are obtained by

$$D = X_2 - X_1$$

The sample mean, $\overline{D}$, and sample variance, s^2, are used to summarize and describe the set of difference scores.

3. The formula for the repeated-measures t statistic is

$$t = \frac{\overline{D} - \mu_D}{s_{\overline{D}}}$$

In the formula, the null hypothesis specifies $\mu_D = 0$, and the estimated standard error is computed by

$$s_{\overline{D}} = \sqrt{\frac{s^2}{n}} \quad \text{or} \quad s_{\overline{D}} = \frac{s}{\sqrt{n}}$$

4. A repeated-measures design may be more useful than an independent-measures study when one wants to observe changes in behavior in the same subjects, as in learning or developmental studies. An important advantage of the repeated-measures design is that it removes or reduces individual differences, which in turn lowers sample variability and tends to increase the chances for obtaining a significant result.

5. A related-samples study may consist of two samples in which subjects have been matched on some variable. The repeated-measures t test may be used in this situation.

KEY TERMS

repeated-measures design

within-subjects design

matched-subjects design

difference scores

repeated-measures t statistic

estimated standard error for $\overline{D}$

individual differences

FOCUS ON PROBLEM SOLVING

1. Once data have been collected, we must then select the appropriate statistical analysis. How can you tell if the data call for a repeated-measures t test? Look at the experiment carefully. Is there only one sample of subjects? Are the same subjects tested a second time? If your answers are yes to both of these questions, then a repeated-measures t test should be done. There is only one situation in which the repeated-measures t test can be used for data from two samples, and that is for *matched-subjects* experiments (page 262).

2. The repeated-measures t test is based on difference scores. In finding difference scores, be sure you are consistent with your method. That is, you may use either $X_2 - X_1$ or $X_1 - X_2$ to find D scores, but you must use the same method for all subjects. We suggest that you use $D = X_2 - X_1$ so that the sign of the difference scores ($+$ or $-$) indicates the direction of difference (increase or decrease).

DEMONSTRATION 11.1

A REPEATED-MEASURES t TEST

A major oil company would like to improve its tarnished image following a large oil spill. Its marketing department develops a short television commercial and tests it on a sample of $n = 7$ subjects. People's attitudes about the company are measured with a short questionnaire, both before and after viewing the commercial. The data are as follows:

Person	X_1 (Before)	X_2 (After)
A	15	15
B	11	13
C	10	18
D	11	12
E	14	16
F	10	10
G	11	19

Was there a significant change?

Note that subjects are being tested twice—once before and once after viewing the commercial. Therefore, we have a repeated-measures design.

STEP 1 State the hypothesis, and select an alpha level.

The null hypothesis states that the commercial has no effect on people's attitudes, or in symbols,

$$H_0: \mu_D = 0 \quad \text{(the mean difference is zero)}$$

The alternative hypothesis states that the commercial does alter attitudes about the company, or

$$H_1: \mu_D \neq 0 \quad \text{(there is a mean change in attitudes)}$$

For this demonstration, we will use an alpha level of .05 for a two-tailed test.

STEP 2 Locate the critical region.

Degrees of freedom for the repeated-measures t test are obtained by the formula

$$df = n - 1$$

For these data, degrees of freedom equal

$$df = 7 - 1 = 6$$

The t-distribution table is consulted for a two-tailed test with $\alpha = .05$ for $df = 6$. The t values for the critical region are $t = \pm 2.447$.

STEP 3 Obtain the sample data, and compute the test statistic.

To compute the repeated-measures t statistic, we will have to determine the values for the difference (D) scores, $\overline{D}$, SS for the D scores, the sample variance for the D scores, and the estimated standard error for $\overline{D}$.

The difference scores. The following table illustrates the computation of the D values for our sample data. Remember, $D = X_2 - X_1$.

X_1	X_2	D
15	15	$15 - 15 =\ \ 0$
11	13	$13 - 11 = +2$
10	18	$18 - 10 = +8$
11	12	$12 - 11 = +1$
14	16	$16 - 14 = +2$
10	10	$10 - 10 =\ \ 0$
11	19	$19 - 11 = +8$

The sample mean of D values. The sample mean for the difference scores is equal to the sum of the D values divided by n. For these data,

$$\Sigma D = 0 + 2 + 8 + 1 + 2 + 0 + 8 = 21$$

$$\overline{D} = \frac{\Sigma D}{n} = \frac{21}{7} = 3$$

Sum of squares for D scores. We will use the computational formula for SS. The following table summarizes the calculations:

D	D^2	
0	0	$\Sigma D = 21$
2	4	$\Sigma D^2 = 0 + 4 + 64 + 1 + 4 + 0 + 64 = 137$
8	64	
1	1	$SS = \Sigma D^2 - \dfrac{(\Sigma D)}{n} = 137 - \dfrac{(21)^2}{7}$
2	4	
0	0	
8	64	$SS = 137 - \dfrac{441}{7} = 137 - 63 = 74$

Variance for the D scores. The variance for the sample of D scores is

$$s^2 = \frac{SS}{df} = \frac{74}{6} = 12.33$$

Estimated standard error for $\bar{D}$. The estimated standard error for the sample mean difference is computed as follows:

$$s_{\bar{D}} = \sqrt{\frac{s^2}{n}} = \sqrt{\frac{12.33}{7}} = \sqrt{1.76} = 1.33$$

The repeated-measures t statistic. We now have the information required to calculate the t statistic.

$$t = \frac{\bar{D} - \mu_D}{s_{\bar{D}}} = \frac{3 - 0}{1.33} = \frac{3}{1.33} = 2.26$$

STEP 4 Make a decision about H_0, and state the conclusion.

The obtained t value is not extreme enough to fall in the critical region. Therefore, we fail to reject the null hypothesis. We conclude that there is no evidence that the commercial will change people's attitudes, $t(6) = 2.26$, $p > .05$, two-tailed. (Note that we state that p is *greater than* .05 because we failed to reject H_0.)

PROBLEMS

1. For the following studies, indicate whether or not a repeated-measures t test is the appropriate analysis. Explain your answers.
   a. A researcher examines the effect of relaxation training on test anxiety. One sample of subjects receives relaxation training for three weeks. A second sample serves as a control group and does not receive the treatment. The researcher then measures anxiety levels for both groups in a test-taking situation.
   b. Another researcher does a similar study. Baseline levels of test anxiety are recorded for a sample of subjects. Then all subjects receive relaxation training for three weeks, and their anxiety levels are measured again.
   c. In a test-anxiety study, two samples are used. Subjects are assigned to groups so that they are matched for self-esteem and for grade-point average. One sample receives relaxation training for three weeks, and the second serves as a no-treatment control group. Test anxiety is measured for both groups at the end of three weeks.

2. What is the primary advantage of a repeated-measures design over an independent-measures design?

3. Explain the difference between a matched-subjects design and a repeated-measures design.

4. A researcher conducts an experiment comparing two treatment conditions and obtains data with 10 scores in each treatment.
   a. If the researcher used an independent-measures design, how many subjects participated in the experiment?
   b. If the researcher used a repeated-measures design, how many subjects participated in the experiment?
   c. If the researcher used a matched-subjects design, how many subjects participated in the experiment?

5. Many people who are trying to quit smoking will switch to a lighter brand of cigarettes to reduce their intake of tar and nicotine. However, there is some evidence that switching to a lighter brand simply results in people smoking more cigarettes. A researcher examining this phenomenon recorded the number of cigarettes smoked each day for a sample of $n = 16$ subjects before and after they switched to a lighter brand. On average, the subjects smoked $\bar{D} = 3.2$ more cigarettes after switching, with $SS = 375$. Do these data indicate a significant change in the number of cigarettes smoked per day? Test at the .05 level of significance, two tails.

6. A college professor performed a study to assess the effectiveness of computerized exercises in teaching mathematics. She decides to use a *matched-subjects design*. One group of subjects is assigned to a regular lecture section

of introductory mathematics. A second group must attend a computer laboratory in addition to the lecture. These students work on computerized exercises for additional practice and instruction. Both samples are matched in terms of general mathematics ability, as measured by mathematical SAT scores. At the end of the semester, both groups are given the same final exam. For $n = 16$ matched pairs of subjects, the professor found that the computer group scored an average of $\overline{D} = 9.3$ points higher. The SS for the D values was 2160. Does the computerized instruction lead to a significant change? Test at the .01 level of significance.

7. The following are two samples of difference scores obtained from repeated-measures experiments:

Sample A: 6,   5,   7,   5,   4,   3,   4,   5,   6

Sample B: −2,   25,   5,   12,   −15,   5,   −5,   15

a. For each sample of D scores, sketch a histogram showing the frequency distribution, and calculate the mean, $\overline{D}$, and SS.
b. According to the null hypothesis, each sample comes from a population of difference scores with $\mu_D = 0$. Locate this value in each sketch from part a. Does sample A appear to have come from a population centered at zero? What about sample B?
c. Perform the repeated-measures hypothesis test for each sample using $\alpha = .05$. You should find that the conclusions from the hypothesis tests agree with your observations in part b.

8. Many over-the-counter cold medications come with a warning that the medicine may cause drowsiness. To evaluate this effect, a researcher measured reaction time for a sample of $n = 36$ subjects. Each subject was tested for reaction time, given a dose of a popular cold medicine, and then tested again for reaction time. For this sample, reaction time increased by an average of $\overline{D} = 24$ milliseconds with $s = 8$ after the medication. Do these data indicate that the cold medicine has a significant effect on reaction time? Use a two-tailed test with $\alpha = .05$.

9. Following are data from two repeated-measures experiments. In each experiment, a sample of $n = 4$ subjects is tested before treatment and again after treatment. For both experiments, the scores after treatment average $\overline{D} = 5$ points higher than the scores before treatment. However, in the first experiment, the treatment effect is fairly consistent across subjects; that is, all four subjects show an increase of about 5 points. In the second experiment, there is no consistent effect: Some subjects show an increase, some a decrease, and some no change at all.

Experiment 1		Experiment 2	
Before treatment	After treatment	Before treatment	After treatment
8	13	8	25
10	14	10	10
14	19	14	21
11	17	11	7

a. For each experiment, find the difference scores, and compute the mean and the standard deviation for the D values.
b. For each experiment, use a repeated-measures t test to determine whether or not the treatment has a significant effect. In each case, use a two-tailed test with $\alpha = .05$.
c. Explain why the two experiments lead to different conclusions about the significance of the treatment effect.

10. A psychologist is testing a new drug for its pain-killing effects. Pain threshold is measured for a sample of $n = 9$ subjects by determining the intensity of an electric shock that causes discomfort. After the initial baseline is established, each subject receives the drug, and pain threshold is measured once again. For this sample, the pain threshold increased by an average of $\overline{D} = 1.6$ milliamperes with $s = 0.4$ after receiving the drug. Do these data indicate that the drug produces a significant increase in pain tolerance? Use a one-tailed test with $\alpha = .05$.

11. The level of lymphocytes (white blood cells) in the blood is associated with susceptibility to disease—lower levels indicate greater susceptibility. In a study of the effects of stress on physical well-being, Schleifer et al. (1983) recorded the lymphocyte counts for men who were going through a period of extreme emotional stress (the death of a spouse). For a sample of $n = 20$ men, the lymphocyte counts dropped an average of $\overline{D} = 0.09$ with $s = 0.08$. Do these data indicate a significant change? Test at the .05 level of significance.

12. A consumer protection agency is testing the effectiveness of a new gasoline additive that claims to improve gas mileage. A sample of 10 cars is obtained, and each car is driven over a standard 100-mile course with and without the additive. The researchers carefully record the miles per gallon for each test drive. The cars in this test averaged $\overline{D} = 2.7$ miles per gallon higher with the additive than without, with $s = 1.4$. Are these data sufficient to conclude that the additive significantly increases gas mileage? Use a one-tailed test with $\alpha = .05$.

13. One of the benefits of aerobic exercise is the release of endorphins, which are natural chemicals in the brain that produce a feeling of general well-being. A sample of $n = 16$ subjects is obtained, and each person's tolerance for pain is tested before and after a 50-minute session of aerobic exercise. On average, the pain tolerance for this sample was $\overline{D} = 10.5$ higher after exercise than it was before. The SS for the sample of difference scores was $SS = 960$. Do these data indicate a significant increase in pain tolerance following exercise? Use a one-tailed test with $\alpha = .01$.

14. A researcher for a dog food manufacturer would like to determine whether animals show any preference between two new food mixes that the company has recently developed. A sample of $n = 9$ dogs is obtained. The dogs are deprived of food overnight and then presented simultaneously with two bowls of food, one with each mix, the next morning. After 10 minutes, the bowls are removed, and the amount of food consumed (in ounces) is determined for each mix. The researcher computes the difference between the two mixes for each dog. The data show that the dogs ate $\overline{D} = 6$ ounces more of the first mix than of the second, with $SS = 288$. Do these data indicate a significant difference between the two food mixes? Test with $\alpha = .05$.

15. Although it is generally assumed that routine exercise will improve overall health, there is some question about how much exercise is necessary to produce the benefits. To address this question, a researcher obtained $n = 7$ pairs of subjects, matched one-to-one for age, sex, and weight. All subjects exercised regularly, but one subject within each pair exercised less than 2 hours per week, and the other spent more than 5 hours per week exercising. Each subject was evaluated by a physician and given an overall health rating. The data from this study are as follows:

Subject pair	2 hours per week	5 hours per week
A	15	18
B	12	14
C	16	12
D	9	11
E	13	14
F	16	16
G	17	16

Do these data indicate that the amount of regular exercise has an effect on health? Test for a significant difference using $\alpha = .05$.

16. It has been demonstrated that memory for a list of items can be improved if you make meaningful associations with each item. In one demonstration of this phenomenon, Craik and Tulving (1975) presented subjects with a list of words to be remembered. Each word was presented in the context of a sentence. For some words, the sentences were very simple, and for others, the sentences were more elaborate. For the word *rabbit,* for example, a simple sentence would be "She cooked the rabbit." An elaborate sentence would be "The great bird swooped down and carried off the struggling rabbit." The researchers recorded the number of words recalled for each type of sentence to determine whether the more elaborate sentences produced richer associations and better memory. Hypothetical data, similar to the experimental results, are as follows:

Subject	Simple sentence	Elaborate sentence
A	14	22
B	15	17
C	19	24
D	12	19
E	17	28
F	15	20

Do these data indicate a significant difference between the two types of sentences? Test at the .01 level of significance.

17. At the Olympic level of competition, even the smallest factors can make the difference between winning and losing. For example, Pelton (1983) has shown that Olympic marksmen shoot much better if they fire between heartbeats, rather than squeezing the trigger during a heartbeat. The small vibration caused by a heartbeat seems to be sufficient to affect the marksman's aim. The following hypothetical data demonstrate this phenomenon. A sample of $n = 6$ Olympic marksmen fires a series of rounds while a researcher records heartbeats. For each marksman, the total score is recorded for shots fired during heartbeats and for shots fired between heartbeats. Do these data indicate a significant difference? Test with $\alpha = .05$.

Subject	During heartbeat	Between heartbeats
A	93	98
B	90	94
C	95	96
D	92	91
E	95	97
F	91	97

18. Sensory isolation chambers are used to examine the effects of mild sensory deprivation. The chamber is a dark, silent tank where subjects float on heavily salted water and are thereby deprived of nearly all external stimulation. Sensory deprivation produces deep relaxation and has been shown to produce temporary increases in sensitivity for vision, hearing, touch, and even taste. The following data represent hearing threshold scores for a group of subjects who were tested before and immediately after one hour of deprivation. A lower score indicates more sensitive hearing. Do these data indicate that deprivation has a significant effect on the hearing threshold? Test at the .05 level of significance.

Subject	Before	After
A	31	30
B	34	31
C	29	29
D	33	29
E	35	32
F	32	34
G	35	28

19. There are two supermarkets in town, and a student would like to determine whether there is any significant difference in the prices they charge. The student identifies a sample of nine basic grocery items and records the prices of the items at each store. Do these sample data indicate a significant difference in prices between the two stores? Test at the .05 level of significance.

Item	Store A	Store B
paper towels	$1.09	$1.05
1/2 gallon milk	.89	1.09
bread	1.29	1.05
5 pounds potatoes	1.79	1.60
corn flakes	2.65	2.65
1 dozen eggs	.99	.85
1 pound coffee	2.29	2.19
1 pound ground beef	1.85	1.79
peanut butter	2.50	2.35

20. A researcher studies the effect of a drug (MAO inhibitor) on the number of nightmares occurring in veterans with post-traumatic stress disorder (PTSD). A sample of PTSD clients records each incident of a nightmare for one month before treatment. Subjects are then given the medication for one month, and they continue to report each occurrence of a nightmare. For the following hypothetical data, determine if the MAO inhibitor significantly reduces nightmares. Use the .05 level of significance and a one-tailed test.

Number of nightmares	
One month before treatment	One month during treatment
6	1
10	2
3	0
5	5
7	2

CHAPTER 12

ESTIMATION

TOOLS YOU WILL NEED

The following items are considered essential background material for this chapter. If you doubt your knowledge of any of these items, you should review the appropriate chapter or section before proceeding.

- Distribution of sample means (Chapter 7)

- Hypothesis tests with z (Chapter 8)

- Single-sample t statistic (Chapter 9)

- Independent-measures t statistic (Chapter 10)

- Related-samples t statistic (Chapter 11)

CONTENTS

12.1   AN OVERVIEW OF ESTIMATION

In Chapter 8, we introduced hypothesis testing as a statistical procedure that allows researchers to use sample data to draw inferences about populations. Hypothesis testing is probably the most frequently used inferential technique, but it is not the only one. In this chapter, we will examine the process of estimation, which provides researchers with an additional method for using samples as the basis for drawing general conclusions about populations.

The basic principle underlying all of inferential statistics is that samples are representative of the populations from which they come. The most direct application of this principle is the use of sample values as estimators of the corresponding population values, that is, using statistics to estimate parameters. This process is called *estimation.*

DEFINITION   The inferential process of using sample statistics to estimate population parameters is called *estimation.*

The use of samples to estimate populations is quite common. For example, you often hear news reports such as "Sixty percent of the general public approves of the president's new budget plan." Clearly, the percentage that is reported was obtained from a sample (they don't ask everyone's opinion), and this sample statistic is being used as an estimate of the population parameter.

We already have encountered estimation in earlier sections of this book. For example, the formula for sample variance (Chapter 4) was developed so that the sample value would give an accurate and unbiased estimate of the population variance. Now we will examine the process of using sample means as the basis for estimating population means.

PRECISION AND CONFIDENCE IN ESTIMATION

Before we begin the actual process of estimation, there are a few general points that should be kept in mind. First, a sample will not give a perfect picture of the whole population. A sample is expected to be representative of the population, but there always will be some differences between the sample and the entire population. These differences are referred to as *sampling error.* Second, there are two distinct ways of making estimates. Suppose, for example, you are asked to estimate the weight of this book. You could pick a single value (say, 2 pounds), or you could choose a range of values (say, between 1.5 pounds and 2.5 pounds). The first estimate, using a single number, is called a *point estimate.* Point estimates have the advantage of being very precise; they specify a particular value. On the other hand, you generally do not have much confidence that a point estimate is correct. You would not bet on it, for example.

DEFINITION   For a *point estimate,* you use a single number as your estimate of an unknown quantity.

The second type of estimate, using a range of values, is called an *interval estimate.* Interval estimates do not have the precision of point estimates, but they do

FIGURE 12.1

The basic research situation for either hypothesis testing or estimation. The goal is to use the sample data to answer questions about the unknown population mean after treatment.

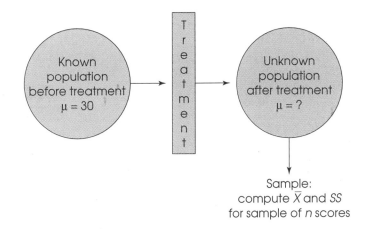

Sample:
compute $\bar{X}$ and SS
for sample of n scores

give you more confidence. You would feel more comfortable, for example, saying that this book weighs "around 2 pounds." At the extreme, you would be very confident in estimating that this book weighs between 0.5 and 10 pounds. Notice that there is a trade-off between precision and confidence. As the interval gets wider and wider, your confidence grows. But, at the same time, the precision of the estimate gets worse. We will be using samples to make both point and interval estimates of a population mean. Because the interval estimates are associated with confidence, they usually are called *confidence intervals*.

DEFINITIONS

For an *interval estimate*, you use a range of values as your estimate of an unknown quantity.

When an interval estimate is accompanied by a specific level of confidence (or probability), it is called a *confidence interval*.

Estimation is used in the same general situations in which we have already used hypothesis testing. In fact, there is an estimation procedure that accompanies each of the hypothesis tests we presented in the preceding four chapters. Figure 12.1 shows an example of a research situation where either hypothesis testing or estimation could be used. The figure shows a population with an unknown mean (the population after treatment). A sample is selected from the unknown population. The goal of estimation is to use the sample data to obtain an estimate of the unknown population mean.

COMPARISON OF HYPOTHESIS TESTS AND ESTIMATION

You should recognize that the situation shown in Figure 12.1 is the same situation in which we have used hypothesis tests in the past. In many ways, hypothesis testing and estimation are similar. They both make use of sample data and either z-scores or t statistics to find out about unknown populations. But these two inferential procedures are designed to answer different questions. Using the situation shown in Figure 12.1 as an example, we could use a hypothesis test to evaluate the effect of the treatment. The test would determine whether or not the treatment has any effect. Notice that this is a yes-no question. The null hypothesis says, "No, there

12.1   HYPOTHESIS TESTING VERSUS ESTIMATION: STATISTICAL SIGNIFICANCE VERSUS PRACTICAL SIGNIFICANCE

AS WE already noted, hypothesis tests tend to involve a yes-no decision. Either we decide to reject H_0, or we fail to reject H_0. The language of hypothesis testing reflects this process. The outcome of the hypothesis test is one of two conclusions:

There is no evidence for a treatment effect (fail to reject H_0)

or

There is a statistically significant effect (H_0 is rejected).

For example, a researcher studies the effect of a new drug on people with high cholesterol. In hypothesis testing, the question is whether or not the drug has a significant effect on cholesterol levels. Suppose that the hypothesis test revealed that the drug did produce a significant decrease in cholesterol. The next question might be, How much of a reduction occurs? This question calls for estimation, in which the size of a treatment effect for the population is estimated.

Estimation can be of great practical importance be-

cause the presence of a "statistically significant" effect does not necessarily mean the results are large enough for use in practical applications. Consider the following possibility: Before drug treatment, the sample of patients had a mean cholesterol level of 225. After drug treatment, their cholesterol reading was 210. When analyzed, this 15-point change reached statistical significance (H_0 was rejected). Although the hypothesis test revealed that the drug produced a *statistically significant* change, it may not be *clinically significant*. That is, a cholesterol level of 210 is still quite high. In estimation, we would estimate the population mean cholesterol level for patients who are treated with the drug. This estimated value may reveal that even though the drug does in fact reduce cholesterol levels, it does not produce a large enough change (notice we are looking at a "how much" question) to make it of any practical value. Thus, the hypothesis test might reveal that an effect occurred, but estimation indicates it is small and of little *practical significance* in real-world applications.

is no treatment effect." The alternative hypothesis says, "Yes, there is a treatment effect."

The goal of estimation, on the other hand, is to determine the value of the population mean after treatment. Essentially, estimation will determine *how much* effect the treatment has (see Box 12.1). If, for example, we obtained a point estimate of $\mu = 38$ for the population after treatment, we could conclude that the effect of the treatment is to increase scores by an average of 8 points (from the original mean of $\mu = 30$ to the post-treatment mean of $\mu = 38$).

WHEN TO USE ESTIMATION

There are three situations where estimation commonly is used:

1. Estimation is used after a hypothesis test where H_0 is rejected. Remember that when H_0 is rejected, the conclusion is that the treatment does have an effect. The next logical question would be, How much effect? This is exactly the question that estimation is designed to answer.

2. Estimation is used when you already know that there is an effect and simply want to find out how much. For example, the city school board probably knows that a special reading program will help students. However, they want to be sure that the effect is big enough to justify the cost. Estimation is used to determine the size of the treatment effect.

TABLE 12.1

A summary of research situations where a sample statistic is used to draw inferences about the corresponding population parameter

Sample statistic	Population parameter	Standard error*	
$\overline{X}$	μ	$\sigma_{\overline{X}}$	(Chapter 8)
$\overline{X}$	μ	$s_{\overline{X}}$	(Chapter 9)
$\overline{X}_1 - \overline{X}_2$	$\mu_1 - \mu_2$	$s_{\overline{X}_1 - \overline{X}_2}$	(Chapter 10)
$\overline{D}$	μ_D	$s_{\overline{D}}$	(Chapter 11)

*In each case, the standard error provides a measure of the standard distance or error between the statistic and the parameter.

3. Estimation is used when you simply want some basic information about an unknown population. Suppose, for example, you want to know about the political attitudes of students at your college. You could use a sample of students as the basis for estimating the population mean.

THE LOGIC OF ESTIMATION

As we have noted, estimation and hypothesis testing are both *inferential* statistical techniques that involve using sample data as the basis for drawing conclusions about an unknown population. More specifically, a researcher begins with a question about an unknown population parameter. To answer the question, a sample is obtained, and a sample statistic is computed. In general, *statistical inference* involves using sample statistics to help answer questions about population parameters. The general logic underlying the processes of estimation and hypothesis testing is based on the fact that each population parameter has a corresponding sample statistic. In addition, you usually can compute a standard error that measures how much discrepancy is expected, on average, between the statistic and the parameter. For example, a sample mean, $\overline{X}$, corresponds to the population mean, μ, with a standard error measured by $\sigma_{\overline{X}}$ or $s_{\overline{X}}$.

In the preceding four chapters, we examined four hypothesis-testing situations involving sample statistics, parameters, and standard errors. The four situations are summarized in Table 12.1.

The statistic, the parameter, and the standard error can all be combined into a single formula to compute a *z*-score or a *t* statistic. For example,

$$z = \frac{\overline{X} - \mu}{\sigma_{\overline{X}}} \qquad \text{or} \qquad t = \frac{\overline{X} - \mu}{s_{\overline{X}}}$$

For estimation and for hypothesis testing, the population parameter is unknown, and the *z*-score or *t* statistic formula has the general structure

$$z \text{ or } t = \frac{\text{sample statistic} - \text{unknown population parameter}}{\text{standard error}}$$

In previous chapters, we used this general formula to test hypotheses about the unknown parameter. Now we will use the same formula as the basis for *estimating* the value of an unknown parameter. Although hypothesis tests and

FIGURE 12.2

For estimation or hypothesis testing, the distribution of z-scores or t statistics is divided into two sections: the middle of the distribution, consisting of high-probability outcomes that are considered "reasonable," and the extreme tails of the distribution, consisting of low-probability, "unreasonable" outcomes.

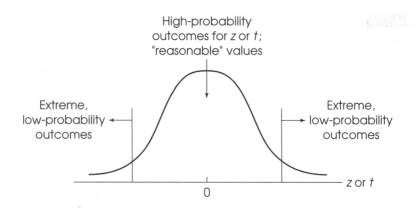

High-probability outcomes for z or t; "reasonable" values

Extreme, low-probability outcomes

Extreme, low-probability outcomes

z or t

0

estimation use the same basic formula, these two inferential procedures follow different logical paths because they have different goals. The two different paths are developed as follows:

Hypothesis test	Estimation
Goal: To test a hypothesis about a population parameter—usually the null hypothesis, which states that the treatment has no effect.	*Goal:* To estimate the value of an unknown population parameter—usually the value for an unknown population mean.
A. For a hypothesis test, you begin by hypothesizing a value for the unknown population parameter. This value is specified in the null hypothesis.	A. For estimation, you do not attempt to calculate z or t. Instead, you begin by estimating what the z or t value ought to be. The strategy for making this estimate is to select a "reasonable" value for z or t. (*Note:* You are not just picking a value for z or t, but rather you are estimating where the sample is located in the distribution.)
B. The hypothesized value is substituted into the formula, and *the value for z or t is computed.*	
C. If the hypothesized value produces a "reasonable" value for z or t, we conclude that the hypothesis was "reasonable," and we fail to reject H_0. If the result is an extreme value for z or t, H_0 is rejected.	B. As with hypothesis testing, a "reasonable" value for z or t is defined as a high-probability outcome located near the center of the distribution (see Figure 12.2).
D. A "reasonable" value for z or t is defined by its location in a distribution. In general, "reasonable" values are high-probability outcomes in the center of the distribution. Extreme values with low probability are considered "unreasonable" (see Figure 12.2).	C. The "reasonable" value for z or t is substituted into the formula, and *the value for the unknown population parameter is computed.*
	D. Because you used a "reasonable" value for z or t in the formula, it is assumed that the computation will produce a "reasonable" estimate of the population parameter.

Because the goal of the estimation process is to compute a value for an unknown population parameter, it usually is easier to regroup the terms in the z or t formula so that the population parameter is isolated on one side of the equation. In algebraic terms, we simply solve the equation for the unknown parameter. The result takes the following form:

$$\begin{matrix} \text{unknown} \\ \text{population parameter} \end{matrix} = \begin{matrix} \text{sample} \\ \text{statistic} \end{matrix} \pm [(z \text{ or } t)(\text{standard error})] \qquad (12.1)$$

This is the general equation that we will use for estimation. Consider the following four points about equation 12.1:

1. In this formula, the values for the sample mean and the standard error can be computed directly from the sample data. Only the value for the z-score (or t statistic) cannot be computed. If we can determine this missing value, then the equation can be used to compute the unknown population parameter.

2. Although the specific value for the z-score (or t statistic) cannot be computed, you do know what the entire distribution of z-scores (or t statistics) looks like. For example, the z-scores may be normally distributed and always have a mean of zero. Likewise, the t distribution is bell-shaped with its exact shape determined by the value of df, and it also will have a mean of zero.

3. While you cannot compute the specific value for the z-score (or t statistic) for the sample data, you can *estimate* the location of the sample data in the sampling distribution. The procedure involves selecting "reasonable" values for the z-scores (or t statistics) to estimate the location of the sample.

4. For a point estimate, your best bet is to estimate that the sample statistic (for example, $\overline{X}$) is located in the exact center of its sampling distribution (distribution of sample means), that is, the location that corresponds to $z = 0$ (or $t = 0$). This is a reasonable estimate for location because sample statistics become increasingly unlikely as you move toward the tails of the distribution.

   For an interval estimate, your best bet is to predict that the sample statistic is in the middle section of the sampling distribution. For example, to be 90% confident that μ is in your interval, you would use the z-score (or t statistic) values that mark off the middle 90% of the distribution of sample means.

   Notice that by estimating the location of a sample statistic ($\overline{X}$) within its sampling distribution (distribution of sample means) and by taking the standard error into account, we can arrive at an estimate of μ using equation 12.1.

LEARNING CHECK

1. Estimation procedures basically address a "yes-no" question: whether or not a treatment effect exists. (True or false?)

2. Estimation is primarily used to estimate the value of sample statistics. (True or false?)

FIGURE 12.3

A population distribution before the treatment is administered and the same population after treatment. Note that the effect of the treatment is to add a constant amount to each score. The goal of estimation is to determine how large the treatment effect is; in other words, what is the new population mean (μ = ?)?

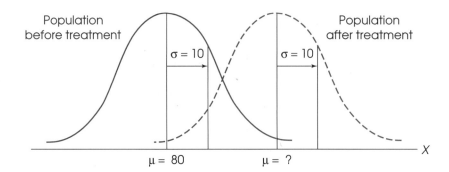

3. In general, as precision of an interval estimate increases, confidence decreases. (True or false?)

4. Describe three common applications of estimation.

ANSWERS   1. False. Estimation asks *how much* difference exists.

2. False. Estimation procedures are used to estimate the value for unknown population parameters.

3. True. There is a trade-off between precision and confidence.

4. **a.** After a hypothesis test results in rejecting H_0, a researcher may want to determine the size of the treatment effect.

   b. If it is already known that a treatment effect exists, estimation can be used to determine how much of an effect is present.

   c. Estimation can be used anytime you want to get information about unknown population parameters.

12.2 ESTIMATION WITH THE *z*-SCORE

The *z*-score statistic is used in situations where the population standard deviation is known, but the population mean is unknown. Often this is a population that has received some treatment. Suppose you are examining the effect of a special summer reading program for grade-school children. Using a standard reading achievement test, you know that the scores for second-graders in the city school district form a normal distribution with μ = 80 and σ = 10. It is reasonable to assume that a special reading program would increase the students' scores. The question is, How much?

The example we are considering is shown graphically in Figure 12.3. Notice that we have assumed that the effect of the treatment (the special program) is to add a constant amount to each student's reading score. As a result, after the summer reading program, the entire distribution would be shifted to a new location with a larger mean. This new mean is what we want to estimate.

Because it would not be reasonable to put all the students in the special program, we cannot measure this mean directly. However, we can get a sample and use the

If the treatment simply adds a constant to each score, the standard deviation will not be changed. Although it is common practice to assume that a treatment will add a constant amount, you should realize that in most real-life situations there is a general tendency for variability to increase when the mean increases.

FIGURE 12.4

The distribution of sample means based on $n = 25$. Samples were selected from the unknown population (after treatment) shown in Figure 12.3.

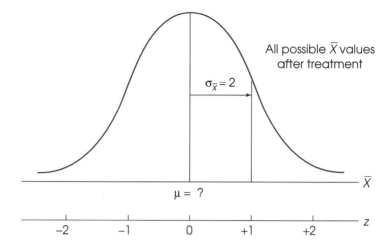

All possible $\overline{X}$ values after treatment

$\sigma_{\overline{X}} = 2$

$\mu = ?$

$\overline{X}$

z

$-2 \quad -1 \quad 0 \quad +1 \quad +2$

sample mean to estimate the population value for μ. For this example, assume that a random sample of $n = 25$ students is selected to participate in the summer program. At the end of the summer, each student takes the reading test, and we compute a mean reading score of $\overline{X} = 88$. Note that this sample represents the population after the special program. The goal of estimation is to use this sample mean as the basis for estimating the unknown population mean.

The procedure for estimating μ is based on the distribution of sample means (see Chapter 7). You should recall that this distribution is the set of all the possible $\overline{X}$ values for a specified sample size (n). The parameters of this distribution are the following:

1. The mean (called expected value) is equal to the population mean.
2. The standard deviation for this distribution (called standard error) is equal to $\sigma/\sqrt{n}$.
3. The distribution of sample means will be normal if either
   a. The population is normal, or
   b. The sample size is at least $n = 30$.

For the example we are considering, the distribution of sample means for $n = 25$ will be normal (because the population is normal), it will have a standard error of $\sigma/\sqrt{n} = 10/\sqrt{25} = 10/5 = 2$, and it will have a mean that is equal to the unknown population mean ($\mu = ?$). This distribution is shown in Figure 12.4.

Our sample mean, $\overline{X} = 88$, is somewhere in this distribution; that is, we have one value out of all the possible sample means. Unfortunately, we do not know where our sample mean is located in the distribution. Nonetheless, we can specify different locations by using z-scores. The z-score values and their locations have been identified in Figure 12.4.

Our sample mean, $\overline{X} = 88$, has a z-score given by the formula

$$z = \frac{\overline{X} - \mu}{\sigma_{\overline{X}}}$$

Because our goal is to find the population mean (μ), we will solve this z-score equation for μ. The algebra in this process is as follows:

$$z\sigma_{\overline{X}} = \overline{X} - \mu \qquad \text{(Multiply both sides of the equation by } \sigma_{\overline{X}}.\text{)}$$

$$\mu + z\sigma_{\overline{X}} = \overline{X} \qquad \text{(Add } \mu \text{ to both sides of the equation.)}$$

$$\mu = \overline{X} - z\sigma_{\overline{X}} \qquad \text{(Subtract } z\sigma_{\overline{X}} \text{ from both sides of the equation.)}$$

To use this equation, we begin with the values we know: $\overline{X} = 88$ and $\sigma_{\overline{X}} = 2$. To complete the equation, we must obtain a value for z. The value we will use for z is determined by estimating the z-score for our sample mean. More precisely, we are estimating the location of our sample mean within the distribution of sample means. The estimated position will determine a z-score value that can be used in the equation to compute μ. It is important to note that the z-score value we will be using is an *estimate;* therefore, the population mean that we compute will also be an estimate.

POINT ESTIMATES

For a point estimate, you must select a single value for the z-score. It should be clear that your best bet is to select the exact middle of the distribution—that is, the most reasonable estimate is $z = 0$. It would be unwise to pick an extreme value, such as $z = 2$, because there are relatively few samples that far away from the population mean. Most of the sample means pile up around $z = 0$, so this is your best choice. When this z-value is used in the equation, we get

$$\mu = \overline{X} - z\sigma_{\overline{X}}$$
$$\mu = 88 - 0(2)$$
$$= 88$$

This is our point estimate of the population mean. You should notice that we simply have used the sample mean, $\overline{X}$, to estimate the population mean, μ. The sample is the *only* information we have about the population, and you should recall from Chapter 7 that sample means tend to approximate μ (central limit theorem). Our conclusion is that the special summer program will increase reading scores from an average of $\mu = 80$ to an average of $\mu = 88$. We are estimating that the program will have an 8-point effect on reading scores.

INTERVAL ESTIMATES

To make an interval estimate, you select a range of z-score values, rather than a single point. Looking again at the distribution of sample means in Figure 12.4, where would you estimate our sample mean is located? Remember, you now can pick a range of values. As before, your best bet is to predict that the sample mean is located somewhere in the center of the distribution. There is a good chance, for example, that our sample mean is located somewhere between $z = +1$ and $z = -1$. You would be almost certain that $\overline{X}$ is between $z = +3$ and $z = -3$. How do you know what range to use? Because several different ranges are possible and each range has its own degree of confidence, the first step is to determine the amount of confidence we want and then use this value to determine the range. Commonly used levels of confidence start at about 60% and go up. For this example, we will use 90%. This means that we want to be 90% confident that our interval estimate of μ is correct.

There are no strict rules for choosing a level of confidence. Researchers must decide how much precision and how much confidence are needed in each specific situation.

To be 90% confident, we simply estimate that our sample mean is somewhere in the middle 90% of the distribution of sample means. This section of the distribution is bounded by z-scores of $z = +1.65$ and $z = -1.65$ (check the unit normal table). We are 90% confident that our particular sample mean ($\overline{X} = 88$) is in this range because 90% of all the possible means are there.

The next step is to use this range of z-score values in the estimation equation. We use the two ends of the z-score range to compute the two ends of the interval estimate for μ.

At one extreme, $z = +1.65$, which gives

$$\mu = \overline{X} - z\sigma_{\overline{X}}$$
$$= 88 - 1.65(2)$$
$$= 88 - 3.30$$
$$= 84.70$$

At the other extreme, $z = -1.65$, which gives

$$\mu = \overline{X} - z\sigma_{\overline{X}}$$
$$= 88 - (-1.65)(2)$$
$$= 88 + 3.30$$
$$= 91.30$$

How did we get $z = \pm1.65$ from the unit normal table? If we are looking for the middle 90% of the distribution, then there must be 5% left in each tail ($90 + 5 + 5 = 100\%$). Look in column C to find the value closest to 5% (.0500). You will find the value .0495. Moving across to column A, the value for z is 1.65.

 Minitab only.

The result is an interval estimate for μ. We are estimating that the population mean after the special summer program is between 84.70 and 91.30. If the mean is as small as 84.70, then the effect of the special program will be to increase reading scores by an average of 4.70 points (from $\mu = 80$ to $\mu = 84.70$). If the mean is as large as 91.30, the program will increase scores by an average of 11.30 points (from $\mu = 80$ to $\mu = 91.30$). Thus, we conclude that the special summer program will increase reading scores, and we estimate that the magnitude of the increase will be between 4.7 and 11.3 points. Because we used a reasonable range of z-score values to generate this estimate, we can assume that the equation has produced a reasonable range of values for μ. Specifically, we can be 90% confident that the true population mean is between 84.7 and 91.3 because we are 90% confident that the z-score for our sample mean is between -1.65 and $+1.65$.

Notice that the confidence interval sets up a range of values with the sample mean in the middle. As with point estimates, we are using the sample mean to estimate the population mean, but now we are saying that the value of μ should be *around* $\overline{X}$, rather than exactly equal to $\overline{X}$. Because the confidence interval is built around $\overline{X}$, adding in one direction and subtracting in the other, we will modify the estimation equation in order to simplify the arithmetic.

$$\mu = \overline{X} \pm z\sigma_{\overline{X}} \tag{12.2}$$

To build the confidence interval, start with the sample mean, and add $z\sigma_{\overline{X}}$ to get the boundary in one direction; then subtract $z\sigma_{\overline{X}}$ to get the other boundary. Translated into words, the formula says that

population mean = sample mean $\pm$ some error

The sample mean is expected to be representative of the population mean with some margin of error. Although it may seem obvious that the sample mean is used as the

FIGURE 12.5

The distribution of sample means for samples of size $n = 25$ selected from a normal population with $\sigma = 10$ and an unknown mean ($\mu = ?$). Note that 90% of all the possible sample means are located between $z = 1.65$ and $z = -1.65$. Also note that any sample mean located within these z-score boundaries will be within $(1.65)(2) = 3.30$ points of the population mean.

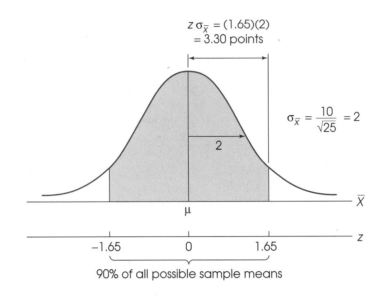

basis for estimating the population mean, you should not overlook the reason for this result. Sample means, on average, provide an accurate, unbiased representation of the population mean. You should recognize this fact as one of the characteristics of the distribution of sample means: The mean (expected value) of the distribution of sample means is μ.

INTERPRETATION OF THE CONFIDENCE INTERVAL

In the preceding example, we computed the 90% confidence interval for μ. The interval ranges from 84.7 to 91.3, and we are 90% confident that the true (unknown) population mean is located somewhere in this interval. The logic that allows us to be 90% confident in this estimate proceeds as follows:

1. The sample mean, $\overline{X} = 88$, is located somewhere within the distribution of sample means. The exact location is determined by a z-score.

2. Although we do not know the exact location of the sample mean, or its z-score, we can be 90% confident that the z-score is between 1.65 and -1.65 because 90% of all the possible z-scores are contained in this range (see Figure 12.5).

3. Therefore, we *estimate* that our sample mean is located between $z = 1.65$ and $z = -1.65$. If this estimate is correct, and it will be correct 90% of the time, then the sample mean is guaranteed to be within a specific distance of the unknown population mean (see Figure 12.5). This distance is determined by z times the standard error: In this example,

In Chapter 5 (page 118), we noted that $z\sigma$ is a deviation score.

$$z\sigma_{\overline{X}} = (1.65)(2) = 3.30 \text{ points}$$

This distance is then used to construct the confidence interval. As long as the sample mean falls within the z-score boundaries, you are guaranteed that the population mean will fall within the limits of the confidence interval.

4. If we took additional samples and computed each sample mean, then we would have a variety of values for $\overline{X}$ and could compute a variety of differ-

ent confidence intervals. However, 90% of all the different sample means would be located between $z = 1.65$ and $z = -1.65$ (see Figure 12.5). Thus, 90% of the sample means would be located within a distance of 3.30 points of the unknown population mean. Because each confidence interval extends 3.30 points in either direction from the sample mean, 90% of the different confidence intervals would contain the true population mean. As long as the sample mean is located in the middle 90% of the distribution, the 90% confidence interval will contain μ.

LEARNING CHECK

1. A cattle rancher is interested in using a newly developed hormone to increase the weight of beef cattle. Before investing in this hormone, the rancher would like to obtain some estimate of its effect. Without the hormone, the cattle weigh an average of $\mu = 1250$ pounds when they are sold at 8 months of age. The distribution of weights is approximately normal with $\sigma = 80$. A sample of 16 calves is selected to test the hormone. At age 8 months, the average weight for this sample is $\overline{X} = 1340$ pounds.

   a. Use these data to make a point estimate of the population mean weight if all the cattle were given the hormone.

   b. Make an interval estimate of the population mean so that you are 95% confident that the true mean is in your interval.

2. A researcher has collected sample data and is ready to compute a confidence interval. What *z*-scores should be used so that the researcher is 80% confident that the interval contains μ? What *z*-scores should be used for 70% confidence?

3. Suppose the 95% confidence interval for adult male weights is from 125 to 175 pounds. This indicates that 95% of all men weigh between these two values. (True or false?)

ANSWERS

1. **a.** For a point estimate, use the sample mean: $\overline{X} = 1340$ pounds.

   b. For the 95% confidence interval, $z = \pm 1.96$ and $\sigma_{\overline{X}} = 20$. The interval would be

   $$\mu = 1340 \pm 39.20$$

   The interval ranges from 1300.80 to 1379.20 pounds.

2. For the 80% confidence interval, the *z*-scores are $+1.28$ and -1.28. For the 70% confidence interval, *z*-scores of $+1.04$ and -1.04 should be used.

3. False. It indicates that you can be 95% confident that the population mean, μ, for men's weight is in the interval from 125 to 175 pounds.

12.3 ESTIMATION WITH THE *t* STATISTIC

In the preceding three chapters, we introduced three different versions of the *t* statistic: the single-sample *t* in Chapter 9, the independent-measures *t* in Chapter 10, and the repeated-measures *t* in Chapter 11. Although the three *t* statistics were introduced in the context of hypothesis testing, they all can be adapted for use in estimation. As we saw in the previous section, the basic z-score formula can be converted to the following form for estimation:

$$\mu = \overline{X} \pm z\sigma_{\overline{X}}$$

Each of the three t statistics can be converted in the same way to produce the general estimation formula

$$\begin{array}{c} \text{population mean} \\ \text{or mean difference} \end{array} = \begin{array}{c} \text{sample mean} \\ \text{or mean difference} \end{array} \pm t\,(\text{standard error})$$

With the single-sample t, we will estimate an unknown population mean, μ, using a sample mean, $\overline{X}$. The estimation formula for the single-sample t is

$$\mu = \overline{X} \pm ts_{\overline{X}} \tag{12.3}$$

With the independent-measures t, we will estimate the size of the difference between two population means, $\mu_1 - \mu_2$, using the difference between two sample means, $\overline{X}_1 - \overline{X}_2$. The estimation formula for the independent-measures t is

$$\mu_1 - \mu_2 = \overline{X}_1 - \overline{X}_2 \pm ts_{\overline{X}_1 - \overline{X}_2} \tag{12.4}$$

Finally, the repeated-measures t statistic will be used to estimate the mean difference for the general population, μ_D, using the mean difference for a sample, $\overline{D}$. The estimation formula for the repeated-measures t is

$$\mu_D = \overline{D} \pm ts_{\overline{D}} \tag{12.5}$$

The procedure for using the t statistic formulas for estimation is essentially the same as the procedure we used with the z-score formula. First, you use the sample data to compute the sample mean (or mean difference) and the standard error. Next, you estimate a value, or a range of values, for t. More precisely, you are estimating where the sample data are located in the t distribution. The most likely value is $t = 0$, and this is the value that will be used for point estimates. For interval estimates, we will use a range of values around zero, with the exact range determined by the level of confidence. For 90% confidence, for example, we will use the range of t values that form the middle 90% of the t distribution. Finally, the sample mean(s) and the standard error (both computed from the sample data) and the estimated value for t (obtained from the t-distribution table) are plugged into the estimation formula. These values complete the right-hand side of the equation and allow us to compute an estimated value for the mean (or the mean difference). The following examples demonstrate the estimation procedure with each of the three t statistics.

ESTIMATION OF μ FOR SINGLE-SAMPLE STUDIES

In Chapter 9, we introduced single-sample studies and hypothesis testing with the t statistic. Now we will use a single-sample study to estimate the value for μ, using point and interval estimates. A t statistic, rather than a z-score, is used because the population standard deviation is not known.

EXAMPLE 12.1

A marketing researcher for a major U.S. jeans manufacturer would like to estimate the mean age for the population of people who buy its products. This information will be valuable in making decisions about how to spend advertising dollars. For example, should the company place more advertisements in

In this example, we are simply trying to estimate the value of μ. Because no particular treatment is involved here, we are not trying to determine the size of a treatment effect.

Seventeen magazine or in *Cosmopolitan?* These represent publications that are directed at different age groups. It would be too costly and time-consuming to record the age of every person in the population of their consumers, so a random sample is taken to estimate the value of μ. A sample of $n = 30$ people is drawn from the consumers who purchase the jeans from several major clothing outlets. The mean age of this sample is $\overline{X} = 30.5$ years with $SS = 709$. The marketing researcher wishes to make a point estimate and to determine the 95% confidence interval for μ.

Notice that nothing is known about the population parameters. Estimation of the value for μ will be based solely on the sample data. Because σ is unknown, a *t* statistic will be used for the estimation. The confidence level has been selected (95%), and the sample data have been collected. Now we can turn our attention to the computational steps of estimation.

Compute s^2 and $s_{\overline{X}}$ The population standard deviation is not known; therefore, to estimate μ, it is necessary to use the estimated standard error. In Chapters 7 and 9, we introduced ways to compute standard errors with variance. To obtain $s_{\overline{X}}$, we must first compute the sample variance. Using the information provided, we obtain

$$s^2 = \frac{SS}{n-1}$$

$$= \frac{709}{29}$$

$$= 24.45$$

For estimated standard error, we obtain

$$s_{\overline{X}} = \sqrt{\frac{s^2}{n}}$$

$$= \sqrt{\frac{24.45}{30}}$$

$$= \sqrt{0.81}$$

$$= 0.90$$

Compute the point estimate   The value for *t* that is used depends on the type of estimate being made. A single *t* value is used for a point estimate, and an interval of values is used for the confidence interval. Just as we observed with the *z*-score distribution, *t* values are symmetrically distributed with a mean of zero. Therefore, we will use $t = 0$, the center of the distribution, as the best choice for the point estimate. Using the sample data, the estimation formula yields a point estimate of

$$\mu = \overline{X} \pm ts_{\overline{X}}$$

$$= 30.5 \pm 0(0.90)$$

$$= 30.5 \pm 0$$

$$= 30.5$$

As noted before, the sample mean is the most appropriate point estimate of the population mean.

FIGURE 12.6

The 95% confidence interval for $df = 29$ will have boundaries that range from $t = -2.045$ to $t = +2.045$. Because the t distribution table presents proportions in both tails of the distribution, for the 95% confidence interval you find the t values under $p = 0.05$ (proportions of t-scores in two tails) for $df = 29$.

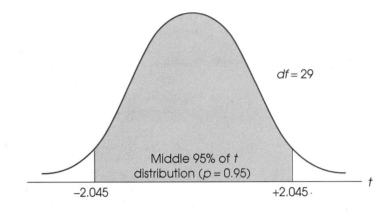

$df = 29$

Middle 95% of t distribution ($p = 0.95$)

-2.045 $+2.045$ t

We do not know the actual t value associated with the $\overline{X}$ we obtained. For that information, we would need the value for μ—which we are trying to estimate. So we use values of t, just as we did with z, to define an interval around $\overline{X}$ that probably contains the value of μ.

Construct an interval estimate   For an interval estimate of μ, we construct an interval around the sample mean in which the value for μ probably falls. We now use a range of t values to define this interval. For example, there is a good chance that the sample has a t value somewhere between $t = +2$ and $t = -2$ and even a much better chance it is between $t = +4$ and $t = -4$. The level of confidence (percentage of confidence) will determine the t values that mark off the boundaries of this interval. However, unlike the normal z distribution, there is a family of t distributions in which the exact shape of the distribution depends on the value of degrees of freedom. Therefore, the value of df that is associated with the sample is another determining factor of the t values to be used in the interval estimate. For this example,

$$df = n - 1 = 30 - 1 = 29$$

The marketing researcher selected the 95% confidence interval. Figure 12.6 depicts the t distribution for $df = 29$. To obtain the t values associated with the 95% confidence interval, we must consult the t distribution table. We look under the heading of proportions in *two tails*. If the middle 95% of the distribution is of interest to us, then both tails outside of the interval together will contain 5% of the t values. Therefore, to find the t values associated with the 95% confidence interval, we look for the entry under $p = .05$, two tails, for $df = 29$. The values of t used for the boundaries of this confidence interval are -2.045 and $+2.045$ (Figure 12.6). Using these values in the formula for μ, we obtain, for one end of the confidence interval,

$$\mu = \overline{X} - ts_{\overline{X}}$$
$$= 30.5 - 2.045(.90)$$
$$= 30.5 - 1.84$$
$$= 28.66$$

Minitab only.

For the other end of the interval,

$$\mu = \overline{X} + ts_{\overline{X}}$$
$$= 30.5 + 2.045(0.90)$$
$$= 30.5 + 1.84$$
$$= 32.34$$

Therefore, the marketing researcher can be 95% confident that the population mean age for consumers of his product is between 28.66 and 32.34 years. The confidence level (%) determines the range of *t* values used in constructing the interval. As long as the obtained sample really does have a *t* value that falls within the estimated range of *t* values, the population mean will be included in the confidence interval.

ESTIMATION OF $\mu_1 - \mu_2$ FOR INDEPENDENT-MEASURES STUDIES

The independent-measures *t* statistic can be used for estimation as well as hypothesis testing. Remember, hypothesis testing (Chapter 10) is used to answer a yes-no question: Is there any mean difference between the two populations? With estimation, we ask, *How much* difference? In this case, the independent-measures *t* statistic is used to estimate the value of $\mu_1 - \mu_2$.

EXAMPLE 12.2

Recent studies have allowed psychologists to establish definite links between specific foods and specific brain functions. For example, lecithin (found in soybeans, eggs, and liver) has been shown to increase the concentration of certain brain chemicals that help regulate memory and motor coordination. This experiment is designed to demonstrate the importance of this particular food substance.

The experiment involves two separate samples of newborn rats (an independent-measures experiment). The 10 rats in the first sample are given a normal diet containing standard amounts of lecithin. The 5 rats in the other sample are fed a special diet, which contains almost no lecithin. After six months, each of the rats is tested on a specially designed learning problem that requires both memory and motor coordination. The purpose of the experiment is to demonstrate the deficit in performance that results from lecithin deprivation. The score for each animal is the number of errors it makes before it solves the learning problem. The data from this experiment are as follows:

Regular diet	No-lecithin diet
$n = 10$	$n = 5$
$\overline{X} = 25$	$\overline{X} = 33$
$SS = 250$	$SS = 140$

Because we fully expect that there will be a significant difference between these two treatments, we will not do the hypothesis test (although you should be able to do it). We want to use these data to obtain an estimate of the size of the difference between the two population means; that is, how much does lecithin affect learning performance? We will use a point estimate and an 80% confidence interval.

The basic equation for estimation with an independent-measures experiment is

$$\mu_1 - \mu_2 = (\overline{X}_1 - \overline{X}_2) \pm ts_{\overline{X}_1 - \overline{X}_2}$$

The first step is to obtain the known values from the sample data. The sample mean difference is easy; one group averaged $\overline{X} = 25$, and the other averaged

FIGURE 12.7

The distribution of t values with $df = 13$. Note that t values pile up around zero and that 80% of the values are between $+1.350$ and -1.350.

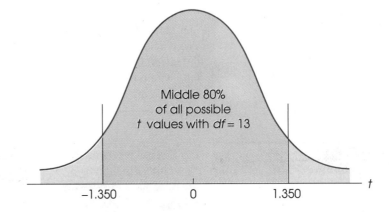

Middle 80% of all possible t values with $df = 13$

−1.350   0   1.350 t

$\overline{X} = 33$, so there is an 8-point difference. Notice that it is not important whether we call this a $+8$ or a -8 difference. In either case, the size of the difference is 8 points, and the regular diet group scored lower. Because it is easier to do arithmetic with positive numbers, we will use

$$\overline{X}_1 - \overline{X}_2 = 8$$

Compute the standard error   To find the standard error, we first must pool the two variances:

$$
\begin{aligned}
s_p^2 &= \frac{SS_1 + SS_2}{df_1 + df_2} \\
&= \frac{250 + 140}{9 + 4} \\
&= \frac{390}{13} \\
&= 30
\end{aligned}
$$

Next, the pooled variance is used to compute the standard error:

$$s_{\overline{X}_1 - \overline{X}_2} = \sqrt{\frac{s_p^2}{n_1} + \frac{s_p^2}{n_2}} = \sqrt{\frac{30}{10} + \frac{30}{5}} = \sqrt{3 + 6} = \sqrt{9} = 3$$

You should recall that this standard error combines the error from the first sample and the error from the second sample. Because the first sample is much larger, $n = 10$, it should have less error. This difference shows up in the formula. The larger sample contributes an error of 3 points, and the smaller sample contributes 6 points, which combine for a total error of 9 points under the square root.

Sample 1 has $df = 9$, and sample 2 has $df = 4$. The t statistic has $df = 9 + 4 = 13$.

The final value needed on the right-hand side of the equation is t. The data from this experiment would produce a t statistic with $df = 13$. With 13 degrees of freedom, we can sketch the distribution of all the possible t values. This distribution is shown in Figure 12.7. The t statistic for our data is somewhere in this distribution. The problem is to estimate where. For a point estimate, the best bet is to use $t = 0$. This is the most likely value, located exactly in the middle of the distribution. To gain more confidence in the estimate, you can select a range of t values. For 80% confidence, for example, you would estimate that the t statistic is somewhere in the middle 80% of the distribution. Checking the table, you find that the middle 80% is bounded by values of $t = +1.350$ and $t = -1.350$.

Using these t values and the sample values computed earlier, we now can estimate the magnitude of the performance deficit caused by lecithin deprivation.

Compute the point estimate   For a point estimate, use the single-value (point) estimate of $t = 0$:

$$\mu_1 - \mu_2 = (\overline{X}_1 - \overline{X}_2) \pm t s_{\overline{X}_1 - \overline{X}_2}$$
$$= 8 \pm 0(3)$$
$$= 8$$

Notice that the result simply uses the sample mean difference to estimate the population mean difference. The conclusion is that lecithin deprivation produces an average of 8 more errors on the learning task. (Based on the fact that the normal animals averaged around 25 errors, an 8-point increase would mean a performance deficit of approximately 30%.)

Construct the interval estimate   For an interval estimate, or confidence interval, use the range of t values. With 80% confidence, at one extreme,

$$\mu_1 - \mu_2 = (\overline{X}_1 - \overline{X}_2) + t s_{\overline{X}_1 - \overline{X}_2}$$
$$= 8 + 1.350(3)$$
$$= 8 + 4.05$$
$$= 12.05$$

and at the other extreme,

Minitab only.

$$\mu_1 - \mu_2 = (\overline{X}_1 - \overline{X}_2) - t s_{\overline{X}_1 - \overline{X}_2}$$
$$= 8 - 1.350(3)$$
$$= 8 - 4.05$$
$$= 3.95$$

This time we are concluding that the effect of lecithin deprivation is to increase errors, with an average increase somewhere between 3.95 and 12.05 errors. We are 80% confident of this estimate because the only thing estimated was the location of the t statistic, and we used the middle 80% of all the possible t values.

Note that the result of the point estimate is to say that lecithin deprivation will increase errors by exactly 8. To gain confidence, you must lose precision and say that errors will increase by around 8 (for 80% confidence, we say that the average increase will be 8 ± 4.05).

ESTIMATION OF μ_D FOR REPEATED-MEASURES STUDIES

Finally, we turn our attention to the repeated-measures study. Remember, this type of study has a single sample of subjects, which is measured in two different treatment conditions. By finding the difference between the score for treatment 1 and the score for treatment 2, we can determine a difference score for each subject.

$$D = X_2 - X_1$$

The mean of the sample of D scores, $\overline{D}$, is used to estimate the value of the mean of the population of D scores, μ_D.

E X A M P L E   1 2 . 3

A school psychologist has determined that a remedial reading course increases scores on a reading comprehension test. The psychologist now would like to estimate how much improvement might be expected for the whole population of students in his city. A random sample of $n = 16$ children is obtained. These children are first tested for level of reading comprehension and then enrolled in the course. At the completion of the remedial reading course, the students are tested again, and the difference between the second score and the first score is recorded for each child. For this sample, the average difference was $\overline{D} = +21$, and the SS for the difference scores was $SS = 1215$. The psychologist would like to use these data to make a point estimate and a 90% confidence interval estimate of μ_D.

The formula for estimation requires that we know the values of $\overline{D}$, $s_{\overline{D}}$, and t. We know that $\overline{D} = +21$ points for this sample, so all that remains is to compute $s_{\overline{D}}$ and look up the value of t in the t-distribution table.

Compute the standard error To find the standard error, we first must compute the sample variance:

$$s^2 = \frac{SS}{n - 1} = \frac{1215}{15} = 81$$

Now the estimated standard error is

$$s_{\overline{D}} = \sqrt{\frac{s^2}{n}} = \sqrt{\frac{81}{16}} = \frac{9}{4} = 2.25$$

To complete the estimate of μ_D, we must identify the value of t. We will consider the point estimate and the interval estimate separately.

Compute the point estimate To obtain a point estimate, a single value of t is selected to approximate the location of $\overline{D}$. Remember that the t distribution is symmetrical and bell-shaped with a mean of zero (see Figure 12.8). Because $t = 0$ is the most frequently occurring value in the distribution, this is the t value that is used for the point estimate. Using this value in the estimation formula gives

$$\mu_D = \overline{D} \pm ts_{\overline{D}}$$
$$= 21 \pm 0(2.25)$$
$$= 21$$

As noted several times before, the sample mean, $\overline{D} = 21$, provides the best point estimate of μ_D.

FIGURE 12.8

The *t* values for the 90% confidence interval are obtained by consulting the *t* table for *df* = 15, *p* = 0.10 for two tails.

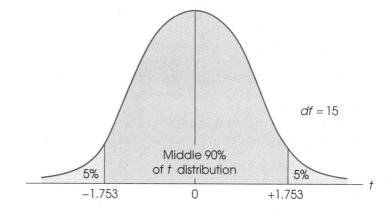

df = 15

Middle 90%
of *t* distribution

5%   5%

−1.753   0   +1.753   *t*

Construct the interval estimate   The psychologist also wanted to make an interval estimate in order to be 90% confident that the interval contains the value of μ_D. To get the interval, it is necessary to determine what *t* values form the boundaries of the middle 90% of the *t* distribution. To use the *t* distribution table, we first must determine the proportion associated with the tails of this distribution. With 90% in the middle, the remaining area in both tails must be 10%, or *p* = .10. Also note that our sample has *n* = 16 scores, so the *t* statistic will have *df* = *n* − 1 = 15. Using *df* = 15 and *p* = 0.10 for two tails, you should find the values +1.753 and −1.753 in the *t* table. These values form the boundaries for the middle 90% of the *t* distribution. (See Figure 12.8.) We are confident that the *t* value for our sample is in this range because 90% of all the possible *t* values are there. Using these values in the estimation formula, we obtain the following: On one end of the interval,

$$\mu_D = \bar{D} - ts_{\bar{D}}$$
$$= 21 - 1.753(2.25)$$
$$= 21 - 3.94$$
$$= 17.06$$

and on the other end of the interval,

Minitab only.

$$\mu_D = 21 + 1.753(2.25)$$
$$= 21 + 3.94$$
$$= 24.94$$

Therefore, the school psychologist can be 90% confident that the average amount of improvement in reading comprehension for the population (μ_D) will be somewhere between 17.06 and 24.94 points.

LEARNING CHECK

1. A professor notices that students who get an A in physics have high grade-point averages (GPAs) in their engineering courses. The professor selects a sample of *n* = 16 engineering majors who have earned As in physics. The mean GPA in engineering courses for this sample is $\bar{X}$ = 3.30. For this sample, *SS* = 10. What is the 99% confidence interval?

2. A psychologist studies the change in mood from the follicular phase (prior to ovulation) to the luteal phase (after ovulation) during the menstrual cycle. In a repeated-measures study, a sample of $n = 9$ women take a mood questionnaire during each phase. On average, the participants show an increase in dysphoria (negative moods) of $D = 18$ points with $SS = 152$. Determine the 95% confidence interval for population mean change in mood.

3. In families with several children, the first-born tend to be more reserved and serious, whereas the last-born tend to be more outgoing and happy-go-lucky. A psychologist is using a standardized personality inventory to measure the magnitude of this difference. Two samples are used: 8 first-born children and 8 last-born children. Each child is given the personality test. The results are as follows:

First-born	Last-born
$\overline{X} = 11.4$	$\overline{X} = 13.9$
$SS = 26$	$SS = 30$

   a. Use these sample statistics to make a point estimate of the population mean difference in personality for first-born versus last-born children.

   b. Make an interval estimate of the population mean difference so that you are 80% confident that the true mean difference is in your interval.

ANSWERS 1. $s^2 = 0.667$, $s_{\overline{X}} = .20$, $df = 15$, $t = \pm 2.947$; estimate that μ is between 2.71 and 3.89.

2. $s^2 = 19$, $s_{\overline{D}} = 1.45$, $df = 8$, $t = \pm 2.306$; estimate that μ_D is between 14.66 and 21.34.

3. a. For a point estimate, use the sample mean difference: $\overline{X}_1 - \overline{X}_2 = 2.5$ points.

   b. Pooled variance $= 4$, estimated standard error $= 1$, $df = 14$, $t = \pm 1.345$. The 80% confidence interval is 1.16 to 3.85.

12.4 FACTORS AFFECTING THE WIDTH OF A CONFIDENCE INTERVAL

Two characteristics of the confidence interval should be noted. First, notice what happens to the width of the interval when you change the level of confidence (the percent confidence). To gain more confidence in your estimate, you must increase the width of the interval. Conversely, to have a smaller interval, you must give up confidence. This is the basic trade-off between precision and confidence that was discussed earlier. In the estimation formula, the percent confidence influences the width of the interval by way of the z-score or t value. The larger the level of confidence (the percentage), the larger the z or t value, and the larger the interval. This relationship can be seen in Figure 12.8. In the figure, we have identified the middle 90% of the t distribution in order to find a 90% confidence interval. It should be obvious that if we were to increase the confidence level to 95%, it would be necessary to increase the range of t values and thereby increase the width of the interval.

   Second, notice what would happen to the interval width if you had a different sample size. This time the basic rule is as follows: The bigger the sample (n), the smaller the interval. This relationship is straightforward if you consider the sample size as a measure of the amount of information. A bigger sample gives you more information about the population and allows you to make a more precise estimate (a narrower interval). The sample size controls the magnitude of the standard error in the estimation formula. As the sample size increases, the standard error decreases, and the interval gets smaller.

With t statistics, the sample size has an additional effect on the width of a confidence interval. Remember that the exact shape of the t distribution depends on degrees of freedom. As the sample size gets larger, df also get larger, and the t values associated with any specific percentage of confidence get smaller. This fact simply enhances the general relationship that the larger a sample, the smaller a confidence interval.

SUMMARY

1. Estimation is a procedure that uses sample data to obtain an estimate of a population mean. The estimate can be either a point estimate (single value) or an interval estimate (range of values). Point estimates have the advantage of precision, but they do not give much confidence. Interval estimates provide confidence, but you lose precision as the interval grows wider.

2. Estimation and hypothesis testing are similar processes: Both use sample data to answer questions about populations. However, these two procedures are designed to answer different questions. Hypothesis testing will tell you whether or not a treatment effect exists (yes or no). Estimation will tell you how much treatment effect there is.

3. The z-score or a t formula can be used to estimate a population mean with data from a single sample. The z-score formula is used when the population standard deviation is known, and the t formula is used when σ is unknown. The two formulas are

$$\mu = \bar{X} \pm z\sigma_{\bar{X}} \quad \text{and} \quad \mu = \bar{X} \pm ts_{\bar{X}}$$

To use either formula, first calculate the sample mean and the standard error ($\sigma_{\bar{X}}$ or $s_{\bar{X}}$) from the sample data. Next, obtain an estimate of the value of z or t by estimating the location of the sample statistic within the appropriate distribution. For a point estimate, use $z = 0$ or $t = 0$. For an interval estimate, select a level of confidence (percent confidence) and then look up the range of z-scores or t values in the appropriate table.

4. For an independent-measures study, the formula for estimation is

$$\mu_1 - \mu_2 = (\bar{X}_1 - \bar{X}_2) \pm ts_{\bar{X}_1 - \bar{X}_2}$$

To use this formula, first decide on a degree of precision and a level of confidence desired for the estimate. If your primary concern is precision, use $t = 0$ to make a point estimate of the mean difference. Otherwise, select a level of confidence (percent confidence) that determines a range of t values to be used in the formula.

5. For a repeated-measures study, estimation of the amount of mean change for the population is accomplished by solving the t statistic formula for μ_D:

$$\mu_D = \bar{D} \pm ts_{\bar{D}}$$

For a point estimate, use a t value of zero. For an interval estimate, use a range of t values to construct an interval around $\bar{D}$. As in previous estimation problems, the t values that mark the interval boundaries are determined by the confidence level selected and by degrees of freedom.

6. The width of a confidence interval is an indication of its precision: A narrow interval is more precise than a wide interval. The interval width is influenced by the sample size and the level of confidence.
   a. As sample size (n) gets larger, the interval width gets smaller (greater precision).
   b. As the percent confidence increases, the interval width gets larger (less precision).

KEY TERMS

estimation   point estimate   interval estimate   confidence interval

FOCUS ON PROBLEM SOLVING

1. Although hypothesis tests and estimation are similar in some respects, you should remember that they are separate statistical techniques. A hypothesis test is used to determine whether or not there is evidence for a treatment effect. Estimation is used to determine how much effect a treatment has.

2. When students perform a hypothesis test and estimation with the same set of data, a common error is to take the z-score or t statistic from the hypothesis test and use it in the estimation formula. For estimation, the z-score or t value is determined by the level of confidence and must be looked up in the appropriate table.

3. Now that you are familiar with several different formulas for hypothesis tests and estimation, one problem will be determining which formula is appropriate for each set of data. When the data consist of a single sample selected from a single population, the appropriate statistic will be either z or the single-sample t, depending on whether σ is known or unknown, respectively. For an independent-measures design, you will always have two separate samples. In a repeated-measures design, there is only one sample, but each individual is measured twice so that difference scores can be computed.

DEMONSTRATION 12.1

ESTIMATION WITH A SINGLE-SAMPLE *t* STATISTIC

A sample of $n = 16$ is randomly selected from a population with unknown parameters. For the following sample data, estimate the value of μ using a point estimate and a 90% confidence interval:

Sample data: 13,   10,   8,   13,   9,   14,   12,   10

11,   10,   15,   13,   7,   6,   15,   10

Note that we have a single sample and we do not know the value for σ. Thus, the single-sample t statistic should be used for these data. The formula for estimation is

$$\mu = \overline{X} \pm ts_{\overline{X}}$$

STEP 1 Compute the sample mean.
The sample mean is the basis for our estimate of μ. For these data,

$$\Sigma X = 13 + 10 + 8 + 13 + 9 + 14 + 12 + 10 +$$
$$11 + 10 + 15 + 13 + 7 + 6 + 15 + 10$$
$$= 176$$

$$\overline{X} = \frac{\Sigma X}{n} = \frac{176}{16} = 11$$

STEP 2 Compute the estimated standard error, $s_{\overline{X}}$.
To compute the estimated standard error, we must first find the value for SS and the sample variance.
Sum of squares. We will use the definitional formula for SS. The following table demonstrates the computations:

X	$X - \overline{X}$	$(X - \overline{X})^2$
13	$13 - 11 = +2$	4
10	$10 - 11 = -1$	1
8	$8 - 11 = -3$	9
13	$13 - 11 = +2$	4
9	$9 - 11 = -2$	4
14	$14 - 11 = +3$	9
12	$12 - 11 = +1$	1
10	$10 - 11 = -1$	1
11	$11 - 11 = 0$	0
10	$10 - 11 = -1$	1
15	$15 - 11 = +4$	16
13	$13 - 11 = +2$	4
7	$7 - 11 = -4$	16
6	$6 - 11 = -5$	25
15	$15 - 11 = +4$	16
10	$10 - 11 = -1$	1

To obtain *SS*, we sum the squared deviation scores in the last column.

$$SS = \Sigma(X - \overline{X})^2 = 112$$

Variance. The sample variance is computed for these data.

$$s^2 = \frac{SS}{n - 1} = \frac{112}{16 - 1} = \frac{112}{15} = 7.47$$

Estimated standard error. The estimated standard error can now be determined.

$$s_{\overline{X}} = \sqrt{\frac{s^2}{n}} = \sqrt{\frac{7.47}{16}} = \sqrt{0.467} = 0.68$$

STEP 3 Compute the point estimate for μ.

For a point estimate, we use $t = 0$. Using the estimation formula, we obtain

$$\mu = \overline{X} \pm ts_{\overline{X}}$$
$$= 11 \pm 0(0.68)$$
$$= 11 \pm 0 = 11$$

The point estimate for the population mean is $\mu = 11$.

STEP 4 Determine the confidence interval for μ.

For these data, we want the 90% confidence interval. Therefore, we will use a range of t values that form the middle 90% of the distribution. For this demonstration, degrees of freedom are

$$df = n - 1 = 16 - 1 = 15$$

If we are looking for the middle 90% of the distribution, then 10% ($p = 0.10$) would lie in both tails outside of the interval. To find the t values, we look up $p = 0.10$, two tails, for $df = 15$ in the t-distribution table. The t values for the 90% confidence interval are $t = \pm 1.753$.

Using the estimation formula, one end of the confidence interval is

$$\mu = \overline{X} - ts_{\overline{X}}$$
$$= 11 - 1.753(0.68)$$
$$= 11 - 1.19 = 9.81$$

For the other end of the confidence interval, we obtain

$$\mu = \overline{X} + ts_{\overline{X}}$$
$$= 11 + 1.753(0.68)$$
$$= 11 + 1.19 = 12.19$$

Thus, the 90% confidence interval for μ is from 9.81 to 12.19.

DEMONSTRATION 12.2

ESTIMATION WITH THE INDEPENDENT-MEASURES t STATISTIC

Samples are taken from two school districts, and knowledge of American history is tested with a short questionnaire. For the following sample data, estimate the amount of mean difference between the students of these two districts. Specifically, provide a point estimate and a 95% confidence interval for $\mu_1 - \mu_2$.

District A scores: 18,   15,   24,   15

District B scores: 9,   12,   13,   6

STEP 1 Compute the sample means.

The estimate of population mean difference $(\mu_1 - \mu_2)$ is based on the sample mean difference $(\overline{X}_1 - \overline{X}_2)$.

For district A,

$$\Sigma X = 18 + 15 + 24 + 15 = 72$$

$$\overline{X}_1 = \frac{\Sigma X}{n} = \frac{72}{4} = 18$$

For district B,

$$\Sigma X = 9 + 12 + 13 + 6 = 40$$

$$\overline{X}_2 = \frac{\Sigma X}{n} = \frac{40}{4} = 10$$

STEP 2 Calculate the estimated standard error for mean difference, $s_{\overline{X}_1 - \overline{X}_2}$.

To compute the estimated standard error, we first need to determine the values of SS for both samples and pooled variance.

Sum of squares. The computations for sum of squares, using the definitional formula, are shown for both samples in the following tables:

District A		
X	$X - \overline{X}$	$(X - \overline{X})^2$
18	$18 - 18 = 0$	0
15	$15 - 18 = -3$	9
24	$24 - 18 = +6$	36
15	$15 - 18 = -3$	9

District B		
X	$X - \overline{X}$	$(X - \overline{X})^2$
9	$9 - 10 = -1$	1
12	$12 - 10 = +2$	4
13	$13 - 10 = +3$	9
6	$6 - 10 = -4$	16

For district A,

$$SS_1 = \Sigma(X - \overline{X})^2 = 0 + 9 + 36 + 9 = 54$$

For district B,

$$SS_2 = \Sigma(X - \overline{X})^2 = 1 + 4 + 9 + 16 = 30$$

Pooled variance. For pooled variance, we use the *SS* and *df* values from both samples. For district A, $df_1 = n_1 - 1 = 3$. For district B, $df_2 = n_2 - 1 = 3$. Pooled variance is

$$s_p^2 = \frac{SS_1 + SS_2}{df_1 + df_2} = \frac{54 + 30}{3 + 3} = \frac{84}{6} = 14$$

Estimated standard error. The estimated standard error for mean difference can now be calculated.

$$s_{\overline{X}_1 - \overline{X}_2} = \sqrt{\frac{s_p^2}{n_1} + \frac{s_p^2}{n_2}} = \sqrt{\frac{14}{4} + \frac{14}{4}} = \sqrt{3.5 + 3.5}$$
$$= \sqrt{7} = 2.65$$

STEP 3 Compute the point estimate for $\mu_1 - \mu_2$.

For the point estimate, we use a *t* value of zero. Using the sample means and estimated standard error from previous steps, we obtain

$$\mu_1 - \mu_2 = (\overline{X}_1 - \overline{X}_2) \pm ts_{\overline{X}_1 - \overline{X}_2}$$
$$= (18 - 10) \pm 0(2.65)$$
$$= 8 \pm 0 = 8$$

STEP 4 Determine the confidence interval for $\mu_1 - \mu_2$.

For the independent-measures *t* statistic, degrees of freedom are determined by

$$df = n_1 + n_2 - 2$$

For these data, *df* is

$$df = 4 + 4 - 2 = 6$$

With a 95% level of confidence, 5% of the distribution falls in the tails outside the interval. Therefore, we consult the *t*-distribution table for $p = 0.05$, two tails, with $df = 6$. The *t* values from the table are $t = \pm2.447$. On one end of the confidence interval, the population mean difference is

$$\mu_1 - \mu_2 = (\overline{X}_1 - \overline{X}_2) - ts_{\overline{X}_1 - \overline{X}_2}$$
$$= (18 - 10) - 2.447(2.65)$$
$$= 8 - 6.48$$
$$= 1.52$$

On the other end of the confidence interval, the population mean difference is

$$\mu_1 - \mu_2 = (\overline{X}_1 - \overline{X}_2) + ts_{\overline{X}_1 - \overline{X}_2}$$
$$= (18 - 10) + 2.447(2.65)$$
$$= 8 + 6.48$$
$$= 14.48$$

Thus, the 95% confidence interval for population mean difference is from 1.52 to 14.48.

PROBLEMS

1. Explain how the purpose of estimation differs from the purpose of a hypothesis test.

2. Explain why it would *not* be reasonable to use estimation after a hypothesis test where the decision was "fail to reject H_0."

3. Explain how each of the following factors affects the width of a confidence interval:
   a. Increasing the sample size
   b. Increasing the sample variability
   c. Increasing the level of confidence (the percent confidence)

4. For the following studies, state whether estimation or hypothesis testing is required. Also, is an independent- or a repeated-measures *t* statistic appropriate?
   a. An educator wants to determine how much mean difference can be expected for the population in SAT scores following an intensive review course. Two samples are selected. The first group takes the review course, and the second receives no treatment. SAT scores are subsequently measured for both groups.
   b. A psychiatrist would like to test the effectiveness of a new antipsychotic medication. A sample of patients is first assessed for the severity of psychotic symptoms.

Then the patients are placed on drug therapy for two weeks. The severity of their symptoms is assessed again at the end of the treatment.

5. In 1985, an extensive survey indicated that fifth-grade students in the city school district spent an average of $\mu = 5.5$ hours per week doing homework. The distribution of homework times was approximately normal, with $\sigma = 2$. Last year, a sample of $n = 100$ fifth-grade students produced a mean of $\overline{X} = 5.1$ hours of homework each week.
   a. Use the data to make a point estimate of the population mean for last year. Assume there was no change in the standard deviation.
   b. Based on your point estimate, how much change has occurred in homework time since 1985?
   c. Make an interval estimate of last year's homework time so that you are 80% confident that the true population mean is in your interval.

6. A researcher has constructed a 90% confidence interval of 87 ± 10, based on a sample of $n = 25$ scores. Note that this interval is 20 points wide (from 77 to 97). How large a sample would be needed to produce a 90% interval that is only 10 points wide?

7. Researchers have developed a filament that should add to the life expectancy of light bulbs. The standard 60-watt bulb burns for an average of $\mu = 750$ hours with $\sigma = 20$. A sample of $n = 100$ bulbs is prepared using the new filament. The average life for this sample is $\overline{X} = 820$ hours.
   a. Use these sample data to make a point estimate of the mean life expectancy for the new filament.
   b. Make an interval estimate so that you are 80% confident that the true mean is in your interval.
   c. Make an interval estimate so that you are 99% confident that the true mean is in your interval.

8. A toy manufacturer asks a developmental psychologist to test children's responses to a new product. Specifically, the manufacturer wants to know how long, on average, the toy captures children's attention. The psychologist tests a sample of $n = 9$ children and measures how long they play with the toy before they get bored. This sample had a mean of $\overline{X} = 31$ minutes with $SS = 648$.
   a. Make a point estimate for μ.
   b. Make an interval estimate for μ using a confidence level of 95%.

9. A random sample of $n = 11$ scores is selected from a population with unknown parameters. The scores in the sample are as follows: 12, 5, 9, 9, 10, 14, 7, 10, 14, 13, 8.
   a. Provide an estimate of the population standard deviation.
   b. Use the sample data to make a point estimate for μ and to construct the 95% confidence interval for μ.

10. A psychologist has developed a new personality questionnaire for measuring self-esteem and would like to estimate the population parameters for the test scores. The questionnaire is administered to a sample of $n = 25$ subjects. This sample has an average score of $\overline{X} = 43$ with $SS = 2400$.
   a. Provide an estimate for the population standard deviation.
   b. Make a point estimate for the population mean.
   c. Make an interval estimate of μ so that you are 90% confident that the value for μ is in your interval.

11. On a test of short-term memory for random strings of digits, the number of digits recalled for a sample of adults is measured. The data are as follows: 7, 9, 8, 10, 8, 6, 7, 8, 7, 6, 5. Make a point estimate for the population mean. Construct the 95% confidence interval.

12. Most adolescents experience a growth spurt when they are between 12 and 15 years old. This period of dramatic growth generally occurs around age 12 for girls and around age 14 for boys. A researcher studying physical development selected a random sample of $n = 9$ girls and a second sample of $n = 16$ boys and recorded the gain in height (in millimeters) between the 14th birthday and 15th birthday for each subject. The girls showed an average gain of $\overline{X} = 40$ millimeters with $SS = 1152$, and the boys gained an average of $\overline{X} = 95$ millimeters with $SS = 2160$.
   a. Estimate the population mean growth in one year for 14-year-old boys. Make a point estimate and an 80% confidence interval estimate.
   b. Estimate the population mean growth in one year for 14-year-old girls. Make a point estimate and an 80% confidence interval estimate.
   c. Estimate the mean difference in growth for boys versus girls during this one-year period. Again, make a point estimate and an 80% confidence interval estimate.

13. A therapist has demonstrated that five sessions of relaxation training significantly reduced anxiety levels for a sample of $n = 16$ clients. However, the therapist is concerned about the long-term effects of the training. Six months after therapy is completed, the patients are recalled, and their anxiety levels are measured again. On average, the anxiety scores for these patients are $\overline{D} = 5.5$ points higher after six months than they had been at the end of therapy. The difference scores had $SS = 960$. Use these data to estimate the mean amount of relapse that occurs after therapy ends. Make a point estimate and an 80% confidence interval estimate of the population mean difference.

14. An educational psychologist has observed that children seem to lose interest and enthusiasm for school as they progress through the elementary grades. To measure the extent of this phenomenon, the psychologist selects a sample of $n = 15$ second-grade children and a sample of $n = 15$ fifth-graders. Each child is given a questionnaire measuring his/her attitude toward school. Higher scores indicate a more positive attitude. The second-grade children average $\overline{X} = 85$ with $SS = 1620$, and the fifth-graders average $\overline{X} = 71$ with $SS = 1740$. Use these data to estimate how much the enthusiasm for school declines from second to fifth grade. Make a point estimate and a 90% confidence interval estimate of the mean difference.

15. The counseling center at the college offers a short course in study skills for students who are having academic difficulty. To evaluate the effectiveness of this course, a sample of $n = 25$ students is selected, and each student's grade-point average is recorded for the semester before the course and for the semester immediately following the course. On average, these students show an increase of $\overline{D} = 0.72$ with $SS = 24$. Use these data to estimate how much effect the course has on grade-point average. Make a point estimate and a 95% confidence interval estimate of the mean difference.

16. For the study in problem 8 of Chapter 10, construct an interval estimate so that you are 99% confident that it contains the population mean difference.

17. For the study in problem 9 of Chapter 10, make a point estimate for $\mu_1 - \mu_2$, and determine the 90% confidence interval.

18. "Encoding specificity" is a psychological principle that states that a person's recall performance will be best if memory is tested under the same conditions that existed when the person originally learned the material. To evaluate the magnitude of this effect, a researcher obtains a sample of 50 subjects. The subjects all listen to a lecture on river pollution and are warned that they will be tested on the information. One week later the subjects are re-assembled. Twenty-five subjects are assigned to the same room where they heard the lecture, and the other 25 are assigned to a different room. All subjects take the same test. The average score for the same-room subjects is $\overline{X} = 38$ with $SS = 1040$, and the different-room subjects average $\overline{X} = 26$ with $SS = 1360$. Use these data to estimate how much memory is affected by having memory tested in the same room as learning. Make a point estimate and a 90% confidence interval estimate of the mean difference.

19. The following data are from two experiments, as previously examined in problem 14 of Chapter 10.

Experiment 1		Experiment 2	
Treatment A	Treatment B	Treatment A	Treatment B
$n = 10$	$n = 10$	$n = 10$	$n = 10$
$\overline{X} = 42$	$\overline{X} = 52$	$\overline{X} = 61$	$\overline{X} = 71$
$SS = 180$	$SS = 120$	$SS = 986$	$SS = 1042$

a. Which experiment has a larger mean difference (1 or 2)? (This is a trick question.)
b. Without doing any computations, which experiment has more variability? Explain your answer.
c. If you were to construct the 95% confidence interval for the population mean difference, in which experiment would the interval width be greater? Why? (*Note:* You do not need to do computations.)

20. Use the data from problem 8 in Chapter 11 to make a point estimate for μ_D. Make an interval estimate of the population mean difference so that you are 90% confident that your interval contains μ_D.

21. Many researchers have reported that exposure to violence on television can result in increased violent or aggressive behavior in children. To evaluate this effect, a researcher obtains a sample of $n = 4$ children in a preschool setting. The group of children is observed for two hours one afternoon, and the number of violent or aggressive acts is recorded for each child. The following morning the children are shown a video cartoon with several violent and aggressive scenes. That afternoon the children's behavior is observed again. On the average, the children exhibit $\overline{D} = 4.2$ more violent/aggressive behaviors after viewing the video than they did the previous day. The sample difference scores had $SS = 12$. Use these data to estimate the effects of viewing television violence. Make a point estimate and a 95% confidence interval estimate of the mean effect.

22. For the study in problem 13 of Chapter 11, determine the 95% confidence interval for population mean difference.

CHAPTER 13

INTRODUCTION TO ANALYSIS OF VARIANCE

TOOLS YOU WILL NEED

The following items are considered essential background material for this chapter. If you doubt your knowledge of any of these items, you should review the appropriate chapter or section before proceeding.

- Variability (Chapter 4)
  - Sum of squares
  - Sample of variance
  - Degrees of freedom
- Introduction to hypothesis testing (Chapter 8)
  - The logic of hypothesis testing
- Independent-measures t statistic (Chapter 10)

CONTENTS

13.1 INTRODUCTION

Analysis of variance (ANOVA) is a hypothesis-testing procedure that is used to evaluate mean differences between two or more treatments (or populations). As with all inferential procedures, ANOVA uses sample data as the basis for drawing general conclusions about populations. It may appear that analysis of variance and *t* tests are simply two different ways of doing exactly the same job: testing for mean differences. In some respects, this is true—both tests use sample data to test hypotheses about population means. However, ANOVA has a tremendous advantage over *t* tests. Specifically, *t* tests are limited to situations where there are only two treatments to compare. The major advantage of ANOVA is that it can be used to compare *two or more treatments*. Thus, ANOVA provides researchers with much greater flexibility in designing experiments and interpreting results (see Figure 13.1).

Like the *t* tests presented in Chapters 10 and 11, ANOVA can be used with either an independent-measures or a repeated-measures design. You should recall that an independent-measures design means that there is a separate sample for each of the treatments (or populations) being compared. In a repeated-measures design, on the other hand, the same sample is tested in all of the different treatment conditions. In addition, ANOVA can be used to evaluate the results from a research study that involves more than one independent variable. For example, an educational psychologist might want to compare the effectiveness of two different teaching methods (in-

FIGURE 13.1

Research designs for which ANOVA would be appropriate. Note that each design involves comparisons of more than two sample means.

(a) Independent variable:
Age

4 Years	5 Years	6 Years
Vocabulary scores for sample 1	Vocabulary scores for sample 2	Vocabulary scores for sample 3

A research study comparing vocabulary skill (dependent variable) for three different age groups. Note that this study could be done as an independent-measures design using three separate samples or as a repeated-measures design testing the same sample at three different times. In either case, the analysis would compare the three sample means.

(b) Independent variable 1:
Class size

		Small class	Medium class	Large class
Independent variable 2: teaching method	Method A	Sample 1	Sample 2	Sample 3
	Method B	Sample 4	Sample 5	Sample 6

The structure of a research design with two independent variables. The effects of two different teaching methods and three different class sizes are evaluated in a single experiment. The dependent variable is the standardized-achievement-test score for each student. Note that this study involves comparing six different sample means.

dependent variable 1) for three different class sizes (independent variable 2). The dependent variable for this study would be each student's score on a standardized achievement test. The structure of this experiment, which involves comparing sample means from six different treatment conditions, is shown in Figure 13.1(b).

As you can see, analysis of variance provides researchers with an extremely flexible data-analysis technique. It can be used to evaluate the significance of mean differences in a wide variety of research situations and is one of the most commonly used hypothesis-testing procedures.

Before we continue, it is necessary to introduce some additional terminology that will help describe the kinds of research studies presented in Figure 13.1. First, you should recall that when a researcher manipulates a variable to create different treatment conditions, the variable is called an *independent variable*. On the other hand, when a researcher uses a non-manipulated variable to differentiate groups of scores, the variable is called a *quasi-independent variable* (see page 13). In the context of analysis of variance, an independent variable or a quasi-independent variable is called a *factor*. For example, the study shown in Figure 13.1(a) compares three different age groups. In this case, age is a quasi-independent variable (it is a preexisting subject variable that is not manipulated by the researcher) and age is the only factor being examined. By contrast, Figure 13.1(b) shows a study where a researcher has manipulated class size and manipulated teaching method. Therefore, this study has two factors and both of them are independent variables.

DEFINITIONS

In analysis of variance, an independent variable or a quasi-independent variable is called a *factor.*

A research study that involves only one factor is called a *single-factor design.*

A study with more than one factor is called a *factorial design.*

In this chapter we will introduce analysis of variance in its simplest form. Specifically, we will only consider *single-factor, independent-measures* designs. That is, we will examine studies that have only one independent variable (or only one quasi-independent variable) and we will limit our discussion to studies that use a separate sample for each of the treatment conditions or populations being compared. In Chapter 14 we will examine ANOVA as it is used for factorial designs involving two factors.

The basic logic and procedures for the analysis of variance that are presented in this chapter form the foundation for more complex applications of ANOVA (for example, repeated-measures or factorial designs). However, the details of the analysis will differ from one situation to another. If you are interested in learning more about ANOVA, we suggest that you consult a more advanced statistics text such as Gravetter and Wallnau (1996).

THE SINGLE-FACTOR, INDEPENDENT-MEASURES DESIGN

A diagram of a single-factor, independent-measures research design is shown in Figure 13.2. Notice that a separate sample is taken for each of the three treatment conditions. Also notice that the three samples have different scores and different means. The goal for ANOVA is to help the researcher decide between the following two interpretations:

1. There really are no differences between the populations (or treatments). The observed differences between samples are simply due to chance (sampling error).

FIGURE 13.2

A typical situation where ANOVA would be used. Three separate samples are obtained to evaluate the mean differences among three populations (or treatments) with unknown means.

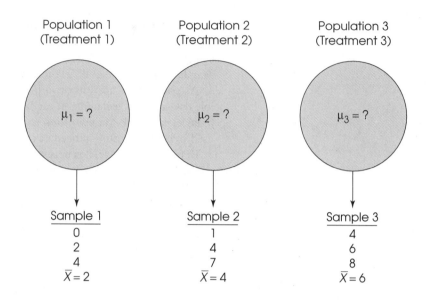

2. The differences between the sample means represent real differences between the populations (or treatments). That is, the populations (or treatments) really do have different means, and the sample data accurately reflect these differences.

You should recognize these two interpretations as corresponding to the two hypotheses (null hypothesis and alternative hypothesis) that are part of the general hypothesis-testing procedure.

STATISTICAL HYPOTHESES FOR ANOVA

The following example will be used to introduce the statistical hypotheses for ANOVA. Suppose a psychologist examined learning performance under three temperature conditions: 50°, 70°, and 90°. Three samples of subjects are selected, one sample for each treatment condition. The purpose of the study is to determine whether room temperature affects learning performance. In statistical terms, we want to decide between two hypotheses: the null hypothesis (H_0), which says temperature has no effect, and the alternative hypothesis (H_1), which states that temperature does affect learning. In symbols, the null hypothesis states

$$H_0: \mu_1 = \mu_2 = \mu_3$$

In words, H_0 states that there is no treatment effect. That is, there are no differences among the means of the populations that receive the three treatments. The population means are all the same. Once again, notice that hypotheses are always stated in terms of population parameters, even though we use sample data to test them.

For the alternative hypothesis, we may state that

$$H_1: \text{At least one population mean is different from the others.}$$

In general, H_1 states that there are some differences among the treatment conditions; that is, there is a real treatment effect. Notice that we have not given any specific alternative hypothesis. This is because many different alternatives are possible, and it would be tedious to list them all. One alternative, for example, would be that the first two populations are identical but that the third is different. Another alternative states that the last two means are the same but that the first is different. Other alternatives might be

$$H_1: \mu_1 \neq \mu_2 \neq \mu_3 \quad \text{(all three means are different)}$$

$$H_1: \mu_1 = \mu_3 \quad \quad \text{(μ_2 is different)}$$

It should be pointed out that a researcher typically entertains only one (or at most a few) of these alternative hypotheses. Usually a theory or the outcomes of previous studies will dictate a specific prediction concerning the treatment effect. For the sake of simplicity, we will state a general alternative hypothesis rather than trying to list all the possible specific alternatives.

THE TEST STATISTIC FOR ANOVA

The test statistic for ANOVA is very similar to the t statistics used in earlier chapters. For the t statistic, we computed a ratio with the following structure:

$$t = \frac{\text{obtained difference between sample means}}{\text{difference expected by chance (error)}}$$

For analysis of variance, the test statistic is called an F-ratio and has the following structure:

$$F = \frac{\text{variance (differences) between sample means}}{\text{variance (differences) expected by chance (error)}}$$

Notice that the F-ratio is based on *variance* instead of sample mean *difference*. The reason for this change is that ANOVA is used in situations where there are more than two sample means. For example, when you have only two means, it is easy to find the difference between them. If two samples have means of $\overline{X} = 20$ and $\overline{X} = 30$, there is a 10-point difference between sample means. However, if we add a third sample mean, $\overline{X} = 35$, it becomes more difficult to describe the "difference" between samples. The solution to this problem is to compute the variance for the set of sample means. If the sample means are all clustered close together (small differences), then the variance will be small. On the other hand, if the sample means are spread over a wide range of values (big differences), then the variance will be large. Thus, the variance in the numerator of the F-ratio provides a single number that describes the differences between all the sample means.

In much the same way, the variance in the denominator of the F-ratio and the standard error in the denominator of the t statistic both measure the differences that would be expected by chance. In the t statistic, the standard error measures standard distance or standard deviation. In the F-ratio, this standard deviation (s) is simply converted to a variance (s^2). In fact, the denominator of the F-ratio is often called *error variance* in order to stress its relationship to the "standard error" in the t statistic.

Finally, you should realize that the t statistic and the F-ratio provide the same basic information. In each case, the numerator of the ratio measures the actual difference obtained from the sample data, and the denominator measures the difference that would be expected by chance. With either the F-ratio or the t statistic, a large

13.1   TYPE I ERRORS AND MULTIPLE HYPOTHESIS TESTS

IF WE already have *t* tests for comparing mean differences, you might wonder why analysis of variance is necessary. Why create a whole new hypothesis-testing procedure that simply duplicates what the *t* tests can already do? The answer to this question is based in a concern about Type I errors.

Remember, each time you do a hypothesis test, you select an alpha level that determines the risk of a Type I error. With $\alpha = .05$, for example, there is a 5%, or a 1-in-20, risk of a Type I error. Thus, for every 20 hypothesis tests, you expect to make one Type I error. The more tests you do, the more risk there is of a Type I error. For this reason, researchers often make a distinction between the *testwise* alpha level and the *experimentwise* alpha level. The testwise alpha level is simply the alpha level you select for each individual hypothesis test. The experimentwise alpha level is the total probability of a Type I error accumulated from all of the separate tests in the experiment. As the number of separate tests increases, so does the experimentwise alpha level.

For an experiment involving three treatments, you would need three separate *t* tests to compare all of the mean differences:

1. Test 1 compares treatment I versus treatment II.
2. Test 2 compares treatment I versus treatment III.
3. Test 3 compares treatment II versus treatment III.

The three separate tests accumulate to produce a relatively large experimentwise alpha level. The advantage of analysis of variance is that it performs all three comparisons simultaneously in the same hypothesis test. Thus, no matter how many different means are being compared, ANOVA uses one test with one alpha level to evaluate the mean differences and thereby avoids the problem of an inflated experimentwise alpha level.

value provides evidence that the sample mean difference is more than chance (see Box 13.1).

13.2   THE LOGIC OF ANALYSIS OF VARIANCE

The formulas and calculations required in ANOVA are somewhat complicated, but the logic that underlies the whole procedure is fairly straightforward. Therefore, this section will give a general picture of analysis of variance before we start looking at the details. We will introduce the logic of ANOVA with the help of the hypothetical data in Table 13.1. These data represent the results of an independent-measures study comparing learning performance under three temperature conditions.

One obvious characteristic of the data in Table 13.1 is that the scores are not all the same. In everyday language, the scores are different; in statistical terms, the scores are variable. Our goal is to measure the amount of variability (the size of the differences) and to explain where it comes from.

The first step is to determine the total variability for the entire set of data. To compute the total variability, we will combine all the scores from all the separate samples to obtain one general measure of variability for the complete experiment. Once we have measured the total variability, we can begin to break it apart into separate components. The word *analysis* means dividing into smaller parts. Because we are going to analyze variability, the process is called *analysis of variance*. This analysis process divides the total variability into two basic components:

TABLE 13.1

Hypothetical data from an experiment examining learning performance under three temperature conditions*

Treatment 1 50° (sample 1)	Treatment 2 70° (sample 2)	Treatment 3 90° (sample 3)
0	4	1
1	3	2
3	6	2
1	3	0
0	4	0
$\overline{X} = 1$	$\overline{X} = 4$	$\overline{X} = 1$

*Note that there are three separate samples, with $n = 5$ in each sample. The dependent variable is the number of problems solved correctly.

1. Between-Treatments Variance. Looking at the data in Table 13.1, we clearly see that much of the variability in the scores is due to general differences between treatment conditions. For example, the scores in the 70° condition tend to be much higher ($\overline{X} = 4$) than the scores in the 50° condition ($\overline{X} = 1$). We will calculate the variability between treatments in order to provide a measure of the overall differences between treatment conditions—that is, the differences between sample means.

2. Within-Treatments Variance. In addition to the general differences between treatment conditions, there is variability within each sample. Looking again at Table 13.1, the scores in the 70° condition are not all the same; they are variable. The within-treatments variance will provide a measure of the variability inside each treatment condition.

Analyzing the total variability into these two components is the heart of analysis of variance. We will now examine each of the components in more detail.

BETWEEN-TREATMENTS VARIANCE

Remember, calculating variance is simply a method for measuring how big the differences are for a set of numbers. When you see the term *variance,* you can automatically translate it into the term *differences.* Thus, the between-treatments variance is simply measuring how much difference exists between the treatment conditions. In addition to measuring the between-treatments differences, we want to explain why the treatments are different. For the data in Table 13.1, for example, why are the scores in treatment 2 higher than the scores in treatment one? In general, why are there differences between treatments?

There are two possible explanations for the differences that exist between treatments:

1. The differences are *caused by the treatments.* For the data we are considering, it is possible that a 70° room produces the best performance and that lowering or raising the temperature causes performance to deteriorate. Thus, changing the temperature (the treatment) has caused the scores to be lower in treatment 1 and treatment 3 than the scores in treatment 2.

2. The differences are *simply due to chance.* If there is no treatment effect at all, you would still expect some differences between samples. Even if all the samples were treated exactly the same, there will still be differences

13.2   DIFFERENCES DUE TO CHANCE

WITHIN ANY set of research data, there will be differences between scores. Some of the differences are systematic and are actually planned and predicted by the researcher. For example, a researcher may test one group of subjects in a 70° room and test a second group in a room where the temperature is 90°. The researcher is hoping to demonstrate that temperature has an effect on performance. If the treatment does have an effect, the scores in one treatment condition will be systematically higher than the scores in the other condition. In this case, the differences are planned and predicted by the researcher and are caused by the change from one temperature to another.

On the other hand, some of the differences that exist within a set of data are not planned and not predicted. Within each treatment condition, all the individuals are treated exactly the same. In particular, the researcher does not do anything to cause the scores to be different. Still, there are differences within each treatment. These unplanned, unpredicted differences that are not caused or explained by any action on the part of the researcher are simply due to chance.

Researchers often analyze chance differences into two general categories:

1. Individual Differences. Subjects enter a research study with different backgrounds, different characteristics, different attitudes, and so on. These individual differences can influence the scores so that differences are obtained between subjects or between groups of subjects. Notice that these differences are not caused or planned by the researcher. The differences simply exist regardless of any treatment and therefore are considered to be the result of chance.

2. Experimental Error. Whenever you make a measurement, there is a chance of error. The error could be caused by poor equipment, lack of attention, or unpredictable changes in the environment or event you are trying to measure. Thus, if you measure exactly the same individual under exactly the same conditions, it is still possible to obtain two different measurements. These differences are uncontrolled and unexplained, and therefore are attributed to chance.

Thus, individual differences and experimental error can cause two subjects to have different scores even though they are treated exactly the same. Also, these factors can cause two groups of subjects to have different means even though the two groups are treated exactly the same. In each case, the differences are simply attributed to chance.

from one sample to another. The samples consist of different individuals with different scores, and it should not be surprising that differences exist between samples just by chance (see Box 13.2).

Thus, when we compute the between-treatments variance, we are measuring differences that could be due to chance or could include some degree of treatment effect. The goal of the hypothesis test is to determine whether or not the treatment has any effect. The problem is that the treatment effect and chance variation are mixed together so that it is impossible to measure the treatment effect by itself. The analysis of variance attempts to solve this problem by measuring chance differences by themselves. This is accomplished by computing the within-treatments variance.

WITHIN-TREATMENTS VARIANCE   Inside each treatment condition, we have a set of individuals who are treated exactly the same; that is, the researcher does not do anything that would cause these individuals to have different scores. In Table 13.1, for example, the data show that five individuals were tested in a 70° room (treatment 2). Although these five individuals were all treated exactly the same, their scores are different. Why are the scores different? The answer is that the differences within a treatment are simply due to chance (again, see Box 13.2).

FIGURE 13.3

The independent-measures analysis of variance partitions, or analyzes, the total variability into two components: variability between treatments and variability within treatments.

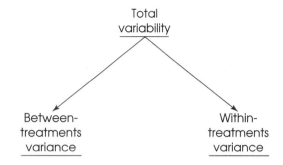

Total
variability

Between-
treatments
variance

Within-
treatments
variance

1. Treatment effect
2. Differences due to chance

1. Differences due to chance

Thus, the within-treatments variance provides a measure of how much difference is reasonable to expect just by chance. In particular, the within-treatments variance measures the differences that exist when there is no treatment that could cause differences.

THE *F*-RATIO: THE TEST STATISTIC FOR ANOVA

Once we have analyzed the total variability into two basic components (between treatments and within treatments), we simply compare them. The comparison is made by computing a statistic called an *F-ratio*. For the independent-measures ANOVA, the *F*-ratio has the following structure:

$$F = \frac{\text{variance between treatments}}{\text{variance within treatments}} \tag{13.1}$$

When we express each component of variability in terms of its sources (see Figure 13.3), the structure of the *F*-ratio is

$$F = \frac{\text{treatment effect} + \text{differences due to chance}}{\text{differences due to chance}} \tag{13.2}$$

The value obtained for the *F*-ratio will help determine whether or not any treatment effects exist. Consider the following two possibilities:

1. When the treatment has no effect, then the differences between treatments (numerator) are entirely due to chance. In this case, the numerator and the denominator of the F-ratio are both measuring chance differences and should be roughly the same size. With the numerator and denominator roughly equal, the *F*-ratio should have a value around 1.00. In terms of the formula, when the treatment effect is zero, we obtain

$$F = \frac{0 + \text{differences due to chance}}{\text{differences due to chance}}$$

Thus, an *F*-ratio near 1.00 indicates that the differences between treatments (numerator) are about the same as the differences that are expected by chance (the denominator). With an *F*-ratio near 1.00, we will conclude that there is no evidence to suggest that the treatment has any effect.

2. When the treatment does have an effect, causing differences between samples, then the between-treatment differences (numerator) should be larger than chance (denominator). In this case, the numerator of the F-ratio should be noticeably larger than the denominator, and we should obtain an F-ratio noticeably larger than 1.00. Thus, a large F-ratio indicates that the differences between treatments are greater than chance; that is, the treatment does have a significant effect.

LEARNING CHECK

1. ANOVA is a statistical procedure that compares two or more treatment conditions for differences in variance. (True or false?)

2. In ANOVA, what value is expected, on average, for the F-ratio when the null hypothesis is true?

3. What happens to the value of the F-ratio if differences between treatments are increased? What happens to the F-ratio if variability inside the treatments is increased?

4. In ANOVA, the total variability is partitioned into two parts. What are these two variability components called, and how are they used in the F-ratio?

ANSWERS

1. False. Although ANOVA uses variability in the computations, the purpose of the test is to evaluate differences in *means* between treatments.

2. When H_0 is true, the expected value for the F-ratio is 1.00 because the top and the bottom of the ratio are both measuring the same variance.

3. As differences between treatments increase, the F-ratio will increase. As variability within treatments increases, the F-ratio will decrease.

4. The two components are between-treatments variance and within-treatments variance. Between treatments variance is the numerator of the F-ratio, and within-treatments variance is the denominator.

13.3 ANOVA VOCABULARY, NOTATION, AND FORMULAS

Before we introduce the notation, we will look at some special terminology that is used for ANOVA. As noted earlier, in analysis of variance an independent variable (or a quasi-independent variable) is called a *factor*. Therefore, for the experiment shown in Table 13.1, the factor is temperature. Because this experiment has only one independent variable, it is called a *single-factor experiment*. The next term you need to know is *levels*. The levels in an experiment consist of the different values used for the independent variable (factor). For example, in the learning experiment (Table 13.1) we are using three values of temperature. Therefore, the temperature factor has three levels.

DEFINITION

The individual treatment conditions that make up a factor are called *levels* of the factor.

TABLE 13.2

Hypothetical data from an experiment examining learning performance under three temperature conditions*

Temperature conditions			
1 50°	2 70°	3 90°	
0	4	1	$\Sigma X^2 = 106$
1	3	2	$G = 30$
3	6	2	$N = 15$
1	3	0	$k = 3$
0	4	0	
$T_1 = 5$	$T_2 = 20$	$T_3 = 5$	
$SS_1 = 6$	$SS_2 = 6$	$SS_3 = 4$	
$n_1 = 5$	$n_2 = 5$	$n_3 = 5$	
$\overline{X}_1 = 1$	$\overline{X}_2 = 4$	$\overline{X}_3 = 1$	

*Summary values and notation for an analysis of variance are also presented.

Because ANOVA most often is used to examine data from more than two treatment conditions (and more than two samples), we will need a notational system to help keep track of all the individual scores and totals. To help introduce this notational system, we will use the hypothetical data from Table 13.1 again. The data are reproduced in Table 13.2 along with some of the notation and statistics that will be described.

1. The letter k is used to identify the number of treatment conditions, that is, the number of levels of the factor. For an independent-measures experiment, k also specifies the number of separate samples. For the data in Table 13.2, there are three treatments, so $k = 3$.

2. The number of scores in each treatment is identified by a lowercase letter n. For the example in Table 13.2, $n = 5$ for all the treatments. If the samples are of different sizes, you can identify a specific sample by using a subscript. For example, n_2 is the number of scores in treatment 2.

3. The total number of scores in the entire experiment is specified by a capital letter N. When all the samples are the same size (n is constant), $N = kn$. For the data in Table 13.2, there are $n = 5$ scores in each of the $k = 3$ treatments, so $N = 3(5) = 15$.

Because ANOVA formulas require ΣX for each treatment and ΣX for the entire set of scores, we have introduced new notation (T and G) to help identify which ΣX is being used. Remember, T stands for *treatment total,* and G stands for *grand total.*

4. The total (ΣX) for each treatment condition is identified by the capital letter T. The total for a specific treatment can be identified by adding a numerical subscript to the T. For example, the total for the second treatment in Table 13.2 is $T_2 = 20$.

5. The sum of all the scores in the experiment (the grand total) is identified by G. You can compute G by adding up all N scores or by adding up the treatment totals: $G = \Sigma T$.

6. Although there is no new notation involved, we also have computed SS and $\overline{X}$ for each sample, and we have calculated ΣX^2 for the entire set of $N = 15$ scores in the experiment. These values are given in Table 13.2 and will be important in the formulas and calculations for ANOVA.

FIGURE 13.4

The structure of ANOVA calculations.

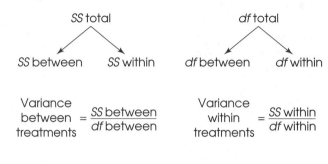

$$F = \frac{\text{Variance between treatments}}{\text{Variance within treatments}}$$

ANOVA FORMULAS

Because analysis of variance requires extensive calculations and many formulas, one common problem for students is simply keeping track of the different formulas and numbers. Therefore, we will examine the general structure of the procedure and look at the organization of the calculations before we introduce the individual formulas.

1. The final calculation for ANOVA is the F-ratio, which is composed of two variances:

$$F = \frac{\text{variance between treatments}}{\text{variance within treatments}}$$

2. You should recall that variance for sample data has been defined as

$$\text{sample variance} = s^2 = \frac{SS}{df}$$

Therefore, we will need to compute an SS and a df for the variance between treatments (numerator of F), and we will need another SS and df for the variance within treatments (denominator of F). To obtain these SS and df values, we must go through two separate analyses: First, compute SS for the total experiment and analyze it into two components (between and within). Then compute df for the total experiment and analyze it into two components (between and within).

Thus, the entire process of analysis of variance will require nine calculations: three values for SS, three values for df, two variances (between and within), and a final F-ratio. However, these nine calculations are all logically related and are all directed toward finding the final F-ratio. Figure 13.4 shows the logical structure of ANOVA calculations.

ANALYSIS OF SUM OF SQUARES (SS)

ANOVA requires that we first compute a total variability and then partition this value into two components: between treatments and within treatments. This analysis is outlined in Figure 13.5. We will examine each of the three components separately.

1. Total Sum of Squares, SS_{total}. As the name implies, SS_{total} is simply the sum of squares for the entire set of N scores. We calculate this value by using the computational formula for SS:

FIGURE 13.5

Partitioning the sum of squares (*SS*) for the independent-measures analysis of variance.

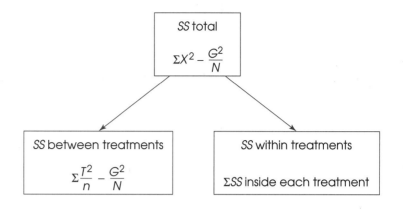

$$SS = \Sigma X^2 - \frac{(\Sigma X)^2}{N}$$

To make this formula consistent with ANOVA notation, we substitute the letter *G* in place of ΣX and obtain

$$SS_{total} = \Sigma X^2 - \frac{G^2}{N} \tag{13.3}$$

Applying this formula to the set of data in Table 13.2, we obtain

$$SS_{total} = 106 - \frac{30^2}{15}$$
$$= 106 - 60$$
$$= 46$$

2. Within-Treatments Sum of Squares, SS_{within}. Now we are looking at the variability inside each of the treatment conditions. We already have computed the *SS* within each of the three treatment conditions (Table 13.2): $SS_1 = 6$, $SS_2 = 6$, and $SS_3 = 4$. To find the overall within-treatment sum of squares, we simply add these values together:

$$SS_{within} = \Sigma SS_{inside\ each\ treatment} \tag{13.4}$$

For the data in Table 13.2, this formula gives

$$SS_{within} = 6 + 6 + 4$$
$$= 16$$

3. Between-Treatments Sum of Squares, $SS_{between}$. Before we introduce the equation for $SS_{between}$, consider what we have found so far. The total variability for the data in Table 13.2 is $SS_{total} = 46$. We intend to partition this total into two parts (see Figure 13.5). One part, SS_{within}, has been found to be equal to 16. This means that $SS_{between}$ must be equal to 30 in order for the two parts (16 and 30) to add up to the total (46). The equation for the between-treatments sum of squares should produce a value of $SS_{between} = 30$. You should recall that the variability between treatments is measuring the differences between treatment means. Conceptually, the most direct way of measuring the amount of variability among the treatment means is to compute the sum of squares for the set of means, but this method usually is awkward, especially when treatment means are not whole numbers. Therefore, we will use a computational formula for $SS_{between}$ that uses the treatment totals (T) instead of the treatment means.

$$SS_{between} = \Sigma \frac{T^2}{n} - \frac{G^2}{N}$$ (13.5)

You probably notice the general similarity between this formula and the one for SS_{total} (formula 13.3). The formula for $SS_{between}$, however, is based on squared treatment totals (T^2) and measures the variability between treatments.

Using this new formula with the data in Table 13.2, we obtain

$$
\begin{aligned}
SS_{between} &= \frac{5^2}{5} + \frac{20^2}{5} + \frac{5^2}{5} - \frac{30^2}{15} \\
&= 5 + 80 + 5 - 60 \\
&= 90 - 60 \\
&= 30
\end{aligned}
$$

At this point of the analysis, the work may be checked to see if total SS equals between-treatments SS plus within-treatments SS.

The formula for each SS and the relationships among these three values are shown in Figure 13.5.

ANALYSIS OF DEGREES OF FREEDOM (df)

The analysis of degrees of freedom (df) follows the same pattern as the analysis of SS (see Figure 13.4). First, we will find df for the total set of N scores, and then we will partition this value into two components: degrees of freedom between treatments and degrees of freedom within treatments. In computing degrees of freedom, there are two important considerations to keep in mind:

1. Each df value is associated with a specific SS value.
2. Normally the value of df is obtained by counting the number of items that were used to calculate SS and then subtracting 1. For example, if you compute SS for a set of n scores, then $df = n - 1$.

With this in mind, we will examine the degrees of freedom for each part of the analysis:

1. Total Degrees of Freedom, df_{total}. To find the df associated with SS_{total}, you must first recall that this SS value measures variability for the entire set of N scores. Therefore, the df value will be

$$df_{total} = N - 1$$ (13.6)

For the data in Table 13.2, the total number of scores is $N = 15$, so the total degrees of freedom would be

$$
\begin{aligned}
df_{total} &= 15 - 1 \\
&= 14
\end{aligned}
$$

2. Within-Treatments Degrees of Freedom, df_{within}. To find the df associated with SS_{within}, we must look at how this SS value is computed. Remember, we first find SS inside each of the treatments and then add these values together. Each of the treatment SS values measures variability for the n scores in the treatment, so each SS will have $df = n - 1$. When all these individual treatment values are added together, we obtain

$$df_{\text{within}} = \Sigma(n - 1) \tag{13.7}$$

For the experiment we have been considering, each treatment has $n = 5$ scores. This means there are $n - 1 = 4$ degrees of freedom inside each treatment. Because there are 3 different treatment conditions, this gives a total of 12 for the within-treatments degrees of freedom. Notice that this formula for df simply adds up the number of scores in each treatment (the n values) and subtracts 1 for each treatment. If these two stages are done separately, you obtain

$$df_{\text{within}} = N - k \tag{13.8}$$

(Adding up all the n values gives N. If you subtract 1 for each treatment, then altogether you have subtracted k because there are k treatments.) For the data in Table 13.2, $N = 15$ and $k = 3$, so

$$\begin{aligned} df_{\text{within}} &= 15 - 3 \\ &= 12 \end{aligned}$$

3. Between-Treatments Degrees of Freedom, df_{between}. The df associated with SS_{between} can be found by considering the SS formula. This SS formula measures the variability among the treatment means or totals. To find df_{between}, simply count the number of T values (or means) and subtract 1. Because the number of treatments is specified by the letter k, the formula for df is

$$df_{\text{between}} = k - 1 \tag{13.9}$$

For the data in Table 13.2, there are three different treatment conditions (three T values), so the between-treatments degrees of freedom are computed as follows:

$$\begin{aligned} df_{\text{between}} &= 3 - 1 \\ &= 2 \end{aligned}$$

Notice that the two parts we obtained from this analysis of degrees of freedom add up to equal the total degrees of freedom:

$$\begin{aligned} df_{\text{total}} &= df_{\text{within}} + df_{\text{between}} \\ 14 &= 12 + 2 \end{aligned}$$

The complete analysis of degrees of freedom is shown in Figure 13.6.

CALCULATION OF VARIANCES (MS) AND THE F-RATIO

The final step in the analysis of variance procedure is to compute the variance between treatments and the variance within treatments in order to calculate the F-ratio (see Figure 13.4). You should recall (from Chapter 4) that variance is defined as the average squared deviation. For a sample, you compute this average by the following formula:

$$\text{variance} = \frac{SS}{n - 1} = \frac{SS}{df}$$

FIGURE 13.6

Partitioning degrees of freedom (*df*) for the independent-measures analysis of variance.

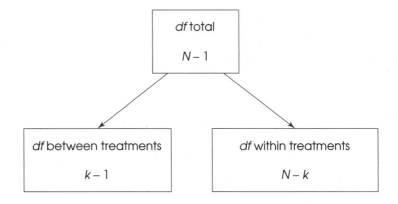

In ANOVA, it is customary to use the term *mean square,* or simply *MS*, in place of the term *variance.* Note that variance is the *mean squared* deviation, so this terminology is quite sensible. For the final *F*-ratio, you will need an *MS* between treatments and an *MS* within treatments. In each case,

$$MS = \frac{SS}{df} \tag{13.10}$$

For the data we have been considering,

$$MS_{\text{between}} = \frac{SS_{\text{between}}}{df_{\text{between}}} = \frac{30}{2} = 15$$

and

$$MS_{\text{within}} = \frac{SS_{\text{within}}}{df_{\text{within}}} = \frac{16}{12} = 1.33$$

We now have a measure of the variance (or differences) between the treatments and a measure of the variance within the treatments. The *F*-ratio simply compares these two variances:

$$F = \frac{MS_{\text{between}}}{MS_{\text{within}}} \tag{13.11}$$

For the experiment we have been examining, the data give an *F*-ratio of

$$F = \frac{15}{1.33} = 11.28$$

For this example, the obtained value of $F = 11.28$ indicates that the numerator of the *F*-ratio is substantially larger than the denominator. If you recall the conceptual structure of the *F*-ratio as presented in formulas 13.1 and 13.2, the *F* value we obtained indicates that the differences between treatments are substantially greater than would be expected by chance, providing evidence that a treatment effect really exists. Stated in terms of the experimental variables, it appears that temperature does have an effect on learning performance. However, to properly evaluate the *F*-ratio, we must examine the *F* distribution.

LEARNING CHECK

1. Can SS_{within} ever be larger than SS_{total}?

2. A researcher conducts a study comparing 3 treatment conditions with a sample of $n = 10$ individuals in each treatment. If the data are evaluated with an ANOVA,
   a. What is the value of df_{total}?
   b. What is the value of df_{within}?
   c. What is the value of $df_{between}$?

3. A researcher computes $SS_{between} = 35$ for a set of data. If the data have $SS_{total} = 50$, then what is the value of SS_{within}?

ANSWERS

1. No. SS_{within} is always a part of SS_{total}.

2. a. $df_{total} = 29$ (There are 30 subjects.)
   b. $df_{within} = 27$ (Each of the three treatments has $df = 10 - 1 = 9$.)
   c. $df_{between} = 2$ (The ANOVA is looking at differences between 3 treatments.)

3. $SS_{within} = 15$. The two components, between and within, must add to equal the total.

13.4 THE DISTRIBUTION OF *F*-RATIOS

In analysis of variance, the *F*-ratio is constructed so that the numerator and the denominator of the ratio are measuring exactly the same variance when the null hypothesis is true (see formula 13.2). In this situation, we expect the value of *F* to be around 1.00. The problem now is to define precisely what we mean by "around 1.00." What values are considered to be close to 1.00, and what values are far away? To answer this question, we need to look at all the possible *F* values, that is, the *distribution of F-ratios*.

Before we examine this distribution in detail, you should note two obvious characteristics:

1. Because *F*-ratios are computed from two variances (the numerator and the denominator of the ratio), *F* values always will be positive numbers. Remember, variance is always positive.

2. When H_0 is true, the numerator and the denominator of the *F*-ratio are measuring the same variance. In this case, the two sample variances should be about the same size, so the ratio should be near 1.00. In other words, the distribution of *F*-ratios should pile up around 1.00.

With these two factors in mind, we can sketch the distribution of *F*-ratios. The distribution is cut off at zero (all positive values), piles up around 1.00, and then tapers off to the right (see Figure 13.7). The exact shape of the *F* distribution depends on the degrees of freedom for the two variances in the *F*-ratio. You should recall that the precision of a sample variance depends on the number of scores or the degrees of freedom. In general, the variance for a large sample (large *df*) provides a more accurate estimate of the population variance. Because the precision of the *MS* values depends on *df*, the shape of the *F* distribution also will depend on the *df* values for the numerator and the denominator of the *F*-ratio. With very large *df* val-

FIGURE 13.7

The distribution of F-ratios with $df = 2$, 12. Of all the values in the distribution, only 5% are larger than $F = 3.88$, and only 1% are larger than $F = 6.93$.

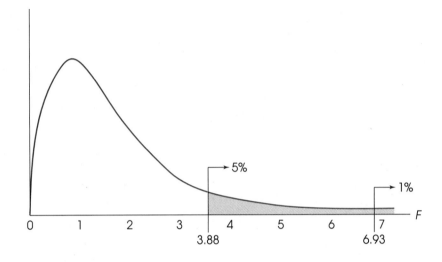

ues, nearly all the F-ratios will be clustered very near to 1.00. With smaller df values, the F distribution is more spread out.

For analysis of variance, we expect F near 1.00 if H_0 is true, and we expect a large value for F if H_0 is not true. In the F distribution, we need to separate those values that are reasonably near 1.00 from the values that are significantly greater than 1.00. These critical values are presented in an F-distribution table in Appendix B, page A-29. To use the table, you must know the df values for the F-ratio (numerator and denominator), and you must know the alpha level for the hypothesis test. It is customary for an F table to have the df values for the numerator of the F-ratio printed across the top of the table. The df values for the denominator of F are printed in a column on the left-hand side. A portion of the F-distribution table is shown in Table 13.3. For the temperature experiment we have been considering,

TABLE 13.3

A portion of the F-distribution table

Degrees of freedom: Denominator	Degrees of freedom: Numerator					
	1	2	3	4	5	6
10	4.96	4.10	3.71	3.48	3.33	3.22
	10.04	**7.56**	**6.55**	**5.99**	**5.64**	**5.39**
11	4.84	3.98	3.59	3.36	3.20	3.09
	9.65	**7.20**	**6.22**	**5.67**	**5.32**	**5.07**
12	4.75	3.88	3.49	3.26	3.11	3.00
	9.33	**6.93**	**5.95**	**5.41**	**5.06**	**4.82**
13	4.67	3.80	3.41	3.18	3.02	2.92
	9.07	**6.70**	**5.74**	**5.20**	**4.86**	**4.62**
14	4.60	3.74	3.34	3.11	2.96	2.85
	8.86	**6.51**	**5.56**	**5.03**	**4.69**	**4.46**

Entries in roman type are critical values for the .05 level of significance, and bold type values are for the .01 level of significance. The critical F-ratios for $df = 2$, 12 have been highlighted (see text).

the numerator of the F-ratio (between treatments) has $df = 2$, and the denominator of the F-ratio (within treatments) has $df = 12$. This F-ratio is said to have "degrees of freedom equal to 2 and 12." The degrees of freedom would be written as $df = 2$, 12. To use the table, you would first find $df = 2$ across the top of the table and $df = 12$ in the first column. When you line up these two values, they point to a pair of numbers in the middle of the table. These numbers give the critical cutoffs for $\alpha = .05$ and $\alpha = .01$. With $df = 2$, 12, for example, the numbers in the table are 3.88 and 6.93. These values indicate that the most unlikely 5% of the distribution ($\alpha = .05$) begins at a value of 3.88. The most extreme 1% of the distribution begins at a value of 6.93 (see Figure 13.7).

In the temperature experiment, we obtained an F-ratio of 11.28. According to the critical cutoffs in Figure 13.7, this value is extremely unlikely (it is in the most extreme 1%). Therefore, we would reject H_0 with alpha set at either .05 or .01 and conclude that temperature does have a significant effect on learning performance.

LEARNING CHECK

1. Calculate SS_{total}, $SS_{between}$, and SS_{within} for the following set of data:

Treatment 1	Treatment 2	Treatment 3	
$n = 10$	$n = 10$	$n = 10$	$N = 30$
$T = 10$	$T = 20$	$T = 30$	$G = 60$
$SS = 27$	$SS = 16$	$SS = 23$	$\Sigma X^2 = 206$

2. A researcher uses ANOVA to compare 3 treatment conditions with a sample of $n = 8$ in each treatment. For this analysis, find df_{total}, $df_{between}$, and df_{within}.

3. With $\alpha = .05$, what value forms the boundary for the critical region in the distribution of F-ratios with $df = 2$, 24?

ANSWERS

1. $SS_{total} = 86$; $SS_{between} = 20$; $SS_{within} = 66$

2. $df_{total} = 23$; $df_{between} = 2$; $df_{within} = 21$

3. The critical value is 3.40.

13.5 EXAMPLES OF HYPOTHESIS TESTING WITH ANOVA

Although we have seen all the individual components of ANOVA, the following example demonstrates the complete ANOVA process using the standard four-step procedure for hypothesis testing.

EXAMPLE 13.1

The data depicted in Table 13.4 were obtained from an independent-measures study designed to measure the effectiveness of three pain relievers (A, B, and C). A fourth group that received a placebo (sugar pill) also was tested.

The purpose of the analysis is to determine whether these sample data provide evidence of any significant differences among the four drugs. The depen-

TABLE 13.4

The effect of drug treatment on the amount of time (in seconds) a stimulus is endured

Placebo	Drug A	Drug B	Drug C	
0	0	3	8	$N = 12$
0	1	4	5	$G = 36$
3	2	5	5	$\Sigma X^2 = 178$
$T = 3$	$T = 3$	$T = 12$	$T = 18$	
$SS = 6$	$SS = 2$	$SS = 2$	$SS = 6$	

dent variable is the amount of time (in seconds) that subjects can withstand a painfully hot stimulus.

Before we begin the hypothesis test, note that we already have computed a variety of summary statistics for the data in Table 13.4. Specifically, the treatment totals (T) and SS values are shown for each sample, and the grand total (G), N, and ΣX^2 are shown for the entire set of data. Having these summary values will simplify the computations in the hypothesis test, and we suggest that you always compute these summary statistics before you begin an analysis of variance.

STEP 1   The first step is to state the hypotheses and select an alpha level:

$$H_0: \mu_1 = \mu_2 = \mu_3 = \mu_4 \quad \text{(no treatment effect)}$$

$$H_1: \text{At least one of the treatment means is different.}$$

We will use $\alpha = .05$.

STEP 2   To locate the critical region for the F-ratio, we first must determine degrees of freedom for $MS_{between}$ and MS_{within} (the numerator and the denominator of F). For these data, the total degrees of freedom would be

Often it is easier to postpone finding the critical region until after step 3, where you compute the df values as part of the calculations for the F-ratio.

$$\begin{aligned} df_{total} &= N - 1 \\ &= 12 - 1 \\ &= 11 \end{aligned}$$

Analyzing this total into two components, we obtain

$$\begin{aligned} df_{between} &= k - 1 \\ &= 4 - 1 \\ &= 3 \\ df_{within} &= N - k \\ &= 12 - 4 \\ &= 8 \end{aligned}$$

The F-ratio for these data will have $df = 3, 8$. The distribution of all the possible F-ratios with $df = 3, 8$ is presented in Figure 13.8. Almost always (95% of the time) we should obtain an F-ratio less than 4.07 if H_0 is true.

STEP 3   We already have the data for this experiment, so it now is time for the calculations.

FIGURE 13.8

The distribution of F-ratios with $df = 3$, 8. The critical value for $\alpha = .05$ is $F = 4.07$.

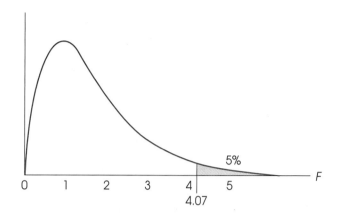

Analysis of SS. First, we will compute the total *SS* and then the two components, as indicated in Figure 13.4.

$$SS_{\text{total}} = \Sigma X^2 - \frac{G^2}{N}$$

$$= 178 - \frac{36^2}{12}$$

$$= 178 - 108$$

$$= 70$$

$$SS_{\text{within}} = \Sigma SS_{\text{inside each treatment}}$$

$$= 6 + 2 + 2 + 6$$

$$= 16$$

$$SS_{\text{between}} = \Sigma \frac{T^2}{n} - \frac{G^2}{N}$$

$$= \frac{3^2}{3} + \frac{3^2}{3} + \frac{12^2}{3} + \frac{18^2}{3} - \frac{36^2}{12}$$

$$= 3 + 3 + 48 + 108 - 108$$

$$= 54$$

Notice that the two components, between and within, add up to the total.

Calculation of mean squares. Now we must compute the variance, or *MS*, for each of the two components:

The *df* values ($df_{\text{between}} = 3$ and $df_{\text{within}} = 8$) were computed in step 2 when we located the critical region.

$$MS_{\text{between}} = \frac{SS_{\text{between}}}{df_{\text{between}}} = \frac{54}{3} = 18$$

$$MS_{\text{within}} = \frac{SS_{\text{within}}}{df_{\text{within}}} = \frac{16}{8} = 2$$

Calculation of F. Finally, we compute the F-ratio:

$$F = \frac{MS_{\text{between}}}{MS_{\text{within}}} = \frac{18}{2} = 9.00$$

STEP 4   Finally, we make the statistical decision. The F value we obtained, $F = 9.00$, is in the critical region (see Figure 13.8). It is very unlikely ($p < .05$) that we will obtain a value this large if H_0 is true. Therefore, we reject H_0 and conclude that there is a significant treatment effect.

IN THE LITERATURE:
REPORTING THE RESULTS OF ANALYSIS OF VARIANCE

The APA format for reporting the results of ANOVA begins with a presentation of the treatment means and standard deviations in the narrative of the article, a table, or a graph. The means and standard deviations are descriptive statistics that summarize the data. Next, the results of the ANOVA are reported. For the study described in Example 13.1, the report might state

> The means and standard deviations are presented in Table 1. The analysis of variance revealed a significant difference, $F(3, 8) = 9.00$, $p < .05$.
>
> **TABLE 1**
> Amount of time (seconds) the stimulus was endured
>
> | | | Treatment condition | | |
> |---|---|---|---|---|
> | | Placebo | Drug A | Drug B | Drug C |
> | M | 1.0 | 1.0 | 4.0 | 6.0 |
> | SD | 1.73 | 1.00 | 1.00 | 1.73 |

Notice how the F-ratio is reported. In this example, degrees of freedom for between and within treatments are $df = 3, 8$, respectively. These values are placed in parentheses immediately following the symbol F. Next, the calculated value for F is reported, followed by the probability of committing a Type I error. Because H_0 has been rejected and alpha was set at .05, p is *less than* .05.

We have rejected the null hypothesis; thus, we are concluding that not all treatments are the same. However, the analysis has not determined which ones are different. Is drug A different from the placebo? Is drug A different from drug B? Is drug B different from drug C, and so on? Unfortunately, these questions remain unanswered. However, we do know that at least one difference exists [H_0 was rejected with $F(3, 8) = 9.00$, $p < .05$]. Additional analysis is necessary to determine exactly which groups differ. This additional analysis is addressed in Section 13.6.

Finally, many years ago APA format required a table that summarizes the results of the ANOVA calculations. The table, called an ANOVA summary table, includes columns for the source of variance, SS, df, and MS. The following table summarizes the ANOVA for Example 13.1.

Source	ANOVA summary SS	df	MS	
Between treatments	54	3	18	$F = 9.00$
Within treatments	16	8	2	
Total	70	11		

These tables took additional page space in journals, along with other tables reporting means and standard deviations or graphs of the results. Consequently, they are not used in most scientific journals anymore. Even though these tables are no longer reported in articles, they are sometimes included in a thesis or a dissertation. However, they are especially useful during the computations of ANOVA because they help organize the steps of the analysis and allow you to keep track of your calculations. At the start of performing an ANOVA, it is a good idea to create a blank ANOVA summary table, showing the headings (SS, df, MS) and the sources of variability (between, within, and total). You start your analysis by doing the computations necessary to fill in the values of the SS column. Then you continue to df, MS, and, finally, F.   ❏

AN EXAMPLE WITH UNEQUAL SAMPLE SIZES

In the previous example, all the samples were exactly the same size (equal ns). However, the formulas for ANOVA can be used when the sample size varies within an experiment. With unequal sample sizes, you must take care to be sure that each value of n is matched with the proper T value in the equations. You also should note that the general ANOVA procedure is most accurate when used to examine experimental data with equal sample sizes. Therefore, researchers generally try to plan experiments with equal ns. However, there are circumstances where it is impossible or impractical to have an equal number of subjects in every treatment condition. In these situations, ANOVA still provides a valid test, especially when the samples are relatively large and when the discrepancy between sample sizes is not extreme.

EXAMPLE 13.2

A psychologist conducts an experiment to compare learning performance for three species of monkeys. The animals are tested individually on a delayed-response task. A raisin is hidden in one of three containers while the animal is viewing from its cage window. A shade is then pulled over the window for 1 minute to block the view. After this delay period, the monkey is allowed to respond by tipping over one container. If its response is correct, the monkey is rewarded with the raisin. The number of trials it takes before the animal makes five consecutive correct responses is recorded. The experimenter used all of the available animals from each species, which resulted in unequal sample sizes (n). The data are summarized in Table 13.5.

TABLE 13.5

The performance of different species of monkeys on a delayed-response task

Vervet	Rhesus	Baboon	
$n = 4$	$n = 10$	$n = 6$	$N = 20$
$\overline{X} = 9$	$\overline{X} = 14$	$\overline{X} = 4$	$G = 200$
$T = 36$	$T = 140$	$T = 24$	$\Sigma X^2 = 3400$
$SS = 200$	$SS = 500$	$SS = 320$	

STEP 1 State the hypotheses, and select the alpha level.

$$H_0:\ \mu_1 = \mu_2 = \mu_3$$

H_1: At least one population is different from the others.

$$\alpha = .05$$

STEP 2 Locate the critical region.

To find the critical region, we first must determine the df values for the F-ratio:

$$df_{\text{total}} = N - 1 = 20 - 1 = 19$$

$$df_{\text{between}} = k - 1 = 3 - 1 = 2$$

$$df_{\text{within}} = N - k = 20 - 3 = 17$$

The F-ratio for these data will have $df = 2, 17$. With $\alpha = .05$, the critical value for the F-ratio is 3.59.

STEP 3 Compute the F-ratio.

First, compute SS for all three parts of the analysis:

$$SS_{\text{total}} = \Sigma X^2 - \frac{G^2}{N}$$

$$= 3400 - \frac{200^2}{20}$$

$$= 3400 - 2000$$

$$= 1400$$

$$SS_{\text{between}} = \Sigma \frac{T^2}{n} - \frac{G^2}{N} = \frac{T_1^2}{n_1} + \frac{T_2^2}{n_2} + \frac{T_3^2}{n_3} - \frac{G^2}{N}$$

$$= \frac{36^2}{4} + \frac{140^2}{10} + \frac{24^2}{6} - \frac{200^2}{20}$$

$$= 324 + 1960 + 96 - 2000$$

$$= 380$$

$$SS_{\text{within}} = \Sigma SS_{\text{inside each treatment}}$$

$$= 200 + 500 + 320$$

$$= 1020$$

Finally, compute the MS values and the F-ratio:

$$MS_{\text{between}} = \frac{SS}{df} = \frac{380}{2} = 190$$

$$MS_{\text{within}} = \frac{SS}{df} = \frac{1020}{17} = 60$$

$$F = \frac{MS_{\text{between}}}{MS_{\text{within}}} = \frac{190}{60} = 3.17$$

STEP 4 Make a decision.

Because the obtained F-ratio is not in the critical region, we fail to reject H_0 and conclude that these data do not provide evidence of significant differences among the three populations of monkeys in terms of average learning performance.

A CONCEPTUAL VIEW OF ANOVA   Because analysis of variance requires relatively complex calculations, students encountering this statistical technique for the first time often tend to be overwhelmed by the formulas and arithmetic and lose sight of the general purpose of the analysis. The following two examples are intended to minimize the role of the formulas and shift your attention back to the conceptual goal of the ANOVA process.

EXAMPLE 13.3   The following data represent the outcome of an experiment using two separate samples to evaluate the mean difference between two treatment conditions. Take a minute to look at the data, and without doing any calculations, try to predict the outcome of an ANOVA for these values. Specifically, predict what values should be obtained for the between-treatments variance ($SS_{between}$, $MS_{between}$) and the F-ratio. If you do not "see" the answer after 20 or 30 seconds, try reading the hints that follow the data.

Treatment I	Treatment II	
4	2	$N = 8$
0	1	
1	0	$G = 16$
3	5	$\Sigma X^2 = 56$
$T = 8$	$T = 8$	
$SS = 10$	$SS = 14$	

If you are having trouble predicting the outcome of the ANOVA, read the following hints, and then go back and look at the data.

Hint 1: Remember, $SS_{between}$ and $MS_{between}$ provide a measure of how much difference there is between treatment conditions.

Hint 2: Find the mean for each treatment, and determine how much difference there is between the two means.

You should realize by now that the data have been constructed so that there is zero difference between treatments. The two sample means (and totals) are identical, so $SS_{between} = 0$, $MS_{between} = 0$, and the F-ratio is zero.

Conceptually, the numerator of the F-ratio always measures how much difference exists between treatments. In Example 13.3 we constructed an extreme set of scores with zero difference. However, you should be able to look at any set of data and quickly compare the means (or totals) to determine whether there are big differences between treatments or small differences between treatments.

Being able to estimate the magnitude of between-treatments differences is a good first step in understanding ANOVA and should help you to predict the outcome of an analysis of variance. However, the *between-treatments* differences are only one part of the analysis. You must also understand the *within-treatments* differences that form the denominator of the *F*-ratio. The following example is intended to demonstrate the concepts underlying SS_{within} and MS_{within}. In addition, the example should give you a better understanding of how the between-treatments differences and the within-treatments differences act together within the ANOVA.

EXAMPLE 13.4   The purpose of this example is to demonstrate how you can look at a set of data and estimate the size of the between-treatments differences and the size of the within-treatment differences. In this example, we will compare two hypothetical outcomes for the same experiment. In each case, the experiment uses two separate samples to evaluate the mean difference between two treatments. The following data represent the two outcomes, which we will call case 1 and case 2.

Case 1 Treatment		Case 2 Treatment	
I	II	I	II
29	34	16	28
30	35	30	31
30	35	36	35
31	36	38	46
$\overline{X} = 30$	$\overline{X} = 35$	$\overline{X} = 30$	$\overline{X} = 35$
$SS = 2$	$SS = 2$	$SS = 296$	$SS = 186$

For the data in case 1, it is easy to see that the between-treatments difference is 5 points: One treatment averages $\overline{X} = 30$, and the other averages $\overline{X} = 35$, for a 5-point difference between treatments. To estimate the within-treatment differences, you simply look at the scores inside each of the treatment conditions. In treatment I, for example, the scores vary from 29 to 30 to 31. Thus, the differences within treatments are only 1 or 2 points. Overall, the data in case 1 show a between-treatments difference (numerator of the *F*-ratio) that is substantially larger than the within-treatments differences (denominator of *F*). As a result, the *F*-ratio for these data should be much larger than 1.00. In fact, the actual *F*-ratio for these data is $F = 75$, which is large enough to reject H_0 and conclude that there is a significant difference between treatments. The point, however, is that you should be able to see the difference even before you compute the *F*-ratio.

Now consider the data in case 2. Again, there is a 5-point difference between the two treatment means. This is the size of the between-treatments difference that will form the numerator of the *F*-ratio. However, when you look inside each treatment for case 2, there are much larger differences from score to score. Within each treatment, it is easy to find differences greater than 5 points or even 10 points. Remember, the within-treatments differences provide a measure of how much difference is reasonable to expect by chance. For these data, the 5-point difference between treatments is *not* greater than chance. The actual F-ratio for the case 2 data is $F = 0.62$, which means that we fail to reject H_0 and we conclude that there is no significant difference between treatments.

The preceding two examples were intended to give you a better understanding of the concepts of between-treatments variance and within-treatments variance. In general, you should be able to look at a set of data and make rough estimates of the between-treatments differences and the within-treatments differences. Although you probably cannot estimate the F-ratio with any great precision, you should be able to judge whether the differences between treatments are more or less than the differences expected by chance.

LEARNING CHECK

1. The following data summarize the results of an experiment using 3 separate samples to compare three treatment conditions:

Treatment 1	Treatment 2	Treatment 3	
$n = 5$	$n = 5$	$n = 5$	
$T = 5$	$T = 10$	$T = 30$	$\Sigma X^2 = 325$
$SS = 45$	$SS = 25$	$SS = 50$	

Do these data provide evidence of any significant mean differences among the treatments? Test with $\alpha = .05$.

2. A researcher reports an F-ratio with $df = 2, 30$ for an independent-measures analysis of variance. How many treatment conditions were compared in the experiment? How many subjects participated in the experiment?

3. The following data summarize the results from two separate experiments. Each experiment compares 3 treatment conditions, and each experiment uses a separate sample of $n = 10$ for each treatment.

	Experiment A Treatment			Experiment B Treatment		
	I	II	III	I	II	III
	$\bar{X} = 1$	$\bar{X} = 10$	$\bar{X} = 20$	$\bar{X} = 1$	$\bar{X} = 3$	$\bar{X} = 5$
	$SS = 8$	$SS = 12$	$SS = 10$	$SS = 40$	$SS = 60$	$SS = 50$

Just looking at the data, without doing any serious calculations, answer each of the following questions.

a. Which experiment will produce the larger $MS_{between}$?
b. Which experiment will produce the larger MS_{within}?
c. Which experiment will produce the larger F-ratio?

ANSWERS **1.** The following summary table presents the results of the analysis:

Source	SS	df	MS	
Between	70	2	35	$F = 3.5$
Within	120	12	10	
Total	190	14		

The critical value for F is 3.88. The obtained value for F is not in the critical region, and we fail to reject H_0.

2. There were 3 treatment conditions ($df_{between} = k - 1 = 2$). A total of $N = 33$ individuals participated ($df_{within} = 30 = N - k$).

3. **a.** Experiment A has larger mean differences between treatments.

b. Experiment B has larger SS values within treatments.

c. Experiment A has larger between-treatments differences and smaller within-treatments differences. This combination will produce a larger F-ratio.

13.6 POST HOC TESTS

In analysis of variance, the null hypothesis states that there is no treatment effect:

$$H_0: \mu_1 = \mu_2 = \mu_3 \ldots$$

When you reject the null hypothesis, you conclude that the means are not all the same. Although this appears to be a simple conclusion, in most cases it actually creates more questions than it answers. When there are only two treatments in an experiment, H_0 will state that $\mu_1 = \mu_2$. If you reject this hypothesis, the conclusion is quite straightforward; that is, the two means are not equal ($\mu_1 \neq \mu_2$). However, when you have more than two treatments, the situation immediately becomes more complex. With $k = 3$, for example, rejecting H_0 indicates that not all the means are the same. Now you must decide which ones are different. Is μ_1 different from μ_2? Is μ_1 different from μ_3? Is μ_2 different from μ_3? Are all three different? The purpose of *post hoc tests* is to answer these questions.

As the name implies, post hoc tests are done after an analysis of variance. More specifically, these tests are done after ANOVA when

1. You reject H_0 and
2. There are 3 or more treatments ($k \geq 3$).

Rejecting H_0 indicates that at least one difference exists among the treatments. With $k = 3$ or more, the problem is to find where the differences are.

In general, a post hoc test enables you to go back through the data and compare the individual treatments two at a time. In statistical terms, this is called making *pairwise comparisons*. For example, with $k = 3$, we would compare μ_1 versus μ_2, then μ_2 versus μ_3, and then μ_1 versus μ_3. In each case, we are looking for a significant mean difference.

The process of conducting pairwise comparisons involves performing a series of separate hypothesis tests, and each of these tests includes the risk of a Type I error. As you do more and more separate tests, the risk of a Type I error accumulates and is called the *experimentwise alpha level* (see Box 13.1)

DEFINITION

The *experimentwise alpha level* is the overall probability of a Type I error that is accumulated over a series of separate hypothesis tests. Typically, the experimentwise alpha level is substantially greater than the value of alpha used for any one of the individual tests.

When conducting a series of post hoc tests, you must always be concerned about the experimentwise alpha level. The more comparisons you make, the greater the experimentwise alpha level is, and the greater the risk of a Type I error is. Fortunately, many post hoc tests have been developed that attempt to control the experimentwise alpha level. We will examine one of the commonly used procedures, the Scheffé test.

THE SCHEFFÉ TEST

Because it uses an extremely cautious method for reducing the risk of a Type I error, the *Scheffé test* has the distinction of being one of the safest of all possible post hoc tests. The Scheffé test uses an *F*-ratio to test for a significant difference between any two treatment conditions. The numerator of the *F*-ratio is an *MS* between treatments that is calculated using *only the two treatments you want to compare.* The denominator is the same *MS* within treatments that was used for the overall ANOVA. The "safety factor" for the Scheffé test comes from the following two considerations:

1. Although you are comparing only two treatments, the Scheffé test uses the value of k from the original experiment to compute df between treatments. Thus, df for the numerator of the *F*-ratio is $k - 1$.
2. The critical value for the Scheffé *F*-ratio is the same as was used to evaluate the *F*-ratio from the overall ANOVA. Thus, Scheffé requires that every post-test satisfy the same criteria used for the complete analysis of variance. The following example uses the data from Example 13.1 (Table 13.6) to demonstrate the Scheffé post-test procedure.

EXAMPLE 13.5

Rather than test all the possible comparisons of the four treatment conditions shown in Table 13.6, we will begin with the largest mean difference and then test progressively smaller differences until we find one that is not significant.

TABLE 13.6

Pain threshold data for four different pain relievers

Placebo	Drug A	Drug B	Drug C
$n = 3$	$n = 3$	$n = 3$	$n = 3$
$T = 3$	$T = 3$	$T = 12$	$T = 18$
$\overline{X} = 1$	$\overline{X} = 1$	$\overline{X} = 4$	$\overline{X} = 6$

The grand total, G, for these two treatments is found by adding the two treatment totals ($G = 3 + 18$), and N is found by adding the number of scores in the two treatments ($N = 3 + 3$).

For these data, the largest difference is between the placebo ($T = 3$) and drug C ($T = 18$). The first step is to compute $SS_{between}$ for these two treatments.

$$SS_{between} = \Sigma \frac{T^2}{n} - \frac{G^2}{N}$$

$$= \frac{3^2}{3} + \frac{18^2}{3} - \frac{21^2}{6}$$

$$= 37.5$$

Although we are comparing only two treatments, these two were selected from an experiment consisting of $k = 4$ treatments. The Scheffé test uses the overall experiment ($k = 4$) to determine the degrees of freedom between treatments. Therefore, $df_{between} = k - 1 = 3$, and the MS between treatments is

$$MS_{between} = \frac{SS_{between}}{df_{between}} = \frac{37.5}{3} = 12.5$$

Scheffé also uses the within-treatments variance from the complete experiment, $MS_{within} = 2.00$ with $df = 8$, so the Scheffé F-ratio is

$$F = \frac{MS_{between}}{MS_{within}} = \frac{12.5}{2} = 6.25$$

With $df = 3, 8$ and $\alpha = .05$, the critical value for F is 4.07 (see Table B.4). Therefore, our F-ratio is in the critical region, and we conclude that there is a significant difference between the placebo and drug C. Because the data for drug A and the placebo are equivalent (both have $T = 3$), we also conclude that drug C is significantly different from drug A.

Next, we consider the second largest mean difference for the data: drug B ($T = 12$) versus the placebo ($T = 3$). Again, we compute $SS_{between}$ using only these two treatment groups.

$$SS_{between} = \Sigma \frac{T^2}{n} - \frac{G^2}{N}$$

$$= \frac{3^2}{3} + \frac{12^2}{3} - \frac{15^2}{6}$$

$$= 13.5$$

As before, the Scheffé test uses the overall experiment ($k = 4$) to determine df between treatments. Thus, $df_{between} = k - 1 = 3$, and $MS_{between}$ for this comparison is

$$MS_{between} = \frac{SS_{between}}{df_{between}} = \frac{13.5}{3} = 4.5$$

Using $MS_{within} = 2$ with $df = 8$, we obtain a Scheffé F-ratio of

$$F = \frac{MS_{between}}{MS_{within}} = \frac{4.5}{2} = 2.25$$

With $df = 3, 8$ and $\alpha = .05$, the obtained F-ratio is less than the critical value of 4.07. Because the F-ratio is not in the critical region, our decision is that these data do not provide sufficient evidence to conclude that there is a significant difference between the placebo and drug B. Because the mean difference between the placebo and drug B is larger than any of the remaining pairs of treatments, we can conclude that none of the other pairwise comparisons would be significant with the Scheffé test.

Thus, the conclusion from the Scheffé post-test is that drug C is significantly different from both the placebo and drug A. These are the only significant differences for this experiment (using Scheffé), and these differences are the source of the significant *F*-ratio obtained for the overall ANOVA.

IN THE LITERATURE:
REPORTING THE RESULTS OF POST HOC TESTS

When reporting the results of analysis of variance, it is necessary to include the outcome of any post hoc tests that were performed. A statement of these results immediately follows the report of the overall *F*-ratio from the analysis (see In the Literature, page 332). Thus, for Example 13.1, we would add to our report of the results the following:

> Post hoc analyses were performed using the Scheffé post-test to identify exactly where significant differences exist. The analyses revealed that drug C differed significantly from the placebo group, $F(3, 8) = 6.25, p < .05$, and from drug A, $F(3, 8) = 6.25, p < .05$. No other differences were found.

When the study has many treatment groups (that is, when *k* is large), the results of post-tests may be incorporated into a table that shows treatment means and standard deviations. The table would contain one or more footnotes indicating which groups differ significantly along with the probabilities for those tests. The report would simply state that the results of the Scheffé post-test are summarized in the table. For Example 13.1, the table could be constructed as follows:

	Treatment condition			
	Placebo	Drug A	Drug B	Drug C
M	1.0	1.0	4.0	6.0*
SD	1.73	1.00	1.00	1.73

*Differs from placebo and drug A, $p < .05$

13.7 THE RELATIONSHIP BETWEEN ANOVA AND *t* TESTS

When you have data from an independent-measures study with only two treatment conditions, you can use either a *t* test (Chapter 10) or an independent-measures ANOVA. In practical terms, it makes no difference which you choose. These two statistical techniques always will result in the same statistical decision. In fact, the two methods use many of the same calculations and are very closely related in several other respects. The basic relationship between t statistics and F-ratios can be stated in an equation:

$$F = t^2$$

This relationship can be explained by first looking at the structure of the formulas for F and t.

The structure of the t statistic compares the actual difference between the samples (numerator) with the standard difference that would be expected by chance (denominator).

$$t = \frac{\text{obtained difference between sample means}}{\text{difference expected by chance (error)}}$$

The structure of the F-ratio also compares differences between samples versus the difference due to chance or error.

$$F = \frac{\text{variance (differences) between sample means}}{\text{variance (differences) expected by chance (error)}}$$

However, the numerator and the denominator of the F-ratio measure variances, or mean squared differences. Therefore, we can express the F-ratio as follows:

$$F = \frac{(\text{differences between samples})^2}{(\text{differences expected by chance})^2}$$

The fact that the t statistic is based on differences and the F-ratio is based on *squared* differences leads to the basic relationship $F = t^2$.

There are several other points to consider in comparing the t statistic to the F-ratio.

1. It should be obvious that you will be testing the same hypotheses whether you choose a t test or an ANOVA. With only two treatments, the hypotheses for either test are

   $$H_0: \mu_1 = \mu_2$$
   $$H_1: \mu_1 \neq \mu_2$$

2. The degrees of freedom for the t statistic and the df for the denominator of the F-ratio (df_{within}) are identical. For example, if you have two samples, each with 6 scores, the independent-measures t statistic will have $df = 10$, and the F-ratio will have $df = 1, 10$. In each case, you are adding the df from the first sample ($n - 1$) and the df from the second sample.

3. The distribution of t statistics and the distribution of F-ratios match perfectly if you take into consideration the relationship $F = t^2$. Consider the t distribution with $df = 18$ and the corresponding F distribution with $df = 1, 18$ that are presented in Figure 13.9. Notice the following relationships:

   a. If each of the t values is squared, then all of the negative values will become positive. As a result, the whole left-hand side of the t distribution (below zero) will be flipped over to the positive side. This creates a nonsymmetrical, positively skewed distribution, that is, the F distribution.

FIGURE 13.9

The distribution of *t* statistics with $df = 18$ and the corresponding distribution of *F*-ratios with $df = 1, 18$. Notice that the critical values for $\alpha = .05$ are $t = \pm 2.101$ and that $F = 2.101^2 = 4.41$.

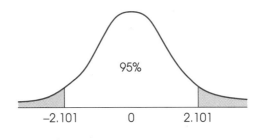

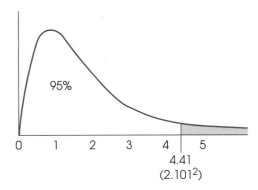

b. For $\alpha = .05$, the critical region for *t* is determined by values greater than $+2.101$ or less than -2.101. When these boundaries are squared, you get

$$\pm 2.101^2 = 4.41$$

Notice that 4.41 is the critical value for $\alpha = .05$ in the *F* distribution. Any value that is in the critical region for *t* will end up in the critical region for *F*-ratios after it is squared.

ASSUMPTIONS FOR THE INDEPENDENT-MEASURES ANOVA

The independent-measures ANOVA requires the same three assumptions that were necessary for the independent-measures *t* hypothesis test:

1. The observations within each sample must be independent (see page 202).
2. The populations from which the samples are selected must be normal.
3. The populations from which the samples are selected must have equal variances (homogeneity of variance).

Ordinarily researchers are not overly concerned with the assumption of normality, especially when large samples are used, unless there are strong reasons to suspect the assumption has not been satisfied. The assumption of homogeneity of variance is an important one. If a researcher suspects it has been violated, it can be tested by Hartley's *F*-max test for homogeneity of variance (Chapter 10, page 253).

LEARNING CHECK

1. The Scheffé post hoc test uses the between-treatments *df* from the original ANOVA even though SS_{between} is calculated for a pair of treatments. (True or false?)

2. An ANOVA produces an *F*-ratio with $df = 1, 34$. Could the data have been analyzed with a *t* test? What would be the degrees of freedom for the *t* statistic?

3. With $k = 2$ treatments, are post hoc tests necessary when the null hypothesis is rejected? Explain why or why not.

ANSWERS   **1.** True

2. If the *F*-ratio has $df = 1, 34$, then the experiment compared only two treatments, and you could use a *t* statistic to evaluate the data. The *t* statistic would have $df = 34$.

3. No. Post hoc tests are used to determine which treatments are different. With only two treatment conditions, there is no uncertainty as to which two treatments are different.

SUMMARY

1. Analysis of variance (ANOVA) is a statistical technique that is used to test for mean differences among two or more treatment conditions or among two or more populations. The null hypothesis for this test states that there are no differences among the population means. The alternative hypothesis states that at least one mean is different from the others. Although analysis of variance can be used with either an independent- or a repeated-measures design, this chapter examined only independent-measures designs, that is, experiments with a separate sample for each treatment condition.

2. The test statistic for analysis of variance is a ratio of two variances called an *F*-ratio. The *F*-ratio is structured so that the numerator and the denominator measure the same variance when the null hypothesis is true. In this way, the existence of a significant treatment effect is apparent if the data produce an "unbalanced" *F*-ratio. The variances in the *F*-ratio are called mean squares, or *MS* values. Each *MS* is computed by

$$MS = \frac{SS}{df}$$

3. For the independent-measures analysis of variance, the *F*-ratio is

$$F = \frac{MS_{between}}{MS_{within}}$$

The $MS_{between}$ measures differences among the treatments by computing the variability of the treatment means or totals. These differences are assumed to be produced by
a. Treatment effects (if they exist)
b. Differences due to chance

The MS_{within} measures variability inside each of the treatment conditions. Because individuals inside a treatment condition are all treated exactly the same, any differences within treatments cannot be caused by treatment effects. Thus, the within-treatments *MS* is produced only by differences due to chance.

With these factors in mind, the *F*-ratio has the following structure:

$$F = \frac{\text{treatment effect} + \text{differences due to chance}}{\text{differences due to chance}}$$

When there is no treatment effect (H_0 is true), the numerator and the denominator of the *F*-ratio are measuring the same variance, and the obtained ratio should be near 1.00. If there is a significant treatment effect, the numerator of the ratio should be larger than the denominator, and the obtained *F* value should be much greater than 1.00.

4. The formulas for computing each *SS*, *df*, and *MS* value are presented in Figure 13.10, which also shows the general structure for the analysis of variance.

5. The *F*-ratio has two values for degrees of freedom, one associated with the *MS* in the numerator and one associated with the *MS* in the denominator. These *df* values are used to find the critical value for the *F*-ratio in the *F*-distribution table.

6. When the decision from an analysis of variance is to reject the null hypothesis and when the experiment contained more than two treatment conditions, it is necessary to continue the analysis with a post hoc test, such as the Scheffé test. The purpose of these tests is to determine exactly which treatments are significantly different and which are not.

FIGURE 13.10

Formulas for ANOVA

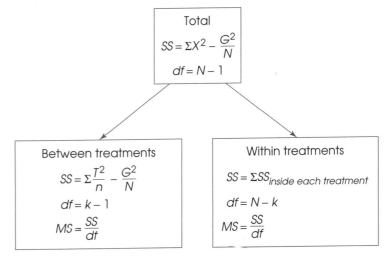

$$F\text{-ratio} = \frac{MS\,\text{between treatments}}{MS\,\text{within treatments}}$$

KEY TERMS

analysis of variance (ANOVA)

between-treatments variability

within-treatments variability

treatment effect

individual differences

experimental error

F-ratio

factor

levels

mean square (MS)

ANOVA summary table

distribution of F-ratios

post hoc tests

Scheffé test

FOCUS ON PROBLEM SOLVING

1. The words and labels used to describe the different components of variance can help you remember the ANOVA formulas. For example, *total* refers to the total experiment. Therefore, the SS_{total} and df_{total} values are based on the whole set of N scores. The word *within* refers to the variability inside (within) the treatment groups. Thus, the value for SS_{within} is based on an SS value from each group, computed from the scores *within* each group. Finally, *between* refers to the variability (or differences) between treatments. The $SS_{between}$ component measures the differences between treatments (T_1 versus T_2, and so on), and $df_{between}$ is simply the number of T values (k) minus one.

2. When you are computing SS and df values, always calculate all three components (total, between, and within) separately and then check your work by making sure that the *between-treatments* and *within-treatments* components add up to the *total*.

3. Because ANOVA requires a fairly lengthy series of calculations, it helps to organize your work. We suggest that you compute all of the SS values first, followed by the df values, then the two MS values, and, finally, the F-ratio. If you use the same system all the time (practice it!), you will be less likely to get lost in the middle of a problem.

4. The previous two focus points are facilitated by using an ANOVA summary table (for example, see page 333). The first column has the heading *source*. Listed below the heading are the three sources of variability (between, within, and total). The second and third columns have the headings *SS* and *df*. These can be filled in as you perform the appropriate computations. The last column is headed *MS* for the mean square values.

5. Remember that an *F*-ratio has two separate values for *df*: one for the numerator and one for the denominator. Properly reported, the $df_{between}$ value is stated first. You will need both *df* values when consulting the *F*-distribution table for the critical *F* value. You should recognize immediately that an error has been made if you see an *F*-ratio reported with a single value for *df*.

6. When you encounter an *F*-ratio and its *df* values reported in the literature, you should be able to reconstruct much of the original experiment. For example, if you see "$F(2, 36) = 4.80$," you should realize that the experiment compared $k = 3$ treatment groups (because $df_{between} = k - 1 = 2$), with a total of $N = 39$ subjects participating in the experiment (because $df_{within} = N - k = 36$).

7. Keep in mind that a large value for *F* indicates evidence for a treatment effect. A combination of factors will yield a large *F* value. One such factor is a large value for $MS_{between}$ in the numerator of the ratio. This will occur when there are large differences between groups, as would be expected when a treatment effect occurs. Another factor that contributes to a large *F* value would be a small value for MS_{within} on the bottom of the *F*-ratio. This will occur when there is little error variability, reflected in low variability within groups.

——— DEMONSTRATION 13.1 ———

ANALYSIS OF VARIANCE

A human factors psychologist studied three computer keyboard designs. Three samples of individuals were given material to type on a particular keyboard, and the number of errors committed by each subject was recorded. The data are as follows:

Keyboard A: 0, 4, 0, 1, 0

Keyboard B: 6, 8, 5, 4, 2

Keyboard C: 6, 5, 9, 4, 6

Does typing performance differ significantly among the three types of keyboards?

STEP 1 State the hypotheses, and specify the alpha level.

The null hypothesis states that there is no difference among the keyboards in terms of number of errors committed. In symbols, we would state

H_0: $\mu_1 = \mu_2 = \mu_3$ (there is no effect of type of keyboard used)

As noted previously in this chapter, there are a number of possible statements for the alternative hypothesis. Here we state the general alternative hypothesis:

H_1: At least one of the treatment means is different.

That is, the type of keyboard has an effect on typing performance. We will set alpha at $\alpha = .05$.

STEP 2   Locate the critical region.

To locate the critical region, we must obtain the values for $df_{between}$ and df_{within}.

$$df_{between} = k - 1 = 3 - 1 = 2$$

$$df_{within} = N - k = 15 - 3 = 12$$

The F-ratio for this problem will have $df = 2, 12$. The F-distribution table is consulted for $df = 2$ in the numerator and $df = 12$ in the denominator. The critical F value for $\alpha = .05$ is $F = 3.88$. The obtained F-ratio must exceed this value to reject H_0.

STEP 3   Perform the analysis.

The analysis involves the following steps:

1. Compute T and SS for each sample, and obtain G and ΣX^2 for all ($N = 15$) scores.
2. Perform the analysis of SS.
3. Perform the analysis of df.
4. Calculate the MS values.
5. Calculate the F-ratio.

Compute T, SS, G, and ΣX^2. We will use the computational formula for the SS of each sample. The calculations are illustrated with the following tables:

Keyboard A		Keyboard B		Keyboard C	
X	X^2	X	X^2	X	X^2
0	0	6	36	6	36
4	16	8	64	5	25
0	0	5	25	9	81
1	1	4	16	4	16
0	0	2	4	6	36
$\Sigma X = 5$	$\Sigma X^2 = 17$	$\Sigma X = 25$	$\Sigma X^2 = 145$	$\Sigma X = 30$	$\Sigma X^2 = 194$

For keyboard A, T and SS are computed using only the $n = 5$ scores of this sample:

$$T_1 = \Sigma X = 0 + 4 + 0 + 1 + 0 = 5$$

$$SS_1 = \Sigma X^2 - \frac{(\Sigma X)^2}{n} = 17 - \frac{(5)^2}{5} = 17 - \frac{25}{5} = 17 - 5$$

$$= 12$$

For keyboard B, T and SS are computed for its $n = 5$ scores:

$$T_2 = \Sigma X = 25$$

$$SS_2 = \Sigma X^2 - \frac{(\Sigma X)^2}{n} = 145 - \frac{(25)^2}{5} = 145 - \frac{625}{5} = 145 - 125$$

$$= 20$$

For the last sample, keyboard C, we obtain

$$T_3 = \Sigma X = 30$$

$$SS_3 = \Sigma X^2 - \frac{(\Sigma X)^2}{n} = 194 - \frac{(30)^2}{5} = 194 - \frac{900}{5} = 194 - 180$$

$$= 14$$

The grand total (G) for $N = 15$ scores is

$$G = \Sigma T = 5 + 25 + 30 = 60$$

The ΣX^2 for all $N = 15$ scores in this study can be obtained by summing the X^2 columns for the three samples. For these data, we obtain

$$\Sigma X^2 = 17 + 145 + 194 = 356$$

Perform the analysis of SS. We will compute SS_{total} followed by its two components.

$$SS_{total} = \Sigma X^2 - \frac{G^2}{N} = 356 - \frac{60^2}{15} = 356 - \frac{3600}{15}$$

$$= 356 - 240 = 116$$

$$SS_{within} = \Sigma SS_{inside \ each \ treatment}$$

$$= 12 + 20 + 14$$

$$= 46$$

$$SS_{between} = \Sigma \frac{T^2}{n} - \frac{G^2}{N}$$

$$= \frac{5^2}{5} + \frac{25^2}{5} + \frac{30^2}{5} - \frac{60^2}{15}$$

$$= \frac{25}{5} + \frac{625}{5} + \frac{900}{5} - \frac{3600}{15}$$

$$= 5 + 125 + 180 - 240$$

$$= 70$$

Perform the analysis of df. We will compute df_{total}. Its components, $df_{between}$ and df_{within}, were previously calculated (step 2).

$$df_{total} \quad = N - 1 = 15 - 1 = 14$$

$$df_{between} = 2$$

$$df_{within} \quad = 12$$

Calculate the MS values. The values for $MS_{between}$ and MS_{within} are determined.

$$MS_{between} = \frac{SS_{between}}{df_{between}} = \frac{70}{2} = 35$$

$$MS_{within} = \frac{SS_{within}}{df_{within}} = \frac{46}{12} = 3.83$$

Compute the F-ratio. Finally, we can compute F.

$$F = \frac{MS_{between}}{MS_{within}} = \frac{35}{3.83} = 9.14$$

STEP 4 Make a decision about H_0, and state a conclusion.

The obtained F of 9.14 exceeds the critical value of 3.88. Therefore, we can reject the null hypothesis. The type of keyboard used has a significant effect on the number of errors committed, $F(2, 12) = 9.14$, $p < .05$. The following table summarizes the results of the analysis:

Source	SS	df	MS	
Between treatments	70	2	35	$F = 9.14$
Within treatments	46	12	3.83	
Total	116	14		

PROBLEMS

1. Explain why the expected value for an F-ratio is equal to 1.00 when there is no treatment effect?

2. Describe the similarities between an F-ratio and a t statistic.

3. What happens to the value of the F-ratio in analysis of variance if the differences between the sample means increase? What happens to the value of the F-ratio if the sample variances increase?

4. Explain why you should use ANOVA instead of several t tests to evaluate mean differences when an experiment consists of three or more treatment conditions.

5. Describe *when* and *why* post-tests are used. Explain why you would not need to do post-tests for an experiment with only $k = 2$ treatment conditions.

6. For the following set of data, without doing any calculations (just look at the data), what value should be obtained for the variance within treatments? ($MS_{within} = ?$)

Treatments		
I	II	
1	3	
1	3	$G = 16$
1	3	$\Sigma X^2 = 40$
1	3	
$T = 4$	$T = 12$	

7. First-born children tend to develop language skills faster than their younger siblings. One possible explanation for this phenomenon is that first-born children have undivided attention from their parents. If this explanation is correct, then it is also reasonable that twins should show slower language development than single children and that triplets should be even slower. Davis (1937) found exactly this result. The following hypothetical data demonstrate the relationship. The dependent variable is a measure of language skill at age 3 years for each child. Do the data provide evidence for significant differences between the three populations? Test at the .05 level.

Single child	Twin	Triplet
8	4	4
7	6	4
10	7	7
6	4	2
9	9	3

8. The following data represent three separate samples tested in three different treatment conditions:

	Treatments	
I	II	III
0	6	4
4	6	0
0	2	4
0	6	4
$T = 4$	$T = 20$	$T = 12$
$SS = 12$	$SS = 12$	$SS = 12$

a. Using only the first two samples (treatments I and II), use an ANOVA to test for a mean difference between the two treatments.

b. Now use all three samples to test for mean differences among the three treatment conditions.

c. You should find that the first two treatments are significantly different (part a) but that there are no significant differences when the third treatment is included in the analysis (part b). How can you explain this outcome? (*Hint:* Compute the three sample means. How large is the mean difference when you are comparing only treatment I versus treatment II? How large are the mean differences, on average, when you are comparing all three treatments?)

9. A psychologist would like to examine the relative effectiveness of three therapy techniques for treating mild phobias. A sample of $N = 15$ individuals who display a moderate fear of spiders is obtained. These individuals are randomly assigned (with $n = 5$) to each of the three therapies. The dependent variable is a measure of reported fear of spiders after therapy. The data are as follows:

Therapy A	Therapy B	Therapy C	
5	3	1	
2	3	0	$G = 30$
2	0	1	$\Sigma X^2 = 86$
4	2	2	
2	2	1	
$T = 15$	$T = 10$	$T = 5$	
$SS = 8$	$SS = 6$	$SS = 2$	

Do these data indicate that there are any significant differences among the three therapies? Test at the .05 level of significance.

10. The following data are from an experiment comparing three treatment conditions, with a separate sample of $n = 4$ in each treatment.

Treatments			
I	II	III	
0	1	8	
4	5	5	$G = 48$
0	4	6	$\Sigma X^2 = 284$
4	2	9	
$T = 8$	$T = 12$	$T = 28$	
$SS = 16$	$SS = 10$	$SS = 10$	

a. Use an ANOVA with $\alpha = .05$ to determine whether there are any significant differences among the three treatments.

b. Use Scheffé post-tests with $\alpha = .05$ to determine exactly which treatment means are significantly different from each other.

11. The following data were obtained from a study using two separate samples to evaluate the mean difference between two treatment conditions.

Treatments		
I	II	
0	10	$G = 40$
1	4	
7	10	$\Sigma X^2 = 298$
4	4	

a. Using an analysis of variance with $\alpha = .05$, determine whether the data indicate a significant difference between the two treatments.

b. Using an independent-measures t test with $\alpha = .05$, determine whether the data indicate a significant difference between the two treatments.

c. Check that your results confirm the general relationship between the F-ratio and the t statistic. Comparing your F-ratio from part a and your t statistic from part b, you should find that $F = t^2$.

12. A pharmaceutical company has developed a drug that is expected to reduce hunger. To test the drug, three samples of rats are selected, with $n = 10$ in each sample. The first sample receives the drug every day. The second sample is given the drug once a week, and the third sample receives no drug at all. The dependent variable is the amount of food eaten by each rat over a one-month period. These data are analyzed by an analysis of variance, and the results are reported in the following summary table. Fill in all missing values in the table. (*Hint:* Start with the df column.)

Source	SS	df	MS	
Between treatments	——	——	——	$F = 12$
Within treatments	54	——	——	
Total	——	——		

13. The following summary table presents the results of an ANOVA from an experiment comparing four treatment conditions with a sample of $n = 10$ in each treatment. Complete all missing values in the table.

Source	SS	df	MS	
Between treatments	___	___	15	$F = $ ___
Within treatments	108	___	___	
Total		___	___	

14. A common science-fair project involves testing the effects of music on the growth of plants. For one of these projects, a sample of 24 newly sprouted bean plants is obtained. These plants are randomly assigned to four treatments, with $n = 6$ in each group. The four conditions are rock, heavy metal, country, and classical music. The dependent variable is the height of each plant after two weeks. The data from this experiment were examined using an ANOVA, and the results are summarized in the following table. Fill in all missing values.

Source	SS	df	MS	
Between treatments	___	___	10	$F = $ ___
Within treatments	40	___	___	
Total		___	___	

15. A researcher reports an F-ratio with $df = 3, 24$ for an independent-measures research study.
  a. How many treatment conditions were compared in the study?
  b. How many subjects participated in the entire study?

16. A developmental psychologist is examining problem-solving ability for grade-school children. Random samples of 5-year-old, 6-year-old, and 7-year-old children are obtained, with $n = 3$ in each sample. Each child is given a standardized problem-solving task, and the psychologist records the number of errors. These data are as follows:

5-year-olds	6-year-olds	7-year-olds	
5	6	0	$G = 30$
4	4	1	$\Sigma X^2 = 138$
6	2	2	
$T = 15$	$T = 12$	$T = 3$	
$SS = 2$	$SS = 8$	$SS = 2$	

  a. Use these data to test whether there are any significant differences among the three age groups. Use $\alpha = .05$.
  b. Use the Scheffé test to determine which groups are different.

17. The extent to which a person's attitude can be changed depends in part on how big a change you are trying to produce. In a classic study on persuasion, Aronson,

Turner, and Carlsmith (1963) obtained three groups of subjects. One group listened to a persuasive message that differed only slightly from the subjects' original attitudes. For the second group, there was a moderate discrepancy between the message and the original attitudes. For the third group, there was a large discrepancy between the message and the original attitudes. For each subject, the amount of attitude change was measured. Hypothetical data, similar to the experimental results, are as follows:

	Size of discrepancy		
Small	Moderate	Large	
1	3	0	
0	4	2	$G = 36$
0	6	0	$\Sigma X^2 = 138$
2	3	4	
3	5	0	
0	3	0	
$T = 6$	$T = 24$	$T = 6$	
$SS = 8$	$SS = 8$	$SS = 14$	

  a. Use an analysis of variance with $\alpha = .01$ to determine whether the amount of discrepancy between the original attitude and the persuasive argument has a significant effect on the amount of attitude change.
  b. For these data, describe how the effectiveness of a persuasive argument is related to the discrepancy between the argument and a person's original attitude.

18. There have been a number of studies on the adjustment problems (that is, "jetlag") people experience when traveling across time zones (see Moore-Ede, Sulzman, and Fuller, 1982). Jetlag always seems to be worse when traveling east. Consider this hypothetical study. A researcher examines how many days it takes a person to adjust after taking a long flight. One group flies east across time zones (California to New York), a second group travels west (New York to California), and a third group takes a long flight within one time zone (San Diego to Seattle). For the following data, create a table that displays the mean and the standard deviation for each sample. Then perform an analysis of variance to determine if jetlag varies for type of travel. Use the .05 level of significance.

Westbound	Eastbound	Same time zone
2	6	1
1	4	0
3	6	1
3	8	1
2	5	0
4	7	0

19. One way to define and measure the capacity of immediate memory is to show subjects a series of objects one at a time and then ask them to report the items they just saw. We begin with only one or two items and gradually increase the number until the subject begins to make errors. Most adults perform perfectly up to seven or eight items before memory begins to fail. (For example, you can recall a 7-digit phone number after hearing it once, but most of us have trouble if presented with a list of 9 or 10 digits to be recalled.) This basic memory ability appears to develop gradually during childhood. The following data represent immediate memory performance for 2-year-old, 6-year-old, and 10-year-old children. Use an analysis of variance with $\alpha = .05$ to test for mean differences among the three age groups. (*Note:* You will need to convert means to totals and convert the standard deviations to *SS* values before you can use the normal ANOVA formulas. Also, these summarized data do not provide enough information for direct calculation of SS_{total}.)

$$2\text{-year-olds:} \quad n = 20, \overline{X} = 2.1, s = 1.3$$

$$6\text{-year-olds:} \quad n = 20, \overline{X} = 4.3, s = 1.5$$

$$10\text{-year-olds:} \quad n = 20, \overline{X} = 6.9, s = 1.8$$

20. Several studies indicate that handedness (left-handed/right-handed) is related to differences in brain function. Because different parts of the brain are specialized for specific behaviors, this means that left- and right-handed people should show different skills or talents. To test this hypothesis, a psychologist tested pitch discrimination (a component of musical ability) for three groups of subjects: left-handed, right-handed, and ambidextrous. The data from this study are as follows:

Right-handed	Left-handed	Ambidextrous	
6	1	2	
4	0	0	$G = 30$
3	1	0	$\Sigma X^2 = 102$
4	1	2	
3	2	1	
$T = 20$	$T = 5$	$T = 5$	
$SS = 6$	$SS = 2$	$SS = 4$	

Each score represents the number of errors during a series of pitch discrimination trials.
a. Do these data indicate any differences among the three groups? Test with $\alpha = .05$.
b. Use the F-max test to determine whether these data satisfy the homogeneity of variance assumption (see Chapter 10).

21. Sheldon (1940) examined the relationship between personality characteristics and physical characteristics (body shape). He identified three categories of body shape: endomorph, ectomorph, and mesomorph, corresponding to rounded, thin, and muscular, respectively. He then evaluated personality characteristics that tend to be associated with each body type. One relationship Sheldon examined involved sociability and body type. Specifically, endomorphs tend to be more social, and ectomorphs tend to be more private. The following hypothetical data represent an attempt to demonstrate this relationship. Sociability scores were obtained for three separate samples representing the three different body types. Do these data indicate significant differences in personality between groups with different physical characteristics? Test with $\alpha = .05$.

Endomorphs	Ectomorphs	Mesomorphs
23	19	18
25	17	14
19	16	15
20	21	11
23	15	17

22. Betz and Thomas (1979) have reported a distinct connection between personality and health. They identified three personality types who differ in their susceptibility to serious, stress-related illness (heart attack, high blood pressure, etc.). The three personality types are Alphas, who are cautious and steady; Betas, who are carefree and outgoing; and Gammas, who tend toward extremes of behavior such as being overly cautious or very careless. Sample data representing general health scores for each of these three groups are as follows. A low score indicates poor health.

Alphas		Betas		Gammas	
43	44	41	52	36	29
41	56	40	57	38	36
49	42	36	48	45	42
52	53	51	55	25	40
41	21	52	39	41	36

a. Compute the mean for each personality type. Do these data indicate a significant difference among the three types? Test with $\alpha = .05$.
b. Use the Scheffé test to determine which groups are different. Explain what happened in this study.

CHAPTER 14

TWO-FACTOR ANALYSIS OF VARIANCE (INDEPENDENT MEASURES)

TOOLS YOU WILL NEED

The following items are considered essential background material for this chapter. If you doubt your knowledge of any of these items, you should review the appropriate chapter or section before proceeding.

- Introduction to analysis of variance (Chapter 13)
  - The logic of analysis of variance
  - ANOVA notation and formulas
  - Distribution of *F*-ratios

CONTENTS

14.1   INTRODUCTION TO FACTORIAL DESIGNS

In most research situations, the goal is to examine the relationship between two variables. Typically, the research study will attempt to isolate the two variables in order to eliminate or reduce the influence of any outside variables that may distort the relationship being studied. A typical experiment, for example, will focus on one independent variable (which is expected to influence behavior) and one dependent variable (which is a measure of the behavior). In real life, however, variables rarely exist in isolation. That is, behavior usually is influenced by a variety of different variables acting and interacting simultaneously. To examine these more complex, real-life situations, researchers often design research studies that include more than one independent variable. That is, researchers will systematically change two (or more) variables and then observe how the changes influence another (dependent) variable.

In the preceding chapter, we examined analysis of variance for *single-factor* research designs, that is, designs that included only one independent variable or only one quasi-independent variable. When a research study involves more than one factor, it is called a *factorial design*. In this chapter, we consider the simplest version of a factorial design. Specifically, we will examine analysis of variance as it applies to research studies with exactly two factors. In addition, we will limit our discussion to studies that use a separate sample for each treatment condition, that is, independent-measures designs. Finally, we will only consider research designs where the sample size (n) is the same for all treatment conditions. In the terminology of ANOVA, this chapter will examine *two-factor, independent-measures, equal n designs*. The following example demonstrates the general structure of this kind of research study.

EXAMPLE 14.1   Most of us find it difficult to think clearly or to work efficiently on hot summer days. If you listen to people discussing this problem, you will occasionally hear comments like "It's not the heat; it's the humidity." To evaluate this claim scientifically, you would need to design a study where both heat and humidity are manipulated within the same experiment and then observe behavior under a variety of different heat and humidity combinations. Figure 14.1 shows the structure of a hypothetical study intended to examine the heat-versus-humidity question. Notice that the study involves two independent variables: The temperature (heat) is varied from 70° to 80° to 90°, and the humidity is varied from 30% to 70%. The two independent variables are used to create a matrix with the different values of temperature defining the columns and the different levels of humidity defining the rows. The resulting two-by-three matrix shows six different combinations of the variables, producing six treatment conditions. Thus, the research study would require six separate samples, one for each of the cells or boxes in the matrix. The dependent variable for our study would be a measure of thinking/working proficiency (such as performance on a problem-solving task) for people observed in each of the six conditions.

FIGURE 14.1

The structure of a two-factor experiment presented as a matrix. The factors are humidity and temperature. There are two levels for the humidity factor (30% and 70%), and there are three levels for the temperature factor (70°, 80°, and 90°).

		Factor B: Temperature		
		70° room	80° room	90° room
Factor A: Humidity	30% humidity	Scores for $n = 15$ subjects tested in a 70° room at 30% humidity	Scores for $n = 15$ subjects tested in an 80° room at 30% humidity	Scores for $n = 15$ subjects tested in a 90° room at 30% humidity
	70% humidity	Scores for $n = 15$ subjects tested in a 70° room at 70% humidity	Scores for $n = 15$ subjects tested in a 80° room at 70% humidity	Scores for $n = 15$ subjects tested in a 90° room at 70% humidity

The two-factor analysis of variance will test for mean differences in research studies that are structured like the heat-and-humidity example in Figure 14.1. For this example, the two-factor ANOVA tests for

1. Mean difference between the two humidity levels.
2. Mean differences between the three temperature levels.
3. Any other mean differences that may result from unique combinations of a specific temperature and a specific humidity level. (For example, high humidity may be especially disruptive when the temperature is also high.)

Thus, the two-factor ANOVA combines three separate hypothesis tests in one analysis. Each of these three tests will be based on its own F-ratio computed from the data. The three F-ratios will all have the same basic structure:

$$F = \frac{\text{variance (differences) between sample means}}{\text{variance (differences) expected by chance}}$$

As always in ANOVA, a large value for the F-ratio indicates that the sample mean differences are greater than chance. To determine whether the obtained F-ratios are *significantly* greater than chance, we will need to compare each F-ratio with the critical values found in the F-distribution table in Appendix B.

14.2 MAIN EFFECTS AND INTERACTIONS

As noted in the previous section, a two-factor ANOVA actually involves three distinct hypothesis tests. In this section, we will examine these three tests in more detail.

Traditionally, the two independent variables in a two-factor experiment are identified as factor A and factor B. For the experiment presented in Figure 14.1, humidity would be factor A, and temperature would be factor B. The goal of the experiment is to evaluate the mean differences that may be produced by either of these factors independently or by the two factors acting together.

MAIN EFFECTS

One purpose of the experiment is to determine whether differences in humidity (factor A) result in differences in performance. To answer this question, we will compare the mean score for all subjects tested with 30% humidity versus the mean score for all subjects tested with 70% humidity. Notice that this process evaluates mean differences between the rows in Figure 14.1.

To make this process more concrete, we have presented a set of hypothetical data in Table 14.1. This table shows the mean score for each of the treatment conditions (cells) as well as the mean for each column (each temperature) and for each row (humidity level). These data indicate that subjects in the low humidity condition (the top row) obtained an average score of $\overline{X} = 80$. This overall mean was obtained by computing the average of the three means in the top row. In contrast, high humidity resulted in a mean score of $\overline{X} = 70$ (the overall mean for the bottom row). The difference between these means constitutes what is called the *main effect* for humidity, or the *main effect for factor A*.

Similarly, the main effect for factor B (temperature) is defined by the mean differences among columns of the matrix. For the data in Table 14.1, the two groups of subjects tested with a temperature of 70° obtained an overall mean score of $\overline{X} = 80$. Subjects tested with the temperature at 80° averaged only $\overline{X} = 75$, and subjects tested at 90° achieved a mean score of $\overline{X} = 70$. The differences among these means constitute the *main effect* for temperature, or the *main effect for factor B*.

DEFINITION

The mean differences among the levels of one factor are referred to as the *main effect* of that factor. When the design of the research study is represented as a matrix with one factor determining the rows and the second factor determining the columns, then the mean differences among the rows would describe the main effect of one factor, and the mean differences among the columns would describe the main effect for the second factor.

You should realize that the mean differences among columns or rows simply *describe* the main effects for a two-factor study. As we have observed in earlier chapters, the existence of sample mean differences does not necessarily imply that the differences are *statistically significant*. In the case of a two-factor study, any main effects that are observed in the data must be evaluated with a hypothesis test to determine whether or not they are statistically significant effects. Unless the hypothesis test demonstrates that the main effects are significant, you must conclude that the observed mean differences are simply the result of sampling error.

The evaluation of main effects will make up two of the three hypothesis tests contained in a two-factor ANOVA. We will state hypotheses concerning the main effect of factor A and the main effect of factor B and then calculate two separate F-ratios to evaluate the hypotheses.

For the example we are considering, factor A involves the comparison of two different levels of humidity. The null hypothesis would state that there is no differ-

TABLE 14.1

Hypothetical data from an experiment examining two different levels of humidity (factor A) and three different levels of temperature (factor B)

	70°	80°	90°	
30% humidity	$\overline{X} = 85$	$\overline{X} = 80$	$\overline{X} = 75$	$\overline{X} = 80$
70% humidity	$\overline{X} = 75$	$\overline{X} = 70$	$\overline{X} = 65$	$\overline{X} = 70$
	$\overline{X} = 80$	$\overline{X} = 75$	$\overline{X} = 70$	

ence between the two levels; that is, humidity has no effect on performance. In symbols,

$$H_0: \mu_{A_1} = \mu_{A_2}$$

The alternative hypothesis is that the two different levels of humidity do produce different scores:

$$H_1: \mu_{A_1} \neq \mu_{A_2}$$

To evaluate these hypotheses, we will compute an F-ratio that compares the actual mean differences between the two humidity levels versus the amount of difference that would be expected by chance or sampling error.

$$F = \frac{\text{variance (differences) between the means for factor } A}{\text{variance (differences) expected by chance}}$$

$$F = \frac{\text{variance (differences) between row means}}{\text{variance (differences) expected by chance}}$$

Similarly, factor B involves the comparison of the three different temperature conditions. The null hypothesis would state that overall there are no differences in mean performance among the three temperatures. In symbols,

$$H_0: \mu_{B_1} = \mu_{B_2} = \mu_{B_3}$$

As always, the alternative hypothesis states that there are differences:

H_1: At least one mean is different from the others.

Again, the F-ratio will compare the obtained mean differences among the three temperature conditions versus the amount of difference that would be expected by chance.

$$F = \frac{\text{variance (differences) between the means for factor } B}{\text{variance (differences) expected by chance}}$$

$$F = \frac{\text{variance (differences) between column means}}{\text{variance (differences) expected by chance}}$$

INTERACTIONS

In addition to evaluating the main effect of each factor individually, the two-factor ANOVA allows you to evaluate other mean differences that may result from unique combinations of the two factors. For example, specific combinations of heat and humidity may have effects that are different from the overall effects of heat or humidity acting alone. Any "extra" mean differences that are not explained by the main effects are called an *interaction between factors*. The real advantage of combining two factors within the same study is the ability to examine the unique effects caused by an interaction.

DEFINITION

An *interaction between two factors* occurs whenever the mean differences between individual treatment conditions, or cells, are different from what would be predicted from the overall main effects of the factors.

To make the concept of an interaction more concrete, we will re-examine the data shown in Table 14.1. For these data, there is no interaction; that is, there are no extra mean differences that are not explained by the main effects. For example, within each temperature condition (each column of the matrix) the subjects scored 10 points higher in the 30% humidity condition than in the 70% condition. This 10-point mean difference is exactly what is predicted by the overall main effect for humidity.

Now consider the data shown in Table 14.2. These new data show exactly the same main effects that existed in Table 14.1 (the column means and the row means have not been changed). But now there is an interaction between the two factors. For example, when the temperature is 90° (third column), there is a 20-point difference between the low-humidity and the high-humidity conditions. This 20-point difference cannot be explained by the 10-point main effect for humidity. Also, when the temperature is 70° (first column), the data show no difference between the two humidity conditions. Again, this zero difference is not what would be expected based on the 10-point main effect for humidity. The extra, unexplained mean differences are an indication that there is an interaction between the two factors.

To evaluate the interaction, the two-factor ANOVA first identifies mean differences that cannot be explained by the main effects. After the extra mean differences are identified, they are evaluated by an F-ratio with the following structure:

$$F = \frac{\text{variance (mean differences) not explained by main effects}}{\text{variance (differences) expected by chance}}$$

The null hypothesis for this F-ratio simply states that there is no interaction:

H_0: There is no interaction between factors A and B. All the mean differences between treatment conditions are explained by the main effects of the two factors.

The alternative hypothesis is that there is an interaction between the two factors:

H_1: There is an interaction between factors. The mean differences between treatment conditions are not what would be predicted from the overall main effects of the two factors.

MORE ABOUT INTERACTIONS

In the previous section, we introduced the concept of an interaction as the unique effect produced by two factors working together. This section will present two alternative definitions of an interaction. These alternatives are intended to help you understand the concept of an interaction and to help you identify an interaction when you encounter one in a set of data. You should realize that the new definitions are

TABLE 14.2

Hypothetical data from an experiment examining two different levels of humidity (factor A) and three different temperature conditions (factor B) (These data show the same main effects as the data in Table 14.1, but the individual treatment means have been modified to produce an interaction.)

	70°	80°	90°	
30% humidity	$\overline{X} = 80$	$\overline{X} = 80$	$\overline{X} = 80$	$\overline{X} = 80$
70% humidity	$\overline{X} = 80$	$\overline{X} = 70$	$\overline{X} = 60$	$\overline{X} = 70$
	$\overline{X} = 80$	$\overline{X} = 75$	$\overline{X} = 70$	

equivalent to the original and simply present slightly different perspectives on the same concept.

The first new perspective on the concept of an interaction focuses on the notion of interdependency between the two factors. More specifically, if the two factors are interdependent, so that one factor does influence the effect of the other, then there will be an interaction. On the other hand, if the two factors are independent, so that the effect of either one is not influenced by the other, then there will be no interaction. You should realize that the notion of interdependence is consistent with our earlier discussion of interactions. In particular, if one factor does influence the effect of the other, then unique combinations of the factors will produce unique effects.

DEFINITION   When the effect of one factor depends on the different levels of a second factor, then there is an *interaction between the factors.*

Returning to the data in Table 14.1, you will notice that the size of the humidity effect (top row versus bottom row) *does not depend on* the temperature. For these data, the change in humidity shows the same 10-point effect for all three levels of temperature. Thus, the humidity effect does not depend on temperature, and there is no interaction. Now consider the data in Table 14.2. This time the effect of changing humidity *depends on* the temperature, and there is an interaction. At 70°, for example, changing humidity has no effect. However, there is a 10-point difference between high and low humidity when the temperature is 80°, and there is a 20-point effect at 90°. Thus, the effect of humidity depends on the temperature, which means that there is an interaction between the two factors (see Box 14.1).

The second alternative definition of an interaction is obtained when the results of a two-factor study are presented in a graph. In this case, the concept of an interaction can be defined in terms of the pattern displayed in the graph. Figure 14.2 shows the two sets of data we have been considering. The original data from Table 14.1, where there is no interaction, are presented in Figure 14.2(a). To construct this figure, we selected one of the factors to be displayed on the horizontal axis; in this case, the different levels of temperature are displayed. The dependent variable, mean level of performance, is shown on the vertical axis. Notice that the figure actually contains two separate graphs: The top line shows the relationship between temperature and mean performance when the humidity is set at 30%, and the bottom line shows the relationship when the humidity is 70%. In general, the picture in the graph matches the structure of the data matrix; the columns of the matrix appear as values along the X-axis, and the rows of the matrix appear as separate lines in the graph.

For this particular set of data, Figure 14.2(a), notice that the two lines are parallel; that is, the distance between lines is constant. In this case, the distance between lines reflects the 10-point difference in mean performance between high and low humidity, and this 10-point difference is the same for all three temperature conditions.

Now look at a graph that is obtained when there is an interaction in the data. Figure 14.2(b) shows that data from Table 14.2. This time notice that the lines in the graph are not parallel. The distance between the lines changes as you scan from left to right. For these data, the distance between the lines corresponds to the humidity effect, that is, the mean difference in performance for 30% humidity versus 70% humidity. The fact that this difference depends on temperature indicates an interaction between factors.

DEFINITION   When the results of a two-factor study are presented in a graph, the existence of nonparallel lines (lines that cross or converge) indicates an *interaction between the two factors.*

14.1   INTERACTIONS IN RESEARCH

THE NOTION that the effect of a treatment may *depend on* other factors is central to the concept of an interaction, and it is often the starting point for a research study. Specifically, a researcher may read about a significant treatment effect and wonder if the same effect would be found with a different set of subjects or under different conditions. Notice that the researcher is wondering whether or not the treatment effect *depends on* other factors. To answer this kind of question, researchers often design two-factor research studies where the primary prediction is that the data will produce a significant interaction.

Consider the following example:

A research report concludes that children who watch violent shows on television tend to be more aggressive than children who watch nonviolent shows. In the study, however, the violent television shows all involved human characters. A psychologist suspects that the children's aggressive behavior occurs because they are modeling their own behavior based on what they see other people doing on television. In addition, the psychologist suspects that the modeling would not occur if the television shows involved nonhuman (cartoon) characters. To examine this suspicion, the psychologist creates the two-factor study shown in the following matrix:

Factor 1:
Amount of violence

		Nonviolent	Violent
	Human	Sample of children who watch nonviolent shows with human characters	Sample of children who watch violent shows with human characters
Factor 2: Type of TV character	Nonhuman	Sample of children who watch nonviolent shows with nonhuman characters	Sample of children who watch violent shows with nonhuman characters

There are two important items to notice about this research study:

1. The top row of the matrix is an exact replication of the study that was presented in the original research report. Thus, the researcher started with a published report and simply expanded the original study by adding a second factor.

2. The researcher is predicting an interaction. The expectation is that the effect of violence (factor 1) will depend on the type of character (factor 2). In particular, the researcher predicts that violence will produce increased aggression when the television shows involve human characters but that violence will have no effect when the characters are nonhuman.

For many students, the concept of an interaction is easiest to understand using the perspective of interdependency; that is, an interaction exists when the effects of one variable *depend on* another factor. However, the easiest way to identify an interaction within a set of data is to draw a graph showing the treatment means. The presence of nonparallel lines is an easy way to spot an interaction.

INDEPENDENCE OF MAIN EFFECTS AND INTERACTIONS

The two-factor ANOVA consists of three hypothesis tests, each evaluating specific mean differences: the A effect, the B effect, and the $A \times B$ interaction. As we have noted, these are three *separate* tests, but you should also realize that the three tests are

FIGURE 14.2

(a) Graph showing the data from Table 14.1, where there is no interaction. (b) Graph showing the data from Table 14.2, where there is an interaction.

(a)

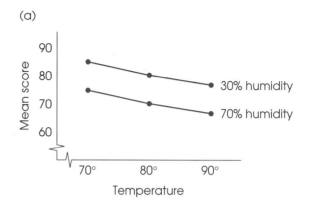

(b)

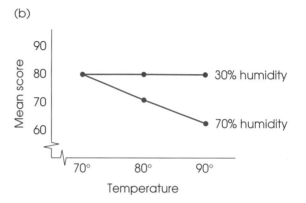

independent. That is, the outcome for any one of the three tests is totally unrelated to the outcome for either of the other two. Thus, it is possible for data from a two-factor study to display any possible combination of significant and/or nonsignificant main effects and interactions. The data sets in Table 14.3 show several possibilities.

Table 14.3(a) shows data with mean differences between levels of factor A (an A effect) but no mean differences for factor B or for an interaction. To identify the A effect, notice that the overall mean for A_1 (the top row) is 10 points higher than the overall mean for A_2 (the bottom row). This 10-point difference is the main effect for factor A. To evaluate the B effect, notice that both columns have exactly the same overall mean, indicating no difference between levels of factor B; hence, there is no B effect. Finally, the absence of an interaction is indicated by the fact that the overall A effect (the 10-point difference) is constant within each column; that is, the A effect *does not depend* on the levels of factor B. (Alternatively, the data indicate that the overall B effect is constant within each row.)

Table 14.3(b) shows data with an A effect and a B effect but no interaction. For these data, the A effect is indicated by the 10-point mean difference between rows, and the B effect is indicated by the 20-point mean difference between columns. The fact that the 10-point A effect is constant within each column indicates no interaction.

Finally, Table 14.3(c) shows data that display an interaction but no main effect for factor A or for factor B. For these data, note that there is no mean difference between rows (no A effect) and no mean difference between columns (no B effect). However, within each row (or within each column), there are mean differences. The

TABLE 14.3

Three sets of data showing different combinations of main effects and interaction for a two-factor study. (The numerical value in each cell of the matrices represents the mean value obtained for the sample in that treatment condition.)

(a) Data showing a main effect for factor A but no B effect and no interaction

	B_1	B_2	
A_1	20	20	A_1 mean = 20
A_2	10	10	A_2 mean = 10

B_1 mean = 15 B_2 mean = 15

10-point difference

No difference

(b) Data showing main effects for both factor A and factor B but no interaction

	B_1	B_2	
A_1	10	30	A_1 mean = 20
A_2	20	40	A_2 mean = 30

B_1 mean = 15 B_2 mean = 35

10-point difference

20-point difference

(c) Data showing no main effect for either factor but an interaction

	B_1	B_2	
A_1	10	20	A_1 mean = 15
A_2	20	10	A_2 mean = 15

B_1 mean = 15 B_2 mean = 15

No difference

No difference

"extra" mean differences within the rows and columns cannot be explained by the overall main effects and therefore indicate an interaction.

LEARNING CHECK

1. Each of the following matrices represents a possible outcome of a two-factor experiment. For each experiment:

   a. Describe the main effect for factor A.

   b. Describe the main effect for factor B.

   c. Does there appear to be an interaction between the two factors?

   Experiment I

	B_1	B_2
A_1	$\overline{X} = 10$	$\overline{X} = 20$
A_2	$\overline{X} = 30$	$\overline{X} = 40$

   Experiment II

	B_1	B_2
A_1	$\overline{X} = 10$	$\overline{X} = 30$
A_2	$\overline{X} = 20$	$\overline{X} = 20$

2. In a graph showing the means from a two-factor experiment, parallel lines indicate that there is no interaction. (True or false?)

3. A two-factor ANOVA consists of three hypothesis tests. What are they?

4. It is impossible to have an interaction unless you also have main effects for at least one of the two factors. (True or false?)

ANSWERS   **1.** For Experiment I:

  a. There is a main effect for factor A; the scores in A_2 average 20 points higher than in A_1.

  b. There is a main effect for factor B; the scores in B_2 average 10 points higher than in B_1.

  c. There is no interaction; there is a constant 20-point difference between A_1 and A_2 that does not depend on the levels of factor B.

  For Experiment II:

  a. There is no main effect for factor A; the scores in A_1 and in A_2 both average 20.

  b. There is a main effect for factor B; on average, the scores in B_2 are 10 points higher than in B_1.

  c. There is an interaction. The difference between A_1 and A_2 depends on the level of factor B. (There is a $+10$ difference in B_1 and a -10 difference in B_2.)

2. True

3. The two-factor ANOVA evaluates the main effects for factor A, the main effects for factor B, and the interaction between the two factors.

4. False. The existence of main effects and interactions is completely independent.

14.3 NOTATION AND FORMULAS

The two-factor ANOVA is composed of three distinct hypothesis tests:

1. The main effect of factor A (often called the A-effect). Assuming that factor A is used to define the rows of the matrix, the main effect of factor A evaluates the mean differences between rows.

2. The main effect of factor B (called the B-effect). Assuming that factor B is used to define the columns of the matrix, the main effect of factor B evaluates the mean differences between columns.

3. The interaction (called the $A \times B$ interaction). The interaction evaluates mean differences between treatment conditions that are not predicted from the overall main effects from factor A or factor B.

For each of these three tests, we are looking for mean differences between treatments that are larger than would be expected by chance. In each case, the magnitude of the treatment effect will be evaluated by an F-ratio. All three F-ratios have the same basic structure:

$$F = \frac{\text{variance (mean differences) between treatments}}{\text{variance (differences) expected by chance}}$$

FORMULAS   The general structure for the analysis of a two-factor experiment is shown in Figure 14.3. Notice that the overall analysis can be divided into two stages. At the first stage of the analysis, the total variability is separated into two components: between-treatments variability and within-treatments variability. You should notice

FIGURE 14.3

Structure of the analysis for a two-factor analysis of variance.

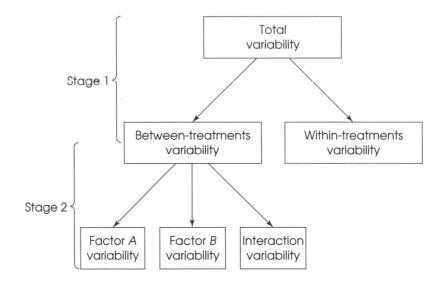

that this first stage is identical to the structure used for the single-factor analysis of variance in Chapter 13 (see Figure 13.4). The second stage of the analysis partitions the between-treatments variability into separate components. With a two-factor experiment, the differences between treatment conditions (cells) could be caused by either of the two factors (*A* or *B*) or by the interaction. These three components are examined individually in the second stage of the analysis.

The goal of this analysis is to compute the variance values needed for the three *F*-ratios. We will need three between-treatments variances (one for factor *A*, one for factor *B*, and one for the interaction), and we will need a within-treatments variance. Each of these variances (or means squares) will be determined by a sum of squares value (*SS*) and a degrees of freedom value (*df*):

Remember, in ANOVA a variance is called a mean square, or *MS*.

$$\text{mean square} = MS = \frac{SS}{df}$$

The actual formulas for the two-factor analysis of variance are almost identical to the formulas for the single-factor analysis (Chapter 13). You may find it useful to refer to this chapter for a more detailed explanation of the formulas. To help demonstrate the use of the formulas, we will use the data in Table 14.4. You should notice that several summary values have been computed for the data and are given in Table 14.4. For example, we have computed the total, *T*, and the *SS* for each sample (each treatment condition or each cell in the matrix). Also, the total is given for each row in the matrix and for each column. Finally, the values of ΣX^2, *N*, and the grand total, *G*, are given for the entire set of scores. All of these summary values are necessary for the two-factor ANOVA.

STAGE 1 OF THE ANALYSIS   The first stage of the two-factor analysis separates the total variability into two basic components: between-treatments and within-treatments. The formulas for this stage are identical to the formulas used in the single-factor ANOVA in

TABLE 14.4

Hypothetical data for a two-factor experiment with two levels of factor A and three levels of factor B. [The individual scores are given for each treatment cell ($n = 5$) along with the cell totals (T values) and the SS for each cell.]

Factor B

		B_1	B_2	B_3	
		5	9	3	
		3	9	8	
	A_1	3	13	3	$T_{A_1} = 90$
		8	6	3	
		6	8	3	
		$T = 25$	$T = 45$	$T = 20$	
		$SS = 18$	$SS = 26$	$SS = 20$	
Factor A		0	0	0	
		2	0	3	
	A_2	0	0	7	$T_{A_2} = 30$
		0	5	5	
		3	0	5	
		$T = 5$	$T = 5$	$T = 20$	
		$SS = 8$	$SS = 20$	$SS = 28$	
		$T_{B_1} = 30$	$T_{B_2} = 50$	$T_{B_3} = 40$	

$N = 30$
$G = 120$
$\Sigma X^2 = 820$

Chapter 13. The formulas and the calculations for the data in Table 14.4 are as follows:

Total variability

$$SS_{\text{total}} = \Sigma X^2 - \frac{G^2}{N} \tag{14.1}$$

For these data,

$$SS_{\text{total}} = 820 - \frac{120^2}{30}$$

$$= 820 - 480$$

$$= 340$$

This SS value measures the variability for all $N = 30$ scores and has degrees of freedom given by

$$df_{\text{total}} = N - 1 \tag{14.2}$$

For the data in Table 14.4, $df_{\text{total}} = 29$.

Between-treatments variability   Remember that each treatment condition corresponds to a cell in the matrix. With this in mind, the between-treatments SS is computed as

$$SS_{\text{between treatments}} = \Sigma\frac{T^2}{n} - \frac{G^2}{N} \tag{14.3}$$

For the data in Table 14.4, there are six treatments (six T values), and the between-treatments SS is

$$SS_{\text{between treatments}} = \frac{25^2}{5} + \frac{45^2}{5} + \frac{20^2}{5} + \frac{5^2}{5} + \frac{5^2}{5} + \frac{20^2}{5} - \frac{120^2}{30}$$
$$= 125 + 405 + 80 + 5 + 5 + 80 - 480$$
$$= 700 - 480$$
$$= 220$$

The between-treatments df value is determined by the number of treatments (or the number of T values) minus one. For a two-factor study, the number of treatments is equal to the number of cells in the matrix. Thus,

$$df_{\text{between treatments}} = \text{number of cells} - 1 \qquad \text{(14.4)}$$

For these data, $df_{\text{between treatments}} = 5$.

Within-treatments variability   To compute the variance within treatments, we will use exactly the same formulas that were introduced in Chapter 13. Specifically, you first compute SS and $df = n - 1$ for each of the individual treatments (each separate sample). Then the within-treatments SS is defined as

$$SS_{\text{within treatments}} = \Sigma SS_{\text{each treatment}}$$

And the within-treatments df is defined as

$$df_{\text{within treatments}} = \Sigma df_{\text{each treatment}}$$

For the data in Table 14.4,

$$SS_{\text{within treatments}} = 18 + 26 + 20 + 8 + 20 + 28$$
$$= 120$$

$$df_{\text{within treatments}} = 4 + 4 + 4 + 4 + 4 + 4$$
$$= 24$$

This completes the first stage of the analysis. You should note that the two components add to equal the total for both SS values and df values.

$$SS_{\text{between treatments}} + SS_{\text{within treatments}} = SS_{\text{total}}$$
$$220 + 120 = 340$$

$$df_{\text{between treatments}} + df_{\text{within treatments}} = df_{\text{total}}$$
$$5 + 24 = 29$$

STAGE 2 OF THE ANALYSIS   The second stage of the analysis will determine the numerators for the three F-ratios. Specifically, this stage will determine the between-treatments variance for factor A, factor B, and the interaction.

1. Factor A. The main effect for factor A evaluates the mean differences between the levels of factor A. For this example, factor A defines the rows of the matrix, so we are evaluating the mean differences between rows. To compute the values for SS and df for factor A, we begin by finding the total for each level of factor A (the total for each row), identified as T_A, and the number of scores for each level of factor A (the number in each row), identified as n_A. With these values, we calculate the between-treatments SS measuring the differences between levels of factor A (between rows) as follows:

$$SS_{\text{between } As} = SS_A = \Sigma \frac{T_A^2}{n_A} - \frac{G^2}{N} \tag{14.5}$$

You should notice that this is the same formula for between-treatments SS that we have used before. Now, however, each treatment corresponds to a row in the matrix, so the totals (T values) and numbers (n values) are determined for each row. For the data in Table 14.4, this formula produces a value of

$$
\begin{aligned}
SS_A &= \frac{90^2}{15} + \frac{30^2}{15} - \frac{120^2}{30} \\
&= 540 + 60 - 480 \\
&= 600 - 480 \\
&= 120
\end{aligned}
$$

The degrees of freedom for factor A are determined by the number of levels of factor A minus one (or the number of rows minus one).

$$df_{\text{between } As} = df_A = (\text{number of levels of } A) - 1 \tag{14.6}$$

For these data, $df_A = 1$.

2. Factor B. The calculations for factor B follow exactly the same pattern that was used for factor A, except for substituting columns in place of rows. The main effect for factor B evaluates the mean differences between the levels of factor B, which define the columns of the matrix. To compute the values for SS and df for factor B, we begin by finding the total for each level of factor B (the total for each column), identified as T_B, and the number of scores for each level of factor B (the number in each column), identified as n_B. With these values, we calculate the between-treatments SS measuring the differences between levels of factor B (between columns) as follows:

$$SS_{\text{between } Bs} = SS_B = \Sigma \frac{T_B^2}{n_B} - \frac{G^2}{N} \tag{14.7}$$

For the data in Table 14.4,

$$
\begin{aligned}
SS_B &= \frac{30^2}{10} + \frac{50^2}{10} + \frac{40^2}{10} - \frac{120^2}{30} \\
&= 90 + 250 + 160 - 480 \\
&= 500 - 480 \\
&= 20
\end{aligned}
$$

The degrees of freedom for factor B are determined by subtracting one from the number of levels of factor B (or the number of rows minus one).

$$df_{\text{between } Bs} = df_B = (\text{number of levels of } B) - 1 \tag{14.8}$$

For these data, $df_B = 2$.

3. The $A \times B$ Interaction: The interaction between A and B is determined by the extra mean differences that are not explained by the main effects for the two factors. This definition will be used to compute the SS and the df for the interaction. Specifically, we begin with the between-treatments values obtained in stage 1 and then subtract out the values computed for the two main effects. The remainder is the variability caused by the interaction. Expressed as formulas,

$$SS_{A \times B} = SS_{\text{between treatments}} - SS_A - SS_B \qquad \text{(14.9)}$$

$$df_{A \times B} = df_{\text{between treatments}} - df_A - df_B \qquad \text{(14.10)}$$

For the data in Table 14.4,

$$SS_{A \times B} = 220 - 120 - 20$$
$$= 80$$

$$df_{A \times B} = 5 - 1 - 2$$
$$= 2$$

This completes the second stage of the two-factor analysis and provides all the elements necessary to compute the variances and the F-ratios for the ANOVA.

The denominator for each of the three F-ratios is intended to measure the variance or differences that would be expected just by chance. As we did in Chapter 13, we will use the within-treatments variance for the denominator. Remember, inside each treatment all the individuals are treated exactly the same, which means that any differences that occur are simply due to chance (see Box 13.2). The within-treatments variance is called a *mean square,* or *MS,* and is computed as follows:

$$MS_{\text{within treatments}} = \frac{SS_{\text{within treatments}}}{df_{\text{within treatments}}}$$

For the data in Table 14.4,

$$MS_{\text{within treatments}} = \frac{120}{24} = 5.00$$

This value will form the denominator for all three F-ratios.

The numerators of the three F-ratios all measure variance or differences between treatments: differences between levels of factor A, or differences between levels of factor B, or extra differences that are attributed to the $A \times B$ interaction. These three variances are computed as follows:

$$MS_A = \frac{SS_A}{df_A} \qquad MS_B = \frac{SS_B}{df_A} \qquad MS_{A \times B} = \frac{SS_{A \times B}}{df_{A \times B}}$$

For the data in Table 14.4, the three MS values are

$$MS_A = \frac{120}{1} = 120.00$$

$$MS_B = \frac{20}{2} = 10.00$$

$$MS_{A \times B} = \frac{80}{2} = 40.00$$

Finally, the three F-ratios are

$$F_A = \frac{MS_A}{MS_{within}} = \frac{120}{5} = 24.00$$

$$F_B = \frac{MS_B}{MS_{within}} = \frac{10}{5} = 2.00$$

$$F_{A \times B} = \frac{MS_{A \times B}}{MS_{within}} = \frac{40}{5} = 8.00$$

A decision concerning the significance of each test would require consulting the F-distribution table for the appropriate alpha level and degrees of freedom for each of the three F-ratios.

LEARNING CHECK   1. The following data summarize the results from a two-factor independent-measures experiment:

		Factor B	
	B_1	B_2	B_3
A_1	$n = 10$ $T = 0$ $SS = 30$	$n = 10$ $T = 10$ $SS = 40$	$n = 10$ $T = 20$ $SS = 50$
A_2	$n = 10$ $T = 40$ $SS = 60$	$n = 10$ $T = 30$ $SS = 50$	$n = 10$ $T = 20$ $SS = 40$

Factor A

a. Calculate the totals for each level of factor A, and compute SS for factor A.

b. Calculate the totals for factor B, and compute SS for this factor. (*Note:* You should find that the totals for B are all the same, so there is no variability for this factor.)

c. Given that the between-treatments (or between-cells) SS is equal to 100, what is the SS for the interaction?

d. Calculate the within-treatments SS, df, and MS for these data. (*Note:* MS_{within} will be the denominator for each of the three F-ratios.)

ANSWERS   1. a. The totals for factor A are 30 and 90, and each total is obtained by summing 30 scores. $SS_A = 60$.

b. All three totals for factor B are equal to 40. Because they are all the same, there is no variability, and $SS_B = 0$.

c. The interaction is determined by differences that remain after the main effects have been accounted for. For these data,

$$SS_{A \times B} = SS_{between \ treatments} - SS_A - SS_B$$
$$= 100 - 60 - 0$$
$$= 40$$

d. $SS_{within} = \Sigma SS_{each\ cell} = 30 + 40 + 50 + 60 + 50 + 40$
$$= 270$$

$df_{within} = \Sigma df_{each\ cell} = 9 + 9 + 9 + 9 + 9 + 9$
$$= 54$$

$$MS_{within} = \frac{SS_{within}}{df_{within}} = \frac{270}{54} = 5.00$$

14.4 AN EXAMPLE OF A TWO-FACTOR ANOVA

Example 14.2 presents the complete hypothesis-testing procedure for a two-factor independent-measures study.

EXAMPLE 14.2

In 1968, Schachter published an article in *Science* reporting a series of experiments on obesity and eating behavior. One of these studies examined the hypothesis that obese individuals do not respond to internal, biological signals of hunger. In simple terms, this hypothesis says that obese individuals tend to eat whether or not their bodies are actually hungry.

In Schachter's study, subjects were led to believe that they were taking part in a "taste test." All subjects were told to come to the experiment without eating for several hours beforehand. The study used two independent variables or factors:

1. Weight (obese versus normal subjects)

2. Full stomach versus empty stomach (half the subjects were given a full meal, as much as they wanted, after arriving at the experiment, and half were left hungry)

All subjects were then invited to taste and rate five different types of crackers. The dependent variable was the number of crackers eaten by each subject.

The prediction for this study was that the obese subjects would eat the same amount of crackers whether or not they were full. The normal subjects were expected to eat more with empty stomachs and less with full stomachs. Notice that the primary prediction of this study is that there will be an interaction between weight and fullness. That is, it is predicted that the effect of fullness *will depend on* the weight of the subjects (normal or obese).

Hypothetical data similar to those obtained by Schachter are presented in Table 14.5.

For this analysis, we will identify weight as factor *A* and fullness as factor *B*.

STEP 1 State the hypotheses, and select α. For factor *A*, the null hypothesis states that there is no difference in the amount eaten for normal versus obese subjects. In symbols,

$$H_0: \mu_{A_1} = \mu_{A_2}$$
$$H_1: \mu_{A_1} \neq \mu_{A_2}$$

TABLE 14.5

Results from an experiment examining the eating behavior of normal and obese individuals who have either a full or an empty stomach*

		Factor *B*: Fullness		
		Empty stomach	Full stomach	
Factor *A*: Weight	Normal	$n = 20$ $\overline{X} = 22$ $T = 440$ $SS = 1540$	$n = 20$ $\overline{X} = 15$ $T = 300$ $SS = 1270$	$T_{normal} = 740$
	Obese	$n = 20$ $\overline{X} = 17$ $T = 340$ $SS = 1320$	$n = 20$ $\overline{X} = 18$ $T = 360$ $SS = 1266$	$T_{obese} = 700$
		$T_{empty} = 780$	$T_{full} = 660$	$G = 1440$ $N = 80$ $\Sigma X^2 = 31{,}836$

*The dependent variable is the number of crackers eaten in a taste test (hypothetical data).

For factor *B*, the null hypothesis states that the amount eaten will be the same for full-stomach subjects as for empty-stomach subjects. In symbols,

$$H_0: \mu_{B_1} = \mu_{B_2}$$

$$H_1: \mu_{B_1} \neq \mu_{B_2}$$

For the $A \times B$ interaction, the null hypothesis can be stated two different ways. First, if there is a difference in eating between the full-stomach and empty-stomach conditions, it will be the same for normal and obese subjects. Second, if there is a difference in eating between the normal and obese subjects, it will be the same for the full-stomach and empty-stomach conditions. In more general terms,

H_0: The effect of factor *A* does not depend on the levels of factor *B* (and *B* does not depend on *A*).

H_1: The effect of one factor does depend on the levels of the other factor.

We will use $\alpha = .05$ for all tests.

STEP 2 Locate the critical region. To locate the critical values for each of the three *F*-ratios, we first must determine the *df* values. For these data (Table 14.5),

$$df_{total} = N - 1 = 79$$

$$df_{between} = (\text{number of cells}) - 1 = 3$$

$$df_{within} = \Sigma(n - 1) = 76$$

$$df_A = (\text{number of levels of } A) - 1 = 1$$

$$df_B = (\text{number of levels of } B) - 1 = 1$$

$$df_{A \times B} = df_{between} - df_A - df_B = 1$$

Notice that the *F*-distribution table has no entry for $df = 1, 76$. A close and conservative estimate of this critical value may be obtained by using $df = 1, 70$ (critical $F = 3.98$). Whenever there is no entry for the *df* value of the error term, use the nearest smaller *df* value in the table.

Thus, all three *F*-ratios will have $df = 1, 76$. With $\alpha = .05$, the critical *F* value is 3.98 for all three tests.

STEP 3   Use the data to compute the F-ratios. First, we will analyze the SS values:

$$SS_{total} = \Sigma X^2 - \frac{G^2}{N} = 31{,}836 - \frac{1440^2}{80}$$

$$= 31{,}836 - 25{,}920$$

$$= 5916$$

$$SS_{between} = \Sigma \frac{T^2}{n} - \frac{G^2}{N}$$

$$= \frac{440^2}{20} + \frac{300^2}{20} + \frac{340^2}{20} + \frac{360^2}{20} - \frac{1440^2}{80}$$

$$= 26{,}440 - 25{,}920$$

$$= 520$$

$$SS_{within} = \Sigma SS_{inside\ each\ cell}$$

$$= 1540 + 1270 + 1320 + 1266$$

$$= 5396$$

For factor A (weight), the two totals are $T_{normal} = 740$ and $T_{obese} = 700$. Each total is obtained by summing 40 scores. Using these values, the SS for factor A is

$$SS_A = \Sigma \frac{T_A^2}{n_A} - \frac{G^2}{N}$$

$$= \frac{740^2}{40} + \frac{700^2}{40} - \frac{1440^2}{80}$$

$$= 25{,}940 - 25{,}920$$

$$= 20$$

For factor B (fullness), the two totals are $T_{empty} = 780$ and $T_{full} = 660$. Each of these totals is the sum of 40 scores. Using these values, the SS for factor B is

$$SS_B = \Sigma \frac{T_B^2}{n_B} - \frac{G^2}{N}$$

$$= \frac{780^2}{40} + \frac{660^2}{40} - \frac{1440^2}{80}$$

$$= 26{,}100 - 25{,}920$$

$$= 180$$

$$SS_{A \times B} = SS_{between} - SS_A - SS_B$$

$$= 520 - 20 - 180$$

$$= 320$$

The MS values needed for the F-ratios are

$$MS_A = \frac{SS_A}{df_A} = \frac{20}{1} = 20$$

$$MS_B = \frac{SS_B}{df_B} = \frac{180}{1} = 180$$

$$MS_{A \times B} = \frac{SS_{A \times B}}{df_{A \times B}} = \frac{320}{1} = 320$$

$$MS_{\text{within}} = \frac{SS_{\text{within}}}{df_{\text{within}}} = \frac{5396}{76} = 71$$

Finally, the F-ratios are

$$F_A = \frac{MS_A}{MS_{\text{within}}} = \frac{20}{71} = 0.28$$

$$F_B = \frac{MS_B}{MS_{\text{within}}} = \frac{180}{71} = 2.54$$

$$F_{A \times B} = \frac{MS_{A \times B}}{MS_{\text{within}}} = \frac{320}{71} = 4.51$$

STEP 4 Make decisions. For these data, factor A (weight) has no significant effect; $F(1, 76) = 0.28$. Statistically, there is no difference in the number of crackers eaten by normal versus obese subjects.

Similarly, factor B (fullness) has no significant effect; $F(1, 76) = 2.54$. Statistically, the number of crackers eaten by full subjects is no different from the number eaten by hungry subjects. (*Note:* This conclusion concerns the combined group of normal and obese subjects. The interaction concerns these two groups separately. See page 374 for information concerning the interpretation of results when there is a significant interaction.)

These data produce a significant interaction; $F(1, 76) = 4.51, p < .05$. This means that the effect of fullness does depend on weight. Specifically, the degree of fullness did affect the normal subjects, but it has no effect on the obese subjects.

As we saw in Chapter 13, the results from an ANOVA can be organized in a summary table, which shows all the components of the analysis (SS, df, etc.) as well as the final F-ratios. For this example, the summary table would be as follows:

Source	SS	df	MS	F
Between treatments	520	3		
Factor A (weight)	20	1	20	0.28
Factor B (fullness)	180	1	180	2.54
$A \times B$ interaction	320	1	320	4.51
Within treatments	5396	76	71	
Total	5916	79		

IN THE LITERATURE
REPORTING THE RESULTS OF A TWO-FACTOR ANOVA

The APA format for reporting the results of a two-factor analysis of variance follows the same basic guidelines as the single-factor report. First, the means and standard deviations are reported. Because a two-factor design typically involves several treatment conditions, these descriptive statistics usually are presented in a table or a graph. Next, the results of all three hypothesis tests (F-ratios) are reported. For the research study in Example 14.2, the report would have the following form:

The means and standard deviations are presented in Table 1. The two-factor analysis of variance showed no significant main effect for the weight factor, $F(1, 76) = 0.28$, $p > .05$; and no significant main effect for the fullness factor, $F(1, 76) = 2.54$, $p > .05$; but the interaction between weight and fullness was significant, $F(1, 76) = 4.51$, $p < .05$.

TABLE 1

Mean number of crackers eaten in each treatment condition

		Fullness	
		Empty stomach	Full stomach
Weight	Normal	$M = 22.0$ $SD = 9.00$	$M = 15.0$ $SD = 8.18$
	Obese	$M = 17.0$ $SD = 8.34$	$M = 18.0$ $SD = 8.16$

You should recognize the elements of this report as being identical to other examples reporting the results from an analysis of variance. Each of the obtained F-ratios is reported with its *df* values in parentheses. The term *significant* is used to indicate that the null hypothesis was rejected (the mean differences are greater than what would be expected by chance), and the term *not significant* indicates that the test failed to reject the null hypothesis. Finally, the *p* value reported with each F-ratio reflects the alpha level used for the test. For example, $p < .05$ indicates that the probability is less than .05 that the obtained mean difference is simply due to chance or sampling error.   ❑

INTERPRETING INTERACTIONS AND MAIN EFFECTS

When a two-factor ANOVA produces a significant interaction, you should be very cautious about accepting the main effects (whether significant or not significant) at face value. In particular, a significant interaction can distort, conceal, or exaggerate the main effects. Therefore, the best advice is to start with the interaction, not the main effects, as the basis for interpreting the results of the study.

The data from Example 14.2 provide a good demonstration of how a significant interaction can influence the interpretation of main effects. The data have been sum-

FIGURE 14.4

A table and a graph showing the mean number of crackers eaten for each of the four groups in Example 14.2.

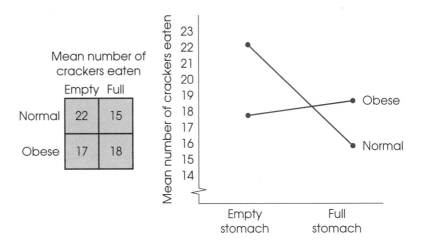

marized in a table and a graph in Figure 14.4. For these data, the ANOVA resulted in a significant interaction but no significant main effect for factor A (normal/obese) or for factor B (full/empty stomach). If you ignore the interaction and focus on the main effects, you might be tempted to conclude that there is no significant difference in eating behavior between normal and obese subjects and that being full versus being hungry has no effect on eating behavior. Looking more closely at the data, however, it is immediately apparent that the main effects do not tell the whole story. Although the obese subjects seem to eat the same amount whether they are full or hungry, the normal subjects show that hunger has a large effect on eating behavior. You should realize that this discrepancy between normal and obese subjects is exactly what Schachter predicted, and it is the source of the significant interaction.

14.5   ASSUMPTIONS FOR THE TWO-FACTOR ANOVA

The validity of the analysis of variance presented in this chapter depends on the same three assumptions we have encountered with other hypothesis tests for independent-measures designs (the *t* test in Chapter 10 and the single-factor ANOVA in Chapter 13):

1. The observations within each sample must be independent (see p. 202).
2. The populations from which the samples are selected must be normal.
3. The populations from which the samples are selected must have equal variances (homogeneity of variance).

As before, the assumption of normality generally is not a cause for concern, especially when the sample size is relatively large. The homogeneity of variance assumption is more important, and if it appears that your data fail to satisfy this requirement, you should conduct a test for homogeneity before you attempt the ANOVA. Hartley's *F*-max test (see p. 253) allows you to use the sample variances from your data to determine whether there is evidence for any differences among the population variances.

HIGHER-ORDER FACTORIAL DESIGNS The basic concepts of the two-factor ANOVA can be extended to more complex designs involving three or more factors. A three-factor design, for example, might look at academic performance scores for two different teaching methods (factor A), for boys versus girls (factor B), and for first-grade versus second-grade classes (factor C). The logic of the analysis and many of the formulas from the two-factor ANOVA are simply extended to the situation with three (or more) factors. In a three-factor experiment, for example, you would evaluate the main effects for each of the three factors, and you would evaluate a set of two-way interactions: $A \times B$, $B \times C$, and $A \times C$. In addition, however, the extra factor introduces the potential for a three-way interaction: $A \times B \times C$.

The general logic for defining and interpreting higher-order interactions follows the pattern set by two-way interactions. For example, a two-way interaction, $A \times B$, means that the effect of factor A depends on the levels of factor B. Extending this definition, a three-way interaction, $A \times B \times C$, indicates that the two-way interaction between A and B depends on the levels of factor C. Although you may have a good understanding of two-way interactions and you may grasp the general idea of a three-way interaction, most people have great difficulty comprehending or interpreting a four-way (or more) interaction. For this reason, factorial experiments involving three or more factors can produce very complex results that are difficult to understand and thus may have limited practical value.

LEARNING CHECK 1. A two-factor experiment has three levels of factor A and two levels of factor B, with $n = 5$ scores in each treatment condition (each cell).

a. For this study, what is the df value for the within-treatment MS?

b. What are the df values for each of the three F-ratios in the ANOVA?

ANSWERS 1. **a.** $df_{within} = 24$

b. The F-ratio for factor A has $df = 2, 24$. For factor B, $df = 1, 24$. The F-ratio for the $A \times B$ interaction has $df = 2, 24$.

SUMMARY

1. A research study with two independent variables (or quasi-independent variables) is called a two-factor design. Such a design can be diagrammed as a matrix by listing the levels of one factor across the top and the levels of the other factor down the side. Each *cell* in the matrix corresponds to a specific combination of the two factors.

2. Traditionally, the two factors are identified as factor A and factor B. The purpose of the analysis of variance is to determine whether there are any significant mean differences among the treatment conditions or cells in the experimental matrix. These treatment effects are classified as follows:
   a. The A-effect: Differential effects produced by the different levels of factor A.

b. The B-effect: Differential effects produced by the different level of factor B.
   c. The $A \times B$ interaction: Differences that are produced by unique combinations of A and B. An interaction exists when the effect of one factor depends on the levels of the other factor.

3. The two-factor analysis of variance produces three F-ratios: one for factor A, one for factor B, and one for the $A \times B$ interaction. Each F-ratio has the same basic structure:

$$F = \frac{MS_{\text{treatment effect}}(\text{either } A \text{ or } B \text{ or } A \times B)}{MS_{\text{within}}}$$

KEY TERMS

two-factor experiment   matrix   cells   main effect   interaction

--- FOCUS ON PROBLEM SOLVING ---

1. Before you begin a two-factor ANOVA, you should take time to organize and summarize the data. It is best if you summarize the data in a matrix with rows corresponding to the levels of one factor and columns corresponding to the levels of the other factor. In each cell of the matrix, show the number of scores (n), the total and mean for the cell, and the SS within the cell. Also compute the row totals and column totals that will be needed to calculate main effects.

2. To draw a graph of the result from a two-factor experiment, first prepare a matrix with each cell containing the mean for that treatment condition.

		Factor B			
		B_1	B_2	B_3	B_4
Factor A	A_1				
	A_2				

Next, list the levels of factor B on the X-axis (B_1, B_2, etc.), and put the scores (dependent variable) on the Y-axis. Starting with the first row of the matrix, place a dot above each B level so that the height of the dot corresponds to the cell mean. Connect the dots with a line, and label the line A_1. In the same way, construct a separate line for each level of factor A (that is, a separate line for each row of the matrix). In general, your graph will be easier to draw and easier to understand if the factor with the larger number of levels is placed on the X-axis (factor B in this example).

3. The concept of an interaction is easier to grasp if you sketch a graph showing the means for each treatment. Remember, parallel lines indicate no interaction. Crossing or converging lines indicate that the effect of one treatment depends on the levels of the other treatment you are examining. This indicates that an interaction exists between the two treatments.

4. For a two-factor ANOVA, there are three separate F-ratios. These three F-ratios use the same measure of error in the denominator (MS_{within}). On the other hand, these F-ratios will have different numerators and may have different df values associated with each of these numerators. Therefore, you must be careful when you look up the critical F values in the table. The two factors and the interaction may have different critical F values.

5. As we have mentioned in the previous ANOVA chapter, it helps tremendously to organize your computations; start with SS and df values, and then compute MS values and F-ratios. Once again, using an ANOVA summary table (see p. 373) can be of great assistance.

DEMONSTRATION 14.1

TWO-FACTOR ANOVA

The following data are from an experiment examining the phenomenon of encoding specificity. According to this psychological principle, recall of information will be best if the testing conditions are the same as the conditions that existed at the time of learning.

The experiment involves presenting a group of students with a lecture on an unfamiliar topic. One week later the students are given a test on the lecture material. To manipulate the conditions at the time of learning, some students receive the lecture in a large classroom, and some hear the lecture in a small classroom. For those students who were lectured in the large room, one-half are tested in the same large room, and the others are changed to the small room for testing. Similarly, one-half of the students who were lectured in the small room are tested in the same small room, and the other half are tested in the large room. Thus, the experiment involves four groups of subjects in a two-factor design, as shown in the following table. The score for each subject is the number of correct answers on the test.

		Testing condition	
		Large testing room	Small testing room
Lecture condition	Large lecture room	15 20 11 18 16	5 8 1 1 5
	Small lecture room	1 4 2 5 8	22 15 20 17 16

Do these data indicate that the size of the lecture room and/or testing room has a significant effect on test performance?

STEP 1 *State the hypotheses, and specify* α. The two-factor ANOVA evaluates three separate sets of hypotheses:

1. The main effect of lecture room size: Is there a significant difference in test performance for students who received the lecture in a large room versus students who received the lecture in a small room? With lecture room size identified as factor A, the null hypothesis states that there is no difference between the two room sizes. In symbols,

$$H_0: \mu_{A_1} = \mu_{A_2}$$

The alternative hypothesis states that there is a difference between the two lecture room sizes.

$$H_1: \mu_{A_1} \neq \mu_{A_2}$$

2. The main effect of testing room size: Is there a significant difference in test performance for students who were tested in a large room versus students who were tested in a small room? With testing room size identified as factor B, the null hypothesis states that there is no difference between the two room sizes. In symbols,

$$H_0: \mu_{B_1} = \mu_{B_2}$$

The alternative hypothesis states that there is a difference between the two testing room sizes.

$$H_1: \mu_{B_1} \neq \mu_{B_2}$$

3. The interaction between lecture room size and testing room size: The null hypothesis states that there is no interaction:

H_0: The effect of testing room size does not depend on the size of the lecture room.

The alternative hypothesis states that there is an interaction:

H_1: The effect of testing room size does depend on the size of the lecture room.

Notice that the researcher is not predicting any main effects for this study: There is no prediction that one room size is better than another for either learning or testing. However, the researcher is predicting that there will be an interaction. Specifically, the small testing room should be better for students who learned in the small room, and the large testing room should be better for students who learned in the large room. Remember, the principle of encoding specificity states that the better the match between testing and learning conditions, the better the recall. We will set alpha at $\alpha = .05$.

STEP 2　*Locate the critical region.* To locate the critical region, we must obtain the *df* values for each of the three *F*-ratios. Specifically, we will need df_A, df_B, and $df_{A \times B}$ for the numerators and df_{within} for the denominator. (Often it is easier to postpone this step until the analysis of the *df* values in step 3.)

$$df_{total} = N - 1 = 19$$

$$df_{between} = (\text{number of cells}) - 1 = 3$$

$$df_{within} = \Sigma(n - 1) = 16$$

$$df_A = (\text{number of levels of } A) - 1 = 1$$

$$df_B = (\text{number of levels of } B) - 1 = 1$$

$$df_{A \times B} = df_{between} - df_A - df_B = 1$$

Thus, all three *F*-ratios will have $df = 1, 16$. With $\alpha = .05$, the critical value for each *F*-ratio is $F = 4.49$. For each test, the obtained *F*-ratio must exceed this critical value to reject H_0.

STEP 3　*Perform the analysis.* The complete two-factor ANOVA can be divided into a series of stages:

1. Compute the summary statistics for the data. This involves calculating the total and *SS* for each treatment condition, finding the *A* totals and *B* totals for the

rows and columns, respectively, and obtaining G and ΣX^2 for the entire set of scores.

2. Perform the first stage of the analysis: Separate the total variability (SS and df) into the between- and within-treatments components.

3. Perform the second stage of the analysis: Separate the between-treatments variability (SS and df) into the A-effect, B-effect, and interaction components.

4. Calculate the mean squares for the F-ratios.

5. Calculate the F-ratios.

Compute the summary statistics. We will use the computational formula to obtain SS for each treatment condition. These calculations will also provide numerical values for the row and column totals as well as G and ΣX^2.

Large testing Large lecture	
X	X^2
15	225
20	400
11	121
18	324
16	256
$\Sigma X = 80$	$\Sigma X^2 = 1326$

$$SS = \Sigma X^2 - \frac{(\Sigma X)^2}{n}$$
$$= 1326 - \frac{80^2}{5}$$
$$= 1326 - 1280$$
$$= 46$$
$$T = \Sigma X = 80$$

Small testing Large lecture	
X	X^2
5	25
8	64
1	1
1	1
5	25
$\Sigma X = 20$	$\Sigma X^2 = 116$

$$SS = \Sigma X^2 - \frac{(\Sigma X)^2}{n}$$
$$= 116 - \frac{20^2}{5}$$
$$= 116 - 80$$
$$= 36$$
$$T = \Sigma X = 20$$

Large testing Small lecture	
X	X^2
1	1
4	16
2	4
5	25
8	64
$\Sigma X = 20$	$\Sigma X^2 = 110$

$$SS = \Sigma X^2 - \frac{(\Sigma X)^2}{n}$$
$$= 110 - \frac{20^2}{5}$$
$$= 110 - 80$$
$$= 30$$
$$T = \Sigma X = 20$$

Small testing Small lecture	
X	X^2
22	484
15	225
20	400
17	289
16	256
$\Sigma X = 90$	$\Sigma X^2 = 1654$

$$SS = \Sigma X^2 - \frac{(\Sigma X)^2}{n}$$
$$= 1654 - \frac{90^2}{5}$$
$$= 1654 - 1620$$
$$= 34$$
$$T = \Sigma X = 90$$

The column totals (factor B) are 100 and 110. The row totals (factor A) are 100 and 110. The grand total for these data is $G = 210$, and ΣX^2 for the entire set can be obtained by summing the ΣX^2 values for the four treatment conditions.

$$\Sigma X^2 = 1326 + 116 + 110 + 1654 = 3206$$

For this study, there are two levels for factor A and for factor B and there are $n = 5$ scores in each of the four conditions, so $N = 20$.

Perform stage 1 of the analysis. We begin by analyzing SS into two basic components:

$$SS_{total} = \Sigma X^2 - \frac{G^2}{N} = 3206 - \frac{210^2}{20} = 3206 - 2205 = 1001$$

$$SS_{between\ cells} = \Sigma \frac{T^2}{n} - \frac{G^2}{N} = \frac{80^2}{5} + \frac{20^2}{5} + \frac{20^2}{5} + \frac{90^2}{5} - \frac{210^2}{20}$$

$$= 1280 + 80 + 80 + 1620 - 2205$$

$$= 3060 - 2205$$

$$= 855$$

$$SS_{within} = \Sigma SS_{each\ cell} = 46 + 36 + 30 + 34 = 146$$

For this stage, the df values are

$$df_{total} = 19$$
$$df_{between\ cells} = 3$$
$$df_{within} = 16$$

Perform stage 2 of the analysis. We begin by analyzing $SS_{between}$.

The row totals (factor A) are 100 and 110, and each total is the sum of 10 scores. Thus, SS for factor A is

$$SS_A = \Sigma \frac{T_A^2}{n_A} - \frac{G^2}{N}$$

$$= \frac{100^2}{10} + \frac{110^2}{10} - \frac{210^2}{20}$$

$$= 1000 + 1210 - 2205$$

$$= 5$$

The column totals (factor B) are 100 and 110, and each total is the sum of 10 scores. Thus, SS for factor B is

$$SS_B = \Sigma \frac{T_B^2}{n_B} - \frac{G^2}{N}$$

$$= \frac{100^2}{10} + \frac{110^2}{10} - \frac{210^2}{20}$$

$$= 1000 + 1210 - 2205$$

$$= 5$$

$$SS_{A \times B} = SS_{between\ cells} - SS_A - SS_B$$

$$= 855 - 5 - 5$$

$$= 845$$

For stage 2, the *df* values are

$$df_A = 1$$

$$df_B = 1$$

$$df_{A \times B} = 1$$

Calculate the MS values.

$$MS_A = \frac{SS_A}{df_A} = \frac{5}{1} = 5$$

$$MS_B = \frac{SS_B}{df_B} = \frac{5}{1} = 5$$

$$MS_{A \times B} = \frac{SS_{A \times B}}{df_{A \times B}} = \frac{845}{1} = 845$$

$$MS_{within} = \frac{SS_{within}}{df_{within}} = \frac{146}{16} = 9.125$$

Calculate the F-ratios. For factor *A* (lecture-room size),

$$F = \frac{MS_A}{MS_{within}} = \frac{5}{9.125} = 0.55$$

For factor *B* (testing-room size),

$$F = \frac{MS_B}{MS_{within}} = \frac{5}{9.125} = 0.55$$

For the *A* × *B* interaction,

$$F = \frac{MS_{A \times B}}{MS_{within}} = \frac{845}{9.125} = 92.60$$

STEP 4 *Make a decision about each H_0, and state conclusions.* For factor *A*, lecture-room size, the obtained *F*-ratio, $F = 0.55$, is not in the critical region. Therefore, we fail to reject the null hypothesis. We conclude that the size of the lecture room does not have a significant effect on test performance, $F(1, 16) = 0.55, p > .05$.

For factor *B*, testing-room size, the obtained *F*-ratio, $F = 0.55$, is not in the critical region. Therefore, we fail to reject the null hypothesis. We conclude that the size of the testing room does not have a significant effect on test performance, $F(1, 16) = 0.55, p > .05$.

For the *A* × *B* interaction, the obtained *F*-ratio, $F = 92.60$, exceeds the critical value of $F = 4.46$. Therefore, we reject the null hypothesis. We conclude that there is a significant interaction between lecture-room size and testing-room size, $F(1, 16) = 92.60, p < .05$.

Notice that the significant interaction means that you must be cautious interpreting the main effects (see p. 374). In this experiment, for example, the size of the testing room (factor *B*) does have an effect on performance, depending on which room was used for the lecture. Specifically, performance is higher when the testing room and lecture room match, and performance is lower when the lecture and testing occur in different rooms.

The following table summarizes the results of the analysis:

Source	SS	df	MS	
Between cells	855	3		
A (lecture room)	5	1	5	F = 0.55
B (testing room)	5	1	5	F = 0.55
A × B interaction	845	1	845	F = 92.60
Within cells	146	16	9.125	
Total	1001	19		

PROBLEMS

1. The structure of a two-factor study can be presented as a matrix with one factor determining the rows and the second factor determining the columns. With this structure in mind, identify the three separate hypothesis tests that make up a two-factor ANOVA, and explain the purpose of each test.

2. The following matrix presents the results of a two-factor experiment with two levels of factor A, two levels of factor B, and $n = 10$ subjects in each treatment condition. Each value in the matrix is the mean score for the subjects in that treatment condition. Notice that one of the mean values is missing.

	Factor B	
	B_1	B_2
Factor A A_1	10	40
A_2	30	?

 a. What value should be assigned to the missing mean so that the resulting data would show no main effect for factor A?
 b. What value should be assigned to the missing mean so that the data would show no main effect for factor B?
 c. What value should be assigned to the missing mean so that the data would show no interaction?

3. Sketch a graph showing each of the following sets of data. Use line graphs with the levels of factor B on the X-axis. In each case, state whether or not the data indicate the presence of a main effect for factor A, a main effect for factor B, and an interaction between A and B. (*Hint:* Main effects are easier to determine by comparing column or row mean differences in the matrix. Interactions are easier to determine by looking at the pattern in the graph.)

a.

	Factor B	
	B_1	B_2
Factor A A_1	$\bar{X} = 40$	$\bar{X} = 10$
A_2	$\bar{X} = 60$	$\bar{X} = 30$

b.

	Factor B	
	B_1	B_2
Factor A A_1	$\bar{X} = 20$	$\bar{X} = 20$
A_2	$\bar{X} = 10$	$\bar{X} = 50$

c.

	Factor B	
	B_1	B_2
Factor A A_1	$\bar{X} = 20$	$\bar{X} = 20$
A_2	$\bar{X} = 10$	$\bar{X} = 10$

4. The results of a two-factor experiment are examined using an ANOVA, and the researcher reports an F-ratio for factor A with $df = 1, 54$ and an F-ratio for factor B with $df = 2, 108$. Explain why this report cannot be correct.

5. A psychologist conducts a two-factor study comparing the effectiveness of two different therapy techniques (factor A) for treating mild and severe phobias (factor B). The dependent variable is a measure of fear for each subject. If the study uses $n = 10$ subjects in each of the four conditions, then identify the df values for each of the three F-ratios.
 a. What are the df values for the F-ratio for factor A?
 b. What are the df values for the F-ratio for factor B?
 c. What are the df values for the F-ratio for the $A \times B$ interaction?

6. The following data are from a two-factor experiment with $n = 10$ subjects in each treatment condition (each cell):

	Factor B	
	B_1	B_2
A_1	$T = 40$, $SS = 70$	$T = 10$, $SS = 80$
A_2	$T = 30$, $SS = 73$	$T = 20$, $SS = 65$

Factor A

$$\Sigma X^2 = 588$$

Test for a significant A-effect, B-effect, and $A \times B$ interaction using $\alpha = .05$ for all tests.

7. The following data are from a study examining the extent to which different personality types are affected by distraction. Individuals were selected to represent two different personality types: introverts and extroverts. Half the individuals in each group were tested on a monotonous task in a relatively quiet, calm room. The individuals in the other half of each group were tested in a noisy room filled with distractions. The dependent variable was the number of errors committed by each individual. The results of this study are as follows:

		Factor B: Personality	
		Introvert	Extrovert
Factor A: Distraction	Quiet	$n = 5$, $T = 10$, $SS = 15$	$n = 5$, $T = 10$, $SS = 25$
	Noisy	$n = 5$, $T = 20$, $SS = 10$	$n = 5$, $T = 40$, $SS = 30$

$$\Sigma X^2 = 520$$

Use a two-factor ANOVA with $\alpha = .05$ to evaluate these results.

8. The following table summarizes the analysis of variance for a two-factor study using two levels of factor A and three levels of factor B. The study used a separate sample of $n = 10$ subjects for each of the six treatment conditions. Fill in all missing values in the table. (*Hint:* Start with the *df* values.)

Source	SS	df	MS	
Between treatments	___	___		
A-effect	20	___	___	$F =$ ___
B-effect	___	___	___	$F =$ ___
$A \times B$ interaction	___	___	___	$F = 5$
Within treatments	108	___	___	
Total	158	___		

9. A researcher studies the effects of need for achievement and task difficulty on problem solving. A two-factor design is used, in which there are two levels of amount of achievement motivation (high versus low need for achievement) and four levels of task difficulty, yielding eight treatment cells. Each cell consists of $n = 6$ subjects. The number of errors each subject made was recorded, and the data were analyzed. The following table summarizes the results of the ANOVA, but it is not complete. Fill in the missing values. (Start with *df* values.)

Source	SS	df	MS	
Between treatments	280	___		
Main effect for achievement motivation	___	___	___	$F =$ ___
Main effect for task difficulty	___	___	48	$F =$ ___
Interaction	120	___	___	$F =$ ___
Within treatments	___	___	___	
Total	600	___		

10. The following data were obtained from an independent-measures experiment using $n = 5$ subjects in each treatment condition:

	Factor B	
	B_1	B_2
A_1	$T = 15$, $SS = 80$	$T = 25$, $SS = 90$
A_2	$T = 5$, $SS = 70$	$T = 55$, $SS = 80$

Factor A

$$\Sigma X^2 = 1100$$

a. Compute the means for each cell, and draw a graph showing the results of this experiment. Your graph should be similar to those shown in Figure 14.2.

b. Just from looking at your graph, does there appear to be a main effect for factor *A*? What about factor *B*? Does there appear to be an interaction?

c. Use an analysis of variance with $\alpha = .05$ to evaluate these data.

11. Many species of animals communicate using odors. A researcher suspects that specific chemicals contained in the urine of male rats can influence the behavior of other males in the colony. The researcher predicts that male rats will become anxious and more active if they think they are in territory that has been marked by another male. Also, it is predicted that these chemicals will have no effect on female rats. To test this theory, the researcher obtains samples of 15 male and 15 female rats. One-third of each group is tested in a sterile cage. Another one-third of each group is tested in a cage that has been painted with a small amount of the chemicals. The rest of the rats are tested in a cage that has been painted with a large amount of the chemicals. The dependent variable is the activity level of each rat. The data from this experiment are as follows:

	Factor *B*: Amount of chemical		
	None	Small	Large
Factor *A*: sex Male	$n = 5$ $T = 10$ $SS = 15$	$n = 5$ $T = 20$ $SS = 19$	$n = 5$ $T = 30$ $SS = 31$
Female	$n = 5$ $T = 10$ $SS = 19$	$n = 5$ $T = 10$ $SS = 21$	$n = 5$ $T = 10$ $SS = 15$
	$\Sigma X^2 = 460$		

Use an ANOVA with $\alpha = .05$ to test the researcher's predictions. Explain the results.

12. It has been demonstrated in a variety of experiments that memory is best when the conditions at the time of testing are identical to the condition at the time of learning. This phenomenon is called *encoding specificity* because the specific cues that you use to learn (or encode) new information are the best possible cues to help you recall the information at a later time. In an experimental demonstration of encoding specificity, Tulving and Osler (1968) prepared a list of words to be memorized. For each word on the list, they selected an associated word to serve as a cue. For example, if the word *queen* were on the list, the word *lady* could be a cue. Four groups of subjects participated in the experiment. One group was given the cues during learning and during the recall test. Another

group received the cues only during recall. A third group received the cues only during learning, and the final group was not given any cues at all. The dependent variable was the number of words correctly recalled. Data similar to Tulving and Osler's results are as follows:

		Cues at learning	
		Yes	No
Cues at recall	Yes	$n = 10$ $\overline{X} = 3$ $SS = 22$	$n = 10$ $\overline{X} = 1$ $SS = 15$
	No	$n = 10$ $\overline{X} = 1$ $SS = 16$	$n = 10$ $\overline{X} = 1$ $SS = 19$
		$\Sigma X^2 = 192$	

Use an ANOVA with $\alpha = .05$ to evaluate the data. Describe the results.

13. The following data show the results of a two-factor experiment with $n = 10$ in each treatment condition (cell):

		Factor *B*	
		B_1	B_2
Factor *A*	A_1	$\overline{X} = 4$ $SS = 40$	$\overline{X} = 2$ $SS = 50$
	A_2	$\overline{X} = 3$ $SS = 50$	$\overline{X} = 1$ $SS = 40$
		$\Sigma X^2 = 480$	

a. Sketch a graph showing the results of this experiment. (See Figure 14.2 for examples.)

b. Looking at your graph, does there appear to be an $A \times B$ interaction? Does factor *A* appear to have any effect? Does factor *B* appear to have any effect?

c. Evaluate these data using an ANOVA with $\alpha = .05$.

14. When subjects are presented with a list of items to be remembered, there are two common methods for testing memory. A *recall* test simply asks each subject to report as many items as he/she can remember. In a *recognition* test, subjects are shown a second list of items (including some items from the original list as well as some new items) and asked to identify which items they remember having seen on the first list. Typically, recall performance is very poor for young children, but improves gradually with age. Recognition performance, on the other hand, is relatively good for all ages. The following data are intended to demonstrate this phenomenon. There are $n = 10$ children in each of the six samples, and the dependent variable is a measure of the number of items correctly remembered for each child. Use an ANOVA with $\alpha = .05$ to evaluate these data.

	Age 2	Age 6	Age 10
Recall test	$\bar{X} = 3$	$\bar{X} = 7$	$\bar{X} = 12$
	$SS = 16$	$SS = 19$	$SS = 14$
Recognition test	$\bar{X} = 15$	$\bar{X} = 16$	$\bar{X} = 17$
	$SS = 21$	$SS = 20$	$SS = 18$

15. Hyperactivity in children usually is treated by counseling, or by drugs, or by both. The following data are from an experiment designed to evaluate the effectiveness of these different treatments. The dependent variable is a measure of attention span (how long each child was able to concentrate on a specific task).

	Drug	No drug
Counseling	$n = 10$	$n = 10$
	$T = 140$	$T = 80$
	$SS = 40$	$SS = 36$
No counseling	$n = 10$	$n = 10$
	$T = 120$	$T = 100$
	$SS = 45$	$SS = 59$

$$\Sigma X^2 = 5220$$

a. Use an ANOVA with $\alpha = .05$ to evaluate these data.
b. Do the data indicate that the drug has a significant effect? Does the counseling have an effect? Describe these results in terms of the effectiveness of the drug and counseling and their interaction.

16. Often two-factor studies occur when researchers want to extend the results of a single-factor study by adding a new population or treatment condition. For example, problem 21 in Chapter 10 demonstrated significant differences in language development among single children, twins, and triplets. However, the significant differences were found when the children were three years old. A researcher would like to determine whether the differences continue into adolescence. The researcher repeats the original experiment with 3-year-olds, but adds a second factor (age) by including samples of 14-year-olds in the study. The dependent variable is a measure of language skill. The resulting data, with $n = 5$ children in each group, are as follows:

	Single child	Twin	Triplet
Age 3	$\bar{X} = 10$	$\bar{X} = 6$	$\bar{X} = 5$
	$SS = 10$	$SS = 9$	$SS = 10$
Age 14	$\bar{X} = 15$	$\bar{X} = 14$	$\bar{X} = 15$
	$SS = 11$	$SS = 11$	$SS = 10$

a. Sketch a graph showing the mean scores for the six groups in this study. Does it appear that language skill is related to the number of siblings for 3-year-old children? Are there still differences at age 14?
b. Evaluate the mean differences with a two-factor ANOVA, using $\alpha = .05$ for all tests.

17. The general relationship between performance and arousal level is described by the Yerkes-Dodson law. This law states that performance is best at a moderate level of arousal. When arousal is too low, people do not care about what they are doing, and performance is poor. At the other extreme, when arousal is too high, people become overly anxious, and performance suffers. In addition, the exact form of the relationship between arousal and performance depends on the difficulty of the task. The following data demonstrate the Yerkes-Dodson law. The dependent variable is a measure of performance.

	Arousal level		
	Low	Medium	High
Easy task	$n = 10$	$n = 10$	$n = 10$
	$T = 80$	$T = 100$	$T = 120$
	$SS = 30$	$SS = 36$	$SS = 45$
Hard task	$n = 10$	$n = 10$	$n = 10$
	$T = 60$	$T = 100$	$T = 80$
	$SS = 42$	$SS = 27$	$SS = 36$

$$\Sigma X^2 = 5296$$

a. Sketch a graph showing the mean level of performance for each treatment condition.
b. Use a two-factor ANOVA to evaluate these data.
c. Describe and explain the main effect for task difficulty.
d. Describe and explain the interaction between difficulty and arousal.

18. The following data are from a study examining the influence of a specific hormone on eating behavior. Three different drug doses were used, including a control condition (no drug), and the study measured eating behavior for males and females. The dependent variable was the amount of food consumed over a 48-hour period.

	No drug	Small dose	Large dose
Males	1	7	3
	6	7	1
	1	11	1
	1	4	6
	1	6	4
Females	0	0	0
	3	0	2
	7	0	0
	5	5	0
	5	0	3

Use an ANOVA with $\alpha = .05$ to evaluate these data, and describe the results (i.e., the drug effect, the sex difference, and the interaction).

19. Shranger (1972) conducted an experiment that examined the effect of an audience on the performance of two different personality types. Hypothetical data from this experiment are as follows. The dependent variable is the number of errors made by each subject.

Self-esteem		Alone	Audience
	High	3	9
		6	4
		2	5
		2	8
		4	4
		7	6
	Low	7	10
		7	14
		2	11
		6	15
		8	11
		6	11

Use an ANOVA with $\alpha = .05$ to evaluate these data. Describe the effect of the audience and the effect of personality on performance.

20. The process of interference is assumed to be responsible for much of forgetting in human memory. New information going into memory interferes with the information that already is there. One demonstration of interference examines the process of forgetting while people are asleep versus while they are awake. Because there should be less interference during sleep, there also should be less forgetting. The following data are the results from an experiment examining four groups of subjects. All subjects were given a list of words to remember. Then half of the subjects went to sleep, and the others stayed awake. Within both the asleep and the awake groups, half of the subjects were tested after 2 hours, and the rest were tested after 8 hours. The dependent variable is the number of words correctly recalled.

	Delay of memory test			
	2 hours		8 hours	
Asleep	5	7	6	4
	6	10	8	4
	4	5	5	7
	3	8	10	8
	6	5	7	6
Awake	3	2	1	2
	4	3	0	1
	5	4	0	2
	3	2	1	0
	2	4	1	1

a. Use a two-factor ANOVA to examine these data.
b. Describe and explain the interaction.

CHAPTER 15

CORRELATION AND REGRESSION

TOOLS YOU WILL NEED

The following items are considered essential background material for this chapter. If you doubt your knowledge of any of these items, you should review the appropriate chapter or section before proceeding.

- Sum of squares (*SS*) (Chapter 4)
  - Computational formula
  - Definitional formula
- *z*-scores (Chapter 5)
- Hypothesis testing (Chapter 8)

CONTENTS

15.1 INTRODUCTION

Correlation is a statistical technique that is used to measure and describe a relationship between two variables. Usually the two variables are simply observed as they exist naturally in the environment—there is no attempt to control or manipulate the variables. For example, a researcher interested in the relationship between nutrition and IQ could observe (and record) the dietary patterns for a group of preschool children and then measure IQ scores for the same group. Notice that the researcher is not trying to manipulate the children's diet or IQ but is simply observing what occurs naturally. You also should notice that a correlation requires two scores for each individual (one score from each of the two variables). These scores normally are identified as *X* and *Y*. The pairs of scores can be listed in a table, or they can be presented graphically in a scatterplot (see Figure 15.1). In the scatterplot, the *X* values are placed on the horizontal axis of a graph, and the *Y* values are placed on the vertical axis. Each individual is then identified by a single point on the graph so that the coordinates of the point (the *X* and *Y* values) match the individual's *X* score and *Y* score. The value of the scatterplot is that it allows you to see the nature of the relationship (see Figure 15.1).

THE CHARACTERISTICS OF A RELATIONSHIP

A correlation measures three characteristics of the relationship between *X* and *Y*. These three characteristics are as follows.

1. The Direction of the Relationship. Correlations can be classified into two basic categories: positive and negative.

DEFINITIONS

In a *positive correlation,* the two variables tend to move in the same direction: When the *X* variable increases, the *Y* variable also increases; if the *X* variable decreases, the *Y* variable also decreases.

In a *negative correlation,* the two variables tend to go in opposite directions. As the *X* variable increases, the *Y* variable decreases. That is, it is an inverse relationship.

FIGURE 15.1

The same set of $n = 6$ pairs of scores (*X* and *Y* values) is shown in a table and in a scatterplot. Notice that the scatterplot allows you to see the relationship between *X* and *Y*.

Person	X	Y
A	1	1
B	1	3
C	3	2
D	4	5
E	6	4
F	7	5

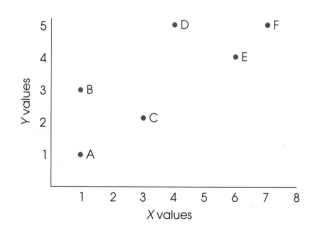

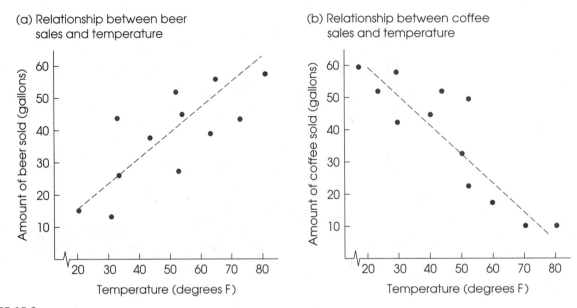

FIGURE 15.2

Examples of positive and negative relationships. Beer sales (gallons) are positively related to temperature, and coffee sales (gallons) are negatively related to temperature.

The direction of a relationship is identified by the sign of the correlation. A positive value ($+$) indicates a positive relationship; a negative value ($-$) indicates a negative relationship. The following example provides a description of positive and negative relationships.

E X A M P L E   1 5 . 1   Suppose you run the drink concession at the football stadium. After several seasons, you begin to notice a relationship between the temperature at game time and the beverages you sell. Specifically, you have noted that when the temperature is high, you tend to sell a lot of beer. When the temperature is low, you sell relatively little beer (see Figure 15.2). This is an example of a positive correlation. At the same time, you have noted a relationship between temperature and coffee sales: On cold days, you sell much more coffee than on hot days (see Figure 15.2). This is an example of a negative relationship.

2. The Form of the Relationship. In the preceding coffee and beer examples, the relationships tend to have a linear form; that is, the points in the scatterplot tend to form a straight line. Notice that we have drawn a line through the middle of the data points in each figure to help show the relationship. The most common use of correlation is to measure straight-line relationships. However, you should note that other forms of relationships do exist and that there are special correlations used to measure them. (We will examine an alternative in Section 15.5.) For example, Figure 15.3(a) shows the relationship between reaction time and age. In this scatterplot, there is a curved relationship. Reaction time improves with age until the late teens, when it reaches a peak; after that, reaction

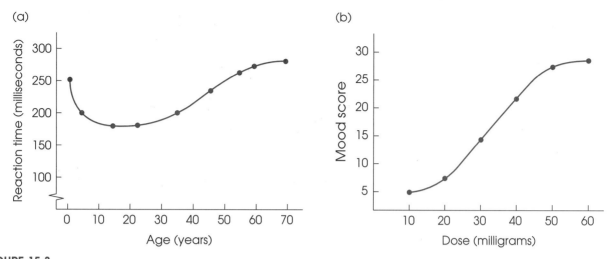

FIGURE 15.3

Examples of relationships that are not linear: (a) relationship between reaction time and age; (b) relationship between mood and drug dose.

time starts to get worse. Figure 15.3(b) shows the typical dose-response relationship. Again, this is not a straight-line relationship. In this graph, the elevation in mood increases rather rapidly with dose increases. However, beyond a certain dose the amount of improvement in mood levels off. Many different types of correlations exist. In general, each type is designed to evaluate a specific form of relationship. In this text, we will concentrate on the correlation that measures linear relationships.

3. The Degree of the Relationship. Finally, a correlation measures how well the data fit the specific form being considered. For example, a linear correlation measures how well the data points fit on a straight line. A *perfect correlation* always is identified by a correlation of 1.00 and indicates a perfect fit. At the other extreme, a correlation of 0 indicates no fit at all. Intermediate values represent the degree to which the data points approximate the perfect fit. The numerical value of the correlation also reflects the degree to which there is a consistent, predictable relationship between the two variables. Again, a correlation of 1.00 (or −1.00) indicates a perfectly consistent relationship.

A correlation of −1.00 also indicates a perfect fit. The direction of the relationship (positive or negative) should be considered separately from the degree of the relationship.

Examples of different values for linear correlations are shown in Figure 15.4. Notice that in each example we have sketched a line around the data points. This line, called an *envelope* because it encloses the data, often helps you to see the overall trend in the data.

WHERE AND WHY CORRELATIONS ARE USED

Although correlations have a number of different applications, a few specific examples are presented next to give an indication of the value of this statistical measure.

1. Prediction. If two variables are known to be related in some systematic way, it is possible to use one of the variables to make accurate predictions about the other. For example, when you applied for admission to college, you were required to submit a great deal of personal information, including your scores on

FIGURE 15.4

Examples of different values for linear correlations: (a) a strong positive relationship, approximately +0.90; (b) a relatively weak negative correlation, approximately −0.40; (c) a perfect negative correlation, −1.00; (d) no linear trend, 0.00.

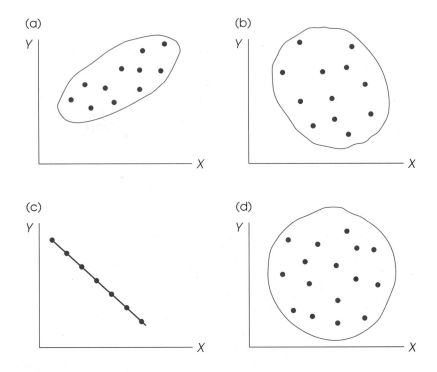

the Scholastic Achievement Test (SAT). College officials want this information so they can predict your chances of success in college. It has been demonstrated over several years that SAT scores and college grade-point averages are correlated. Students who do well on the SAT tend to do well in college; students who have difficulty with the SAT tend to have difficulty in college. Based on this relationship, the college admissions officers can make a prediction about the potential success of each applicant. You should note that this prediction is not perfectly accurate. Not everyone who does poorly on the SAT will have trouble in college. That is why you also submit letters of recommendation, high school grades, and other information with your application.

2. Validity. Suppose a psychologist develops a new test for measuring intelligence. How could you show that this test truly is measuring what it claims; that is, how could you demonstrate the validity of the test? One common technique for demonstrating validity is to use a correlation. If the test actually is measuring intelligence, then the scores on the test should be related to other measures of intelligence, for example, standardized IQ tests, performance on learning tasks, problem-solving ability, and so on. The psychologist could measure the correlation between the new test and each of these other measures of intelligence in order to demonstrate that the new test is valid.

3. Reliability. In addition to evaluating the validity of a measurement procedure, correlations are used to determine reliability. A measurement procedure is considered reliable to the extent that it produces stable, consistent measurements. That is, a reliable measurement procedure will produce the same (or nearly the same) scores when the same individuals are measured under the same conditions. For example, if your IQ were measured as 113 last week, you would expect to obtain nearly the same score if your IQ were measured again this week. One way to evaluate reliability is to use correlations to determine the relationship between two sets of measurements. When reliability is high, the correlation between two measurements should be strong and positive.

4. Theory Verification. Many psychological theories make specific predictions about the relationship between two variables. For example, a theory may predict a relationship between brain size and learning ability; a developmental theory may predict a relationship between the parents' IQs and the child's IQ; a social psychologist may have a theory predicting a relationship between personality type and behavior in a social situation. In each case, the prediction of the theory could be tested by determining the correlation between the two variables.

LEARNING CHECK

1. If the world were fair, would you expect a positive or negative relationship between grade-point average (X) and weekly studying hours (Y) for college students?

2. Data suggest that, on average, children from large families have lower IQs than children from small families. Do these data indicate a positive or a negative relationship between family size and average IQ?

3. If you are measuring linear relationships, correlations of +0.50 and −0.50 are equally good in terms of how well the data fit on a straight line. (True or false?)

4. It is impossible to have a correlation greater than +1.00 or less than −1.00. (True or false?)

ANSWERS

1. Positive. More hours studying should be associated with higher grade-point averages.

2. Negative.

3. True. The degree of fit is measured by the magnitude of the correlation independent of sign.

4. True. Correlations are always from +1.00 to −1.00.

15.2 THE PEARSON CORRELATION

By far the most common correlation is the Pearson correlation (or the Pearson product-moment correlation).

DEFINITION

The *Pearson correlation* measures the degree and the direction of the linear relationship between two variables.

The Pearson correlation is identified by the letter r. Conceptually, this correlation is computed by

$$r = \frac{\text{degree to which } X \text{ and } Y \text{ vary together}}{\text{degree to which } X \text{ and } Y \text{ vary separately}}$$

$$= \frac{\text{covariability of } X \text{ and } Y}{\text{variability of } X \text{ and } Y \text{ separately}}$$

When there is a perfect linear relationship, every change in the X variable is accompanied by a corresponding change in the Y variable. In Figure 15.4(c), for example, every time the value of X increases, there is a perfectly predictable decrease in the value of Y. The result is a perfect linear relationship, with X and Y always varying together. In this case, the covariability (X and Y together) is identical to the variability of X and Y separately, and the formula produces a correlation with a magnitude of 1.00 or −1.00. At the other extreme, when there is no linear relationship, a change in the X variable does not correspond to any predictable change in the Y variable. In this case, there is no covariability, and the resulting correlation is zero.

THE SUM OF PRODUCTS OF DEVIATIONS

To calculate the Pearson correlation, it is necessary to introduce one new concept: the sum of products of deviations. In the past, we have used a similar concept, SS (the sum of squared deviations), to measure the amount of variability for a single variable. The *sum of products,* or *SP,* provides a parallel procedure for measuring the amount of covariability between two variables. The value for SP can be calculated with either a definitional formula or a computational formula.

The *definitional formula* for the sum of products of deviations is

$$SP = \Sigma(X - \overline{X})(Y - \overline{Y}) \tag{15.1}$$

The definitional formula instructs you to perform the following sequence of operations:

1. Find the X deviation and the Y deviation for each individual.
2. Find the product of the deviations for each individual.
3. Sum the products.

Notice that this process "defines" the value being calculated: the sum of the products of the deviations.

The *computational formula* for the sum of products of deviations is

Caution: The n in this formula refers to the number of pairs of scores.

$$SP = \Sigma XY - \frac{\Sigma X \Sigma Y}{n} \tag{15.2}$$

Because the computational formula uses the original scores (X and Y values), it usually results in easier calculations than those required with the definitional formula. However, both formulas will always produce the same value for SP.

You may have noted that the formulas for SP are similar to the formulas you have learned for SS (sum of squares). The relationship between the two sets of formulas is described in Box 15.1. The following example demonstrates the calculation of SP with both formulas.

EXAMPLE 15.2 The set of n = 4 pairs of scores shown in the following table will be used to calculate SP, using first the definitional formula and then the computational formula.

For the definitional formula, you need deviation scores for each of the X values and each of the Y values. Note that the mean for the Xs is $\overline{X} = 3$ and

15.1 COMPARING THE *SP* AND *SS* FORMULAS

IT WILL help you to learn the formulas for *SP* if you note the similarity between the two *SP* formulas and the corresponding formulas for *SS* that were presented in Chapter 4. The definitional formula for *SS* is

$$SS = \Sigma(X - \bar{X})^2$$

In this formula, you must square each deviation, which is equivalent to multiplying it by itself. With this in mind, the formula can be rewritten as

$$SS = \Sigma(X - \bar{X})(X - \bar{X})$$

The similarity between the *SS* formula and the *SP* formula should be obvious—the *SS* formula uses squares and the *SP* formula uses products. This same relation-ship exists for the computational formulas. For *SS*, the computational formula is

$$SS = \Sigma X^2 - \frac{(\Sigma X)^2}{n}$$

As before, each squared value can be rewritten so that the formula becomes

$$SS = \Sigma XX - \frac{\Sigma X \Sigma X}{n}$$

Again, you should note the similarity in structure between the *SS* formula and the *SP* formula. If you remember that *SS* uses squares and *SP* uses products, the two new formulas for the sum of products should be easy to learn.

that the mean for the *Y*s is $\bar{Y} = 5$. The deviations and the products of deviations also are shown in the following table:

Caution: The signs (+ and −) are critical in determining the sum of products, *SP*.

Scores		Deviations		Products
X	Y	$X - \bar{X}$	$Y - \bar{Y}$	$(X - \bar{X})(Y - \bar{Y})$
1	3	−2	−2	+4
2	6	−1	+1	−1
4	4	+1	−1	−1
5	7	+2	+2	+4
				+6 = *SP*

For these scores, the sum of the products of the deviations is *SP* = +6.

For the computational formula, you need the sum of the *X* values, the sum of the *Y* values, and the sum of the *XY* products for each pair. These values are as follows:

X	Y	XY	
1	3	3	
2	6	12	
4	4	16	
5	7	35	
12	20	66	Totals

FIGURE 15.5

Scatterplot of the data from Example 15.3.

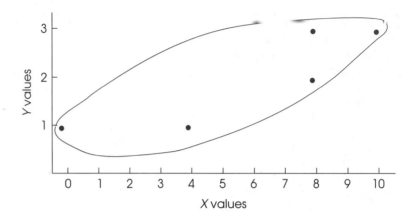

Substituting the sums in the formula gives

$$SP = \Sigma XY - \frac{\Sigma X\, \Sigma Y}{n}$$

$$= 66 - \frac{12(20)}{4}$$

$$= 66 - 60$$

$$= 6$$

Note that both formulas produce the same result, $SP = 6$.

CALCULATION OF THE PEARSON CORRELATION

As noted earlier, the Pearson correlation consists of a ratio comparing the covariability of X and Y (the numerator) with the variability of X and Y separately (the denominator). In the formula for the Pearson r, we will use SP to measure the covariability of X and Y. The variability of X and Y will be measured by computing SS for the X scores and SS for the Y scores separately. With these definitions, the formula for the Pearson correlation becomes

Note that you *multiply* SS for X and SS for Y in the denominator of the Pearson formula.

$$r = \frac{SP}{\sqrt{SS_X SS_Y}}$$ (15.3)

The following example demonstrates the use of this formula with a simple set of scores.

EXAMPLE 15.3

X	Y
0	1
10	3
4	1
8	2
8	3

The Pearson correlation is computed for the set of $n = 5$ pairs of scores shown in the margin.

Before starting any calculations, it is useful to put the data in a scatterplot and make a preliminary estimate of the correlation. These data have been graphed in Figure 15.5. Looking at the scatterplot, it appears that there is a very good (but not perfect) positive correlation. You should expect an approximate value of $r = +.8$ or $+.9$. To find the Pearson correlation, we will need SP, SS for X, and SS for Y. Each of these values is calculated using the definitional formula. The mean for the X values is $\overline{X} = 6$ and the mean for the Y scores is $\overline{Y} = 2$.

Scores		Deviations		Squared deviations		Products
X	Y	$X - \overline{X}$	$Y - \overline{Y}$	$(X - \overline{X})^2$	$(Y - \overline{Y})^2$	$(X - \overline{X})(Y - \overline{Y})$
0	1	−6	−1	36	1	+6
10	3	+4	+1	16	1	+4
4	1	−2	−1	4	1	+2
8	2	+2	0	4	0	0
8	3	+2	+1	4	1	+2
				64 = SS for X	4 = SS for Y	+14 = SP

 By using these values, the Pearson correlation is

$$r = \frac{SP}{\sqrt{SS_X SS_Y}} = \frac{14}{\sqrt{64(4)}}$$

$$= \frac{14}{16} = +0.875$$

Note that the value we obtained is in agreement with the prediction based on the scatterplot.

THE PEARSON CORRELATION AND z-SCORES

The Pearson correlation measures the relationship between an individual's location in the X distribution and his or her location in the Y distribution. For example, a positive correlation means that individuals who score high on X also tend to score high on Y. Similarly, a negative correlation indicates that individuals with high X scores tend to have low Y scores.

You should recall from Chapter 5 that z-scores provide a precise way to identify the location of an individual score within a distribution. Because the Pearson correlation measures the relationship between locations and because z-scores are used to specify locations, the formula for the Pearson correlation can be expressed entirely in terms of z-scores:

$$r = \frac{\sum z_X z_Y}{n} \tag{15.4}$$

In this formula, z_X identifies each individual's position within the X distribution, and z_Y identifies his or her position within the Y distribution. The product of the z-scores (like the product of the deviation scores) determines the strength and direction of the correlation.

Because z-scores are considered to be the best way to describe a location within a distribution, formula 15.4 often is considered to be the best way to define the Pearson correlation. However, you should realize that this formula requires a lot of tedious calculations (changing each score to a z-score), so it rarely is used to calculate a correlation.

LEARNING CHECK

1. Describe what is measured by a Pearson correlation.

2. Can SP ever have a value less than zero?

3. Calculate the sum of products of deviations (*SP*) for the following set of scores. Use the definitional formula and then the computational formula. Verify that you get the same answer with both formulas.

X	Y
1	0
3	1
7	6
5	2
4	1

Remember, it is useful to sketch a scatter-plot and make an estimate of the correlation before you begin calculations.

4. Compute the Pearson correlation for the following data:

X	Y
2	9
1	10
3	6
0	8
4	2

ANSWERS **1.** The Pearson correlation measures the degree and the direction of linear relationship between two variables.

2. Yes. *SP* can be positive, negative, or zero depending on the relationship between *X* and *Y*.

3. *SP* = 19

4. $r = -\dfrac{16}{20} = -.80$

15.3 UNDERSTANDING AND INTERPRETING THE PEARSON CORRELATION

When you encounter correlations, there are four additional considerations that you should bear in mind:

1. Correlation simply describes a relationship between two variables. It does not explain why the two variables are related. Specifically, a correlation should not and cannot be interpreted as proof of a cause-and-effect relationship between the two variables.

2. The value of a correlation can be affected greatly by the range of scores represented in the data.

3. One or two extreme data points, often called *outriders,* can have a dramatic effect on the value of a correlation.

4. When judging how "good" a relationship is, it is tempting to focus on the numerical value of the correlation. For example, a correlation of +.5 is

FIGURE 15.6

Hypothetical data showing the logical relationship between the number of churches and the number of serious crimes for a sample of U.S. cities.

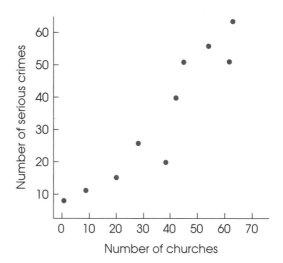

halfway between 0 and 1.00 and therefore appears to represent a moderate degree of relationship. However, a correlation should not be interpreted as a proportion. Although a correlation of 1.00 does mean that there is a 100% perfectly predictable relationship between X and Y, a correlation of .5 does not mean that you can make predictions with 50% accuracy. To describe how accurately one variable predicts the other, you must square the correlation. Thus, a correlation of $r = .5$ provides only $r^2 = .5^2 = 0.25$, or 25% accuracy.

Each of these four points will now be discussed in detail.

CORRELATION AND CAUSATION

One of the most common errors in interpreting correlations is to assume that a correlation necessarily implies a cause-and-effect relationship between the two variables. We constantly are bombarded with reports of relationships: Cigarette smoking is related to heart disease; alcohol consumption is related to birth defects; carrot consumption is related to good eyesight. Do these relationships mean that cigarettes cause heart disease or carrots cause good eyesight? The answer is *no*. Although there may be a causal relationship, the simple existence of a correlation does not prove it. This point should become clear in the following hypothetical example.

EXAMPLE 15.4

Suppose we select a variety of different cities and towns throughout the United States and measure the number of churches (X variable) and the number of serious crimes (Y variable) for each. A scatterplot showing hypothetical data for this study is presented in Figure 15.6. Notice that this scatterplot shows a strong, positive correlation between churches and crime. You also should note that these are realistic data. It is reasonable that the small towns would have less crime and fewer churches and that the large cities would have large values for both variables. Does this relationship mean that churches cause crime? Does it mean that crime causes churches? It should be clear that both answers are no. Although a strong correlation exists between churches and crime, the real cause of the relationship is the size of the population.

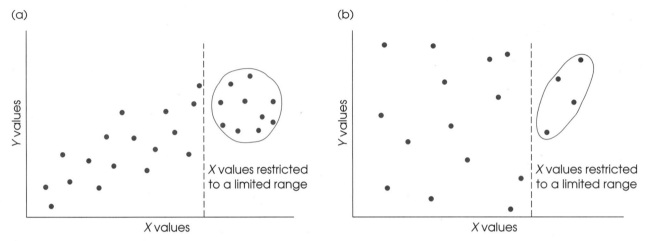

FIGURE 15.7

(a) An example where the full range of X and Y values shows a strong, positive correlation but the restricted range of scores produces a correlation near zero. (b) An example where the full range of X and Y values shows a correlation near zero but the scores in the restricted range produce a strong, positive correlation.

CORRELATION AND RESTRICTED RANGE

Whenever a correlation is computed from scores that do not represent the full range of possible values, you should be cautious in interpreting the correlation. Suppose, for example, you are interested in the relationship between IQ and creativity. If you select a sample of your fellow college students, your data probably would represent only a limited range of IQ scores (most likely from 110 to 130). The correlation within this restricted range could be completely different from the correlation that would be obtained from a full range of IQ scores. Two extreme examples are shown in Figure 15.7.

Figure 15.7(a) shows an example where there is strong, positive relationship between X and Y when the entire range of scores is considered. However, this relationship is obscured when the data are limited to a *restricted range*. In Figure 15.7(b), there is no consistent relationship between X and Y for the full range of scores. However, when the range of X values is restricted, the data show a strong, positive relationship.

To be safe, you should not generalize any correlation beyond the range of data represented in the sample. For a correlation to provide an accurate description for the general population, there should be a wide range of X and Y values in the data.

OUTRIDERS

An outrider is an individual with X and/or Y values that are substantially different (larger or smaller) than the values obtained for the other individuals in the data set. The data point of a single outrider can have a dramatic influence on the value obtained for the correlation. This effect is illustrated in Figure 15.8. Figure 15.8(a) shows a set of $n = 5$ data points where the correlation between the X and Y variables is nearly zero (actually $r = -0.08$). In Figure 15.8(b), one extreme data point (14, 12) has been added to the original data set. When this outrider is included in the analysis, a strong, positive correlation emerges (now $r = +0.85$). Notice that the single outrider drastically alters the value for the correlation and thereby can affect

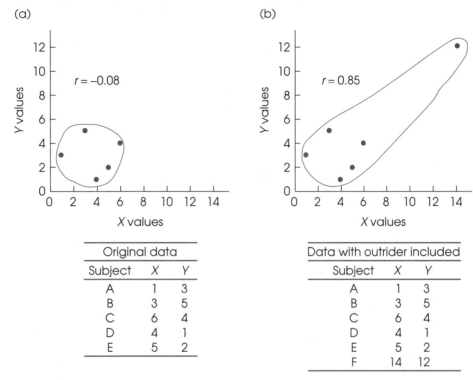

FIGURE 15.8

A demonstration of how one extreme data point (an outrider) can influence the value of a correlation.

one's interpretation of the relationship between variables X and Y. Without the outrider, one would conclude there is no relationship between the two variables. With the extreme data point, $r = +0.85$ implies that as X increases, Y will increase—and do so consistently. The problem of outriders is a good reason why you should always look at a scatterplot, instead of simply basing your interpretation on the numerical value of the correlation. If you only "go by the numbers," you might overlook the fact that one extreme data point inflated the size of the correlation.

CORRELATION AND THE STRENGTH OF THE RELATIONSHIP

A correlation measures the degree of relationship between two variables on a scale from 0 to 1.00. Although this number provides a measure of the degree of relationship, many researchers prefer to square the correlation and use the resulting value to measure the strength of the relationship.

One of the common uses of correlation is for prediction. If two variables are correlated, you can use the value of one variable to predict the other. For example, college admissions officers do not just guess which applicants are likely to do well; they use other variables (SAT scores, high school grades, etc.) to predict which students are most likely to be successful. These predictions are based on correlations. By using correlations, the admissions officers expect to make more-accurate predictions than would be obtained by just guessing. In general, the squared correlation (r^2) measures the gain in accuracy that is obtained from using the correlation for prediction instead of just guessing.

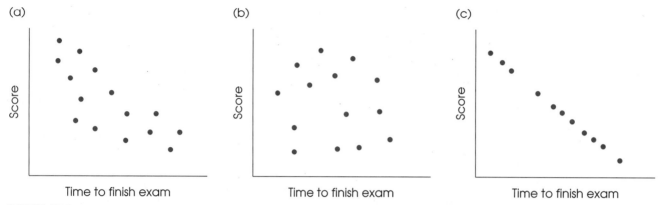

(a)   (b)   (c)

Score   Score   Score

Time to finish exam   Time to finish exam   Time to finish exam

FIGURE 15.9

Three sets of hypothetical data showing the relationship between the time needed to complete a statistics exam and the score on the exam: (a) a moderate, negative relationship ($r = -0.80$); (b) no relationship ($r = 0$); (c) a perfect linear relationship ($r = -1.00$).

DEFINITION

> The value r^2 is called the *coefficient of determination* because it measures the proportion of variability in one variable that can be determined from the relationship with the other variable. A correlation of $r = .80$ (or $-.80$), for example, means that $r^2 = 0.64$ (or 64%) of the variability in the Y scores can be predicted from the relationship with X.

Although a detailed discussion of the coefficient of determination is beyond the scope of this book, the general notion is that whenever two variables are consistently related, it is possible to use one variable to predict the values of the second variable. The amount of variability that can be predicted is determined by r^2. The following example demonstrates this phenomenon.

EXAMPLE 15.5   Several years ago, as students were leaving a statistics exam, we recorded the amount of time each student spent on the exam and the exam grade for each student. Three sets of hypothetical data for this study are shown in Figure 15.9. Figure 15.9(a) is similar to the actual results and shows a moderately strong, negative correlation between time and grade ($r = -0.80$). Figure 15.9(b) shows data that would be obtained if there were no relationship between time and grade. Finally, Figure 15.9(c) shows a perfect linear relationship between time and grade.

Notice that the data in Figure 15.9(a) (the actual results) indicate a fairly consistent, predictable relationship between time and grade: Students who finish early tend to have higher grades than students who hold their papers until the bitter end. From this relationship, it is possible to predict a student's grade based on when he/she turns in the exam. However, you should realize that this prediction is not perfect. Although students who finish early *tend* to have high grades, this is not always true. Thus, knowing how much time a student spent provides some information about the student's grade, or knowing a student's grade provides some information about how much time was spent on the exam. Although we are measuring two separate variables (time and grade), the two scores for each subject actually provide much the same information. In other words, there is some overlap or redundancy between the two measures: To some extent, the same information is provided by knowing either a student's grade or a student's time. With a correlation of $r = -0.80$, we obtain $r^2 = 0.64$, which means that there is 64% overlap between the two measures.

Now consider the data in Figure 15.9(b). Here there is no relationship between time and grade. With $r = 0$ and $r^2 = 0$, you have no ability to predict a student's grade from the amount of time spent on the exam. Knowing how much time a student spent on the exam provides no information (0%) about the student's grade. In this case, the two measurements are completely independent, and there is no overlap in the information they provide. Finally, consider the perfect correlation shown in Figure 15.9(c). In this case, $r = -1.00$ and $r^2 = 1.00$ or 100%. For these data, the student who finished first is guaranteed to have the highest grade, and the student who finished last is guaranteed the lowest grade. With $r = 1.00$ (or -1.00), a student's grade is 100% predictable if you know how much time was spent on the exam. In this case, there is complete overlap between the two measures because a student's time and his/her grade provide *exactly* the same information.

As a final note, you should realize that this example provides an additional demonstration of the fact that a correlation does not necessarily imply a cause-and-effect relationship. In particular, you should not expect to improve your exam grade by turning in your paper earlier. Although there is a relationship between early papers and high grades, you cannot conclude that spending less time on an exam will *cause* your grade to go up.

15.4 HYPOTHESIS TESTS WITH THE PEARSON CORRELATION

The Pearson correlation is generally computed for sample data. As with most sample statistics, however, a sample correlation often is used to answer questions about the general population. That is, the sample correlation is used as the basis for drawing inferences about the corresponding population correlation. For example, a psychologist would like to know whether there is a relationship between IQ and creativity. This is a general question concerning a population. To answer the question, a sample would be selected, and the sample data would be used to compute the correlation value. You should recognize this process as an example of inferential statistics: using samples to draw inferences about populations. In the past, we have been concerned primarily with using sample means as the basis for answering questions about population means. In this section, we will examine the procedures for using a sample correlation as the basis for testing hypotheses about the corresponding population correlation.

The basic question for this hypothesis test is whether or not a correlation exists in the population. The null hypothesis is "No, there is no correlation in the population" or "The population correlation is zero." The alternative hypothesis is "Yes,

FIGURE 15.10

Scatterplot of a population of X and Y values with a near-zero correlation. However, a small sample of $n = 3$ data points from this population shows a relatively strong, positive correlation. These data points in the sample are circled.

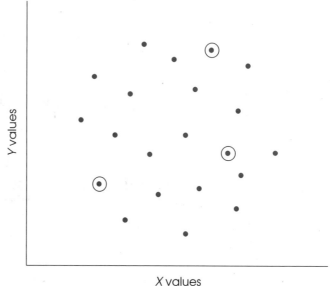

X values

there is a real, nonzero correlation in the population." Because the population correlation is traditionally represented by ρ (the Greek letter rho), these hypotheses would be stated in symbols as

Directional hypotheses for a one-tailed test would specify either a positive correlation ($\rho > 0$) or a negative correlation ($\rho < 0$).

$$H_0: \rho = 0 \quad \text{(no population correlation)}$$

$$H_1: \rho \neq 0 \quad \text{(there is a real correlation)}$$

The correlation from the sample data (r) will be used to evaluate these hypotheses. As always, samples are not expected to be identical to the populations from which they come; there will be some discrepancy (sampling error) between a sample statistic and the corresponding population parameter. Specifically, you should always expect some error between a sample correlation (r) and the population correlation (ρ) it represents. One implication of this fact is that even when there is no correlation in the population ($\rho = 0$), you are still likely to obtain a nonzero value for the sample correlation. This is particularly true for small samples. Figure 15.10 illustrates how a small sample from a population with a near-zero correlation could result in a correlation that deviates from zero. Also relevant here is our previous discussion of outriders (page 400). Just by chance, you might obtain an individual in your sample with an extreme data point (X and/or Y). As a result, you may obtain a nonzero sample correlation even though there is no correlation for the population. The purpose of the hypothesis test is to decide between the following two alternatives:

1. The nonzero sample correlation is simply due to chance. That is, there is no correlation in the population, and the sample value is simply the result of sampling error. This is the alternative specified by H_0.

2. The nonzero sample correlation accurately represents a real, nonzero correlation in the population. This is the alternative stated in H_1.

The table lists critical values in terms of degrees of freedom: $df = n - 2$. Remember to subtract 2 when using this table.

Although it is possible to conduct the hypothesis test by computing either a t statistic or an F-ratio, the computations have been completed and are summarized in Table B.5 in Appendix B. To use this table, you must know the sample size (n), the magnitude of the sample correlation (independent of sign), and the alpha level. The values given in the table indicate how large a sample correlation must be before it is *significantly* greater than chance (significantly more than sampling error). That is, in order to reject H_0, the magnitude of the sample correlation must *equal or exceed* the value given in the table. The following examples demonstrate the use of the table.

EXAMPLE 15.6   A researcher is using a regular, two-tailed test with $\alpha = .05$ to determine whether or not a nonzero correlation exists in the population. A sample of $n = 30$ individuals is obtained. With $\alpha = .05$ and $n = 30$, the table lists a value of 0.361. Thus, the sample correlation (independent of sign) must have a value greater than or equal to 0.361 to reject H_0 and conclude that there is a significant correlation in the population. Any sample correlation between 0.361 and -0.361 is considered within the realm of sampling error and therefore not significant.

EXAMPLE 15.7   This time the researcher is using a directional, one-tailed test to determine whether or not there is a positive correlation in the population.

$$H_0: \rho \leq 0 \quad \text{(not positive)}$$

$$H_1: \rho > 0 \quad \text{(positive)}$$

With $\alpha = .05$ and a sample of $n = 30$, the table lists a value of 0.306 for a one-tailed test. Thus, the researcher must obtain a sample correlation that is positive (as predicted) and has a value greater than or equal to 0.306 to reject H_0 and conclude that there is a significant positive correlation in the population.

IN THE LITERATURE:
REPORTING CORRELATIONS

When correlations are computed, the results are reported using APA format. The statement should include the sample size, the calculated value for the correlation, whether or not it is a statistically significant relationship, the probability level, and the type of test used (one- or two-tailed). For example, a correlation might be reported as follows:

> A correlation for the data revealed that amount of education and annual income were significantly related, $r = +.65$, $n = 30$, $p < .01$, two tails.

Sometimes a study might look at several variables, and correlations between all possible variable pairings are computed. Suppose, for example, a study measured people's annual income, amount of education, age, and intelligence. With four variables, there are six possible pairings. Correlations were computed for every pair of variables. These results are most easily reported in a table called a *correlation matrix,* using footnotes to indicate which correlations are significant. For example, the report might state:

The analysis examined the relationships among income, amount of education, age, and intelligence for $n = 30$ subjects. The correlations between pairs of variables are reported in Table 1. Significant correlations are noted in the table.

TABLE 1

Correlation matrix for income, amount of education, age, and intelligence

	Education	Age	IQ
Income	+.65*	+.41**	+.27
Education		+.11	+.38**
Age			−.02

$n = 30$
*$p < .01$, two tails
**$p < .05$, two tails

LEARNING CHECK

1. A researcher obtains a correlation of $r = -.41$ for a sample of $n = 25$ individuals. Does this sample provide sufficient evidence to conclude that there is a significant, nonzero correlation in the population? Assume a nondirectional test with $\alpha = .05$.

2. For a sample of $n = 20$, how large a correlation is needed to conclude at the .05 level that there is a nonzero correlation in the population? Assume a nondirectional test.

3. As sample size gets smaller, what happens to the magnitude of the correlation necessary for significance? Explain why this occurs.

ANSWERS

1. Yes. For $n = 25$, the critical value is $r = .396$. The sample value is in the critical region.

2. For $n = 20$, the critical value is $r = .444$.

3. As the sample size gets smaller, the magnitude of the correlation needed for significance gets larger. With a small sample, it is easy to get a relatively good correlation just by chance (see Figure 15.10). Therefore, a small sample requires a very large correlation before you can be confident that there is a real (nonzero) relationship in the population.

FIGURE 15.11

Hypothetical data showing the relationship between practice and performance. Although this relationship is not linear, there is a consistent positive relationship. An increase in performance tends to accompany an increase in practice.

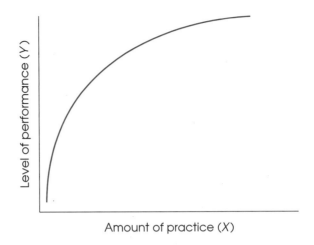

Level of performance (*Y*)

Amount of practice (*X*)

15.5 THE SPEARMAN CORRELATION

The Pearson correlation specifically measures the degree of linear relationship between two variables. It is the most commonly used measure of relationship and is used with data from an interval or a ratio scale of measurement. However, other correlation measures have been developed for nonlinear relationships and for other types of data (scales of measurement). One of these useful measures is called the *Spearman correlation.* The Spearman correlation is used in two situations.

First, the Spearman correlation is designed to measure the relationship between variables measured on an ordinal scale of measurement. Recall that in Chapter 1 we noted that an ordinal scale of measurement involves placing observations into rank order. Rank-order data are fairly common because they are often easier to obtain than interval or ratio scale data. For example, a teacher may feel confident about rank-ordering students' leadership abilities but would find it difficult to measure leadership on some other scale.

In addition to measuring relationships for ordinal data, the Spearman correlation can be used as a valuable alternative to the Pearson correlation, even when the original raw scores are on an interval or a ratio scale. As we have noted, the Pearson correlation measures the degree of linear relationship between two variables—that is, how well the data points fit on a straight line. However, a researcher often expects the data to show a consistent relationship but not necessarily a linear relationship (for example, see Figure 15.3 and the discussion on page 390). Consider the situation where a researcher is investigating the relationship between amount of practice (*X*) and level of performance (*Y*). For these data, the researcher expects a strong, positive relationship: More practice leads to better performance. However, the expected relationship probably does not fit a linear form, so a Pearson correlation would not be appropriate (see Figure 15.11). In this situation, the Spearman correlation can be used to obtain a measure of the consistency of relationship, independent of its specific form.

The reason that the Spearman correlation measures consistency, rather than form, comes from a simple observation: When two variables are consistently related, their ranks will be linearly related. For example, a perfectly consistent positive relation-

(a) Scores

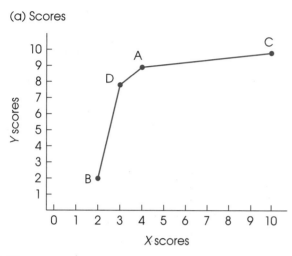

(b) Ranks

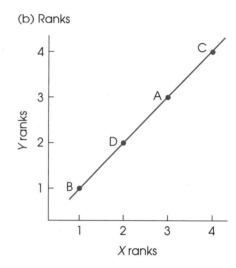

FIGURE 15.12

Scatterplots showing (a) the scores and (b) the ranks for the data in Example 15.8. Notice that there is a consistent, positive relationship between the X and Y scores, although it is not a linear relationship. Also notice that the scatterplot of the ranks shows a perfect linear relationship.

ship means that every time the X variable increases, the Y variable also increases. Thus, the smallest value of X is paired with the smallest value of Y, the second-smallest value of X is paired with the second-smallest value of Y, and so on. This phenomenon is demonstrated in the following example.

EXAMPLE 15.8

TABLE 15.1

Scores for Example 15.8

Person	X	Y
A	4	9
B	2	2
C	10	10
D	3	8

TABLE 15.2

Ranks for Example 15.8

Person	X Rank	Y Rank
A	3	3
B	1	1
C	4	4
D	2	2

The data in Table 15.1 represent X and Y scores for a sample of $n = 4$ people. Note that person B has the lowest X score and the lowest Y score. Similarly, person D has the second-lowest score for both X and Y, person A has the third-lowest scores, and person C has the highest scores. These data show a perfectly consistent relationship: Each increase in X is accompanied by an increase in Y. However, the relationship is not linear, as can be seen in the graph of the data in Figure 15.12(a).

Now we convert the raw scores to ranks: The lowest X is assigned a rank of 1, the next lowest X a rank of 2, and so on. This procedure is repeated for the Y scores. What happens when the X and Y scores are converted to ranks? Again, person B has the lowest X and Y scores, so this individual is ranked first on both variables. Similarly, person D is ranked second on both variables, person A is ranked third, and person C is ranked fourth (Table 15.2). When the ranks are plotted on a graph [see Figure 15.12(b)], the result is a perfect linear relationship.

The preceding example has demonstrated that a consistent relationship among scores produces a linear relationship when the scores are converted to ranks. Thus, if you want to measure the consistency of a relationship for a set of scores, you can simply convert the scores to ranks and then use the Spearman correlation to measure the correlation for the ranked data. The degree of relationship for the ranks (the Spearman correlation) provides a measure of the degree of consistency for the original scores.

To summarize, the Spearman correlation measures the relationship between two variables when both are measured on ordinal scales (ranks). There are two general situations where the Spearman correlation is used:

1. Spearman is used when the original data are ordinal; that is, when the X and Y values are ranks.

2. Spearman is used when a researcher wants to measure the consistency of a relationship between X and Y, independent of the specific form of the relationship. In this case, the original scores are first converted to ranks; then the Spearman correlation is used to measure the relationship for the ranks. Incidentally, when there is a consistently one-directional relationship between two variables, the relationship is said to be *monotonic*. Thus, the Spearman correlation can be used to measure the degree of monotonic relationship between two variables.

The word *monotonic* describes a sequence that is consistently increasing (or decreasing). Like the word *monotonous,* it means constant and unchanging.

CALCULATION OF THE SPEARMAN CORRELATION

The calculation of the Spearman correlation is remarkably simple, provided you know how to compute a Pearson correlation. First, be sure that you have ordinal data (ranks) for the X and the Y scores. If necessary, you may have to place X scores and/or Y scores in rank order if they are not already. Ranking is accomplished as follows: The smallest X is assigned a rank of 1, the next smallest a rank of 2, and so on. Then the same is done for the Y scores. (Note that the X and Y scores are ranked separately.) Finally, using the same formula from the Pearson correlation, the Spearman correlation is computed *for the ranks* of X and Y.

That's all there is to it. When you use the Pearson correlation formula for ordinal data, the result is called a Spearman correlation. The Spearman correlation is identified by the symbol r_S to differentiate it from the Pearson correlation. The complete process of computing the Spearman correlation, including ranking scores, is demonstrated in Example 15.9.

EXAMPLE 15.9

The following data show a nearly perfect monotonic relationship between X and Y. When X increases, Y tends to decrease, and there is only one reversal in this general trend. To compute the Spearman correlation, we first rank the X and Y values, and we then compute the Pearson correlation for the ranks.

We have listed the X values in order so that the trend is easier to recognize.

| Original data | | | Ranks | | |
X	Y		X	Y	XY
3	12		1	5	5
4	5		2	3	6
5	6		3	4	12
10	4		4	2	8
13	3		5	1	5
					$36 = \Sigma XY$

FIGURE 15.13

Scatterplots showing (a) the scores and (b) the ranks for the data in Example 15.9.

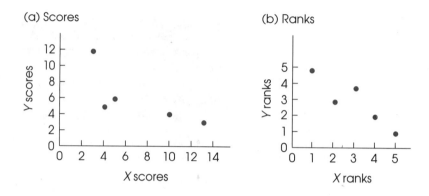

The scatterplots for the original data and the ranks are shown in Figure 15.13. To compute the correlation, we will need SS for X, SS for Y, and SP. Remember, all these values are computed with the ranks, not the original scores.

The X ranks are simply the integers 1, 2, 3, 4, and 5. These values have $\Sigma X = 15$ and $\Sigma X^2 = 55$. The SS for the X ranks is

$$SS_X = \Sigma X^2 - \frac{(\Sigma X)^2}{n}$$

$$= 55 - \frac{(15)^2}{5}$$

$$= 55 - 45$$

$$= 10$$

You should note that the ranks for Y are identical to the ranks for X; that is, they are the integers 1, 2, 3, 4, and 5. Therefore, the SS for Y will be identical to the SS for X:

$$SS_Y = 10$$

To compute the SP value, we need ΣX, ΣY, and ΣXY for the ranks. The XY values are listed in the table with the ranks, and we already have found that both the Xs and the Ys have a sum of 15. Using these values, we obtain

$$SP = \Sigma XY - \frac{(\Sigma X)(\Sigma Y)}{n}$$

$$= 36 - \frac{(15)(15)}{5}$$

$$= 36 - 45$$

$$= -9$$

The final Spearman correlation is

$$r_S = \frac{SP}{\sqrt{(SS_X)(SS_Y)}}$$

$$= \frac{-9}{\sqrt{10(10)}}$$

$$= -0.9$$

The Spearman correlation indicates that the data show a strong (nearly perfect) negative trend.

RANKING TIED SCORES

When you are converting scores into ranks for the Spearman correlation, you may encounter two (or more) identical scores. Whenever two scores have exactly the same value, their ranks should also be the same. This is accomplished by the following procedure:

1. List the scores in order from smallest to largest. Include tied values in this list.
2. Assign a rank (first, second, etc.) to each position in the ordered list.
3. When two (or more) scores are tied, compute the mean of their ranked positions, and assign this mean value as the final rank for each score.

The process of finding ranks for tied scores is demonstrated here. These scores have been listed in order from smallest to largest.

Scores	Rank position	Final rank	
3	1	1.5	Mean of 1 and 2
3	2	1.5	
5	3	3	
6	4	5	
6	5	5	Mean of 4, 5, and 6
6	6	5	
12	7	7	

Note that this example has seven scores and uses all seven ranks. For $X = 12$, the largest score, the appropriate rank is 7. It cannot be given a rank of 6 because that rank has been used for the tied scores.

SPECIAL FORMULA FOR THE SPEARMAN CORRELATION

After the original X values and Y values have been ranked, the calculations necessary for SS and SP can be greatly simplified. First, you should note that the X ranks and the Y ranks are really just a set of integers: 1, 2, 3, 4, . . . , n. To compute the mean for these integers, you can locate the midpoint of the series by $\overline{X} = (n + 1)/2$. Similarly, the SS for this series of integers can be computed by

$$SS = \frac{n(n^2 - 1)}{12} \quad \text{(Try it out.)}$$

Also, because the X ranks and the Y ranks are the same values, the SS for X will be identical to the SS for Y.

Because calculations with ranks can be simplified and because the Spearman correlation uses ranked data, these simplifications can be incorporated into the final calculations for the Spearman correlation. Instead of using the Pearson formula after ranking the data, you can put the ranks directly into a simplified formula:

Caution: In this formula, you compute the value of the fraction and then subtract from 1. The 1 is not part of the fraction.

$$r_S = 1 - \frac{6\Sigma D^2}{n(n^2 - 1)} \quad (15.5)$$

where D is the difference between the X rank and the Y rank for each individual. This special formula will produce the same result that would be obtained from the Pearson formula. However, you should note that this special formula can be used only after the scores have been converted to ranks and only when there are no ties among the ranks. If there are relatively few tied ranks, the formula still may be used, but it loses accuracy as the number of ties increases. The application of this formula is demonstrated in the following example.

EXAMPLE 15.10 To demonstrate the special formula for the Spearman correlation, we will use the same data that were presented in Example 15.9. The ranks for these data are shown again here:

Ranks		Difference	
X	Y	D	D^2
1	5	4	16
2	3	1	1
3	4	1	1
4	2	−2	4
5	1	−4	16
			38 $= \Sigma D^2$

Using the special formula for the Spearman correlation, we obtain

$$r_S = 1 - \frac{6\Sigma D^2}{n(n^2 - 1)}$$

$$= 1 - \frac{6(38)}{5(25 - 1)}$$

$$= 1 - \frac{228}{120}$$

$$= 1 - 1.90$$

$$= -0.90$$

Notice that this is exactly the same answer that we obtained in Example 15.9, using the Pearson formula on the ranks.

LEARNING CHECK 1. Describe what is measured by a Spearman correlation, and explain how this correlation is different from the Pearson correlation.

2. Identify the two procedures that can be used to compute the Spearman correlation.

3. Compute the Spearman correlation for the following set of scores:

FIGURE 15.14

Hypothetical data showing the relationship between SAT scores and GPA with a regression line drawn through the data points. The regression line defines a precise, one-to-one relationship between each *X* value (SAT score) and its corresponding *Y* value (GPA).

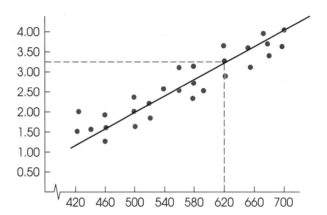

X	Y
2	7
12	38
9	6
10	19

ANSWERS

1. The Spearman correlation measures the consistency of the direction of the relationship between two variables. The Spearman correlation does not depend on the form of the relationship, whereas the Pearson correlation measures how well the data fit a linear form.

2. After the *X* and *Y* values have been ranked, you can compute the Spearman correlation by using either the special formula or the Pearson formula.

3. $r_S = 0.80$

15.6 INTRODUCTION TO REGRESSION

Earlier in this chapter, we introduced the Pearson correlation as a technique for describing and measuring the linear relationship between two variables. Figure 15.14 presents hypothetical data showing the relationship between SAT scores and college grade-point average (GPA). Note that the figure shows a good, but not perfect, positive relationship. Also note that we have drawn a line through the middle of the data points. This line serves several purposes:

1. The line makes the relationship between SAT and GPA easier to see.

2. The line identifies the center, or *central tendency*, of the relationship, just as the mean describes central tendency for a set of scores. Thus, the line provides a simplified description of the relationship. For example, if the data points were removed, the straight line would still give a general picture of the relationship between SAT and GPA.

FIGURE 15.15

Relationship between total cost and number of hours playing tennis. The tennis club charges a $25 membership fee plus $5 per hour. The relationship is described by a linear equation:

total cost = $5(number of hours) + $25

$$Y = bX + a$$

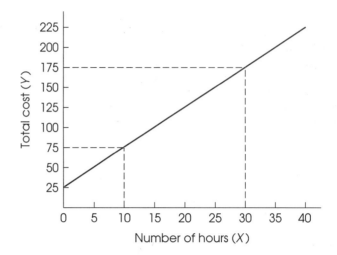

3. Finally, the line can be used for prediction. The line establishes a precise relationship between each X value (SAT score) and a corresponding Y value (GPA). For example, an SAT score of 620 corresponds to a GPA of 3.25 (see Figure 15.14). Thus, the college admissions officers could use the straight-line relationship to predict that a student entering college with an SAT score of 620 should achieve a college GPA of approximately 3.25.

Our goal in this section is to develop a procedure that identifies and defines the straight line that provides the best fit for any specific set of data. You should realize that this straight line does not have to be drawn on a graph; it can be presented in a simple equation. Thus, our goal is to find the equation for the line that best describes the relationship for a set of X and Y data.

LINEAR EQUATIONS   In general, a *linear relationship* between variables X and Y can be expressed by the equation $Y = bX + a$, where b and a are fixed constants.

For example, a local tennis club charges a fee of $5 per hour plus an annual membership fee of $25. With this information, the total cost of playing tennis can be computed using a *linear equation* that describes the relationship between the total cost (Y) and the number of hours (X):

$$Y = 5X + 25$$

Note that a positive slope means that Y increases when X increases, and a negative slope indicates that Y decreases when X increases.

In the general linear equation, the value of b is called the *slope*. The slope determines how much the Y variable will change when X is increased by one point. For the tennis club example, the slope is $b = \$5$ and indicates that your total cost will increase by $5 for each hour you play. The value of a in the general equation is called the *Y-intercept* because it determines the value of Y when $X = 0$. (On a graph, the a value identifies the point where the line intercepts the Y-axis.) For the tennis club example, $a = \$25$; there is a $25 charge even if you never play tennis.

Figure 15.15 shows the general relationship between cost and number of hours for the tennis club example. Notice that the relationship results in a straight line. To obtain this graph, we picked any two values of X and then used the equation to compute the corresponding values for Y. For example,

$$\text{when } X = 10:\qquad\qquad \text{when } X = 30:$$

$$
\begin{aligned}
Y &= bX + a & Y &= bX + a \\
&= \$5(10) + \$25 & &= \$5(30) + \$25 \\
&= \$50 + \$25 & &= \$150 + \$25 \\
&= \$75 & &= \$175
\end{aligned}
$$

When drawing a graph of a linear equation, it is wise to compute and plot at least three points to be certain you have not made a mistake.

Next, these two points are plotted on the graph: one point at $X = 10$ and $Y = 75$, the other point at $X = 30$ and $Y = 175$. Because two points completely determine a straight line, we simply drew the line so that it passed through these two points.

Because a straight line can be extremely useful for describing a relationship between two variables, a statistical technique has been developed that provides a standardized method for determining the best-fitting straight line for any set of data. The statistical procedure is regression, and the resulting straight line is called the regression line.

DEFINITION

The statistical technique for finding the best-fitting straight line for a set of data is called *regression,* and the resulting straight line is called the *regression line.*

The goal for regression is to find the best-fitting straight line for a set of data. To accomplish this goal, however, it is first necessary to define precisely what is meant by "best fit." For any particular set of data, it is possible to draw lots of different straight lines that all appear to pass through the center of the data points. Each of these lines can be defined by a linear equation of the form

$$Y = bX + a$$

where b and a are constants that determine the slope and Y-intercept of the line, respectively. Each individual line has its own unique values for b and a. The problem is to find the specific line that provides the best fit to the actual data points.

LEARNING CHECK

1. Identify the slope and Y-intercept for the following linear equation:

$$Y = -3X + 7$$

2. Use the linear equation $Y = 2X - 7$ to determine the value of Y for each of the following values of X: 1, 3, 5, 10.

3. If the slope constant (b) in a linear equation is positive, then a graph of the equation will be a line tilted from lower left to upper right. (True or false?)

ANSWERS

1. Slope $= -3$ and Y-intercept $= +7$.

2.

X	Y
1	-5
3	-1
5	3
10	13

3. True. A positive slope indicates that Y increases (goes up in the graph) when X increases (goes to the right in the graph).

FIGURE 15.16

The distance between the actual data point (Y) and the predicted point on the line ($\hat{Y}$) is defined as $Y - \hat{Y}$. The goal of regression is to find the equation for the line that minimizes these distances.

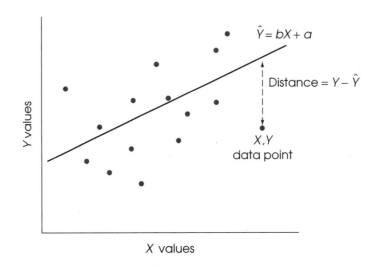

$\hat{Y} = bX + a$

Distance $= Y - \hat{Y}$

X, Y
data point

Y values

X values

THE LEAST-SQUARES SOLUTION

To determine how well a line fits the data points, the first step is to define mathematically the distance between the line and each data point. For every X value in the data, the linear equation will determine a Y value on the line. This value is the predicted Y and is called $\hat{Y}$ ("Y hat"). The distance between this predicted value and the actual Y value in the data is determined by

$$\text{distance} = Y - \hat{Y}$$

Notice that we simply are measuring the vertical distance between the actual data point (Y) and the predicted point on the line. This distance measures the error between the line and the actual data (see Figure 15.16).

Because some of these distances will be positive and some will be negative, the next step is to square each distance in order to obtain a uniformly positive measure of error. Finally, to determine the total error between the line and the data, we sum the squared errors for all of the data points. The result is a measure of overall squared error between the line and the data:

$$\text{total squared error} = \Sigma(Y - \hat{Y})^2$$

Now we can define the *best-fitting* line as the one that has the smallest total squared error. For obvious reasons, the resulting line is commonly called the *least-squared-error* solution.

In symbols, we are looking for a linear equation of the form

$$\hat{Y} = bX + a$$

For each value of X in the data, this equation will determine the point on the line ($\hat{Y}$) *that gives the best prediction of Y*. The problem is to find the specific values for a and b that will make this the best-fitting line.

The calculations that are needed to find this equation require calculus and some sophisticated algebra, so we will not present the details of the solution. The results, however, are relatively straightforward, and the solutions for b and a are as follows:

A commonly used alternative formula for the slope is

$$b = r\frac{s_Y}{s_X}$$

where s_X and s_Y are the standard deviations for X and Y, respectively.

$$b = \frac{SP}{SS_X}$$

(15.6)

where SP is the sum of products and SS_X is the sum of squares for the X scores.

$$a = \overline{Y} - b\overline{X}$$ (15.7)

Note that these two formulas determine the linear equation that provides the best prediction of Y values. This equation is called the regression equation for Y.

DEFINITION   The *regression equation for Y* is the linear equation

$$\hat{Y} = bX + a$$

where the constants b and a are determined by equations 15.6 and 15.7, respectively. This equation results in the least squared error between the data points and the line.

You should notice that the values of SS and SP are needed in the formulas for b and a, just as they are needed to compute the Pearson correlation. An example demonstrating the calculation and use of this best-fitting line is presented now.

EXAMPLE 15.11   The following table presents X and Y scores for a sample of $n = 5$ individuals. These data will be used to demonstrate the procedure for determining the linear regression equation for predicting Y values.

X	Y	$X - \overline{X}$	$Y - \overline{Y}$	$(X - \overline{X})(Y - \overline{Y})$	$(X - \overline{X})^2$
7	11	2	5	10	4
4	3	-1	-3	3	1
6	5	1	-1	-1	1
3	4	-2	-2	4	4
5	7	0	1	0	0
				$16 = SP$	$10 = SS_X$

    For these data, $\Sigma X = 25$, so $\overline{X} = 5$. Also, $\Sigma Y = 30$, so $\overline{Y} = 6$. These means have been used to compute the deviation scores for each X and Y value. The final two columns show the products of the deviation scores and the squared deviations for X. Based on these values,

$$SP = \Sigma(X - \overline{X})(Y - \overline{Y}) = 16$$
$$SS_X = \Sigma(X - \overline{X})^2 = 10$$

Our goal is to find the values for b and a in the linear equation so that we obtain the best-fitting straight line for these data.

By using formulas 15.6 and 15.7, the solutions for b and a are

$$b = \frac{SP}{SS_X} = \frac{16}{10} = 1.6$$

$$a = \overline{Y} - b\overline{X}$$
$$= 6 - 1.6(5)$$
$$= 6 - 8$$
$$= -2$$

FIGURE 15.17

The scatterplot for the data in Example 15.11 is shown with the best-fitting straight line. The predicted Y values ($\hat{Y}$) are on the regression line. Unless the correlation is perfect ($+1.00$ or -1.00), there will be some error between the actual Y values and the predicted Y values. The larger the correlation is, the smaller the error will be.

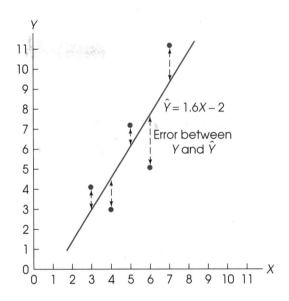

The resulting regression equation is

$$\hat{Y} = 1.6X - 2$$

The original data and the regression line are shown in Figure 15.17.

As we noted at the beginning of this section, one common use of regression equations is for prediction. For any given value of X, we can use the equation to compute a predicted value for Y. For the equation from Example 15.11, an individual with an X score of $X = 5$ would be predicted to have a Y score of

$$\hat{Y} = 1.6X - 2$$
$$= 1.6(5) - 2$$
$$= 8 - 2$$
$$= 6$$

Although regression equations can be used for prediction, there are a few cautions that should be considered whenever you are interpreting the predicted values:

1. The predicted value is not perfect (unless $r = +1.00$ or -1.00). If you examine Figure 15.17, it should be clear that the data points do not fit perfectly on the line. In general, there will be some error between the predicted Y values (on the line) and the actual data. Although the amount of error will vary from point to point, on average the errors will be directly related to the magnitude of the correlation. With a correlation near 1.00 (or -1.00), the data points will generally be close to the line (small error), but as the correlation gets nearer to zero, the magnitude of the error will increase.

2. The regression equation should not be used to make predictions for X values that fall outside the range of values covered by the original data. For Example 15.11, the X values ranged from $X = 3$ to $X = 7$, and the regres-

sion equation was calculated as the best-fitting line within this range. Because you have no information about the *X-Y* relationship outside this range, the equation should not be used to predict *Y* for any *X* value lower than 3 or greater than 7.

LEARNING CHECK

1. Sketch a scatterplot for the following data, that is, a graph showing the *X*, *Y* data points:

X	Y
1	4
3	9
5	8

a. Find the regression equation for predicting *Y* and *X*. Draw this line on your graph. Does it look like the best-fitting line?

b. Use the regression equation to find the predicted *Y* value corresponding to each *X* in the data.

ANSWERS

1. a. $SS_X = 8$, $SP = 8$, $b = 1$, $a = 4$.
The equation is

$$\hat{Y} = X + 4$$

b. The predicted *Y* values are 5, 7, and 9.

SUMMARY

1. A correlation measures the relationship between two variables, *X* and *Y*. The relationship is described by three characteristics:

   a. *Direction.* A relationship can be either positive or negative. A positive relationship means that *X* and *Y* vary in the same direction. A negative relationship means that *X* and *Y* vary in opposite directions. The sign of the correlation (+ or −) specifies the direction.

   b. *Form.* The most common form for a relationship is a straight line. However, special correlations exist for measuring other forms. The form is specified by the type of correlation used. For example, the Pearson correlation measures linear form.

   c. *Degree.* The magnitude of the correlation measures the degree to which the data points fit the specified form. A correlation of 1.00 indicates a perfect fit, and a correlation of 0 indicates no degree of fit.

2. The most commonly used correlation is the Pearson correlation, which measures the degree of linear relationship. The Pearson correlation is identified by the letter *r* and is computed by

$$r = \frac{SP}{\sqrt{SS_X SS_Y}}$$

In this formula, *SP* is the sum of products of deviations and can be calculated with either a definitional formula or a computational formula:

definitional formula: $SP = \Sigma(X - \bar{X})(Y - \bar{Y})$

computational formula: $SP = \Sigma XY - \dfrac{\Sigma X \Sigma Y}{n}$

3. The Pearson correlation and *z*-scores are closely related because both are concerned with the location of individuals within a distribution. When *X* and *Y* scores are transformed into *z*-scores, the Pearson correlation can be computed by

$$r = \frac{\Sigma z_X z_Y}{n}$$

4. A correlation between two variables should not be interpreted as implying a causal relationship. Simply because *X*

and Y are related does not mean that X causes Y or that Y causes X.

5. When the X or the Y values used to compute a correlation are limited to a relatively small portion of the potential range, you should exercise caution in generalizing the value of the correlation. Specifically, a limited range of values can either obscure a strong relationship or exaggerate a poor relationship. Also, one or two extreme data points, called *outriders*, can distort the value of a correlation.

6. To evaluate the strength of a relationship, you should square the value of the correlation. The resulting value, r^2, is called the *coefficient of determination* because it measures the portion of the variability in one variable that can be predicted using the relationship with the second variable.

7. The Spearman correlation (r_S) measures the consistency of direction in the relationship between X and Y, that is, the degree to which the relationship is one-directional, or monotonic. The Spearman correlation is computed by a two-stage process:
   a. Rank the X scores and the Y scores.
   b. Compute the Pearson correlation using the ranks.

Note: After the X and Y values are ranked, you may use a special formula to determine the Spearman correlation:

$$r_S = 1 - \frac{6\Sigma D^2}{n(n^2 - 1)}$$

where D is the difference between the X rank and the Y rank for each individual. The formula is accurate only when there are no tied scores in the data.

8. When there is a general linear relationship between two variables, X and Y, it is possible to construct a linear equation that allows you to predict the Y value corresponding to any known value of X:

$$\text{predicted } Y \text{ value} = \hat{Y} = bX + a$$

The technique for determining this equation is called regression. By using a *least-squares* method to minimize the error between the predicted Y values and the actual Y values, the best-fitting line is achieved when the linear equation has

$$b = \frac{SP}{SS_X} \quad \text{and} \quad a = \bar{Y} - b\bar{X}$$

KEY TERMS

correlation	sum of products (SP)	monotonic relationship	regression
positive correlation	restricted range	Spearman correlation	regression line
negative correlation	coefficient of determination	linear equation	predicted Y
perfect correlation	linear relationship	slope	least-squared error
Pearson correlation	ranks	Y-intercept	regression equation for Y

FOCUS ON PROBLEM SOLVING

1. A correlation always has a value from $+1.00$ to -1.00. If you obtain a correlation outside this range, then you have made a computational error.

2. When interpreting a correlation, do not confuse the sign ($+$ or $-$) with its numerical value. The sign and the numerical value must be considered separately. Remember, the sign indicates the direction of the relationship between X and Y. On the other hand, the numerical value reflects the strength of the relationship or how well the points approximate a linear (straight-line) relationship. Therefore, a correlation of -0.90 is just as strong as a correlation of $+.90$. The signs tell us that the first correlation is an inverse relationship.

3. Before you begin to calculate a correlation, you should sketch a scatterplot of the data and make an estimate of the correlation. (Is it positive or negative? Is it near 1 or near 0?) After computing the correlation, compare your final answer with your original estimate.

4. The definitional formula for the sum of products (SP) should be used only when you have a small set (n) of scores and the means for X and Y are both whole numbers. Otherwise, the computational formula will produce quicker, easier, and more accurate results.

5. For computing a correlation, n is the number of individuals (and therefore the number of *pairs* of X and Y values).

6. When using the special formula for the Spearman correlation, remember that the fraction is computed separately and then subtracted from 1. Students often include the 1 as a part of the numerator, or they get so absorbed in computing the fractional part of the equation that they forget to subtract it from 1. Be careful using this formula.

7. When computing a Spearman correlation, be sure that both X and Y values have been ranked. Sometimes the data will consist of one variable already ranked with the other variable on an interval or a ratio scale. If one variable is ranked, do not forget to rank the other. When interpreting a Spearman correlation, remember that it measures how monotonic (consistent) the relationship is between X and Y.

8. To draw a graph from a linear equation, choose any three values for X, put each value in the equation, and calculate the corresponding values for Y. Then plot the three X, Y points on the graph. It is a good idea to use $X = 0$ for one of the three values because this will give you the Y-intercept. You can get a quick idea of what the graph should look like if you know the Y-intercept and the slope. Remember, the Y-intercept is the point where the line crosses the Y-axis, and the slope identifies the tilt of the line. For example, suppose the Y-intercept is 5 and the slope is -3. The line would pass through the point $(0, 5)$, and its slope indicates that the Y value goes down 3 points each time X increases by 1.

9. Rather than memorizing the formula for the Y-intercept in the regression equation, simply remember that the graphed line of the regression equation always goes through the point $\overline{X}, \overline{Y}$. Therefore, if you plug the mean value for X ($\overline{X}$) into the regression equation, the result equals the mean value for Y ($\overline{Y}$).

$$\overline{Y} = b\overline{X} + a$$

If you simply solve this equation for a, you get the formula for the Y-intercept.

$$a = \overline{Y} - b\overline{X}$$

DEMONSTRATION 15.1

CORRELATION AND REGRESSION

For the following data, calculate the Pearson correlation and find the regression equation:

Person	X	Y
A	0	4
B	2	1
C	8	10
D	6	9
E	4	6

FIGURE 15.18

The scatterplot for the data of Demonstration 15.1. An envelope is drawn around the points to estimate the magnitude of the correlation. A line is drawn through the middle of the envelope to roughly estimate the Y-intercept for the regression equation.

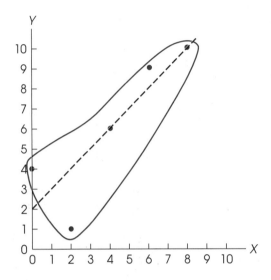

STEP 1   Sketch a scatterplot.

We have constructed a scatterplot for the data (Figure 15.18) and placed an envelope around the data points to make a preliminary estimate of the correlation. Note that the envelope is narrow and elongated. This indicates that the correlation is large—perhaps 0.80 to 0.90. Also, the correlation is positive because increases in X are generally accompanied by increases in Y.

We can sketch a straight line through the middle of the envelope and data points. Now we can roughly approximate the slope and Y-intercept of the best-fit line. This is only an educated guess, but it will tell us what values are reasonable when we actually compute the regression line. The line has a positive slope (as X increases, Y increases), and it intersects the Y-axis in the vicinity of $+2$.

STEP 2   Obtain the values for SS and SP.

To compute the Pearson correlation, we must find the values for SS_X, SS_Y, and SP. These values are needed for the regression equation as well. The following table illustrates these calculations with the computational formulas for SS and SP:

X	Y	X^2	Y^2	XY
0	4	0	16	0
2	1	4	1	2
8	10	64	100	80
6	9	36	81	54
4	6	16	36	24
$\Sigma X = 20$	$\Sigma Y = 30$	$\Sigma X^2 = 120$	$\Sigma Y^2 = 234$	$\Sigma XY = 160$

For SS_X, we obtain

$$SS_X = \Sigma X^2 - \frac{(\Sigma X)^2}{n} = 120 - \frac{20^2}{5} = 120 - \frac{400}{5} = 120 - 80$$

$$= 40$$

For Y, the sum of squares is

$$SS_Y = \Sigma Y^2 - \frac{(\Sigma Y)^2}{n} = 234 - \frac{30^2}{5} = 234 - \frac{900}{5} = 234 - 180$$

$$= 54$$

The sum of products equals

$$SP = \Sigma XY - \frac{\Sigma X \Sigma Y}{n} = 160 - \frac{20(30)}{5} = 160 - \frac{600}{5} = 160 - 120$$

$$= 40$$

STEP 3 Compute the Pearson correlation.
For these data, the Pearson correlation is

$$r = \frac{SP}{\sqrt{SS_X SS_Y}} = \frac{40}{\sqrt{40(54)}} = \frac{40}{\sqrt{2160}} = \frac{40}{46.48}$$

$$= 0.861$$

In step 1, our preliminary estimate for the correlation was between $+0.80$ and $+0.90$. The calculated correlation is consistent with this estimate.

STEP 4 Compute the values for the regression equation.
The general form of the regression equation is

$$\hat{Y} = bX + a$$

We will need to compute the values for the slope (b) of the line and the Y-intercept (a). For slope, we obtain

$$b = \frac{SP}{SS_X} = \frac{40}{40} = +1$$

The formula for the Y-intercept is

$$a = \bar{Y} - b\bar{X}$$

Thus, we will need the values for the sample means. For these data, the sample means are

$$\bar{X} = \frac{\Sigma X}{n} = \frac{20}{5} = 4$$

$$\bar{Y} = \frac{\Sigma Y}{n} = \frac{30}{5} = 6$$

Now we can compute the Y-intercept.

$$a = 6 - 1(4) = 6 - 4 = 2$$

Finally, the regression equation is

$$\hat{Y} = bX + a$$
$$= 1X + 2$$

or

$$\hat{Y} = X + 2$$

DEMONSTRATION 15.2

THE SPEARMAN CORRELATION

The following data will be used to demonstrate the calculation of the Spearman correlation. Both X and Y values are measurements on interval scales.

X	Y
5	12
7	18
2	9
15	14
10	13

STEP 1 Rank the X and Y values.
Remember, the X values and Y values are ranked separately.

X score	Y score	X rank	Y rank
5	12	2	2
7	18	3	5
2	9	1	1
15	14	5	4
10	13	4	3

STEP 2 Use the special Spearman formula to compute the correlation.
The special Spearman formula requires that you first find the difference (D) between the X rank and the Y rank for each individual and then square the differences and find the sum of the squared differences.

X rank	Y rank	D	D^2
2	2	0	0
3	5	2	4
1	1	0	0
5	4	1	1
4	3	1	1
			$6 = \Sigma D^2$

Using this value in the Spearman formula, we obtain

$$r_S = 1 - \frac{6\Sigma D^2}{n(n^2 - 1)}$$

$$= 1 - \frac{6(6)}{5(24)}$$

$$= 1 - \frac{36}{120}$$

$$= 1 - 0.30$$

$$= 0.70$$

There is a positive relationship between X and Y for these data. The Spearman correlation is fairly high, which indicates a very consistent positive relationship.

PROBLEMS

1. For each of the following sets of scores, calculate SP using the definitional formula and then using the computational formula:

Set 1		Set 2		Set 3	
X	Y	X	Y	X	Y
1	3	0	7	1	5
2	6	4	3	2	0
4	4	0	5	3	1
5	7	4	1	2	6

2. For the following set of data,

Data	
X	Y
8	2
9	2
2	4
1	5
5	2

a. Sketch a graph showing the location of the five X, Y points.
b. Just looking at your graph, estimate the value of the Pearson correlation.
c. Compute the Pearson correlation for this data set.

3. For this problem, we have used the same X and Y values that appeared in Problem 2, but we have changed the X, Y pairings:

Data	
X	Y
8	4
9	5
2	2
1	2
5	2

a. Sketch a graph showing these reorganized data.
b. Estimate the Pearson correlation just by looking at your graph.
c. Compute the Pearson correlation. (*Note:* Much of the calculation for this problem was done already in Problem 2.)

If you compare the results of Problem 2 and Problem 3, you will see that the correlation measures the relationship between X and Y. These two problems use the same X and Y values, but they differ in the way X and Y are related.

4. With a very small sample, a single point can have a large influence on the magnitude of a correlation. For the following data set,

X	Y
0	1
10	3
4	1
8	2
8	3

a. Sketch a graph showing the X, Y points.
b. Estimate the value of the Pearson correlation.
c. Compute the Pearson correlation.
d. Now we will change the value of one of the points. For the first individual in the sample ($X = 0$ and $Y = 1$), change the Y value to $Y = 6$. What happens to the graph of the X, Y points? What happens to the Pearson correlation? Compute the new correlation.

5. In the following data, there are three scores (X, Y, and Z) for each of the $n = 5$ individuals:

X	Y	Z
3	5	5
4	3	2
2	4	6
1	1	3
0	2	4

a. Sketch a graph showing the relationship between X and Y. Compute the Pearson correlation between X and Y.
b. Sketch a graph showing the relationship between Y and Z. Compute the Pearson correlation between Y and Z.
c. Given the results of parts a and b, what would you predict for the correlation between X and Z?
d. Sketch a graph showing the relationship between X and Z. Compute the Pearson correlation for these data.
e. What general conclusion can you make concerning relationships among correlations? If X is related to Y and Y is related to Z, does this necessarily mean that X is related to Z?

6. a. Compute the Pearson correlation for the following set of data:

X	Y
2	8
3	10
3	7
5	6
6	7
8	4
9	2
10	3

b. Add 5 points to each X value, and compute the Pearson correlation again.

c. When you add a constant to each score, what happens to SS for X and Y? What happens to SP? What happens to the correlation between X and Y?
d. Now multiply each X in the original data by 3, and calculate the Pearson correlation once again.
e. When you multiply by a constant, what happens to SS for X and Y? What happens to SP? What happens to the correlation between X and Y?

7. A psychology instructor asked each student to report the number of hours he or she had spent preparing for the final exam. In addition, the instructor recorded the number of incorrect answers on each student's exam. These data are as follows:

Hours	Number wrong
4	8
0	6
1	3
2	2
4	5

What is the Pearson correlation between study hours and number wrong?

8. It is well known that similarity in attitudes, beliefs, and interests plays an important role in interpersonal attraction (see Byrne, 1971, for example). Thus, correlations for attitudes between married couples should be strong. Suppose a researcher developed a questionnaire that measures how liberal or conservative one's attitudes are. Low scores indicate that the person has liberal attitudes, while high scores indicate conservatism. The following hypothetical data are scores for married couples.

Couple	Wife	Husband
A	11	14
B	6	7
C	18	15
D	4	7
E	1	3
F	10	9
G	5	9
H	3	3

Compute the Pearson correlation for these data, and determine whether or not there is a significant correlation between attitudes for husbands and wives. Set alpha at .05, two tails.

9. To measure the relationship between anxiety level and test performance, a psychologist obtains a sample of $n = 6$ college students from an introductory statistics course. The students are asked to come to the laboratory 15 minutes before the final exam. In the lab, the psychologist records physiological measures of anxiety (heart rate, skin resistance, blood pressure, etc.) for each subject. In addition, the psychologist obtains the exam score for each subject. Compute the Pearson correlation for the following data:

Student	Anxiety rating	Exam score
A	5	80
B	2	88
C	7	80
D	7	79
E	4	86
F	5	85

10. A high school counselor would like to know if there is a relationship between mathematical skill and verbal skill. A sample of $n = 25$ students is selected, and the counselor records achievement test scores in mathematics and English for each student. The Pearson correlation for this sample is $r = +0.50$. Do these data provide sufficient evidence for a real relationship in the population? Test at the .05 level, two tails.

11. Researchers who measure reaction time for human subjects often observe a relationship between the reaction time scores and the number of errors that the subjects commit. This relationship is known as the *speed-accuracy trade-off*. The following data are from a reaction time study where the researcher recorded the average reaction time (milliseconds) and the total number of errors for each individual in a sample of $n = 8$ subjects.

Subject	Reaction time	Errors
A	184	10
B	213	6
C	234	2
D	197	7
E	189	13
F	221	10
G	237	4
H	192	9

a. Compute the Pearson correlation for the data.
b. In words, describe the speed-accuracy trade-off.

12. Studies have suggested that the stress of major life changes is related to subsequent physical illness. Holmes and Rahe (1967) devised the Social Readjustment Rating Scale (SRRS) to measure the amount of stressful change in one's life. Each event is assigned a point value, which measures its severity. For example, at the top of the list, death of a spouse is assigned 100 life change units (LCU). Divorce is 73 LCUs, retirement is 45, change of career is 36, the beginning or end of school is 26, and so on. The more life change units one has accumulated in the past year, the more likely he or she is to have an illness. The following hypothetical data show the number of life change units during the past year for a group of people. The data also show the number of doctor visits each person made during the year.

Person	Total LCUs	Number of visits
A	73	6
B	15	1
C	204	9
D	31	2
E	63	5
F	278	11
G	15	5
H	36	3
I	110	8
J	116	4
K	162	7

a. For these data, compute the Pearson correlation.
b. Compute the Spearman correlation.

13. Rank the following scores, and compute the Spearman correlation between X and Y:

X	Y
7	19
2	4
11	34
15	28
32	104

14. While grading essay exams, a professor noticed huge differences in the writing skills of the students. To investigate a possible cause for the differences, the professor asked the students how many English courses they had completed. The number of courses completed and the

professor's ranking of the essay are reported for each student in the following table.

Quality of essay (professor's ranking)	Number of English courses
1 (best)	7
2	4
3	1
4	3
5	1
6	1
7	0
8	2

Compute the Spearman correlation between writing ability and English background. (*Note:* You must convert the data before computing the correlation.)

15. A common concern for students (and teachers) is the assignment of grades for essays or term papers. Because there are no absolute right or wrong answers, these grades must be based on a judgment of quality. To demonstrate that these judgments actually are reliable, an English instructor asked a colleague to rank-order a set of term papers. The ranks and the instructor's grades for these papers are as follows:

Rank	Grade
1	A
2	B
3	A
4	B
5	B
6	C
7	D
8	C
9	C
10	D
11	E

a. Calculate the Spearman correlation for these data. (*Note:* You must convert the letter grades to ranks.)
b. Based on this correlation, does it appear that there is reasonable agreement between these two instructors in their judgment of the papers?

16. In the following data, X and Y are related by the equation $Y = X^2$:

X	Y
0	0
1	1
2	4
3	9
4	16
5	25

a. Sketch a graph showing the relationship between X and Y. Describe the relationship shown in your graph.
b. Compute the Spearman correlation for these data.

17. Sketch a graph showing the linear equation $Y = 3X - 2$.

18. Two major companies supply laboratory animals for psychologists. Company A sells laboratory rats for $6 each and charges a $10 fee for delivery. Company B sells rats for only $5 each but has a $20 delivery charge. In each case, the delivery fee is a one-time charge and does not depend on the number of rats in the order.
a. For each company, what is the linear equation that defines the total cost (Y) as a function of the number of rats (X)? Each equation should be of the form

$$Y = bX + a$$

b. What would the total cost be for an order of 10 rats from company A? From company B?
c. If you were buying 20 rats, which company would give you the better deal?

19. For the following set of data, find the linear regression equation for predicting Y from X:

X	Y
0	9
2	9
4	7
6	3

20. a. Find the regression equation for the following data:

X	Y
1	2
4	7
3	5
2	1
5	14
3	7

b. Compute the predicted Y value for each X in the data.

21. Find the regression equation for the following data:

X	Y
3	12
0	8
4	18
2	12
1	8

22. In a study examining the relationship between temperature and violent aggression, Anderson and Anderson (1984) recorded the temperature and the number of murders and rapes per day in a major southwestern city. Data similar to the results of the research study are as follows:

Temperature (degrees F)	Number of murders and rapes
85	6
78	5
97	9
92	9
75	6
62	5
83	6
90	7
88	7
72	5

a. Compute the correlation for these data.
b. What does this correlation indicate?
c. Find the regression equation for these data.

23. You probably have read for years about the relationship between years of education and salary potential. The following hypothetical data represent a sample of $n = 10$ men who have been employed for five years. For each person, we report the total number of years of higher education (high school plus college) and current annual salary.

Salary (in $1000s)	Years of higher education
21.4	4
18.7	4
17.5	2
32.0	8
12.6	0
25.3	5
35.5	10
17.3	4
33.8	12
14.0	0

a. Find the regression equation for predicting salary from education.
b. How would you interpret the slope constant (b) in the regression equation?
c. How would you interpret the Y-intercept (a) in the equation?

CHAPTER 16

THE CHI-SQUARE STATISTIC: TESTS FOR GOODNESS OF FIT AND INDEPENDENCE

CONTENTS

16.1 PARAMETRIC AND NONPARAMETRIC STATISTICAL TESTS

All the statistical tests we have examined thus far are designed to test hypotheses about specific population parameters. For example, we used t tests to assess hypotheses about μ and later about $\mu_1 - \mu_2$. In addition, these tests typically make assumptions about the shape of the population distribution and about other population parameters. Recall that for analysis of variance the population distributions are assumed to be normal and homogeneity of variance is required. Because these tests all concern parameters and require assumptions about parameters, they are called *parametric tests*.

Another general characteristic of parametric tests is that they require a numerical score for each individual in the sample. The scores then are added, squared, averaged, and otherwise manipulated using basic arithmetic. In terms of measurement scales, parametric tests require data from an interval or a ratio scale (see Chapter 1).

Often researchers are confronted with experimental situations that do not conform to the requirements of parametric tests. In these situations, it may not be appropriate to use a parametric test. Remember, when the assumptions of a test are violated, the test may lead to an erroneous interpretation of the data. Fortunately, there are several hypothesis-testing techniques that provide alternatives to parametric tests. These alternatives are called *nonparametric tests*.

In this chapter, we introduce some commonly used nonparametric tests. You should notice that these nonparametric tests usually do not state hypotheses in terms of a specific parameter and that they make few (if any) assumptions about the population distribution. For the latter reason, nonparametric tests sometimes are called *distribution-free tests*. Another distinction is that nonparametric tests are well suited for data that are measured on nominal or ordinal scales. Finally, you should be warned that nonparametric tests generally are not as sensitive as parametric tests; nonparametric tests are more likely to fail in detecting a real difference between two treatments. Therefore, whenever the experimental data give you a choice between a parametric and a nonparametric test, you always should choose the parametric alternative.

16.2 THE CHI-SQUARE TEST FOR GOODNESS OF FIT

Parameters such as the mean and the standard deviation are the most common way to describe a population, but there are situations where a researcher has questions about the proportions or relative frequencies for a distribution. For example,

How does the number of women lawyers compare with the number of men in the profession?

Of the three leading brands of soft drinks, which is preferred by most Americans? Which brands are second and third, and how big are the differences in popularity among the three?

To what extent are different ethnic groups represented in the population of your city?

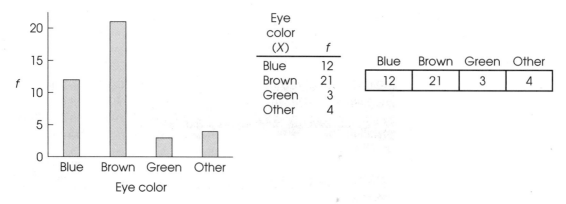

FIGURE 16.1

Distribution of eye colors for a sample of $n = 40$ individuals. The same frequency distribution is shown as a bar graph, as a table, and with the frequencies written in a series of boxes.

The name of the test comes from the Greek letter χ (chi, pronounced "kye"), which is used to identify the test statistic.

Notice that each of the preceding examples asks a question about proportions—in other words, these are all questions about relative frequencies. The chi-square test for goodness of fit is specifically designed to answer this type of question. In general terms, this chi-square test is a hypothesis-testing procedure that uses the proportions obtained for a sample distribution to test hypotheses about the corresponding proportions in the population distribution.

DEFINITION

The *chi-square test for goodness of fit* uses sample data to test hypotheses about the shape or proportions of a population distribution. The test determines how well the obtained sample proportions fit the population proportions specified by the null hypothesis.

You should recall from Chapter 2 that a frequency distribution is defined as a tabulation of the number of individuals located in each category of the scale of measurement. In a frequency distribution graph, the categories that make up the scale of measurement are listed on the X-axis. In a frequency distribution table, the categories are listed in the first column. With chi-square tests, however, it is customary to present the scale of measurement as a series of boxes, with each box corresponding to a separate category on the scale. The frequency corresponding to each category is simply presented as a number written inside the box. Figure 16.1 shows how a distribution of eye colors for a set of $n = 40$ students can be presented as a graph, a table, or a series of boxes. Notice that the scale of measurement for this example consists of four categories of eye color (blue, brown, green, other).

THE NULL HYPOTHESIS FOR THE GOODNESS-OF-FIT TEST

For the chi-square test of goodness of fit, the null hypothesis specifies the proportion (or percentage) of the population in each category. For example, a hypothesis might state that 90% of all lawyers are men and only 10% are women. The simplest way of presenting this hypothesis is to put the hypothesized proportions in the series of boxes representing the scale of measurement:

	Men	Women
H_0:	90%	10%

Although it is conceivable that a researcher could choose any proportions for the null hypothesis, there usually is some well-defined rationale for stating a null hypothesis. Generally H_0 will fall into one of the following categories:

1. No Preference. The null hypothesis often states that there is no preference among the different categories. In this case, H_0 states that the population is divided equally among the categories. For example, a hypothesis stating that there is no preference among the three leading brands of soft drinks would specify a population distribution as follows:

	Brand X	Brand Y	Brand Z
H_0:	$\frac{1}{3}$	$\frac{1}{3}$	$\frac{1}{3}$

(Preferences in the population are equally divided among the three soft drinks.)

The no-preference hypothesis is used in situations where a researcher wants to determine whether there are any preferences among categories or any variations from one category to another.

2. No Difference from a Comparison Population. The null hypothesis can state that the frequency distribution for one population is not different from the distribution that is known to exist for another population. For example, suppose it is known that 60% of Americans favor the president's foreign policy and 40% are in opposition. A researcher might wonder if this same pattern of attitudes exists among Europeans. The null hypothesis would state that there is no difference between the two populations and would specify that the Europeans are distributed as follows:

	Favor	Oppose
H_0:	60%	40%

(Proportions for the European population are not different from the American proportions.)

The no-difference hypothesis is used in situations where a specific population distribution is already known. You may have a known distribution from an earlier time, and the question is whether there has been any change in the distribution. Or you may have a known distribution for one population, and the question is whether a second population has the same distribution.

Because the null hypothesis for the goodness-of-fit test specifies an exact distribution for the population, the alternative hypothesis (H_1) simply states that the population distribution has a different shape from that specified in H_0. If the null hypothesis states that the population is equally divided among three categories, the alternative hypothesis will say that the population is not divided equally.

THE DATA FOR THE GOODNESS-OF-FIT TEST

The data for a chi-square test are remarkably simple to obtain. There is no need to calculate a sample mean or SS; you just select a sample of n individuals and count how many are in each category. The resulting values are called *observed frequencies*. The symbol for observed frequency is f_o. For example, the following data represent observed frequencies for a sample of $n = 40$ subjects. Each person was given a personality questionnaire and classified into one of three personality categories: A, B, or C.

Category A	Category B	Category C	
15	19	6	$n = 40$

Notice that each individual in the sample is classified into one and only one of the categories. Thus, the frequencies in this example represent three completely separate groups of individuals: 15 who were classified as category A, 19 classified as B, and 6 classified as C. Also note that the observed frequencies add up to the total sample size: $\Sigma f_o = n$.

DEFINITION   The *observed frequency* is the number of individuals from the sample who are classified in a particular category. Each individual is counted in one and only one category.

EXPECTED FREQUENCIES   The general goal of the chi-square test for goodness of fit is to compare the data (the observed frequencies) with the null hypothesis. The problem is to determine how well the data fit the distribution specified in H_0—hence the name *goodness of fit*.

The first step in the chi-square test is to determine how the sample distribution would look if the sample was in perfect agreement with the proportions stated in the null hypothesis. Suppose, for example, the null hypothesis states that the population is distributed in three categories with the following proportions:

Category A	Category B	Category C
25%	50%	25%

If this hypothesis is correct, how would you expect a random sample of $n = 40$ individuals to be distributed among the three categories? It should be clear that your best strategy is to predict that 25% of the sample would be in category A, 50% would be in category B, and 25% would be in category C. To find the exact frequency expected for each category, multiply the sample size (n) by the proportion (or percentage) from the null hypothesis. For this example, you would expect

25% of 40 = 0.25(40) = 10 individuals in category A

50% of 40 = 0.50(40) = 20 individuals in category B

25% of 40 = 0.25(40) = 10 individuals in category C

The frequency values predicted from the null hypothesis are called *expected frequencies*. The symbol for expected frequency is f_e, and the expected frequency for each category is computed by

$$\text{expected frequency} = f_e = pn \qquad\qquad (16.1)$$

where p is the proportion stated in the null hypothesis and n is the sample size.

DEFINITION   The *expected frequency* for each category is the frequency value that is predicted from the null hypothesis and the sample size (n). The expected frequencies define an ideal, *hypothetical* sample distribution that would be obtained if the sample proportions were in perfect agreement with the proportions specified in the null hypothesis.

Note that the no-preference null hypothesis will always produce equal f_e values for all categories because the proportions (p) are the same for all categories. That is, the

16.1   THE CHI-SQUARE FORMULA

WE HAVE seen that the chi-square formula compares observed frequencies to expected frequencies in order to assess how well the sample data match the hypothesized data. This function of the chi-square statistic is easy to spot in the numerator of the equation, $(f_o - f_e)^2$. The difference between the observed and the expected frequencies is found first. The greater this difference is, the more discrepancy there is between what is observed and what is expected. The difference is then squared to remove the negative signs (large discrepancies may have negative signs as well as positive signs). The summation sign in front of the equation indicates that we must examine the amount of discrepancy for every category. Why, then, must we divide the squared differences by f_e for each category before we sum the category values? The answer to this question is that the obtained discrepancy between f_o and f_e is viewed as *relatively* large or *relatively* small depending on the size of the expected frequency. This point is demonstrated in the following analogy.

Suppose you were going to throw a party and you *expected* 1000 people to show up. However, at the party you counted the number of guests and *observed* that 1040 actually showed up. Forty more guests than expected are no major problem when all along you were planning for 1000. There will still probably be enough beer and potato chips for everyone. On the other hand, suppose you had a party and you expected 10 people to attend but instead 50 actually showed up. Forty more guests in this case spell big trouble. How "significant" the discrepancy is depends in part on what you were originally expecting. With very large expected frequencies, allowances are made for more error between f_o and f_e. This is accomplished in the chi-square formula by dividing the squared discrepancy for each category, $(f_o - f_e)^2$, by its expected frequency.

probability of observing a response in any category is the same for all categories. On the other hand, the no-difference null hypothesis typically will not produce equal values for the expected frequencies because the probability of observing a response in any category is not the same across categories.

THE CHI-SQUARE STATISTIC

The general purpose of any hypothesis test is to determine whether the sample data support or refute a hypothesis about the population. In the chi-square test for goodness of fit, the sample is expressed as a set of observed frequencies (f_o values), and the null hypothesis is used to generate a set of expected frequencies (f_e values). The *chi-square statistic* simply measures how well the data (f_o) fit the hypothesis (f_e). The symbol for the chi-square statistic is χ^2. The formula for the chi-square statistic is

$$\text{chi-square} = \chi^2 = \Sigma \frac{(f_o - f_e)^2}{f_e} \tag{16.2}$$

As the formula indicates, the value of chi-square is computed by the following steps:

1. Find the difference between f_o (the data) and f_e (the hypothesis) for each category.
2. Square the difference. This ensures that all values are positive.
3. Next, divide the squared difference by f_e. A justification for this step is given in Box 16.1.
4. Finally, sum the values from all the categories.

FIGURE 16.2

Chi-square distributions are positively skewed. The critical region is placed in the extreme tail, which reflects large chi-square values.

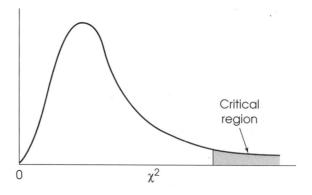

Critical region

0 χ^2

THE CHI-SQUARE DISTRIBUTION AND DEGREES OF FREEDOM

It should be clear from the chi-square formula that the value of chi-square is measuring the discrepancy between the observed frequencies (data) and the expected frequencies (H_0). When there are large differences between f_o and f_e, the value of chi-square will be large, and we will conclude that the data do not fit the hypothesis. Thus, a large value for chi-square will lead us to reject H_0. On the other hand, when the observed frequencies are very close to the expected frequencies, chi-square will be small, and we will conclude that there is a very good fit between the data and the hypothesis. Thus, a small chi-square value indicates that we should fail to reject H_0. To decide whether a particular chi-square value is "large" or "small," we must refer to a *chi-square distribution*. This distribution is the set of chi-square values for all the possible random samples when H_0 is true. Much like other distributions we have examined (t distribution, F distribution), the chi-square distribution is a theoretical distribution with well-defined characteristics. Some of these characteristics are easy to infer from the chi-square formula.

1. The formula for chi-square involves adding squared values, so you can never obtain a negative value. Thus, all chi-square values are zero or larger.

2. When H_0 is true, you expect the data (f_o values) to be close to the hypothesis (f_e values). Thus, we expect chi-square values to be small when H_0 is true.

These two factors suggest that the typical chi-square distribution will be positively skewed (see Figure 16.2). Note that small values, near zero, are expected when H_0 is true and large values (in the right-hand tail) are very unlikely. Thus, unusually large values of chi-square will form the critical region for the hypothesis test.

Although the typical chi-square distribution is positively skewed, there is one other factor that plays a role in the exact shape of the chi-square distribution—the number of categories. You should recall that the chi-square formula requires that you sum values from every category. The more categories you have, the more likely it is that you will obtain a large sum for the chi-square value. On average, chi-square will be larger when you are summing over 10 categories than when you are summing over only 3 categories. As a result, there is a whole family of chi-square distributions, with the exact shape of each distribution determined by the number of categories used in the study. Technically, each specific chi-square distribution is identified by degrees of freedom (*df*), rather than the number of categories. For the goodness-of-fit test, the degrees of freedom are determined by

$$df = C - 1 \qquad (16.3)$$

16.2 A CLOSER LOOK AT DEGREES OF FREEDOM

DEGREES OF freedom for the chi-square test literally measure the number of free choices that exist when you are determining the null hypothesis or the expected frequencies. For example, when you are classifying individuals into three categories, you have exactly two free choices in stating the null hypothesis. You may select any two proportions for the first two categories, but then the third proportion is determined. If you hypothesize 25% in the first category and 50% in the second category, then the third category must be 25% in order to account for 100% of the population. In general, you are free to select proportions for all but one of the categories, but then the final proportion is determined by the fact that the entire set must total 100%. Thus, you have $C - 1$ free choices, where C is the number of categories: Degrees of freedom, df, equal $C - 1$.

The same restriction holds when you are determining the expected frequencies. Again, suppose you have three categories and a sample of $n = 40$ individuals. If you specify expected frequencies of $f_e = 10$ for the first category and $f_e = 20$ for the second category, then you must use $f_e = 10$ for the final category.

Category A	Category B	Category C	
10	20	???	$n = 40$

As before, you may distribute the sample freely among the first $C - 1$ categories, but then the final category is determined by the total number of individuals in the sample.

Caution: The df for a chi-square test is *not* related to sample size (n), as it is in most other tests.

where C is the number of categories. A brief discussion of this df formula is presented in Box 16.2. Figure 16.3 shows the general relationship between df and the shape of the chi-square distribution. Note that the typical chi-square value (the mode) gets larger as the number of categories is increased.

LOCATING THE CRITICAL REGION FOR A CHI-SQUARE TEST

To evaluate the results of a chi-square test, we must determine whether the chi-square statistic is large or small. Remember, an unusually large value indicates a big discrepancy between the data and the hypothesis and suggests that we reject H_0. To

---

FIGURE 16.3

The shape of the chi-square distribution for different values of df. As the number of categories increases, the peak (mode) of the distribution has a larger chi-square value.

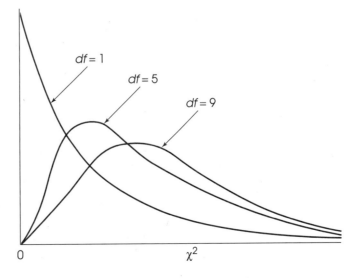

TABLE 16.1

A portion of the table of critical values for the chi-square distribution

df	Proportion in critical region				
	0.10	0.05	0.025	0.01	0.005
1	2.71	3.84	5.02	6.63	7.88
2	4.61	5.99	7.38	9.21	10.60
3	6.25	7.81	9.35	11.34	12.84
4	7.78	9.49	11.14	13.28	14.86
5	9.24	11.07	12.83	15.09	16.75
6	10.64	12.59	14.45	16.81	18.55
7	12.02	14.07	16.01	18.48	20.28
8	13.36	15.51	17.53	20.09	21.96
9	14.68	16.92	19.02	21.67	23.59

determine whether or not a particular chi-square value is significantly large, you first select an alpha level, typically .05 or .01. Then you consult the table entitled The Chi-Square Distribution (Appendix B). A portion of the chi-square distribution table is shown in Table 16.1. The first column in the table lists df values for chi-square. The top row of the table lists proportions of area in the extreme right-hand tail of the distribution. The numbers in the body of the table are the critical values of chi-square. The table shows, for example, in a chi-square distribution with $df = 3$, only 5% (.05) of the values are larger than 7.81, and only 1% (.01) are larger than 11.34.

EXAMPLE OF THE CHI-SQUARE TEST FOR GOODNESS OF FIT

We will use the same step-by-step process for testing hypotheses with chi-square as we used for other hypothesis tests. In general, the steps consist of stating the hypotheses, locating the critical region, computing the test statistic, and making a decision about H_0. The following example demonstrates the complete process of hypothesis testing with the goodness-of-fit test.

EXAMPLE 16.1   A researcher is interested in the factors that are involved in course selection. A sample of 50 students is asked, "Which of the following factors is most important to you when selecting a course?" Students must choose one and only one of the following alternatives:

1. Interest in the course topic

2. Ease of passing the course

3. Instructor for the course

4. Time of day the course is offered

The frequency distribution of responses for this sample (the set of observed frequencies) is as follows:

Interest in topic	Ease of passing	Course instructor	Time of day
18	17	7	8

FIGURE 16.4

For Example 16.1, the critical region begins at a chi-square value of 7.81.

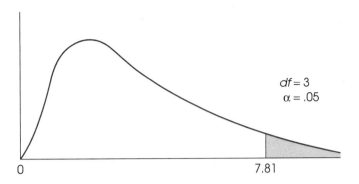

The question for the hypothesis test is whether any of the factors plays a greater role than others for course selection.

STEP 1 The hypotheses must be stated and a level of significance selected. The hypotheses may be stated as follows:

> H_0: The population of students shows no preference in selecting any one of the four factors over the others. Thus, the four factors are named equally often, and the population distribution has the following proportions:

Interest in topic	Ease of passing	Course instructor	Time of day
$\frac{1}{4}$	$\frac{1}{4}$	$\frac{1}{4}$	$\frac{1}{4}$

> H_1: In the population of students, one or more of these factors play a greater role in course selection (the factor is named more frequently by students).

The level of significance is set at a standard value, $\alpha = .05$.

STEP 2 The value for degrees of freedom is determined, and then the critical region is located. For this example, the value for degrees of freedom is

$$df = C - 1 = 4 - 1 = 3$$

For $df = 3$ and $\alpha = .05$, the table of critical values for chi-square indicates that the critical χ^2 has a value of 7.81. The critical region is sketched in Figure 16.4.

STEP 3 The expected frequencies for all categories must be determined, and then the chi-square statistic can be calculated. If H_0 were true and the students displayed no response preference for any of the four alternatives, then the proportion of the population responding to each category would be $\frac{1}{4}$. Because the sample size (n) is 50, the null hypothesis predicts expected frequencies of 12.5 for all categories:

Expected frequencies are computed and may be decimal values. Observed frequencies are always whole numbers.

$$f_e = pn = \tfrac{1}{4}(50) = 12.5$$

The observed frequencies and the expected frequencies are presented in Table 16.2. Using these values, the chi-square statistic may now be calculated.

$$\chi^2 = \Sigma \frac{(f_o - f_e)^2}{f_e}$$

$$= \frac{(18 - 12.5)^2}{12.5} + \frac{(17 - 12.5)^2}{12.5} + \frac{(7 - 12.5)^2}{12.5} + \frac{(8 - 12.5)^2}{12.5}$$

$$= \frac{30.25}{12.5} + \frac{20.25}{12.5} + \frac{30.25}{12.5} + \frac{20.25}{12.5}$$

$$= 2.42 + 1.62 + 2.42 + 1.62$$

$$= 8.08$$

STEP 4 The obtained chi-square value is in the critical region. Therefore, H_0 is rejected, and the researcher may conclude that the subjects mentioned some of the factors significantly more than others in response to the question about course selection.

IN THE LITERATURE:
REPORTING THE RESULTS FOR CHI-SQUARE

The APA style specifies the format for reporting the chi-square statistic in scientific journals. For the results of Example 16.1, the report might state:

> The students showed a significant preference on the question concerning factors involved in course selection, $\chi^2(3, n = 50) = 8.08$, $p < .05$.

Note that the form of the report is similar to that of other statistical tests we have examined. Degrees of freedom are indicated in parentheses following the chi-square symbol. Also contained in the parentheses is the sample size (n). This additional information is important because the degrees of freedom value is based on the number of categories (C), not sample size. Next, the calculated value of chi-square is presented, followed by the probability that a Type I error

TABLE 16.2

(a) The most important factor in course selection
(observed frequencies)

	Interest in topic	Ease of passing	Course instructor	Time of day
f_o	18	17	7	8

(b) The expected frequencies for Example 16.1

	Interest in topic	Ease of passing	Course instructor	Time of day
f_e	12.5	12.5	12.5	12.5

has been committed. Because the null hypothesis was rejected, the probability is *less than* the alpha level.

Additionally, the report should provide the observed frequencies (f_o) for each category. This information may be presented in a simple sentence or in a table [see, for example, Table 16.2(a)].   ❏

LEARNING CHECK

1. A researcher for an insurance company would like to know if high-performance, overpowered automobiles are more likely to be involved in accidents than other types of cars. For a sample of 50 insurance claims, the investigator classifies the automobiles as high-performance, subcompact, mid-size, or full-size. The observed frequencies are as follows:

Observed frequencies of insurance claims

High-performance	Subcompact	Midsize	Full-size	Total
20	14	7	9	50

In determining the f_e values, assume that only 10% of the cars in the population are the high-performance variety. However, subcompacts, midsize cars, and full-size cars make up 40%, 30%, and 20%, respectively. Can the researcher conclude that the observed pattern of accidents does not fit the predicted (f_e) values? Test with $\alpha = .05$.

   a. In a few sentences, state the hypotheses.

   b. Determine the value for *df*, and locate the critical region.

   c. Determine the f_e values, and compute chi-square.

   d. Make a decision regarding H_0.

ANSWERS

1. **a.** H_0: In the population, no particular type of car shows a disproportionate number of accidents. H_1: In the population, a disproportionate number of the accidents occur with certain types of cars.

   b. $df = 3$; the critical χ^2 value is 7.81.

   c. The f_e values for high-performance, subcompact, midsize, and full-size cars are 5, 20, 15, and 10, respectively. The obtained chi-square is 51.17.

   d. Reject H_0.

3

16.3 THE CHI-SQUARE TEST FOR INDEPENDENCE

The chi-square statistic may also be used to test whether or not there is a relationship between two variables. In this situation, each individual in the sample is measured or classified on two separate variables. For example, a group of students could be classified in terms of personality (introvert, extrovert) and in terms of color preference (red, yellow, green, or blue). Usually the data from this classification are presented in the form of a matrix, where the rows correspond to the categories of one variable and the columns correspond to the categories of the second variable. Table

16.3 presents some hypothetical data for a sample of $n = 400$ students who have been classified by personality and color preference. The number in each box, or cell, of the matrix depicts the frequency of that particular group. In Table 16.3, for example, there are 20 introverted students who selected red as their preferred color and 180 extroverted students who preferred red. To obtain these data, the researcher first selects a random sample of $n = 400$ students. Each student is then given a personality test, and each student is asked to select a preferred color from among the four choices. Notice that the classification is based on the measurements for each student; the researcher does not assign students to categories. Also notice that the data consist of frequencies, not scores, from a sample. These sample data will be used to test a hypothesis about the corresponding population frequency distribution. Once again, we will be using the chi-square statistic for the test, but in this case, the test is called the chi-square *test for independence*.

THE NULL HYPOTHESIS FOR THE TEST FOR INDEPENDENCE

The null hypothesis for the chi-square test for independence states that the two variables being measured are independent; that is, for each individual, the value obtained for one variable is not related to (or influenced by) the value for the second variable. This general hypothesis can be expressed in two different conceptual forms, each viewing the data and the test from slightly different perspectives. The data in Table 16.3 describing color preference and personality will be used to present both versions of the null hypothesis.

H_0 version 1   The first version of H_0 focuses on the two variables in the study and simply states that there is no relationship between them. For the color preference and personality study, the null hypothesis would state

H_0: For the general population of students, there is no relationship between color preference and personality.

This version of H_0 demonstrates the similarity between the chi-square test and a correlation. Specifically, the data are viewed as a single sample, with each individual measured on two variables. For this example, the data in Table 16.3 would be viewed as a single sample of $n = 400$ individuals, with measurements of each person's personality and color preference. The goal of the correlation or the chi-square test is to evaluate the relationship between the two variables. For a correlation, the data consist of numerical scores, and you compute means, SS, SP, and so on to measure the correlation between variables. For the chi-square test, on the other hand, the data consist of frequencies, and the test uses the frequency data to determine whether or not there is a relationship between the two variables (H_0 says there is no relationship).

For this version of H_0, the alternative hypothesis, H_1, states that there is a relationship between the two variables. For this example, H_1 states that color preference does depend on personality.

TABLE 16.3

Color preferences according to personality types

	Red	Yellow	Green	Blue	
Introvert	20	6	30	44	100
Extrovert	180	34	50	36	300
	200	40	80	80	$n = 400$

H_0 version 2   The second version of the null hypothesis demonstrates the similarity between the chi-square test and an independent-measures t test (or ANOVA), that is, a hypothesis test that is intended to compare two (or more) separate samples. For this version of H_0, the data in Table 16.3 are viewed as consisting of two separate samples: a sample of $n = 100$ introverts (the top row) and a sample of $n = 300$ extroverts (the bottom row). The goal of the chi-square test is now to determine whether the distribution of color preferences for the introverts (sample 1) is significantly different from the distribution of color preferences for the extroverts (sample 2). The null hypothesis now states that there is no difference between the two distributions: That is, the distribution of color preferences has the same shape (same proportions) for introverts as it does for extroverts. Using this alternative perspective, the null hypothesis is restated as follows:

> H_0: In the general population, the distribution of color preferences has the same shape (same proportions) for both categories of personality.

Stated in this new form, the null hypothesis focuses on the difference between two samples, much the same as an independent-measures t test (or ANOVA). However, the t test (or ANOVA) requires numerical scores (you must compute $\overline{X}$, SS, and so on for each sample), and the test evaluates a hypothesis about the population means. For the chi-square test, on the other hand, the data consist of frequencies, and the test uses the sample proportions to evaluate a hypothesis about the corresponding population proportions (H_0 states that there is no difference between population proportions).

For this second version of H_0, the alternative hypothesis, H_1, would state that there is a difference between the populations: For the example we are considering, H_1 would state that the distribution of color preferences for introverts has a different shape (different proportions) than the distribution for extroverts.

Equivalence of H_0 version 1 and H_0 version 2   Although we have presented two different statements of the null hypothesis, you should realize that these two versions are equivalent. The first version of H_0 states that color preference is not related to personality. If this hypothesis is correct, then the distribution of color preferences should not depend on personality. In other words, the distribution of color preferences should be the same for introverts and extroverts, which is the second version of H_0.

For example, if we found that 60% of the introverts preferred red, then H_0 would predict that we also should find that 60% of the extroverts prefer red. In this case, knowing that an individual prefers red does not help you predict his/her personality. Notice that finding the *same proportions* indicates *no relationship*.

On the other hand, if the proportions were different, it would suggest that there is a relationship. For example, if red were preferred by 60% of the extroverts, but only 10% of the introverts, then there is a clear, predictable relationship between personality and color preference. (If I know your personality, I can predict your color preference.) Thus, finding *different proportions* means that there is *a relationship* between the two variables.

In general, when two variables are independent, the distribution for one variable will not depend on the categories of the second variable. In other words, the frequency distribution for one variable will have the *same shape* (same proportions) for all categories of the second variable.

DEFINITION

Two variables are *independent* when the frequency distribution for one variable is not related to (or dependent on) the categories of the second variable. As a result, the frequency distribution for one variable will have the same shape for all categories of the second variable.

Thus, stating that the two variables are independent (version 1 of H_0) is equivalent to stating that the distributions have equal proportions (version 2 of H_0).

OBSERVED AND EXPECTED FREQUENCIES

The chi-square test for independence uses the same basic logic that was used for the goodness-of-fit test. First, a sample is selected, and each individual is classified or categorized. Because the test for independence considers two variables, every individual is classified on both variables, and the resulting frequency distribution is presented as a two-dimensional matrix (see Table 16.3). As before, the frequencies in the sample distribution are called observed frequencies and are identified by the symbol f_o.

The next step is to find the expected frequencies, or f_e values, for this chi-square test. As before, the expected frequencies define an ideal hypothetical distribution that is in perfect agreement with the null hypothesis. Once the expected frequencies are obtained, we will compute a chi-square statistic to determine how well the data (observed frequencies) fit the null hypothesis (expected frequencies).

Although you can use either version of the null hypothesis to find the expected frequencies, the logic of the process is much easier when you use H_0 stated in terms of equal proportions. For the example we are considering, the null hypothesis states

H_0: The frequency distribution of color preference has the same shape (same proportions) for both categories of personality.

Now we must determine how the sample *should* be distributed according to this hypothesis. The first step is to look at the general characteristics of the original sample. The data in Table 16.3 show an overall sample of $n = 400$ individuals. However, the total sample is divided into a set of subgroups. For example, the sample contains 100 introverts and 300 extroverts. Similarly, the sample consists of 200 people who selected red as their favorite color, 40 people who picked yellow, 80 people who chose green, and 80 who preferred blue. These pre-existing subgroups are defined by the row totals and the column totals in the original data and will be crucial to the calculation of the expected frequencies. Table 16.4(a) presents an empty frequency distribution table showing only the row and column totals for the sample. To compute the expected frequencies, we will use the null hypothesis to create an ideal distribution for the sample. In other words, we will use H_0 to fill in the empty spaces in Table 16.4(a).

Once again, the null hypothesis states that the distribution of color preferences has the same proportions for introverts as for extroverts. But what are the proportions? The best information we have comes from the sample data [see Table 16.4(a)]. Specifically, the column totals identify how many people preferred each color. We simply need to convert these frequencies into proportions. For example:

TABLE 16.4

Expected frequencies for color preferences and personality types

(a) An empty frequency distribution matrix showing only the row totals and column totals (These numbers describe the basic characteristics of the sample from Table 16.3.)

	Red	Yellow	Green	Blue	
Introvert					100
Extrovert					300
	200	40	80	80	

(b) Expected frequencies (This is the distribution that is predicted by the null hypothesis.)

	Red	Yellow	Green	Blue	
Introvert	50	10	20	20	100
Extrovert	150	30	60	60	300
	200	40	80	80	

$$200 \text{ out of } 400 \text{ people preferred red:} \quad \frac{200}{400} = 50\% \text{ prefer red}$$

$$40 \text{ out of } 400 \text{ people preferred yellow:} \quad \frac{40}{400} = 10\% \text{ prefer yellow}$$

$$80 \text{ out of } 400 \text{ people preferred green:} \quad \frac{80}{400} = 20\% \text{ prefer green}$$

$$80 \text{ out of } 400 \text{ people preferred blue:} \quad \frac{80}{400} = 20\% \text{ prefer blue}$$

Thus, the overall proportions for the distribution of color preferences are as follows:

Red	Yellow	Green	Blue
50%	10%	20%	20%

According to the null hypothesis, we should get these same proportions for both the introverts and the extroverts. Therefore, to compute the expected frequencies, we simply use these proportions for the sample of 100 introverts and then use the proportions again for the 300 extroverts.

For the 100 introverts, we expect

$$50\% \text{ choose red:} \quad f_e = 0.50(100) = 50$$

$$10\% \text{ choose yellow:} \quad f_e = 0.10(100) = 10$$

$$20\% \text{ choose green:} \quad f_e = 0.20(100) = 20$$

$$20\% \text{ choose blue:} \quad f_e = 0.20(100) = 20$$

For the 300 extroverts, we expect

$$50\% \text{ choose red:}\quad f_e = 0.50(300) = 150$$
$$10\% \text{ choose yellow:}\quad f_e = 0.10(300) = 30$$
$$20\% \text{ choose green:}\quad f_e = 0.20(300) = 60$$
$$20\% \text{ choose blue:}\quad f_e = 0.20(300) = 60$$

These expected frequencies are shown in Table 16.4(b). Notice that the row totals and the column totals for the expected frequencies are the same as those for the original data in Table 16.3.

A SIMPLE FORMULA FOR DETERMINING EXPECTED FREQUENCIES

Although you should understand that expected frequencies are derived directly from the null hypothesis and the sample characteristics, it is not necessary to go through extensive calculations in order to find f_e values. In fact, there is a simple formula that determines f_e for any cell in the frequency distribution table:

$$f_e = \frac{f_c f_r}{n} \tag{16.4}$$

where f_c is the frequency total for the column (column total), f_r is the frequency total for the row (row total), and n is the number of individuals in the entire sample. To demonstrate this formula, we will compute the expected frequency for introverts selecting yellow in Table 16.4. First, note that this cell is located in the top row and second column in the table. The column total is $f_c = 40$, the row total is $f_r = 100$, and the sample size is $n = 400$. Using these values in formula 16.4, we obtain

$$f_e = \frac{f_c f_r}{n} = \frac{40(100)}{400} = 10$$

Notice that this is identical to the expected frequency we obtained using percentages from the overall distribution.

THE CHI-SQUARE STATISTIC AND DEGREES OF FREEDOM

The chi-square test of independence uses exactly the same chi-square formula as the test for goodness of fit:

$$\chi^2 = \Sigma \frac{(f_o - f_e)^2}{f_e}$$

As before, the formula measures the discrepancy between the data (f_o values) and the hypothesis (f_e values). A large discrepancy will produce a large value for chi-square and will indicate that H_0 should be rejected. To determine whether a particular chi-square statistic is significantly large, you must first determine degrees of freedom (df) for the statistic and then consult the chi-square distribution in the appendix. For the chi-square test of independence, degrees of freedom are based on the number of cells for which you can freely choose expected frequencies. You should recall that the f_e values are partially determined by the sample size (n) and by the row totals and column totals from the original data. These various totals restrict your freedom in selecting expected frequencies. This point is illustrated in

TABLE 16.5

Degrees of freedom and expected frequencies (Once three values have been selected, all the remaining expected frequencies are determined by the row totals and the column totals. This example has only three free choices, so $df = 3$.)

	Red	Yellow	Green	Blue	
	50	10	20	?	100
	?	?	?	?	300
	200	40	80	80	

Table 16.5. Once three of the f_e values have been selected, all the other f_e values in the table are also determined. In general, the row totals and the column totals restrict the final choices in each row and column. Thus, we may freely choose all but one f_e in each row and all but one f_e in each column. The total number of f_e values that you can freely choose is $(R - 1)(C - 1)$, where R is the number of rows and C is the number of columns. The degrees of freedom for the chi-square test of independence are given by the formula

$$df = (R - 1)(C - 1) \tag{16.5}$$

AN EXAMPLE OF THE CHI-SQUARE TEST FOR INDEPENDENCE

The steps for the chi-square test of independence should be familiar by now. First, the hypotheses are stated, and an alpha level is selected. Second, the value for degrees of freedom is computed, and the critical region is located. Third, expected frequencies are determined, and the chi-square statistic is computed. Finally, a decision is made regarding the null hypothesis. The following example demonstrates the complete hypothesis-testing procedure.

EXAMPLE 16.2

Darley and Latané (1968) did a study that examined the relationship between the number of observers and aid-giving behaviors. The group sizes consisted of two people (subject and victim), three people, and six people. The investigators categorized the response of a subject in terms of whether or not the observer exhibited any helping behaviors when the victim (actually another laboratory worker) staged an epileptic seizure. The data are presented in Table 16.6. Do aid-giving behaviors depend on group size?

STEP 1 State the hypotheses, and select a level of significance.

One version of H_0 states that group size and helping behavior are independent of each other in the population. The alternative hypothesis (H_1) states that helping behavior does depend on group size.

The second version of H_0 views the data as three separate samples (one sample representing the group size of two, one the group size of three, and one the group size of six). The null hypothesis states that in the population the proportion of people helping (or not helping) will be the same for each of the three group sizes. The alternative hypothesis (H_1) states that the proportion of people helping will be different from one group size to another. The level of significance is set at $\alpha = .05$.

STEP 2 Calculate the degrees of freedom, and locate the critical region.

For the chi-square test of independence,

$$df = (R - 1)(C - 1)$$

TABLE 16.6

The relationship between size of group and type of response to the victim.

Based on data from "Bystander Intervention in Emergencies: Diffusion of Responsibility," by J. M. Darley and B. Latané, 1968, *Journal of Personality and Social Psychology, 8,* 377–383, American Psychological Association.

(a) Observed frequencies

		Assistance	No assistance	Totals
	2	11	2	13
Group size	3	16	10	26
	6	4	9	13
Totals		31	21	

(b) Expected frequencies

		Assistance	No assistance	Totals
	2	7.75	5.25	13
Group size	3	15.50	10.50	26
	6	7.75	5.25	13
Totals		31	21	

Therefore, for this study,

$$df = (2 - 1)(3 - 1) = 1(2) = 2$$

With two degrees of freedom and a level of significance of .05, the critical value for χ^2 is 5.99 (see Table B.6 in Appendix B for critical values of chi-square).

STEP 3   Determine the expected frequencies, and calculate the chi-square statistic.

As noted before, it is quicker to use the computational formula to determine the f_e values, rather than the percentage method. The expected frequency for each cell is as follows:

1. Group size 2—showed aid-giving behavior:

$$f_e = \frac{f_c f_r}{n} = \frac{13(31)}{52} = 7.75$$

2. Group size 3—showed aid-giving behavior:

$$f_e = \frac{26(31)}{52} = 15.5$$

3. Group size 6—showed aid-giving behavior:

$$f_e = \frac{13(31)}{52} = 7.75$$

4. Group size 2—showed no aid-giving behavior:

$$f_e = \frac{13(21)}{52} = 5.25$$

5. Group size 3—showed no aid-giving behavior:

$$f_e = \frac{26(21)}{52} = 10.5$$

6. Group size 6—showed no aid-giving behavior:

$$f_e = \frac{13(21)}{52} = 5.25$$

Note that the row totals and the column totals for the expected frequencies are the same as the totals for the observed frequencies.

The expected frequencies are summarized in Table 16.6(b). Using these expected frequencies along with the observed frequencies [Table 16.6(a)], we can now calculate the value for the chi-square statistic:

$$\chi^2 = \Sigma \frac{(f_o - f_e)^2}{f_e}$$
$$= \frac{(11 - 7.75)^2}{7.75} + \frac{(16 - 15.5)^2}{15.5} + \frac{(4 - 7.75)^2}{7.75}$$
$$+ \frac{(2 - 5.25)^2}{5.25} + \frac{(10 - 10.5)^2}{10.5} + \frac{(9 - 5.25)^2}{5.25}$$
$$= 1.363 + 0.016 + 1.815 + 2.012 + 0.024 + 2.679$$
$$= 7.91$$

STEP 4 Make a decision regarding the null hypothesis.

The obtained chi-square value exceeds the critical value (5.99). Therefore, the decision is to reject H_0 and conclude that there is a relationship between the group size and the likelihood that someone will aid another person in trouble. For purposes of reporting the data, the researchers could state that there is a significant relationship between size of group and helping behavior, $\chi^2(2, n = 52) = 7.91, p < .05$. By examining the observed frequencies in Table 16.6(a), we see that the likelihood of aid-giving behavior decreases as group size increases.

The results of the test for independence typically are reported using APA style (page 440). Reports also should include a table showing the observed frequencies for each cell (for example, Table 16.6[a]).

LEARNING CHECK 1. A researcher suspects that color blindness is inherited by a sex-linked gene. This possibility is examined by looking for a relationship between gender and color vision. A sample of 1000 people is tested for color blindness, and then they are classified according to their sex and color vision status (normal, red-green blind, other color blindness). Is color blindness related to gender? The data are as follows:

Observed frequencies of color vision status according to sex

	Normal color vision	Red-green color blindness	Other color blindness	Totals
Male	320	70	10	400
Female	580	10	10	600
Totals	900	80	20	

a. State the hypotheses.
b. Determine the value for *df*, and locate the critical region.
c. Compute the f_e values and then chi-square.
d. Make a decision regarding H_0.

A N S W E R S **1. a.** H_0: In the population, there is no relationship between gender and color vision.

H_1: In the population, gender and color vision are related.

b. $df = 2$; critical $\chi^2 = 5.99$ for $\alpha = .05$

c. f_e values are as follows:

Expected frequencies

	Normal	Red-green	Other
Male	360	32	8
Female	540	48	12

Obtained $\chi^2 = 83.44$

d. Reject H_0.

16.4 ASSUMPTIONS AND RESTRICTIONS FOR CHI-SQUARE TESTS

To use a chi-square test for goodness of fit or for independence, several conditions must be satisfied. For any statistical test, violation of assumptions and restrictions will cast doubt on the results. For example, the probability of committing a Type I error may be distorted when assumptions of statistical tests are not satisfied. Some important assumptions and restrictions for using chi-square tests are the following:

1. Random Sampling. As we have seen with other inferential techniques, it is assumed that the sample under study is selected randomly from the population of interest.

2. Independence of Observations. This is *not* to be confused with the concept of independence between *variables,* as seen in the test for independence (Section 16.3). One consequence of independent observations is that each observed frequency is generated by a different subject. A chi-square test would be inappropriate if a person could produce responses that can be classified in more than one category or contribute more than one frequency count to a single category. (See page 202 for more information on independence.)

3. Size of Expected Frequencies. A chi-square test should not be performed when the expected frequency of any cell is less than 5. The chi-square statistic can be distorted when f_e is very small. Consider the chi-square computations for a single cell. Suppose the cell has values of $f_e = 1$ and $f_o = 5$. The contribution of this cell to the total chi-square value is

$$\text{cell} = \frac{(f_o - f_e)^2}{f_e} = \frac{(5-1)^2}{1} = \frac{4^2}{1} = 16$$

Now consider another instance, where $f_e = 10$ and $f_o = 14$. The difference between the observed and the expected frequencies is still 4, but the contribution of this cell to the total chi-square value differs from that of the first case:

$$\text{cell} = \frac{(f_o - f_e)^2}{f_e} = \frac{(14-10)^2}{10} = \frac{4^2}{10} = 1.6$$

It should be clear that a small f_e value can have a great influence on the chi-square value. This problem becomes serious when f_e values are less than 5. When f_e is very small, what would otherwise be a minor discrepancy between f_o and f_e will now result in large chi-square values. The test is too sensitive when f_e values are extremely small. One way to avoid small expected frequencies is to use large samples.

SUMMARY

1. Chi-square tests are a type of nonparametric technique that tests hypotheses about the form of the entire frequency distribution. Two types of chi-square tests are the test for goodness of fit and the test for independence. The data for these tests consist of the frequencies observed for various categories of a variable.

2. The test for goodness of fit compares the frequency distribution for a sample to the frequency distribution that is predicted by H_0. The test determines how well the observed frequencies (sample data) fit the expected frequencies (predicted by H_0).

3. The expected frequencies for the goodness-of-fit test are determined by

   $$\text{expected frequency} = f_e = pn$$

   where p is the hypothesized proportion (according to H_0) of observations falling into a category and n is the size of the sample.

4. The chi-square statistic is computed by

   $$\text{chi-square} = \chi^2 = \Sigma \frac{(f_o - f_e)^2}{f_e}$$

   where f_o is the observed frequency for a particular category and f_e is the expected frequency for that category. Large values for χ^2 indicate that there is a large discrepancy between the observed (f_o) and the expected (f_e) frequencies, which may warrant rejection of the null hypothesis.

5. Degrees of freedom for the test for goodness of fit are

   $$df = C - 1$$

   where C is the number of categories in the variable. Degrees of freedom measure the number of categories for which f_e values can be freely chosen. As can be seen from the formula, all but the last f_e value to be determined are free to vary.

6. The chi-square distribution is positively skewed and begins at the value of zero. Its exact shape is determined by degrees of freedom.

7. The test for independence is used to assess the relationship between two variables. The null hypothesis states that the two variables in question are independent of each other. That is, the frequency distribution for one variable does not depend on the categories of the second variable. On the other hand, if a relationship does exist, then the form of the distribution for one variable will depend on the categories of the other variable.

8. For the test for independence, the expected frequencies for H_0 can be directly calculated from the marginal frequency totals:

   $$f_e = \frac{f_c f_r}{n}$$

   where f_c is the total column frequency and f_r is the total row frequency for the cell in question.

9. Degrees of freedom for the test for independence are computed by

   $$df = (R - 1)(C - 1)$$

   where R is the number of row categories and C is the number of column categories.

10. For the test for independence, a large chi-square value means there is a large discrepancy between the f_o and the f_e values. Rejecting H_0 in this test provides support for a relationship between the two variables.

11. Both chi-square tests (for goodness of fit and independence) are based on the assumption that the sample is randomly selected from the population. It is also assumed that each observation is independent of the others. That is, each observed frequency reflects a different individual, and no individual can produce a response that would be classified in more than one category or more than one frequency in a single category.

12. The chi-square statistic is distorted when f_e values are small. Chi-square tests therefore are restricted to situations where f_e values are 5 or greater. The test should not be performed when the expected frequency of any cell is less than 5.

KEY TERMS

goodness-of-fit test   expected frequencies   distribution of chi-square   test for independence

observed frequencies   chi-square statistic

——— FOCUS ON PROBLEM SOLVING ———

1. The expected frequencies that you calculate must satisfy the constraints of the sample. For the goodness-of-fit test, $\Sigma f_e = \Sigma f_o = n$. For the test for independence, the row totals and the column totals for the expected frequencies should be identical to the corresponding totals for the observed frequencies.

2. It is entirely possible to have fractional (decimal) values for expected frequencies. Observed frequencies, however, are always whole numbers.

3. Whenever $df = 1$, the difference between observed and expected frequencies $(f_o - f_e)$ will be identical (the same value) for all cells. This makes the calculation of chi-square easier.

4. Although you are advised to compute expected frequencies for all categories (or cells), you should realize that it is not essential to calculate all f_e values separately. Remember, df for chi-square identifies the number of f_e values that are free to vary. Once you have calculated that number of f_e values, the remaining f_e values are determined. You can get these remaining values by subtracting the calculated f_e values from their corresponding row or column total.

5. Remember, unlike previous statistical tests, the degrees of freedom (df) for a chi-square test are *not* determined by the sample size (n). Be careful!

——— DEMONSTRATION 16.1 ———

TEST FOR INDEPENDENCE

A manufacturer of watches would like to examine preferences for digital versus analog watches. A sample of $n = 200$ people is selected, and these individuals are classified by age and preference. The manufacturer would like to know if there is a relationship between age and watch preference. The observed frequencies (f_o) are as follows:

		Preference		
		Digital	Analog	Undecided
Age	Under 30	90	40	10
	30 or over	10	40	10

STEP 1   State the hypotheses, and select an alpha level.

The null hypothesis states that there is no relationship between the two variables.

H_0: Preference is independent of age. That is, the frequency distribution of preference has the same form for people under 30 as for people 30 or over.

The alternative hypothesis states that there is a relationship between the two variables.

H_1: Preference is related to age. That is, the type of watch preferred depends on a person's age.

We will set alpha to $\alpha = .05$.

STEP 2 Locate the critical region.
Degrees of freedom for the chi-square test for independence are determined by

$$df = (C - 1)(R - 1)$$

For these data,

$$df = (3 - 1)(2 - 1) = 2(1) = 2$$

For $df = 2$ with $\alpha = .05$, the critical chi-square value is 5.99. Thus, our obtained chi-square must exceed 5.99 to be in the critical region and to reject H_0.

STEP 3 Compute the test statistic.
Computing the chi-square statistic requires the following preliminary calculations:

1. Obtain the row and the column totals.
2. Calculate the expected frequencies.

Row and column totals. We start by determining the row and the column totals from the original observed frequencies, f_o.

	Digital	Analog	Undecided	Row totals
Under 30	90	40	10	140
30 or over	10	40	10	60
Column totals	100	80	20	$n = 200$

Expected frequencies, f_e. For the test for independence, the following formula is used to obtain expected frequencies:

$$f_e = \frac{f_c f_r}{n}$$

For people under 30, we obtain the following expected frequencies:

$$f_e = \frac{100(140)}{200} = \frac{14{,}000}{200} = 70 \text{ for digital}$$

$$f_e = \frac{80(140)}{200} = \frac{11{,}200}{200} = 56 \text{ for analog}$$

$$f_e = \frac{20(140)}{200} = \frac{2800}{200} = 14 \text{ for undecided}$$

For individuals 30 or over, the expected frequencies are as follows:

$$f_e = \frac{100(60)}{200} = \frac{6000}{200} = 30 \text{ for digital}$$

$$f_e = \frac{80(60)}{200} = \frac{4800}{200} = 24 \text{ for analog}$$

$$f_e = \frac{20(60)}{200} = \frac{1200}{200} = 6 \text{ for undecided}$$

The following table summarizes the expected frequencies:

	Digital	Analog	Undecided
Under 30	70	56	14
30 or over	30	24	6

The chi-square statistic. The chi-square statistic is computed from the formula

$$\chi^2 = \Sigma \frac{(f_o - f_e)^2}{f_e}$$

That is, we must

1. Find the $f_o - f_e$ difference for each cell.
2. Square these differences.
3. Divide the squared differences by f_e.
4. Sum the results.

The following table summarizes these calculations:

Cell	f_o	f_e	$(f_o - f_e)$	$(f_o - f_e)^2$	$(f_o - f_e)^2/f_e$
Under 30—digital	90	70	20	400	5.71
Under 30—analog	40	56	−16	256	4.57
Under 30—undecided	10	14	−4	16	1.14
30 or over—digital	10	30	−20	400	13.33
30 or over—analog	40	24	16	256	10.67
30 or over—undecided	10	6	4	16	2.67

Finally, we can sum the last column to get the chi-square value.

$$\chi^2 = 5.71 + 4.57 + 1.14 + 13.33 + 10.67 + 2.67$$
$$= 38.09$$

STEP 4 Make a decision about H_0, and state the conclusion.

The chi-square value is in the critical region. Therefore, we can reject the null hypothesis. There is a relationship between watch preference and age, $\chi^2(2, n = 200) = 38.09, p < .05.$

PROBLEMS

1. Parametric tests (such as *t* or ANOVA) differ from non-parametric tests (such as chi-square) primarily in terms of the assumptions they require and the data they use. Explain these differences.

2. A psychology professor is trying to decide which textbook to use for next year's introductory class. To help make the decision, the professor asks the current students to review three texts and identify which one is preferred. The distribution of preferences for the current class is as follows:

Book 1	Book 2	Book 3
52	41	27

Do these data indicate any significant preferences among the three books? Test with $\alpha = .05$.

3. Automobile insurance is much more expensive for teenage drivers than for older drivers. To justify this cost difference, insurance companies claim that the younger drivers are much more likely to be involved in costly accidents. To test this claim, a researcher obtains information about registered drivers from the department of motor vehicles and selects a sample of $n = 300$ accident reports from the police department. The motor vehicle department reports the percentage of registered drivers in each age category as follows: 16% are under age 20; 28% are 20 to 29 years old, and 56% are age 30 or older. The number of accident reports for each age group is as follows:

Under age 20	Age 20 to 29	Age 30 or older
68	92	140

Do these data demonstrate a significantly disproportionate number of accidents among the younger drivers? Test with $\alpha = .05$.

4. A developmental psychologist would like to determine whether infants display any color preferences. A stimulus consisting of four color patches (red, green, blue, and yellow) is projected onto the ceiling above a crib. Infants are placed in the crib, one at a time, and the psychologist records how much time an infant spends looking at each of the four colors. The color that receives the most attention during a 100-second test period is identified as the preferred color for that infant. The preferred colors for a sample of 60 infants are shown in the following table:

Red	Green	Blue	Yellow
21	11	18	10

Do these data indicate any significant preferences among the four colors? Test at the .05 level of significance.

5. A psychologist examining art appreciation selected an abstract painting that had no obvious top or bottom. Hangers were placed on the painting so that it could be hung with any one of the four sides at the top. The painting was shown to a sample of $n = 50$ subjects, and each was asked to hang the painting in whatever orientation looked best. The following data indicate how many times each of the four sides was placed at the top:

Top up (correct)	Left side up	Right side up	Bottom side up
19	7	9	15

Do the data indicate any preferences among the four possible orientations? Test with $\alpha = .05$.

6. A questionnaire given to last year's freshman class indicated that 30% intended to be science majors, 50% intended to major in the social sciences or humanities, and 20% were interested in professional programs. A random sample of 100 students from the current freshman class yielded the following frequency distribution:

Intended major

Sciences	Social sciences or humanities	Professional
35	40	25

a. On the basis of these data, should the university officials conclude that there has been a significant change in student interests? Test at the .05 level of significance.

b. If twice as many students had been sampled with the result that the observed frequencies were doubled in each of the three categories, would there be evidence for a significant change? Again, test with $\alpha = .05$.

c. How do you explain the different conclusions for parts a and b?

7. A researcher is investigating the physical characteristics that influence whether or not a person's face is judged as beautiful. The researcher selects a photograph of a woman and then creates two modifications of the photo by (1) moving the eyes slightly farther apart and (2) moving the eyes slightly closer together. The original photograph and the two modifications are then shown to a sample of $n = 150$ college students, and each student is asked to select the "most beautiful" of the three faces. The distribution of responses was as follows:

Original photo	Eyes moved apart	Eyes moved together
51	72	27

Do these data indicate any significant preferences among the three versions of the photograph? Test at the .05 level of significance.

8. It is known that blood type varies among different populations of people. In the United States, for example, types O, A, B, and AB blood make up 45%, 41%, 10%, and 4% of the population, respectively. Suppose blood type is determined for a sample of $n = 136$ individuals from a foreign country. The resulting frequency distribution is as follows:

Blood type

	O	A	B	AB
f_o	43	38	41	14

Is there a significant difference between this distribution and what we would expect for the United States? Set alpha at .05.

9. Suppose an opinion poll taken in 1970 revealed the following data regarding the legalization of marijuana: 15% in favor of, 79% against, and 6% with no opinion regarding legalization. Suppose also that you took a random sample of $n = 220$ people today and obtained the following data:

Attitude toward legalization of marijuana

	For	Against	No opinion
f_o	38	165	17

Is there a significant difference between the current data and the data obtained in 1970? Use the .05 level of significance.

10. A researcher would like to determine whether mental disorders can be reliably detected using the results of a Rorschach test. The researcher obtains Rorschach results for a sample of 20 normal adults and a sample of 20 psychotic patients diagnosed with paranoid schizophrenia.

The two samples are then mixed to create a total set of 40 Rorschach results that are given to a team of psychologists to be classified as either "normal" or "abnormal." The results of the classification are as follows:

		Psychologists' classification	
		Normal	Abnormal
Subjects	Normal	16	4
	Psychotic	12	8

Do the data indicate a significant relationship between an actual disorder and the Rorschach results? Test at the .05 level of significance.

11. In Problem 2, a professor asked students to rate three textbooks to determine whether there were any preferences among them. Although the data appear to indicate an overall preference for book 1, the professor would like to know whether this opinion is shared by students with different levels of academic ability. To answer this question, the median grade was used to separate the students into two groups: the upper half and the lower half of the class. The distribution of preferences for these two groups is as follows:

	Book 1	Book 2	Book 3	
Upper half	17	31	12	60
Lower half	35	10	15	60
	52	41	27	

Do these data indicate that the distribution of preferences for students in the upper half of the class is significantly different from the distribution for students in the lower half? Test at the .05 level of significance.

12. Sociologists often point to negative childhood experiences as possible causal factors in the criminal behavior of adults. A sample of 25 individuals convicted of violent crimes and a matching sample of 25 college students are obtained. A sociologist interviews each individual to determine whether there is any history of childhood abuse (physical or sexual). The resulting frequency distribution is as follows:

	No abuse	History of abuse
Criminals	9	16
Students	19	6

Do the data indicate that a history of abuse is significantly more common for violent criminals than it is for noncriminal college students? Test with $\alpha = .05$.

13. A psychologist would like to examine the factors that are involved in the development of romantic relationships. The psychologist suspects that people at different ages will place emphasis on different factors when looking for a romantic partner. A sample of 30 adolescent males (ages 13–15) and a sample of 30 male college students (ages 19–32) are obtained. Each subject is asked to identify the single most important factor determining whether or not he is likely to pursue a romantic relationship with a new female acquaintance. The frequency data for this study are as follows:

	Physical appearance	Popularity	Personality
Adolescents	15	10	5
College students	12	8	10

Do these data indicate a significant difference in the relative importance of different factors for adolescents versus college students? Test with $\alpha = .05$.

14. Recent studies have demonstrated that naltrexone, a drug that reduces the craving for heroin, can prevent relapse of drinking in alcoholics (Volpicelli et al., 1992). Suppose you studied a group of alcoholics who were receiving counseling and were currently abstaining from drinking. One-third of the people also received a small dose of naltrexone, one-third received a larger dose, and the remainder got a placebo. You obtained the following data, which show the number of people in each group that relapsed (began drinking heavily again):

	Placebo	Small dose	Large dose
Relapse	20	10	6
No relapse	10	20	24

Does naltrexone have an effect on relapse? Set alpha at .05.

15. Gender differences in dream content are well documented (see Winget & Kramer, 1979). Suppose a researcher studies aggression content in the dreams of men and women. Each subject reports his or her most recent dream. Then each dream is judged by a panel of experts to have low, medium, or high aggression content. The observed frequencies are shown in the following matrix:

		Aggression content		
		Low	Medium	High
Gender	Female	18	4	2
	Male	4	17	15

Is there a relationship between gender and the aggression content of dreams? Test with $\alpha = .01$.

16. Cialdini, Reno, and Kallgren (1990) conducted an experiment examining how people conform to norms concerning littering. They wanted to determine whether a person's tendency to litter depends on the amount of litter already in the area. People were handed a handbill as they entered a parking garage that already had 1, 2, 4, or 8 handbills lying on the ground. Then the people were observed to determine whether or not they dropped the handbills on the ground. Hypothetical data similar to the experimental results are as follows:

	Number of pieces of litter			
	1	2	4	8
Littered	5	10	12	20
Did not litter	45	40	38	30

Is there a relationship between amount of pre-existing litter and a person's tendency to litter?

17. Friedman and Rosenman (1974) have suggested that personality type is related to heart disease. Specifically, type A people, who are competitive, driven, pressured, and impatient, are more prone to heart disease. On the other hand, type B individuals, who are less competitive and more relaxed, are less likely to have heart disease. Suppose an investigator would like to examine the relationship between personality type and disease. For a random sample of individuals, personality type is assessed with a standardized test. These individuals are then examined and categorized according to the type of disorder they have. The observed frequencies are as follows:

	Type of disorder			
	Heart	Vascular	Hypertension	None
Type A personality	38	29	43	60
Type B personality	18	22	14	126

Is there a relationship between personality and disorder? Test at the .05 level of significance.

18. Numerous studies have shown that a person's decision whether or not to respond to a survey depends on *who* presents the survey. To examine this phenomenon, 100 surveys were mailed to students at a local college. Half of the surveys were distributed with a cover letter signed by the president of the student body. The other half of the surveys included a cover letter signed by a fictitious administrator (the "Vice President for Student Information"). Of the surveys signed by the student president, 28 were returned and 22 were discarded. For the surveys signed by the unknown administrator, only 11 were returned and 39 were discarded. Do these data indicate that the individual requesting a survey can have a significant effect on the proportion of surveys returned? Test at the .01 level of significance.

19. McClelland (1961) suggested that the strength of a person's need for achievement can predict behavior in a number of situations, including risk-taking situations. This experiment is patterned after his work. A random sample of college students is given a standardized test that measures the need for achievement. On the basis of their test scores, they are classified into high achievers and low achievers. They are then confronted with a task for which they can select the level of difficulty. Their selections are classified as "cautious" (low risk of failure), "moderate" risk of failure, or "high" risk of failure. The observed frequencies for this study are as follows:

Risk taken by subject

	Cautious	Moderate	High
High achiever	8	24	6
Low achiever	17	7	16

Can you conclude there is a relationship between the need for achievement and risk-taking behavior? Set alpha to .05. Describe the outcome of the study.

20. A researcher believes that people with low self-esteem will avoid situations that will focus attention on them. A random sample of $n = 72$ people is selected. Each person is given a standardized test that measures self-esteem and is classified as high, medium, or low in self-esteem. The subjects are then placed in a situation in which they must choose between performing a task in front of other people and performing the same task by themselves. The researcher notes which task setting is chosen. The observed frequencies are as follows:

Task setting chosen

	Audience	No audience
Low self-esteem	4	16
Medium self-esteem	14	14
High self-esteem	18	6

Is there a relationship between self-esteem and the task chosen? Use $\alpha = .05$.

APPENDIX A BASIC MATHEMATICS REVIEW

CONTENTS

PREVIEW

This appendix reviews some of the basic math skills that are necessary for the statistical calculations presented in this book. Many students already will know some or all of this material. Others will need to do extensive work and review. To help you assess your own skills, we are including a skills assessment exam here. You should allow approximately 30 minutes to complete the test. When you finish, grade your test using the answer key on page A-22.

Notice that the test is divided into five sections. If you miss more than three questions in any section of the test, you probably need help in that area. Turn to the section of this appendix that corresponds to your problem area. In each sec-

tion, you will find a general review, some examples, and some additional practice problems. After reviewing the appropriate section and doing the practice problems, turn to the end of the appendix. You will find another version of the skills assessment exam. If you still miss more than three questions in any section of the exam, continue studying. Get assistance from an instructor or a tutor if necessary. At the end of this appendix is a list of recommended books for individuals who need a more extensive review than can be provided here. We must stress that mastering this material now will make the rest of the course much easier.

SKILLS ASSESSMENT PREVIEW EXAM

SECTION 1

(corresponding to Section A.1 of this appendix)

1. $3 + 2 \times 7 = ?$
2. $(3 + 2) \times 7 = ?$
3. $3 + 2^2 - 1 = ?$
4. $(3 + 2)^2 - 1 = ?$
5. $12/4 + 2 = ?$
6. $12/(4 + 2) = ?$
7. $12/(4 + 2)^2 = ?$
8. $2 \times (8 - 2^2) = ?$
9. $2 \times (8 - 2)^2 = ?$
10. $3 \times 2 + 8 - 1 \times 6 = ?$
11. $3 \times (2 + 8) - 1 \times 6 = ?$
12. $3 \times 2 + (8 - 1) \times 6 = ?$

SECTION 2

(corresponding to Section A.2 of this appendix)

1. The fraction ¾ corresponds to a percentage of _____.
2. Express 30% as a fraction.
3. Convert ¹²⁄₄₀ to a decimal.
4. $\frac{2}{13} + \frac{8}{13} = ?$
5. $1.375 + 0.25 = ?$
6. $\frac{2}{5} \times \frac{1}{4} = ?$
7. $\frac{1}{8} + \frac{2}{3} = ?$
8. $3.5 \times 0.4 = ?$
9. $\frac{1}{5} \div \frac{3}{4} = ?$
10. $3.75/0.5 = ?$
11. In a group of 80 students, 20% are psychology majors. How many psychology majors are in this group?
12. A company reports that two-fifths of its employees are women. If there are 90 employees, how many are women?

SECTION 3

(corresponding to Section A.3 of this appendix)

1. $3 + (-2) + (-1) + 4 = ?$
2. $6 - (-2) = ?$
3. $-2 - (-4) = ?$
4. $6 + (-1) - 3 - (-2) - (-5) = ?$
5. $4 \times (-3) = ?$
6. $-2 \times (-6) = ?$
7. $-3 \times 5 = ?$
8. $-2 \times (-4) \times (-3) = ?$
9. $12 \div (-3) = ?$
10. $-18 \div (-6) = ?$
11. $-16 \div 8 = ?$
12. $-100 \div (-4) = ?$

SECTION 4

(corresponding to Section A.4 of this appendix)

For each equation, find the value of X.

1. $X + 6 = 13$
2. $X - 14 = 15$
3. $5 = X - 4$
4. $3X = 12$
5. $72 = 3X$
6. $X/5 = 3$
7. $10 = X/8$
8. $3X + 5 = -4$
9. $24 = 2X + 2$
10. $(X + 3)/2 = 14$
11. $(X - 5)/3 = 2$
12. $17 = 4X - 11$

SECTION 5

(corresponding to Section A.5 of this appendix)

1. $4^3 = ?$
2. $\sqrt{25 - 9} = ?$
3. If $X = 2$ and $Y = 3$, then $XY^3 = ?$
4. If $X = 2$ and $Y = 3$, then $(X + Y)^2 = ?$
5. If $a = 3$ and $b = 2$, then $a^2 + b^2 = ?$
6. $-3^3 = ?$
7. $-4^4 = ?$
8. $\sqrt{4} \times 4 = ?$
9. $36/\sqrt{9} = ?$
10. $(9 + 2)^2 = ?$
11. $5^2 + 2^3 = ?$
12. If $a = 3$ and $b = -1$, then $a^2 b^3 = ?$

The answers to the skills assessment exam are at the end of the appendix (p. A-22).

A.1 SYMBOLS AND NOTATION

Table A.1 presents the basic mathematical symbols that you should know, and it provides examples of their use. Statistical symbols and notation will be introduced and explained throughout this book as they are needed. Notation for exponents and square roots is covered separately at the end of this appendix.

Parentheses are a useful notation because they specify and control the order of computations. Everything inside the parentheses is calculated first. For example,

$$(5 + 3) \times 2 = 8 \times 2 = 16$$

Changing the placement of the parentheses also changes the order of calculations. For example,

$$5 + (3 \times 2) = 5 + 6 = 11$$

ORDER OF OPERATIONS   Often a formula or a mathematical expression will involve several different arithmetic operations, such as adding, multiplying, squaring, and so on. When you encounter these situations, you must perform the different operations in the correct sequence. Following is a list of mathematical operations, showing the order in which they are to be performed.

1. Any calculation contained within parentheses is done first.
2. Squaring (or raising to other exponents) is done second.
3. Multiplying and/or dividing is done third. A series of multiplication and/or division operations should be done in order from left to right.
4. Adding and/or subtracting is done fourth.

The following examples demonstrate how this sequence of operations is applied in different situations.

To evaluate the expression

$$(3 + 1)^2 - 4 \times 7/2$$

first, perform the calculation within parentheses:

$$(4)^2 - 4 \times 7/2$$

TABLE A.1

Symbol	Meaning	Example
+	Addition	$5 + 7 = 12$
−	Subtraction	$8 - 3 = 5$
×, ( )	Multiplication	$3 \times 9 = 27$, $3(9) = 27$
÷, /	Division	$15 \div 3 = 5$, $15/3 = 5$, $^{15}/_3 = 5$
>	Greater than	$20 > 10$
<	Less than	$7 < 11$
≠	Not equal to	$5 \neq 6$

Next, square the value as indicated:

$$16 - 4 \times 7/2$$

Then perform the multiplication and division:

$$16 - 14$$

Finally, do the subtraction:

$$16 - 14 = 2$$

A sequence of operations involving multiplication and division should be performed in order from left to right. For example, to compute $12/2 \times 3$, you divide 12 by 2 and then multiply the result by 3:

$$12/2 \times 3 = 6 \times 3 = 18$$

Notice that violating the left-to-right sequence can change the result. For this example, if you multiply before dividing, you will obtain

$$12/2 \times 3 = 12/6 = 2 \qquad \text{(This is wrong.)}$$

A sequence of operations involving only addition and subtraction can be performed in any order. For example, to compute $3 + 8 - 5$, you can add 3 and 8 and then subtract 5:

$$(3 + 8) - 5 = 11 - 5 = 6$$

Or you can subtract 5 from 8 and then add the result to 3:

$$3 + (8 - 5) = 3 + 3 = 6$$

A mathematical expression or formula is simply a concise way to write a set of instructions. When you evaluate an expression by performing the calculation, you simply follow the instructions. For example, assume you are given the instructions that follow:

1. First, add 3 and 8.
2. Next, square the result.
3. Next, multiply the resulting value by 6.
4. Finally, subtract 50 from the value you have obtained.

You can write these instructions as a mathematical expression.

1. The first step involved addition. Because addition is normally done last, use parentheses to give this operation priority in the sequence of calculations:

$$(3 + 8)$$

2. The instruction to square a value is noted by using the exponent 2 beside the value to be squared:

$$(3 + 8)^2$$

3. Because squaring has priority over multiplication, you can simply introduce the multiplication into the expression:

$$6 \times (3 + 8)^2$$

4. Addition and subtraction are done last, so simply write in the requested subtraction:

$$6 \times (3 + 8)^2 - 50$$

To calculate the value of the expression, you work through the sequence of operations in the proper order:

$$
\begin{aligned}
6 \times (3 + 8)^2 - 50 &= 6 \times (11)^2 - 50 \\
&= 6 \times (121) - 50 \\
&= 726 - 50 \\
&= 676
\end{aligned}
$$

As a final note, you should realize that the operation of squaring (or raising to any exponent) applies only to the value that immediately precedes the exponent. For example,

$$2 \times 3^2 = 2 \times 9 = 18 \qquad \text{(only the 3 is squared)}$$

If the instructions require multiplying values and then squaring the product, you must use parentheses to give the multiplication priority over squaring. For example, to multiply 2 times 3 and then square the product, you would write

$$(2 \times 3)^2 = (6)^2 = 36$$

LEARNING CHECK 1. Evaluate each of the following expressions:

a. $4 \times 8/2^2$
b. $4 \times (8/2)^2$
c. $100 - 3 \times 12/(6 - 4)^2$
d. $(4 + 6) \times (3 - 1)^2$
e. $(8 - 2)/(8 - 9)^2$
f. $6 + (4 - 1)^2 - 3 \times 4^2$
g. $4 \times (8 - 3) + 8 - 3$

ANSWERS 1. a. 8   b. 64   c. 91   d. 40   e. 6   f. −33   g. 25

A.2 PROPORTIONS: FRACTIONS, DECIMALS, AND PERCENTAGES

A proportion is a part of a whole and can be expressed as a fraction, or a decimal, or a percentage. For example, in a class of 40 students, only 3 failed the final exam.

The proportion of the class that failed can be expressed as a fraction

$$\text{fraction} = \tfrac{3}{40}$$

or as a decimal value

$$\text{decimal} = 0.075$$

or as a percentage

$$\text{percentage} = 7.5\%$$

In a fraction, such as ¾, the bottom value (the denominator) indicates the number of equal pieces into which the whole is split. Here the "pie" is split into 4 equal pieces:

If the denominator has a larger value—say, 8—then each piece of the whole pie is smaller:

A larger denominator indicates a smaller fraction of the whole.

The value on top of the fraction (the numerator) indicates how many pieces of the whole are being considered. Thus, the fraction ¾ indicates that the whole is split evenly into 4 pieces and that 3 of them are being used:

A fraction is simply a concise way of stating a proportion: "Three out of four" is equivalent to ¾. To convert the fraction to a decimal, you divide the numerator by the denominator:

$$\tfrac{3}{4} = 3 \div 4 = 0.75$$

To convert the decimal to a percentage, simply multiply by 100, and place a percent sign (%) after the answer:

$$0.75 \times 100 = 75\%$$

The U.S. money system is a convenient way of illustrating the relationship between fractions and decimals. "One quarter," for example, is one-fourth (¼) of a dollar, and its decimal equivalent is 0.25. Other familiar equivalencies are as follows:

	Dime	Quarter	50 cents	75 cents
Fraction	$\frac{1}{10}$	$\frac{1}{4}$	$\frac{1}{2}$	$\frac{3}{4}$
Decimal	0.10	0.25	0.50	0.75
Percentage	10%	25%	50%	75%

FRACTIONS

1. Finding Equivalent Fractions. The same proportional value can be expressed by many equivalent fractions. For example,

$$\tfrac{1}{2} = \tfrac{2}{4} = \tfrac{10}{20} = \tfrac{50}{100}$$

To create equivalent fractions, you can multiply the numerator and denominator by the same value. As long as both the numerator and the denominator of the fraction are multiplied by the same value, the new fraction will be equivalent to the original. For example,

$$\tfrac{3}{10} = \tfrac{9}{30}$$

because both the numerator and the denominator of the original fraction have been multiplied by 3. Dividing the numerator and denominator of a fraction by the same value will also result in an equivalent fraction. By using division, you can reduce a fraction to a simpler form. For example,

$$\tfrac{40}{100} = \tfrac{2}{5}$$

because both the numerator and the denominator of the original fraction have been divided by 20.

You can use these rules to find specific equivalent fractions. For example, find the fraction that has a denominator of 100 and is equivalent to ¾. That is,

$$\tfrac{3}{4} = \tfrac{?}{100}$$

Notice that the denominator of the original fraction must be multiplied by 25 to produce the denominator of the desired fraction. For the two fractions to be equal, both the numerator and the denominator must be multiplied by the same number. Therefore, we also multiply the top of the original fraction by 25 and obtain

$$\tfrac{3 \times 25}{4 \times 25} = \tfrac{75}{100}$$

2. Multiplying Fractions. To multiply two fractions, you first multiply the numerators and then multiply the denominators. For example,

$$\tfrac{3}{4} \times \tfrac{5}{7} = \tfrac{3 \times 5}{4 \times 7} = \tfrac{15}{28}$$

3. Dividing Fractions. To divide one fraction by another, you invert the second fraction and then multiply. For example,

$$\tfrac{1}{2} \div \tfrac{1}{4} = \tfrac{1}{2} \times \tfrac{4}{1} = \tfrac{1 \times 4}{2 \times 1} = \tfrac{4}{2}$$

4. Adding and Subtracting Fractions. Fractions must have the same denominator before you can add or subtract them. If the two fractions already have a common denominator, you simply add (or subtract as the case may be) *only* the values in the numerators. For example,

$$\tfrac{2}{5} + \tfrac{1}{5} = \tfrac{3}{5}$$

Suppose you divided a pie into five equal pieces (fifths). If you first ate two-fifths of the pie and then another one-fifth, the total amount eaten would be three-fifths of the pie:

If the two fractions do not have the same denominator, you must first find equivalent fractions with a common denominator before you can add or subtract. The product of the two denominators will always work as a common denominator for equivalent fractions (although it may not be the lowest common denominator). For example,

$$\tfrac{2}{3} + \tfrac{1}{10} = ?$$

Because these two fractions have different denominators, it is necessary to convert each into an equivalent fraction and find a common denominator. We will use $3 \times 10 = 30$ as the common denominator. Thus, the equivalent fraction of each is

$$\tfrac{2}{3} = \tfrac{20}{30} \quad \text{and} \quad \tfrac{1}{10} = \tfrac{3}{30}$$

Now the two fractions can be added:

$$\tfrac{20}{30} + \tfrac{3}{30} = \tfrac{23}{30}$$

5. Comparing the Size of Fractions. When comparing the size of two fractions with the same denominator, the larger fraction will have the larger numerator. For example,

$$\tfrac{5}{8} > \tfrac{3}{8}$$

The denominators are the same, so the whole is partitioned into pieces of the same size. Five of these pieces are more than three of them:

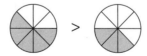

When two fractions have different denominators, you must first convert them to fractions with a common denominator to determine which is larger. Consider the following fractions:

$$\tfrac{3}{8} \quad \text{and} \quad \tfrac{7}{16}$$

If the numerator and denominator of $\frac{3}{8}$ are multiplied by 2, the resulting equivalent fraction will have a denominator of 16:

$$\frac{3}{8} = \frac{3 \times 2}{8 \times 2} = \frac{6}{16}$$

Now a comparison can be made between the two fractions:

$$\frac{6}{16} < \frac{7}{16}$$

Therefore,

$$\frac{3}{8} < \frac{7}{16}$$

DECIMALS

1. Converting Decimals to Fractions. Like a fraction, a decimal represents part of the whole. The first decimal place to the right of the decimal point indicates how many tenths are used. For example,

$$0.1 = \frac{1}{10} \qquad 0.7 = \frac{7}{10}$$

The next decimal place represents $\frac{1}{100}$, the next $\frac{1}{1000}$, the next $\frac{1}{10,000}$, and so on. To change a decimal to a fraction, just use the number without the decimal point for the numerator. Use the denominator that the last (on the right) decimal place represents. For example,

$$0.32 = \frac{32}{100} \qquad 0.5333 = \frac{5333}{10,000} \qquad 0.05 = \frac{5}{100} \qquad 0.001 = \frac{1}{1000}$$

2. Adding and Subtracting Decimals. To add and subtract decimals, the only rule is that you must keep the decimal points in a straight vertical line. For example,

$$
\begin{array}{r}
0.27 \\
+1.326 \\
\hline
1.596
\end{array}
\qquad
\begin{array}{r}
3.595 \\
-0.67 \\
\hline
2.925
\end{array}
$$

3. Multiplying Decimals. To multiply two decimal values, you first multiply the two numbers, ignoring the decimal points. Then you position the decimal point in the answer so that the number of digits to the right of the decimal point is equal to the total number of decimal places in the two numbers being multiplied. For example,

$$
\begin{array}{rl}
1.73 & \text{(two decimal places)} \\
\times 0.251 & \text{(three decimal places)} \\
\hline
173 & \\
865 & \\
346 & \\
\hline
0.43423 & \text{(five decimal places)}
\end{array}
\qquad
\begin{array}{rl}
0.25 & \text{(two decimal places)} \\
\times 0.005 & \text{(three decimal places)} \\
\hline
125 & \\
00 & \\
00 & \\
\hline
0.00125 & \text{(five decimal places)}
\end{array}
$$

4. Dividing Decimals. The simplest procedure for dividing decimals is based on the fact that dividing two numbers is identical to expressing them as a fraction:

$$0.25 \div 1.6 \text{ is identical to } \frac{0.25}{1.6}$$

You now can multiply both the numerator and the denominator of the fraction by 10, 100, 1000, or whatever number is necessary to remove the decimal places. Remember, multiplying both the numerator and the denominator of a fraction by the *same* value will create an equivalent fraction. Therefore,

$$\frac{0.25}{1.6} = \frac{0.25 \times 100}{1.6 \times 100} = \frac{25}{160}$$

The result is a division problem without any decimal places in the two numbers.

PERCENTAGES

1. Converting a Percentage to a Fraction or a Decimal. To convert a percentage to a fraction, remove the percent sign, place the number in the numerator, and use 100 for the denominator. For example,

$$52\% = \frac{52}{100} \qquad 5\% = \frac{5}{100}$$

To convert a percentage to a decimal, remove the percent sign, and divide by 100, or simply move the decimal point two places to the left. For example,

$$83\% = 83. = 0.83$$
$$14.5\% = 14.5 = 0.145$$
$$5\% = 5. = 0.05$$

2. Performing Arithmetic Operations With Percentages. There are situations when it is best to express percent values as decimals in order to perform certain arithmetic operations. For example, what is 45% of 60? This question may be stated as

$$45\% \times 60 = ?$$

The 45% should be converted to decimal form to find the solution to this question. Therefore,

$$0.45 \times 60 = 27$$

LEARNING CHECK

1. Convert $3/25$ to a decimal.

2. Convert $3/8$ to a percentage.

3. Next to each set of fractions, write "True" if they are equivalent and "False" if they are not:
   a. $\frac{3}{8} = \frac{9}{24}$ _____   b. $\frac{7}{9} = \frac{17}{19}$ _____
   c. $\frac{2}{7} = \frac{4}{14}$ _____

4. Compute the following:
   a. $\frac{1}{6} \times \frac{7}{10}$ b. $\frac{7}{8} - \frac{1}{2}$ c. $\frac{9}{10} \div \frac{2}{3}$ d. $\frac{7}{22} + \frac{2}{3}$

5. Identify the larger fraction of each pair:
   a. $\frac{7}{10}, \frac{21}{100}$ b. $\frac{3}{4}, \frac{7}{12}$ c. $\frac{22}{3}, \frac{19}{3}$

6. Convert the following decimals into fractions:
   a. 0.012   b. 0.77   c. 0.005

7. $2.59 \times 0.015 = ?$

8. $1.8 \div 0.02 = ?$

9. What is 28% of 45?

ANSWERS **1.** 0.12   **2.** 37.5%   **3. a.** True   **b.** False   **c.** True

4. a. $\frac{7}{60}$ **b.** $\frac{3}{8}$ **c.** $\frac{27}{20}$ **d.** $\frac{65}{66}$ **5. a.** $\frac{7}{10}$ **b.** $\frac{3}{4}$ **c.** $\frac{22}{3}$

6. a. $\frac{12}{1000}$ **b.** $\frac{77}{100}$ **c.** $\frac{5}{1000}$ **7.** 0.03885   **8.** 90   **9.** 12.6

A.3 NEGATIVE NUMBERS

Negative numbers are used to represent values less than zero. Negative numbers may occur when you are measuring the difference between two scores. For example, a researcher may want to evaluate the effectiveness of a propaganda film by measuring people's attitudes with a test both before and after viewing the film:

	Before	After	Amount of change
Person A	23	27	+4
Person B	18	15	−3
Person C	21	16	−5

Notice that the negative sign provides information about the direction of the difference: A plus sign indicates an increase in value, and a minus sign indicates a decrease.

Because negative numbers are frequently encountered, you should be comfortable working with these values. This section reviews basic arithmetic operations using negative numbers. You should also note that any number without a sign (+ or −) is assumed to be positive.

1. Adding Negative Numbers. When adding numbers that include negative values, simply interpret the negative sign as subtraction. For example,

$$3 + (-2) + 5 = 3 - 2 + 5 = 6$$

When adding a long string of numbers, it often is easier to add all the positive values to obtain the positive sum and then to add all of the negative values to obtain the negative sum. Finally, you subtract the negative sum from the positive sum. For example,

$$-1 + 3 + (-4) + 3 + (-6) + (-2)$$

positive sum = 6   negative sum = 13

Answer: $6 - 13 = -7$

2. Subtracting Negative Numbers. To subtract a negative number, change it to a positive number, and add. For example,

$$4 - (-3) = 4 + 3 = 7$$

This rule is easier to understand if you think of positive numbers as financial gains and negative numbers as financial losses. In this context, taking away a debt is equivalent to a financial gain. In mathematical terms, taking away a negative number is equivalent to adding a positive number. For example, suppose you are meeting a friend for lunch. You have $7, but you owe your friend $3. Thus, you really have only $4 to spend for lunch. But your friend forgives (takes away) the $3 debt. The result is that you now have $7 to spend. Expressed as an equation,

$$\$4 \text{ minus a } \$3 \text{ debt} = \$7$$

$$4 - (-3) = 4 + 3 = 7$$

3. Multiplying and Dividing Negative Numbers. When the two numbers being multiplied (or divided) have the same sign, the result is a positive number. When the two numbers have different signs, the result is negative. For example,

$$3 \times (-2) = -6$$

$$-4 \times (-2) = +8$$

The first example is easy to explain by thinking of multiplication as repeated addition. In this case,

$$3 \times (-2) = (-2) + (-2) + (-2) = -6$$

You add three negative 2s, which results in a total of negative 6. In the second example, we are multiplying by a negative number. This amounts to repeated subtraction. That is,

$$-4 \times (-2) = -(-2) - (-2) - (-2) - (-2)$$
$$= 2 + 2 + 2 + 2 = 8$$

By using the same rule for both multiplication and division, we ensure that these two operations are compatible. For example,

$$-6 \div 3 = -2$$

which is compatible with

$$3 \times (-2) = -6$$

Also,

$$8 \div (-4) = -2$$

which is compatible with

$$-4 \times (-2) = +8$$

LEARNING CHECK

1. Complete the following calculations:

   a. $3 + (-8) + 5 + 7 + (-1) + (-3)$
   b. $5 - (-9) + 2 - (-3) - (-1)$
   c. $3 - 7 - (-21) + (-5) - (-9)$
   d. $4 - (-6) - 3 + 11 - 14$
   e. $9 + 8 - 2 - 1 - (-6)$
   f. $9 \times (-3)$
   g. $-7 \times (-4)$
   h. $-6 \times (-2) \times (-3)$
   i. $-12 \div (-3)$
   j. $18 \div (-6)$

ANSWERS

1. **a.** 3   **b.** 20   **c.** 21   **d.** 4   **e.** 20
   f. -27 **g.** 28   **h.** -36 **i.** 4   **j.** -3

A.4 BASIC ALGEBRA: SOLVING EQUATIONS

An equation is a mathematical statement that indicates two quantities are identical. For example,

$$12 = 8 + 4$$

Often an equation will contain an unknown (or variable) quantity that is identified with a letter or symbol, rather than a number. For example,

$$12 = 8 + X$$

In this event, your task is to find the value of X that makes the equation "true," or balanced. For this example, an X value of 4 will make a true equation. Finding the value of X is usually called *solving the equation.*

To solve an equation, there are two points to be kept in mind:

1. Your goal is to have the unknown value (X) isolated on one side of the equation. This means that you need to remove all of the other numbers and symbols that appear on the same side of the equation as the X.
2. The equation will remain balanced, provided you treat both sides exactly the same. For example, you could add 10 points to *both* sides, and the solution (the X value) for the equation would be unchanged.

FINDING THE SOLUTION FOR AN EQUATION

We will consider four basic types of equations and the operations needed to solve them.

1. When *X* Has a Value Added to It. An example of this type of equation is

$$X + 3 = 7$$

Your goal is to isolate X on one side of the equation. Thus, you must remove the $+3$ on the left-hand side. The solution is obtained by subtracting 3 from *both* sides of the equation:

$$X + 3 - 3 = 7 - 3$$
$$X = 4$$

The solution is $X = 4$. You should always check your solution by returning to the original equation and replacing X with the value you obtained for the solution. For this example,

$$X + 3 = 7$$
$$4 + 3 = 7$$
$$7 = 7$$

2. When X Has a Value Subtracted From It. An example of this type of equation is

$$X - 8 = 12$$

In this example, you must remove the -8 from the left-hand side. Thus, the solution is obtained by adding 8 to *both* sides of the equation:

$$X - 8 + 8 = 12 + 8$$
$$X = 20$$

Check the solution:

$$X - 8 = 12$$
$$20 - 8 = 12$$
$$12 = 12$$

3. When X Is Multiplied by a Value. An example of this type of equation is

$$4X = 24$$

In this instance, it is necessary to remove the 4 that is multiplied by X. This may be accomplished by dividing both sides of the equation by 4:

$$\frac{4X}{4} = \frac{24}{4}$$
$$X = 6$$

Check the solution:

$$4X = 24$$
$$4(6) = 24$$
$$24 = 24$$

4. When X Is Divided by a Value. An example of this type of equation is

$$\frac{X}{3} = 9$$

Now the X is divided by 3, so the solution is obtained by multiplying by 3. Multiplying both sides yields

$$3\left(\frac{X}{3}\right) = 9(3)$$

$$X = 27$$

For the check,

$$\frac{X}{3} = 9$$

$$\frac{27}{3} = 9$$

$$9 = 9$$

SOLUTIONS FOR MORE-COMPLEX EQUATIONS

More-complex equations can be solved by using a combination of the preceding simple operations. Remember, at each stage you are trying to isolate X on one side of the equation. For example,

$$3X + 7 = 22$$

$$3X + 7 - 7 = 22 - 7 \quad \text{(remove + 7 by subtracting 7 from both sides)}$$

$$3X = 15$$

$$\frac{3X}{3} = \frac{15}{3} \quad \text{(remove 3 by dividing both sides by 3)}$$

$$X = 5$$

To check this solution, return to the original equation, and substitute 5 in place of X:

$$3X + 7 = 22$$

$$3(5) + 7 = 22$$

$$15 + 7 = 22$$

$$22 = 22$$

Following is another type of complex equation that is frequently encountered in statistics:

$$\frac{X + 3}{4} = 2$$

First, remove the 4 by multiplying both sides by 4:

$$4\left(\frac{X+3}{4}\right) = 2(4)$$

$$X + 3 = 8$$

Now remove the + 3 by subtracting 3 from both sides:

$$X + 3 - 3 = 8 - 3$$

$$X = 5$$

To check this solution, return to the original equation, and substitute 5 in place of X:

$$\frac{X+3}{4} = 2$$

$$\frac{5+3}{4} = 2$$

$$\frac{8}{4} = 2$$

$$2 = 2$$

LEARNING CHECK 1. Solve for X, and check the solutions:

    a. $3X = 18$　　**b.** $X + 7 = 9$　　**c.** $X - 4 = 18$　　**d.** $5X - 8 = 12$

    e. $\dfrac{X}{9} = 5$　　**f.** $\dfrac{X+1}{6} = 4$　　**g.** $X + 2 = -5$　　**h.** $\dfrac{X}{5} = -5$

    i. $\dfrac{2X}{3} = 12$　　**j.** $\dfrac{X}{3} + 1 = 3$

ANSWERS 1. **a.** $X = 6$　　**b.** $X = 2$　　**c.** $X = 22$　　**d.** $X = 4$　　**e.** $X = 45$

    f. $X = 23$　　**g.** $X = -7$　　**h.** $X = -25$　　**i.** $X = 18$　　**j.** $X = 6$

A.5 EXPONENTS AND SQUARE ROOTS

EXPONENTIAL NOTATION A simplified notation is used whenever a number is being multiplied by itself. The notation consists of placing a value, called an *exponent,* on the right-hand side of and raised above another number, called a *base.* For example,

$$7^3 \leftarrow \text{exponent}$$
$$\uparrow$$
$$\text{base}$$

The exponent indicates how many times the base is multiplied by itself. Following are some examples:

$7^3 = 7(7)(7)$ (read "7 cubed" or "7 raised to the third power")

$5^2 = 5(5)$ (read "5 squared")

$2^5 = 2(2)(2)(2)(2)$ (read "2 raised to the fifth power")

There are a few basic rules about exponents that you will need to know for this course. They are outlined here.

1. Numbers Raised to One or Zero. Any number raised to the first power equals itself. For example,

$$6^1 = 6$$

Any number (except zero) raised to the zero power equals 1. For example,

$$9^0 = 1$$

2. Exponents for Multiple Terms. The exponent applies only to the base that is just in front of it. For example,

$$XY^2 = XYY$$

$$a^2b^3 = aabbb$$

3. Negative Bases Raised to an Exponent. If a negative number is raised to a power, then the result will be positive for exponents that are even and negative for exponents that are odd. For example,

$$-4^3 = -4(-4)(-4)$$
$$= 16(-4)$$
$$= -64$$

and

$$-3^4 = -3(-3)(-3)(-3)$$
$$= 9(-3)(-3)$$
$$= 9(9)$$
$$= 81$$

4. Exponents and Parentheses. If an exponent is present outside of parentheses, then the computations within the parentheses are done first, and the exponential computation is done last:

$$(3 + 5)^2 = 8^2 = 64$$

Notice that the meaning of the expression is changed when each term in the parentheses is raised to the exponent individually:

$$3^2 + 5^2 = 9 + 25 = 34$$

Therefore,

$$X^2 + Y^2 \neq (X + Y)^2$$

5. Fractions Raised to a Power. If the numerator and denominator of a fraction are each raised to the same exponent, then the entire fraction can be raised to that exponent. That is,

$$\frac{a^2}{b^2} = \left(\frac{a}{b}\right)^2$$

For example,

$$\frac{3^2}{4^2} = \left(\frac{3}{4}\right)^2$$

$$\frac{9}{16} = \frac{3}{4}\left(\frac{3}{4}\right)$$

$$\frac{9}{16} = \frac{9}{16}$$

SQUARE ROOTS The square root of a value equals a number that when multiplied by itself yields the original value. For example, the square root of 16 equals 4 because 4 times 4 equals 16. The symbol for the square root is called a *radical*, $\sqrt{}$. The square root is taken for the number under the radical. For example,

$$\sqrt{16} = 4$$

Finding the square root is the inverse of raising a number to the second power (squaring). Thus,

$$\sqrt{a^2} = a$$

For example,

$$\sqrt{3^2} = \sqrt{9} = 3$$

Also,

$$\left(\sqrt{b}\right)^2 = b$$

For example,

$$\left(\sqrt{64}\right)^2 = 8^2 = 64$$

Computations under the same radical are performed *before* the square root is taken. For example,

$$\sqrt{9 + 16} = \sqrt{25} = 5$$

Note that with addition (or subtraction) separate radicals yield a different result:

$$\sqrt{9} + \sqrt{16} = 3 + 4 = 7$$

Therefore,

$$\sqrt{X} + \sqrt{Y} \neq \sqrt{X + Y}$$
$$\sqrt{X} - \sqrt{Y} \neq \sqrt{X - Y}$$

If the numerator and denominator of a fraction each have a radical, then the entire fraction can be placed under a single radical:

$$\frac{\sqrt{16}}{\sqrt{4}} = \sqrt{\frac{16}{4}}$$

$$\frac{4}{2} = \sqrt{4}$$

$$2 = 2$$

Therefore,

$$\frac{\sqrt{X}}{\sqrt{Y}} = \sqrt{\frac{X}{Y}}$$

Also, if the square root of one number is multiplied by the square root of another number, then the same result would be obtained by taking the square root of the product of both numbers. For example,

$$\sqrt{9} \times \sqrt{16} = \sqrt{9 \times 16}$$

$$3 \times 4 = \sqrt{144}$$

$$12 = 12$$

Therefore,

$$\sqrt{a} \times \sqrt{b} = \sqrt{ab}$$

LEARNING CHECK   **1.** Perform the following computations:

    a. $(-6)^3$

    b. $(3 + 7)^2$

    c. $a^3 b^2$ when $a = 2$ and $b = -5$

    d. $a^4 b^3$ when $a = 2$ and $b = 3$

    e. $(XY)^2$ when $X = 3$ and $Y = 5$

    f. $X^2 + Y^2$ when $X = 3$ and $Y = 5$

    g. $(X + Y)^2$ when $X = 3$ and $Y = 5$

    h. $\sqrt{5 + 4}$

    i. $\left(\sqrt{9}\right)^2$

    j. $\dfrac{\sqrt{16}}{\sqrt{4}}$

ANSWERS   **1. a.** −216   **b.** 100   **c.** 200   **d.** 432   **e.** 225

         f. 34   **g.** 64   **h.** 3   **i.** 9   **j.** 2

PROBLEMS FOR APPENDIX A   Basic Mathematics Review

1. $50/(10 - 8) = ?$

2. $(2 + 3)^2 = ?$

3. $20/10 \times 3 = ?$

4. $12 - 4 \times 2 + 6/3 = ?$

5. $24/(12 - 4) + 2 \times (6 + 3) = ?$

6. Convert $^7\!/_{20}$ to a decimal.

7. Express $^9\!/_{25}$ as a percentage.

8. Convert 0.91 to a fraction.

9. Express 0.0031 as a fraction.

10. Next to each set of fractions, write "True" if they are equivalent and "False" if they are not:

   a. $\dfrac{4}{1000} = \dfrac{2}{100}$ _____

   b. $\dfrac{5}{6} = \dfrac{52}{62}$ _____

   c. $\dfrac{1}{8} = \dfrac{7}{56}$ _____

11. Perform the following calculations:

   a. $\dfrac{4}{5} \times \dfrac{2}{3} = ?$

   b. $\dfrac{7}{9} \div \dfrac{2}{3} = ?$

   c. $\dfrac{3}{8} + \dfrac{1}{5} = ?$

   d. $\dfrac{5}{18} - \dfrac{1}{6} = ?$

12. $2.51 \times 0.017 = ?$

13. $3.88 \times 0.0002 = ?$

14. $3.17 + 17.0132 = ?$

15. $5.55 + 10.7 + 0.711 + 3.33 + 0.031 = ?$

16. $2.04 \div 0.2 = ?$

17. $0.36 \div 0.4 = ?$

18. $5 + 3 - 6 - 4 + 3 = ?$

19. $9 - (-1) - 17 + 3 - (-4) + 5 = ?$

20. $5 + 3 - (-8) - (-1) + (-3) - 4 + 10 = ?$

21. $8 \times (-3) = ?$

22. $-22 \div (-2) = ?$

23. $-2(-4) \times (-3) = ?$

24. $84 \div (-4) = ?$

Solve the equations in Problems 25–32 for X.

25. $X - 7 = -2$

26. $9 = X + 3$

27. $\dfrac{X}{4} = 11$

28. $-3 = \dfrac{X}{3}$

29. $\dfrac{X + 3}{5} = 2$

30. $\dfrac{X + 1}{3} = -8$

31. $6X - 1 = 11$

32. $2X + 3 = -11$

33. $-5^2 = ?$

34. $-5^3 = ?$

35. If $a = 4$ and $b = 3$, then $a^2 + b^4 = ?$

36. If $a = -1$ and $b = 4$, then $(a + b)^2 = ?$

37. If $a = -1$ and $b = 5$, then $ab^2 = ?$

38. $\dfrac{18}{\sqrt{4}} = ?$

39. $\sqrt{\dfrac{20}{5}} = ?$

SKILLS ASSESSMENT FINAL EXAM

SECTION 1

1. $4 + 8/4 = ?$

2. $(4 + 8)/4 = ?$

3. $4 \times 3^2 = ?$

4. $(4 \times 3)^2 = ?$

5. $10/5 \times 2 = ?$

6. $10/(5 \times 2) = ?$

7. $40 - 10 \times 4/2 = ?$

8. $(5 - 1)^2/2 = ?$

9. $3 \times 6 - 3^2 = ?$

10. $2 \times (6 - 3)^2 = ?$

11. $4 \times 3 - 1 + 8 \times 2 = ?$

12. $4 \times (3 - 1 + 8) \times 2 = ?$

SECTION 2

1. Express $^{14}/_{80}$ as a decimal.

2. Convert $^6/_{25}$ to a percentage.

3. Convert 18% to a fraction.

4. $\frac{3}{5} \times \frac{2}{3} = ?$

5. $\frac{5}{24} + \frac{5}{6} = ?$

6. $\frac{7}{12} \div \frac{5}{6} = ?$

7. $\frac{5}{9} - \frac{1}{3} = ?$

8. $6.11 \times 0.22 = ?$

9. $0.18 \div 0.9 = ?$

10. $8.742 + 0.76 = ?$

11. In a statistics class of 72 students, three-eighths of the students received a *B* on the first test. How many *B*s were earned?

12. What is 15% of 64?

SECTION 3

1. $3 - 1 - 3 + 5 - 2 + 6 = ?$

2. $-8 - (-6) = ?$

3. $2 - (-7) - 3 + (-11) - 20 = ?$

4. $-8 - 3 - (-1) - 2 - 1 = ?$

5. $8(-2) = ?$

6. $-7(-7) = ?$

7. $-3(-2)(-5) = ?$

8. $-3(5)(-3) = ?$

9. $-24 \div (-4) = ?$

10. $36 \div (-6) = ?$

11. $-56/7 = ?$

12. $-7/(-1) = ?$

SECTION 4

Solve for *X*.

1. $X + 5 = 12$

2. $X - 11 = 3$

3. $10 = X + 4$

4. $4X = 20$

5. $\frac{X}{2} = 15$

6. $18 = 9X$

7. $\frac{X}{5} = 35$

8. $2X + 8 = 4$

9. $\frac{X + 1}{3} = 6$

10. $4X + 3 = -13$

11. $\frac{X + 3}{3} = -7$

12. $23 = 2X - 5$

SECTION 5

1. $5^3 = ?$

2. $-4^3 = ?$

3. $-2^5 = ?$

4. $-2^6 = ?$

5. If $a = 4$ and $b = 2$, then $ab^2 = ?$

6. If $a = 4$ and $b = 2$, then $(a + b)^3 = ?$

7. If $a = 4$ and $b = 2$, then $a^2 + b^2 = ?$

8. $(11 + 4)^2 = ?$

9. $\sqrt{7^2} = ?$

10. If $a = 36$ and $b = 64$, then $\sqrt{a + b} = ?$

11. $\frac{25}{\sqrt{25}} = ?$

12. If $a = -1$ and $b = 2$, then $a^3 b^4 = ?$

ANSWER KEY   Skills Assessment Exams

PREVIEW EXAM

FINAL EXAM

SECTION 1

1. 17
2. 35
3. 6
4. 24
5. 5
6. 2
7. $\dfrac{1}{3}$
8. 8
9. 72
10. 8
11. 24
12. 48

SECTION 1

1. 6
2. 3
3. 36
4. 144
5. 4
6. 1
7. 20
8. 8
9. 9
10. 18
11. 27
12. 80

SECTION 2

1. 75%
2. $\dfrac{30}{100}$, or $\dfrac{3}{10}$
3. 0.3
4. $\dfrac{10}{13}$
5. 1.625
6. $\dfrac{2}{20}$
7. $\dfrac{19}{24}$
8. 1.4
9. $\dfrac{4}{15}$
10. 7.5
11. 16
12. 36

SECTION 2

1. 0.175
2. 24%
3. $\dfrac{18}{100}$, or $\dfrac{9}{50}$
4. $\dfrac{6}{15}$
5. $\dfrac{25}{24}$
6. $\dfrac{42}{60}$, or $\dfrac{7}{10}$
7. $\dfrac{2}{9}$
8. 1.3442
9. 0.2
10. 9.502
11. 27
12. 9.6

SECTION 3

1. 4
2. 8
3. 2
4. 9
5. −12
6. 12
7. −15
8. −24
9. −4
10. 3
11. −2
12. 25

SECTION 3

1. 8
2. −2
3. −25
4. −13
5. −16
6. 49
7. −30
8. 45
9. 6
10. −6
11. −8
12. 7

SECTION 4

1. $X = 7$	7. $X = 80$
2. $X = 29$	8. $X = -3$
3. $X = 9$	9. $X = 11$
4. $X = 4$	10. $X = 25$
5. $X = 24$	11. $X = 11$
6. $X = 15$	12. $X = 7$

SECTION 5

1. 64	7. 256
2. 4	8. 8
3. 54	9. 12
4. 25	10. 121
5. 13	11. 33
6. -27	12. -9

SECTION 4

1. $X = 7$	7. $X = 175$
2. $X = 14$	8. $X = -2$
3. $X = 6$	9. $X = 17$
4. $X = 5$	10. $X = -4$
5. $X = 30$	11. $X = -24$
6. $X = 2$	12. $X = 14$

SECTION 5

1. 125	7. 20
2. -64	8. 225
3. -32	9. 7
4. 64	10. 10
5. 16	11. 5
6. 216	12. -16

SOLUTIONS TO SELECTED PROBLEMS FOR APPENDIX A   Basic Mathematics Review

1. 25

3. 6

5. 21

6. 0.35

7. 36%

9. $\dfrac{31}{10,000}$

10. b. False

11. a. $\dfrac{8}{15}$ b. $\dfrac{21}{18}$ c. $\dfrac{23}{40}$

12. 0.04267

14. 20.1832

17. 0.9

19. 5

21. -24

22. 11

25. $X = 5$

28. $X = -9$

30. $X = -25$

31. $X = 2$

34. -125

36. 9

37. -25

39. 2

SUGGESTED REVIEW BOOKS

There are many basic mathematics review books available if you need a more extensive review than this appendix can provide. The following books are but a few of the many that you may find helpful:

Barker, V. C., & Aufmann, R. N. (1982). *Essential mathematics.* Boston: Houghton Mifflin.

Bloomfield, D. I. (1994). *Introductory algebra.* St. Paul, Minn.: West.

Falstein, L. D. (1986). *Basic mathematics,* 2nd ed. Reading, Mass.: Addison-Wesley.

Goodman, A., & Hirsch, L. (1994). *Understanding elementary algebra.* St. Paul, Minn.: West.

Washington, A. J. (1984). *Arithmetic and beginning algebra.* Menlo Park, Calif.: Benjamin/Cummings.

STATISTICAL TABLES

TABLE B.1   **THE UNIT NORMAL TABLE***

*Column A lists z-score values. A vertical line drawn through a normal distribution at a z-score location divides the distribution into two sections.
Column B identifies the proportion in the larger section, called the *body*.
Column C identifies the proportion in the smaller section, called the *tail*.

Note: Because the normal distribution is symmetrical, the proportions for negative z-scores are the same as those for positive z-scores.

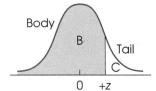 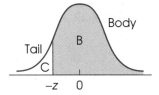

(A)	(B)	(C)	(A)	(B)	(C)	(A)	(B)	(C)
z	Proportion in body	Proportion in tail	z	Proportion in body	Proportion in tail	z	Proportion in body	Proportion in tail
0.00	.5000	.5000	0.20	.5793	.4207	0.40	.6554	.3446
0.01	.5040	.4960	0.21	.5832	.4168	0.41	.6591	.3409
0.02	.5080	.4920	0.22	.5871	.4129	0.42	.6628	.3372
0.03	.5120	.4880	0.23	.5910	.4090	0.43	.6664	.3336
0.04	.5160	.4840	0.24	.5948	.4052	0.44	.6700	.3300
0.05	.5199	.4801	0.25	.5987	.4013	0.45	.6736	.3264
0.06	.5239	.4761	0.26	.6026	.3974	0.46	.6772	.3228
0.07	.5279	.4721	0.27	.6064	.3936	0.47	.6808	.3192
0.08	.5319	.4681	0.28	.6103	.3897	0.48	.6844	.3156
0.09	.5359	.4641	0.29	.6141	.3859	0.49	.6879	.3121
0.10	.5398	.4602	0.30	.6179	.3821	0.50	.6915	.3085
0.11	.5438	.4562	0.31	.6217	.3783	0.51	.6950	.3050
0.12	.5478	.4522	0.32	.6255	.3745	0.52	.6985	.3015
0.13	.5517	.4483	0.33	.6293	.3707	0.53	.7019	.2981
0.14	.5557	.4443	0.34	.6331	.3669	0.54	.7054	.2946
0.15	.5596	.4404	0.35	.6368	.3632	0.55	.7088	.2912
0.16	.5636	.4364	0.36	.6406	.3594	0.56	.7123	.2877
0.17	.5675	.4325	0.37	.6443	.3557	0.57	.7157	.2843
0.18	.5714	.4286	0.38	.6480	.3520	0.58	.7190	.2810
0.19	.5753	.4247	0.39	.6517	.3483	0.59	.7224	.2776

TABLE B.1   **continued**

(A) z	(B) Proportion in body	(C) Proportion in tail	(A) z	(B) Proportion in body	(C) Proportion in tail	(A) z	(B) Proportion in body	(C) Proportion in tail
0.60	.7257	.2743	1.05	.8531	.1469	1.50	.9332	.0668
0.61	.7291	.2709	1.06	.8554	.1446	1.51	.9345	.0655
0.62	.7324	.2676	1.07	.8577	.1423	1.52	.9357	.0643
0.63	.7357	.2643	1.08	.8599	.1401	1.53	.9370	.0630
0.64	.7389	.2611	1.09	.8621	.1379	1.54	.9382	.0618
0.65	.7422	.2578	1.10	.8643	.1357	1.55	.9394	.0606
0.66	.7454	.2546	1.11	.8665	.1335	1.56	.9406	.0594
0.67	.7486	.2514	1.12	.8686	.1314	1.57	.9418	.0582
0.68	.7517	.2483	1.13	.8708	.1292	1.58	.9429	.0571
0.69	.7549	.2451	1.14	.8729	.1271	1.59	.9441	.0559
0.70	.7580	.2420	1.15	.8749	.1251	1.60	.9452	.0548
0.71	.7611	.2389	1.16	.8770	.1230	1.61	.9463	.0537
0.72	.7642	.2358	1.17	.8790	.1210	1.62	.9474	.0526
0.73	.7673	.2327	1.18	.8810	.1190	1.63	.9484	.0516
0.74	.7704	.2296	1.19	.8830	.1170	1.64	.9495	.0505
0.75	.7734	.2266	1.20	.8849	.1151	1.65	.9505	.0495
0.76	.7764	.2236	1.21	.8869	.1131	1.66	.9515	.0485
0.77	.7794	.2206	1.22	.8888	.1112	1.67	.9525	.0475
0.78	.7823	.2177	1.23	.8907	.1093	1.68	.9535	.0465
0.79	.7852	.2148	1.24	.8925	.1075	1.69	.9545	.0455
0.80	.7881	.2119	1.25	.8944	.1056	1.70	.9554	.0446
0.81	.7910	.2090	1.26	.8962	.1038	1.71	.9564	.0436
0.82	.7939	.2061	1.27	.8980	.1020	1.72	.9573	.0427
0.83	.7967	.2033	1.28	.8997	.1003	1.73	.9582	.0418
0.84	.7995	.2005	1.29	.9015	.0985	1.74	.9591	.0409
0.85	.8023	.1977	1.30	.9032	.0968	1.75	.9599	.0401
0.86	.8051	.1949	1.31	.9049	.0951	1.76	.9608	.0392
0.87	.8078	.1922	1.32	.9066	.0934	1.77	.9616	.0384
0.88	.8106	.1894	1.33	.9082	.0918	1.78	.9625	.0375
0.89	.8133	.1867	1.34	.9099	.0901	1.79	.9633	.0367
0.90	.8159	.1841	1.35	.9115	.0885	1.80	.9641	.0359
0.91	.8186	.1814	1.36	.9131	.0869	1.81	.9649	.0351
0.92	.8212	.1788	1.37	.9147	.0853	1.82	.9656	.0344
0.93	.8238	.1762	1.38	.9162	.0838	1.83	.9664	.0336
0.94	.8264	.1736	1.39	.9177	.0823	1.84	.9671	.0329
0.95	.8289	.1711	1.40	.9192	.0808	1.85	.9678	.0322
0.96	.8315	.1685	1.41	.9207	.0793	1.86	.9686	.0314
0.97	.8340	.1660	1.42	.9222	.0778	1.87	.9693	.0307
0.98	.8365	.1635	1.43	.9236	.0764	1.88	.9699	.0301
0.99	.8389	.1611	1.44	.9251	.0749	1.89	.9706	.0294
1.00	.8413	.1587	1.45	.9265	.0735	1.90	.9713	.0287
1.01	.8438	.1562	1.46	.9279	.0721	1.91	.9719	.0281
1.02	.8461	.1539	1.47	.9292	.0708	1.92	.9726	.0274
1.03	.8485	.1515	1.48	.9306	.0694	1.93	.9732	.0268
1.04	.8508	.1492	1.49	.9319	.0681	1.94	.9738	.0262

TABLE B.1   continued

(A)	(B) Proportion in body	(C) Proportion in tail	(A)	(B) Proportion in body	(C) Proportion in tail	(A)	(B) Proportion in body	(C) Proportion in tail
z			z			z		
1.95	.9744	.0256	2.42	.9922	.0078	2.88	.9980	.0020
1.96	.9750	.0250	2.43	.9925	.0075	2.89	.9981	.0019
1.97	.9756	.0244	2.44	.9927	.0073	2.90	.9981	.0019
1.98	.9761	.0239	2.45	.9929	.0071	2.91	.9982	.0018
1.99	.9767	.0233	2.46	.9931	.0069	2.92	.9982	.0018
2.00	.9772	.0228	2.47	.9932	.0068	2.93	.9983	.0017
2.01	.9778	.0222	2.48	.9934	.0066	2.94	.9984	.0016
2.02	.9783	.0217	2.49	.9936	.0064	2.95	.9984	.0016
2.03	.9788	.0212	2.50	.9938	.0062	2.96	.9985	.0015
2.04	.9793	.0207	2.51	.9940	.0060	2.97	.9985	.0015
2.05	.9798	.0202	2.52	.9941	.0059	2.98	.9986	.0014
2.06	.9803	.0197	2.53	.9943	.0057	2.99	.9986	.0014
2.07	.9808	.0192	2.54	.9945	.0055	3.00	.9987	.0013
2.08	.9812	.0188	2.55	.9946	.0054	3.01	.9987	.0013
2.09	.9817	.0183	2.56	.9948	.0052	3.02	.9987	.0013
2.10	.9821	.0179	2.57	.9949	.0051	3.03	.9988	.0012
2.11	.9826	.0174	2.58	.9951	.0049	3.04	.9988	.0012
2.12	.9830	.0170	2.59	.9952	.0048	3.05	.9989	.0011
2.13	.9834	.0166	2.60	.9953	.0047	3.06	.9989	.0011
2.14	.9838	.0162	2.61	.9955	.0045	3.07	.9989	.0011
2.15	.9842	.0158	2.62	.9956	.0044	3.08	.9990	.0010
2.16	.9846	.0154	2.63	.9957	.0043	3.09	.9990	.0010
2.17	.9850	.0150	2.64	.9959	.0041	3.10	.9990	.0010
2.18	.9854	.0146	2.65	.9960	.0040	3.11	.9991	.0009
2.19	.9857	.0143	2.66	.9961	.0039	3.12	.9991	.0009
2.20	.9861	.0139	2.67	.9962	.0038	3.13	.9991	.0009
2.21	.9864	.0136	2.68	.9963	.0037	3.14	.9992	.0008
2.22	.9868	.0132	2.69	.9964	.0036	3.15	.9992	.0008
2.23	.9871	.0129	2.70	.9965	.0035	3.16	.9992	.0008
2.24	.9875	.0125	2.71	.9966	.0034	3.17	.9992	.0008
2.25	.9878	.0122	2.72	.9967	.0033	3.18	.9993	.0007
2.26	.9881	.0119	2.73	.9968	.0032	3.19	.9993	.0007
2.27	.9884	.0116	2.74	.9969	.0031	3.20	.9993	.0007
2.28	.9887	.0113	2.75	.9970	.0030	3.21	.9993	.0007
2.29	.9890	.0110	2.76	.9971	.0029	3.22	.9994	.0006
2.30	.9893	.0107	2.77	.9972	.0028	3.23	.9994	.0006
2.31	.9896	.0104	2.78	.9973	.0027	3.24	.9994	.0006
2.32	.9898	.0102	2.79	.9974	.0026	3.30	.9995	.0005
2.33	.9901	.0099	2.80	.9974	.0026	3.40	.9997	.0003
2.34	.9904	.0096	2.81	.9975	.0025	3.50	.9998	.0002
2.35	.9906	.0094	2.82	.9976	.0024	3.60	.9998	.0002
2.36	.9909	.0091	2.83	.9977	.0023	3.70	.9999	.0001
2.37	.9911	.0089	2.84	.9977	.0023	3.80	.99993	.00007
2.38	.9913	.0087	2.85	.9978	.0022	3.90	.99995	.00005
2.39	.9916	.0084	2.86	.9979	.0021	4.00	.99997	.00003
2.40	.9918	.0082	2.87	.9979	.0021			
2.41	.9920	.0080						

Generated by the Minitab statistical program using the CDL command.

TABLE B.2   **THE _t_ DISTRIBUTION**

df	Proportion in one tail					
	0.25	0.10	0.05	0.025	0.01	0.005
	Proportion in two tails					
	0.50	0.20	0.10	0.05	0.02	0.01
1	1.000	3.078	6.314	12.706	31.821	63.657
2	0.816	1.886	2.920	4.303	6.965	9.925
3	0.765	1.638	2.353	3.182	4.541	5.841
4	0.741	1.533	2.132	2.776	3.747	4.604
5	0.727	1.476	2.015	2.571	3.365	4.032
6	0.718	1.440	1.943	2.447	3.143	3.707
7	0.711	1.415	1.895	2.365	2.998	3.499
8	0.706	1.397	1.860	2.306	2.896	3.355
9	0.703	1.383	1.833	2.262	2.821	3.250
10	0.700	1.372	1.812	2.228	2.764	3.169
11	0.697	1.363	1.796	2.201	2.718	3.106
12	0.695	1.356	1.782	2.179	2.681	3.055
13	0.694	1.350	1.771	2.160	2.650	3.012
14	0.692	1.345	1.761	2.145	2.624	2.977
15	0.691	1.341	1.753	2.131	2.602	2.947
16	0.690	1.337	1.746	2.120	2.583	2.921
17	0.689	1.333	1.740	2.110	2.567	2.898
18	0.688	1.330	1.734	2.101	2.552	2.878
19	0.688	1.328	1.729	2.093	2.539	2.861
20	0.687	1.325	1.725	2.086	2.528	2.845
21	0.686	1.323	1.721	2.080	2.518	2.831
22	0.686	1.321	1.717	2.074	2.508	2.819
23	0.685	1.319	1.714	2.069	2.500	2.807
24	0.685	1.318	1.711	2.064	2.492	2.797
25	0.684	1.316	1.708	2.060	2.485	2.787
26	0.684	1.315	1.706	2.056	2.479	2.779
27	0.684	1.314	1.703	2.052	2.473	2.771
28	0.683	1.313	1.701	2.048	2.467	2.763
29	0.683	1.311	1.699	2.045	2.462	2.756
30	0.683	1.310	1.697	2.042	2.457	2.750
40	0.681	1.303	1.684	2.021	2.423	2.704
60	0.679	1.296	1.671	2.000	2.390	2.660
120	0.677	1.289	1.658	1.980	2.358	2.617
∞	0.674	1.282	1.645	1.960	2.326	2.576

From Table III of R. A. Fisher and F. Yates, _Statistical Tables for Biological, Agricultural and Medical Research_, 6th ed. London: Longman Group Ltd., 1974 (previously published by Oliver and Boyd Ltd., Edinburgh). Adapted and reprinted with permission of Addison Wesley Longman Ltd.

TABLE B.3   CRITICAL VALUES FOR THE *F*-MAX STATISTIC*

*The critical values for α = .05 are in lightface type, and for α = .01, they are in boldface type.

n − 1	k = Number of samples										
	2	3	4	5	6	7	8	9	10	11	12
4	9.60	15.5	20.6	25.2	29.5	33.6	37.5	41.4	44.6	48.0	51.4
	23.2	**37.**	**49.**	**59.**	**69.**	**79.**	**89.**	**97.**	**106.**	**113.**	**120.**
5	7.15	10.8	13.7	16.3	18.7	20.8	22.9	24.7	26.5	28.2	29.9
	14.9	**22.**	**28.**	**33.**	**38.**	**42.**	**46.**	**50.**	**54.**	**57.**	**60.**
6	5.82	8.38	10.4	12.1	13.7	15.0	16.3	17.5	18.6	19.7	20.7
	11.1	**15.5**	**19.1**	**22.**	**25.**	**27.**	**30.**	**32.**	**34.**	**36.**	**37.**
7	4.99	6.94	8.44	9.70	10.8	11.8	12.7	13.5	14.3	15.1	15.8
	8.89	**12.1**	**14.5**	**16.5**	**18.4**	**20.**	**22.**	**23.**	**24.**	**26.**	**27.**
8	4.43	6.00	7.18	8.12	9.03	9.78	10.5	11.1	11.7	12.2	12.7
	7.50	**9.9**	**11.7**	**13.2**	**14.5**	**15.8**	**16.9**	**17.9**	**18.9**	**19.8**	**21.**
9	4.03	5.34	6.31	7.11	7.80	8.41	8.95	9.45	9.91	10.3	10.7
	6.54	**8.5**	**9.9**	**11.1**	**12.1**	**13.1**	**13.9**	**14.7**	**15.3**	**16.0**	**16.6**
10	3.72	4.85	5.67	6.34	6.92	7.42	7.87	8.28	8.66	9.01	9.34
	5.85	**7.4**	**8.6**	**9.6**	**10.4**	**11.1**	**11.8**	**12.4**	**12.9**	**13.4**	**13.9**
12	3.28	4.16	4.79	5.30	5.72	6.09	6.42	6.72	7.00	7.25	7.48
	4.91	**6.1**	**6.9**	**7.6**	**8.2**	**8.7**	**9.1**	**9.5**	**9.9**	**10.2**	**10.6**
15	2.86	3.54	4.01	4.37	4.68	4.95	5.19	5.40	5.59	5.77	5.93
	4.07	**4.9**	**5.5**	**6.0**	**6.4**	**6.7**	**7.1**	**7.3**	**7.5**	**7.8**	**8.0**
20	2.46	2.95	3.29	3.54	3.76	3.94	4.10	4.24	4.37	4.49	4.59
	3.32	**3.8**	**4.3**	**4.6**	**4.9**	**5.1**	**5.3**	**5.5**	**5.6**	**5.8**	**5.9**
30	2.07	2.40	2.61	2.78	2.91	3.02	3.12	3.21	3.29	3.36	3.39
	2.63	**3.0**	**3.3**	**3.5**	**3.6**	**3.7**	**3.8**	**3.9**	**4.0**	**4.1**	**4.2**
60	1.67	1.85	1.96	2.04	2.11	2.17	2.22	2.26	2.30	2.33	2.36
	1.96	**2.2**	**2.3**	**2.4**	**2.4**	**2.5**	**2.5**	**2.6**	**2.6**	**2.7**	**2.7**

From Table 31 of E. Pearson and H. O. Hartley, *Biometrika Tables for Statisticians,* 3rd ed. New York: Cambridge University Press, 1966. Adapted and reprinted with permission of the Biometrika trustees.

TABLE B.4 **THE *F* DISTRIBUTION***

*Table entries in lightface type are critical values for the .05 level of significance.
Boldface type values are for the .01 level of significance.

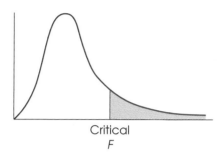

Critical
F

Degrees of freedom: Denominator	Degrees of freedom: Numerator														
	1	2	3	4	5	6	7	8	9	10	11	12	14	16	20
1	161	200	216	225	230	234	237	239	241	242	243	244	245	246	248
	4052	**4999**	**5403**	**5625**	**5764**	**5859**	**5928**	**5981**	**6022**	**6056**	**6082**	**6106**	**6142**	**6169**	**6208**
2	18.51	19.00	19.16	19.25	19.30	19.33	19.36	19.37	19.38	19.39	19.40	19.41	19.42	19.43	19.44
	98.49	**99.00**	**99.17**	**99.25**	**99.30**	**99.33**	**99.34**	**99.36**	**99.38**	**99.40**	**99.41**	**99.42**	**99.43**	**99.44**	**99.45**
3	10.13	9.55	9.28	9.12	9.01	8.94	8.88	8.84	8.81	8.78	8.76	8.74	8.71	8.69	8.66
	34.12	**30.92**	**29.46**	**28.71**	**28.24**	**27.91**	**27.67**	**27.49**	**27.34**	**27.23**	**27.13**	**27.05**	**26.92**	**26.83**	**26.69**
4	7.71	6.94	6.59	6.39	6.26	6.16	6.09	6.04	6.00	5.96	5.93	5.91	5.87	5.84	5.80
	21.20	**18.00**	**16.69**	**15.98**	**15.52**	**15.21**	**14.98**	**14.80**	**14.66**	**14.54**	**14.45**	**14.37**	**14.24**	**14.15**	**14.02**
5	6.61	5.79	5.41	5.19	5.05	4.95	4.88	4.82	4.78	4.74	4.70	4.68	4.64	4.60	4.56
	16.26	**13.27**	**12.06**	**11.39**	**10.97**	**10.67**	**10.45**	**10.27**	**10.15**	**10.05**	**9.96**	**9.89**	**9.77**	**9.68**	**9.55**
6	5.99	5.14	4.76	4.53	4.39	4.28	4.21	4.15	4.10	4.06	4.03	4.00	3.96	3.92	3.87
	13.74	**10.92**	**9.78**	**9.15**	**8.75**	**8.47**	**8.26**	**8.10**	**7.98**	**7.87**	**7.79**	**7.72**	**7.60**	**7.52**	**7.39**
7	5.59	4.47	4.35	4.12	3.97	3.87	3.79	3.73	3.68	3.63	3.60	3.57	3.52	3.49	3.44
	12.25	**9.55**	**8.45**	**7.85**	**7.46**	**7.19**	**7.00**	**6.84**	**6.71**	**6.62**	**6.54**	**6.47**	**6.35**	**6.27**	**6.15**
8	5.32	4.46	4.07	3.84	3.69	3.58	3.50	3.44	3.39	3.34	3.31	3.28	3.23	3.20	3.15
	11.26	**8.65**	**7.59**	**7.01**	**6.63**	**6.37**	**6.19**	**6.03**	**5.91**	**5.82**	**5.74**	**5.67**	**5.56**	**5.48**	**5.36**
9	5.12	4.26	3.86	3.63	3.48	3.37	3.29	3.23	3.18	3.13	3.10	3.07	3.02	2.98	2.93
	10.56	**8.02**	**6.99**	**6.42**	**6.06**	**5.80**	**5.62**	**5.47**	**5.35**	**5.26**	**5.18**	**5.11**	**5.00**	**4.92**	**4.80**
10	4.96	4.10	3.71	3.48	3.33	3.22	3.14	3.07	3.02	2.97	2.94	2.91	2.86	2.82	2.77
	10.04	**7.56**	**6.55**	**5.99**	**5.64**	**5.39**	**5.21**	**5.06**	**4.95**	**4.85**	**4.78**	**4.71**	**4.60**	**4.52**	**4.41**
11	4.84	3.98	3.59	3.36	3.20	3.09	3.01	2.95	2.90	2.86	2.82	2.79	2.74	2.70	2.65
	9.65	**7.20**	**6.22**	**5.67**	**5.32**	**5.07**	**4.88**	**4.74**	**4.63**	**4.54**	**4.46**	**4.40**	**4.29**	**4.21**	**4.10**
12	4.75	3.88	3.49	3.26	3.11	3.00	2.92	2.85	2.80	2.76	2.72	2.69	2.64	2.60	2.54
	9.33	**6.93**	**5.95**	**5.41**	**5.06**	**4.82**	**4.65**	**4.50**	**4.39**	**4.30**	**4.22**	**4.16**	**4.05**	**3.98**	**3.86**
13	4.67	3.80	3.41	3.18	3.02	2.92	2.84	2.77	2.72	2.67	2.63	2.60	2.55	2.51	2.46
	9.07	**6.70**	**5.74**	**5.20**	**4.86**	**4.62**	**4.44**	**4.30**	**4.19**	**4.10**	**4.02**	**3.96**	**3.85**	**3.78**	**3.67**
14	4.60	3.74	3.34	3.11	2.96	2.85	2.77	2.70	2.65	2.60	2.56	2.53	2.48	2.44	2.39
	8.86	**6.51**	**5.56**	**5.03**	**4.69**	**4.46**	**4.28**	**4.14**	**4.03**	**3.94**	**3.86**	**3.80**	**3.70**	**3.62**	**3.51**
15	4.54	3.68	3.29	3.06	2.90	2.79	2.70	2.64	2.59	2.55	2.51	2.48	2.43	2.39	2.33
	8.68	**6.36**	**5.42**	**4.89**	**4.56**	**4.32**	**4.14**	**4.00**	**3.89**	**3.80**	**3.73**	**3.67**	**3.56**	**3.48**	**3.36**
16	4.49	3.63	3.24	3.01	2.85	2.74	2.66	2.59	2.54	2.49	2.45	2.42	2.37	2.33	2.28
	8.53	**6.23**	**5.29**	**4.77**	**4.44**	**4.20**	**4.03**	**3.89**	**3.78**	**3.69**	**3.61**	**3.55**	**3.45**	**3.37**	**3.25**

TABLE B.4   continued

Degrees of freedom: Denominator	Degrees of freedom: Numerator														
	1	2	3	4	5	6	7	8	9	10	11	12	14	16	20
17	4.45	3.59	3.20	2.96	2.81	2.70	2.62	2.55	2.50	2.45	2.41	2.38	2.33	2.29	2.23
	8.40	**6.11**	**5.18**	**4.67**	**4.34**	**4.10**	**3.93**	**3.79**	**3.68**	**3.59**	**3.52**	**3.45**	**3.35**	**3.27**	**3.16**
18	4.41	3.55	3.16	2.93	2.77	2.66	2.58	2.51	2.46	2.41	2.37	2.34	2.29	2.25	2.19
	8.28	**6.01**	**5.09**	**4.58**	**4.25**	**4.01**	**3.85**	**3.71**	**3.60**	**3.51**	**3.44**	**3.37**	**3.27**	**3.19**	**3.07**
19	4.38	3.52	3.13	2.90	2.74	2.63	2.55	2.48	2.43	2.38	2.34	2.31	2.26	2.21	2.15
	8.18	**5.93**	**5.01**	**4.50**	**4.17**	**3.94**	**3.77**	**3.63**	**3.52**	**3.43**	**3.36**	**3.30**	**3.19**	**3.12**	**3.00**
20	4.35	3.49	3.10	2.87	2.71	2.60	2.52	2.45	2.40	2.35	2.31	2.28	2.23	2.18	2.12
	8.10	**5.85**	**4.94**	**4.43**	**4.10**	**3.87**	**3.71**	**3.56**	**3.45**	**3.37**	**3.30**	**3.23**	**3.13**	**3.05**	**2.94**
21	4.32	3.47	3.07	2.84	2.68	2.57	2.49	2.42	2.37	2.32	2.28	2.25	2.20	2.15	2.09
	8.02	**5.78**	**4.87**	**4.37**	**4.04**	**3.81**	**3.65**	**3.51**	**3.40**	**3.31**	**3.24**	**3.17**	**3.07**	**2.99**	**2.88**
22	4.30	3.44	3.05	2.82	2.66	2.55	2.47	2.40	2.35	2.30	2.26	2.23	2.18	2.13	2.07
	7.94	**5.72**	**4.82**	**4.31**	**3.99**	**3.76**	**3.59**	**3.45**	**3.35**	**3.26**	**3.18**	**3.12**	**3.02**	**2.94**	**2.83**
23	4.28	3.42	3.03	2.80	2.64	2.53	2.45	2.38	2.32	2.28	2.24	2.20	2.14	2.10	2.04
	7.88	**5.66**	**4.76**	**4.26**	**3.94**	**3.71**	**3.54**	**3.41**	**3.30**	**3.21**	**3.14**	**3.07**	**2.97**	**2.89**	**2.78**
24	4.26	3.40	3.01	2.78	2.62	2.51	2.43	2.36	2.30	2.26	2.22	2.18	2.13	2.09	2.02
	7.82	**5.61**	**4.72**	**4.22**	**3.90**	**3.67**	**3.50**	**3.36**	**3.25**	**3.17**	**3.09**	**3.03**	**2.93**	**2.85**	**2.74**
25	4.24	3.38	2.99	2.76	2.60	2.49	2.41	2.34	2.28	2.24	2.20	2.16	2.11	2.06	2.00
	7.77	**5.57**	**4.68**	**4.18**	**3.86**	**3.63**	**3.46**	**3.32**	**3.21**	**3.13**	**3.05**	**2.99**	**2.89**	**2.81**	**2.70**
26	4.22	3.37	2.98	2.74	2.59	2.47	2.39	2.32	2.27	2.22	2.18	2.15	2.10	2.05	1.99
	7.72	**5.53**	**4.64**	**4.14**	**3.82**	**3.59**	**3.42**	**3.29**	**3.17**	**3.09**	**3.02**	**2.96**	**2.86**	**2.77**	**2.66**
27	4.21	3.35	2.96	2.73	2.57	2.46	2.37	2.30	2.25	2.20	2.16	2.13	2.08	2.03	1.97
	7.68	**5.49**	**4.60**	**4.11**	**3.79**	**3.56**	**3.39**	**3.26**	**3.14**	**3.06**	**2.98**	**2.93**	**2.83**	**2.74**	**2.63**
28	4.20	3.34	2.95	2.71	2.56	2.44	2.36	2.29	2.24	2.19	2.15	2.12	2.06	2.02	1.96
	7.64	**5.45**	**4.57**	**4.07**	**3.76**	**3.53**	**3.36**	**3.23**	**3.11**	**3.03**	**2.95**	**2.90**	**2.80**	**2.71**	**2.60**
29	4.18	3.33	2.93	2.70	2.54	2.43	2.35	2.28	2.22	2.18	2.14	2.10	2.05	2.00	1.94
	7.60	**5.42**	**4.54**	**4.04**	**3.73**	**3.50**	**3.33**	**3.20**	**3.08**	**3.00**	**2.92**	**2.87**	**2.77**	**2.68**	**2.57**
30	4.17	3.32	2.92	2.69	2.53	2.42	2.34	2.27	2.21	2.16	2.12	2.09	2.04	1.99	1.93
	7.56	**5.39**	**4.51**	**4.02**	**3.70**	**3.47**	**3.30**	**3.17**	**3.06**	**2.98**	**2.90**	**2.84**	**2.74**	**2.66**	**2.55**
32	4.15	3.30	2.90	2.67	2.51	2.40	2.32	2.25	2.19	2.14	2.10	2.07	2.02	1.97	1.91
	7.50	**5.34**	**4.46**	**3.97**	**3.66**	**3.42**	**3.25**	**3.12**	**3.01**	**2.94**	**2.86**	**2.80**	**2.70**	**2.62**	**2.51**
34	4.13	3.28	2.88	2.65	2.49	2.38	2.30	2.23	2.17	2.12	2.08	2.05	2.00	1.95	1.89
	7.44	**5.29**	**4.42**	**3.93**	**3.61**	**3.38**	**3.21**	**3.08**	**2.97**	**2.89**	**2.82**	**2.76**	**2.66**	**2.58**	**2.47**
36	4.11	3.26	2.86	2.63	2.48	2.36	2.28	2.21	2.15	2.10	2.06	2.03	1.98	1.93	1.87
	7.39	**5.25**	**4.38**	**3.89**	**3.58**	**3.35**	**3.18**	**3.04**	**2.94**	**2.86**	**2.78**	**2.72**	**2.62**	**2.54**	**2.43**
38	4.10	3.25	2.85	2.62	2.46	2.35	2.26	2.19	2.14	2.09	2.05	2.02	1.96	1.92	1.85
	7.35	**5.21**	**4.34**	**3.86**	**3.54**	**3.32**	**3.15**	**3.02**	**2.91**	**2.82**	**2.75**	**2.69**	**2.59**	**2.51**	**2.40**
40	4.08	3.23	2.84	2.61	2.45	2.34	2.25	2.18	2.12	2.07	2.04	2.00	1.95	1.90	1.84
	7.31	**5.18**	**4.31**	**3.83**	**3.51**	**3.29**	**3.12**	**2.99**	**2.88**	**2.80**	**2.73**	**2.66**	**2.56**	**2.49**	**2.37**
42	4.07	3.22	2.83	2.59	2.44	2.32	2.24	2.17	2.11	2.06	2.02	1.99	1.94	1.89	1.82
	7.27	**5.15**	**4.29**	**3.80**	**3.49**	**3.26**	**3.10**	**2.96**	**2.86**	**2.77**	**2.70**	**2.64**	**2.54**	**2.46**	**2.35**
44	4.06	3.21	2.82	2.58	2.43	2.31	2.23	2.16	2.10	2.05	2.01	1.98	1.92	1.88	1.81
	7.24	**5.12**	**4.26**	**3.78**	**3.46**	**3.24**	**3.07**	**2.94**	**2.84**	**2.75**	**2.68**	**2.62**	**2.52**	**2.44**	**2.32**
46	4.05	3.20	2.81	2.57	2.42	2.30	2.22	2.14	2.09	2.04	2.00	1.97	1.91	1.87	1.80
	7.21	**5.10**	**4.24**	**3.76**	**3.44**	**3.22**	**3.05**	**2.92**	**2.82**	**2.73**	**2.66**	**2.60**	**2.50**	**2.42**	**2.30**
48	4.04	3.19	2.80	2.56	2.41	2.30	2.21	2.14	2.08	2.03	1.99	1.96	1.90	1.86	1.79
	7.19	**5.08**	**4.22**	**3.74**	**3.42**	**3.20**	**3.04**	**2.90**	**2.80**	**2.71**	**2.64**	**2.58**	**2.48**	**2.40**	**2.28**

TABLE B.4   continued

Degrees of freedom: Denominator	Degrees of freedom: Numerator														
	1	2	3	4	5	6	7	8	9	10	11	12	14	16	20
50	4.03	3.18	2.79	2.56	2.40	2.29	2.20	2.13	2.07	2.02	1.98	1.95	1.90	1.85	1.78
	7.17	**5.06**	**4.20**	**3.72**	**3.41**	**3.18**	**3.02**	**2.88**	**2.78**	**2.70**	**2.62**	**2.56**	**2.46**	**2.39**	**2.26**
55	4.02	3.17	2.78	2.54	2.38	2.27	2.18	2.11	2.05	2.00	1.97	1.93	1.88	1.83	1.76
	7.12	**5.01**	**4.16**	**3.68**	**3.37**	**3.15**	**2.98**	**2.85**	**2.75**	**2.66**	**2.59**	**2.53**	**2.43**	**2.35**	**2.23**
60	4.00	3.15	2.76	2.52	2.37	2.25	2.17	2.10	2.04	1.99	1.95	1.92	1.86	1.81	1.75
	7.08	**4.98**	**4.13**	**3.65**	**3.34**	**3.12**	**2.95**	**2.82**	**2.72**	**2.63**	**2.56**	**2.50**	**2.40**	**2.32**	**2.20**
65	3.99	3.14	2.75	2.51	2.36	2.24	2.15	2.08	2.02	1.98	1.94	1.90	1.85	1.80	1.73
	7.04	**4.95**	**4.10**	**3.62**	**3.31**	**3.09**	**2.93**	**2.79**	**2.70**	**2.61**	**2.54**	**2.47**	**2.37**	**2.30**	**2.18**
70	3.98	3.13	2.74	2.50	2.35	2.23	2.14	2.07	2.01	1.97	1.93	1.89	1.84	1.79	1.72
	7.01	**4.92**	**4.08**	**3.60**	**3.29**	**3.07**	**2.91**	**2.77**	**2.67**	**2.59**	**2.51**	**2.45**	**2.35**	**2.28**	**2.15**
80	3.96	3.11	2.72	2.48	2.33	2.21	2.12	2.05	1.99	1.95	1.91	1.88	1.82	1.77	1.70
	6.96	**4.88**	**4.04**	**3.56**	**3.25**	**3.04**	**2.87**	**2.74**	**2.64**	**2.55**	**2.48**	**2.41**	**2.32**	**2.24**	**2.11**
100	3.94	3.09	2.70	2.46	2.30	2.19	2.10	2.03	1.97	1.92	1.88	1.85	1.79	1.75	1.68
	6.90	**4.82**	**3.98**	**3.51**	**3.20**	**2.99**	**2.82**	**2.69**	**2.59**	**2.51**	**2.43**	**2.36**	**2.26**	**2.19**	**2.06**
125	3.92	3.07	2.68	2.44	2.29	2.17	2.08	2.01	1.95	1.90	1.86	1.83	1.77	1.72	1.65
	6.84	**4.78**	**3.94**	**3.47**	**3.17**	**2.95**	**2.79**	**2.65**	**2.56**	**2.47**	**2.40**	**2.33**	**2.23**	**2.15**	**2.03**
150	3.91	3.06	2.67	2.43	2.27	2.16	2.07	2.00	1.94	1.89	1.85	1.82	1.76	1.71	1.64
	6.81	**4.75**	**3.91**	**3.44**	**3.14**	**2.92**	**2.76**	**2.62**	**2.53**	**2.44**	**2.37**	**2.30**	**2.20**	**2.12**	**2.00**
200	3.89	3.04	2.65	2.41	2.26	2.14	2.05	1.98	1.92	1.87	1.83	1.80	1.74	1.69	1.62
	6.76	**4.71**	**3.88**	**3.41**	**3.11**	**2.90**	**2.73**	**2.60**	**2.50**	**2.41**	**2.34**	**2.28**	**2.17**	**2.09**	**1.97**
400	3.86	3.02	2.62	2.39	2.23	2.12	2.03	1.96	1.90	1.85	1.81	1.78	1.72	1.67	1.60
	6.70	**4.66**	**3.83**	**3.36**	**3.06**	**2.85**	**2.69**	**2.55**	**2.46**	**2.37**	**2.29**	**2.23**	**2.12**	**2.04**	**1.92**
1000	3.85	3.00	2.61	2.38	2.22	2.10	2.02	1.95	1.89	1.84	1.80	1.76	1.70	1.65	1.58
	6.66	**4.62**	**3.80**	**3.34**	**3.04**	**2.82**	**2.66**	**2.53**	**2.43**	**2.34**	**2.26**	**2.20**	**2.09**	**2.01**	**1.89**
∞	3.84	2.99	2.60	2.37	2.21	2.09	2.01	1.94	1.88	1.83	1.79	1.75	1.69	1.64	1.57
	6.64	**4.60**	**3.78**	**3.32**	**3.02**	**2.80**	**2.64**	**2.51**	**2.41**	**2.32**	**2.24**	**2.18**	**2.07**	**1.99**	**1.87**

From Table A14 of *Statistical Methods,* 7th ed. by George W. Snedecor and William G. Cochran. Copyright © 1980 by the Iowa State University Press, 2121 South State Avenue, Ames, Iowa 50010. Reprinted with permission of the Iowa State University Press.

TABLE B.5   CRITICAL VALUES FOR THE PEARSON CORRELATION*

*To be significant, the sample correlation, r, must be greater than or equal to the critical value in the table.

	Level of significance for one-tailed test			
	.05	.025	.01	.005
	Level of significance for two-tailed test			
$df = n - 2$	.10	.05	.02	.01
1	.988	.997	.9995	.9999
2	.900	.950	.980	.990
3	.805	.878	.934	.959
4	.729	.811	.882	.917
5	.669	.754	.833	.874
6	.622	.707	.789	.834
7	.582	.666	.750	.798
8	.549	.632	.716	.765
9	.521	.602	.685	.735
10	.497	.576	.658	.708
11	.476	.553	.634	.684
12	.458	.532	.612	.661
13	.441	.514	.592	.641
14	.426	.497	.574	.623
15	.412	.482	.558	.606
16	.400	.468	.542	.590
17	.389	.456	.528	.575
18	.378	.444	.516	.561
19	.369	.433	.503	.549
20	.360	.423	.492	.537
21	.352	.413	.482	.526
22	.344	.404	.472	.515
23	.337	.396	.462	.505
24	.330	.388	.453	.496
25	.323	.381	.445	.487
26	.317	.374	.437	.479
27	.311	.367	.430	.471
28	.306	.361	.423	.463
29	.301	.355	.416	.456
30	.296	.349	.409	.449
35	.275	.325	.381	.418
40	.257	.304	.358	.393
45	.243	.288	.338	.372
50	.231	.273	.322	.354
60	.211	.250	.295	.325
70	.195	.232	.274	.302
80	.183	.217	.256	.283
90	.173	.205	.242	.267
100	.164	.195	.230	.254

From Table VI of R. A. Fisher and F. Yates, *Statistical Tables for Biological, Agricultural and Medical Research,* 6th ed. London: Longman Group Ltd., 1974 (previously published by Oliver and Boyd Ltd., Edinburgh). Adapted and reprinted with permission of Addison Wesley Longman Ltd.

TABLE B.6   **THE CHI-SQUARE DISTRIBUTION***

*The table entries are critical values of χ^2.

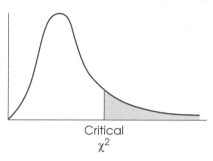

Critical
χ^2

df	Proportion in critical region				
	0.10	0.05	0.025	0.01	0.005
1	2.71	3.84	5.02	6.63	7.88
2	4.61	5.99	7.38	9.21	10.60
3	6.25	7.81	9.35	11.34	12.84
4	7.78	9.49	11.14	13.28	14.86
5	9.24	11.07	12.83	15.09	16.75
6	10.64	12.59	14.45	16.81	18.55
7	12.02	14.07	16.01	18.48	20.28
8	13.36	15.51	17.53	20.09	21.96
9	14.68	16.92	19.02	21.67	23.59
10	15.99	18.31	20.48	23.21	25.19
11	17.28	19.68	21.92	24.72	26.76
12	18.55	21.03	23.34	26.22	28.30
13	19.81	22.36	24.74	27.69	29.82
14	21.06	23.68	26.12	29.14	31.32
15	22.31	25.00	27.49	30.58	32.80
16	23.54	26.30	28.85	32.00	34.27
17	24.77	27.59	30.19	33.41	35.72
18	25.99	28.87	31.53	34.81	37.16
19	27.20	30.14	32.85	36.19	38.58
20	28.41	31.41	34.17	37.57	40.00
21	29.62	32.67	35.48	38.93	41.40
22	30.81	33.92	36.78	40.29	42.80
23	32.01	35.17	38.08	41.64	44.18
24	33.20	36.42	39.36	42.98	45.56
25	34.38	37.65	40.65	44.31	46.93
26	35.56	38.89	41.92	45.64	48.29
27	36.74	40.11	43.19	46.96	49.64
28	37.92	41.34	44.46	48.28	50.99
29	39.09	42.56	45.72	49.59	52.34
30	40.26	43.77	46.98	50.89	53.67
40	51.81	55.76	59.34	63.69	66.77
50	63.17	67.50	71.42	76.15	79.49
60	74.40	79.08	83.30	88.38	91.95
70	85.53	90.53	95.02	100.42	104.22
80	96.58	101.88	106.63	112.33	116.32
90	107.56	113.14	118.14	124.12	128.30
100	118.50	124.34	129.56	135.81	140.17

From Table 8 of E. Pearson and H. Hartley, *Biometrika Tables for Statisticians,* 3d ed. New York: Cambridge University Press, 1966. Adapted and reprinted with permission of the Biometrika trustees.

SOLUTIONS FOR ODD-NUMBERED PROBLEMS IN THE TEXT

Note: Many of the problems in the text require several stages of computation. At each stage, there is an opportunity for rounding answers. Depending on the exact sequence of operations used to solve a problem, different individuals will round their answers at different times and in different ways.

As a result, you may obtain answers that are slightly different from those presented here. As long as those differences are small, they probably can be attributed to rounding error and should not be a matter for concern.

CHAPTER 1 INTRODUCTION TO STATISTICS

1. Descriptive statistics simplify and summarize data. Inferential statistics use sample data to make general conclusions about populations.

3. The distinguishing characteristics of the experimental method are *manipulation* and *control*. In the experimental method, the researcher manipulates one variable and observes a second variable. All other variables are controlled to prevent them from affecting the outcome.

5. In an experiment, the researcher manipulates one variable and observes a second variable for changes. The manipulated variable is called the *independent variable,* and the observed variable is the *dependent variable.*

7. This is a quasi-experimental study. It is not an experiment because the researcher is not manipulating a variable to create the groups. Instead, the study is comparing pre-existing groups (boys versus girls).

9. **a.** The dependent variable is whether or not each subject has a cold.
   b. Discrete
   c. Nominal scale
   d. Experimental (amount of vitamin C is manipulated)

11. A discrete variable consists of separate, indivisible categories. A continuous variable is divisible into an infinite number of fractional parts.

13. **a.** The independent variable is the brand of medication, brand X versus brand Y, which is measured on a nominal scale.
    b. The dependent variable is the degree of indigestion, which is measured on an ordinal scale; a series of ordered categories.

15. **a.** Discrete
    b. Ratio scale (zero means no errors)

17. A construct is a hypothetical concept. An operational definition defines a construct in terms of a measurement procedure.

19. **a.** $\Sigma X = 15$
    b. $\Sigma Y = 10$
    c. $\Sigma XY = 40$

21. **a.** $\Sigma X + 3$
    b. $\Sigma (X - 2)^2$
    c. $\Sigma X^2 - 10$

23. **a.** $\Sigma X = 9$
    b. $\Sigma X^2 = 35$
    c. $\Sigma (X + 1) = 13$
    d. $\Sigma (X + 1)^2 = 57$

25. **a.** $\Sigma X = -3$
    b. $\Sigma Y = 14$
    c. $\Sigma XY = -14$

CHAPTER 2 **FREQUENCY DISTRIBUTIONS**

1. a.

X	f	p	%
5	3	.15	15%
4	4	.20	20%
3	8	.40	40%
2	3	.15	15%
1	2	.10	10%

  b. (1) Highest score is 5; lowest score is 1. Most frequent score is $X = 3$.

    (2) There are 5 people, or 25% of the class, with scores of 2 or lower.

3. A grouped frequency distribution is used when the original scale of measurement consists of more categories than can be listed simply in a regular table. A total of 20 or more categories is generally considered too many for a regular table.

5. a.

X	f
10	2
9	7
8	2
7	1
6	0
5	2
4	0
3	2
2	8
1	2

  b.

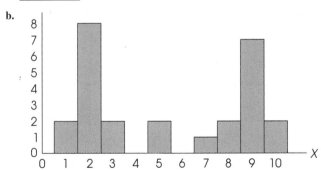

  c. (1) The distribution is symmetrical, with half the class scoring high and half scoring low.

    (2) The class is sharply divided into two distinct groups. For some, the quiz was easy, and for some, the quiz was hard.

7. a. $N = 11$
  b. $\Sigma X = 37$
  c. $\Sigma X^2 = 139$

9. a.

X	f
30–31	1
28–29	0
26–27	2
24–25	1
22–23	4
20–21	5
18–19	2
16–17	3
14–15	1
12–13	1

b.

X	f
30–34	1
25–29	2
20–24	10
15–19	5
10–14	2

11. a.

Set I X	f	Set II X	f
5	1	14–15	1
4	2	12–13	0
3	4	10–11	1
2	2	8–9	2
1	1	6–7	1
		4–5	2
		2–3	1
		0–1	1

  b. Set I

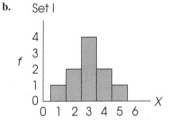

    Set II

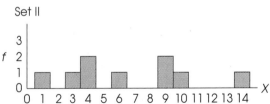

  c. Set I forms a symmetrical distribution, centered at $X = 3$, with most of the scores clustered close to the center. Set II forms a flat distribution with a center near $X = 7$ or $X = 8$, but the scores are spread evenly across the entire scale.

13. a.

X	f
10	1
9	3
8	2
7	1
6	1
5	2
4	3
3	2
2	1

b.

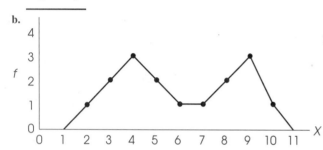

c. (1) The distribution is roughly symmetrical with scores piled up at each end of the scale.
(2) The center is near $X = 6$.
(3) The scores are spread across the entire scale.

15. The scores range from 112 to 816 and should be presented in a grouped table.

X	f
800–899	1
700–799	3
600–699	4
500–599	6
400–499	5
300–399	3
200–299	1
100–199	1

17.

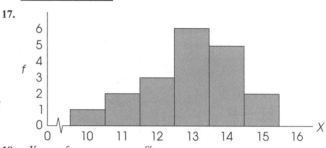

19.

X	f	p	%
6	2	.08	8%
5	3	.12	12%
4	3	.12	12%
3	4	.16	16%
2	5	.20	20%
1	8	.32	32%

The distribution is positively skewed.

CHAPTER 3 **CENTRAL TENDENCY**

1. The purpose of central tendency is to identify the single score that serves as the best representative for an entire distribution, usually a score from the center of the distribution.

3. a. The mean is the *balance point* of a distribution because the sum of the distances above the mean is exactly equal to the sum of the distances below the mean.

b. The median is the *midpoint* of a distribution because exactly 50% of the scores are greater than the median and 50% are less than the median.

5. The mean, median, and mode are identical for a symmetrical distribution with one mode.

7. With a skewed distribution, the extreme scores in the tail can displace the mean. Specifically, the mean is displaced away from the main pile of scores and moved out toward the tail. The result is that the mean is often not a very representative value.

9. a.

Median (8.5)
Mode (9.0)
Mean (7.8)

b. Mean = 78/10 = 7.8, median = 8.5, mode = 9

11. a.

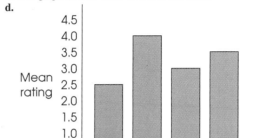

b. Mean = 30/10 = 3

c. The median is still $X = 3$ (unchanged). The new mean is 70/10 = 7.

13. $\Sigma X = 1300$

15. a. $\overline{X} = 120$ (the old mean is multiplied by 6)

b. $\overline{X} = 25$ (the old mean is increased by 5)

17. a. The mean will increase because the new score is greater than the average for the original sample.

b. The mean will decrease because the new score is lower than the average for the original sample.

c. The mean will stay the same because the new score is exactly equal to the average for the original sample.

19. For the original sample, $\Sigma X = 5(21) = 105$. When the new score is added, $\Sigma X = 6(25) = 150$. The new score is $X = 45$.

21. The original samples have $\Sigma X = 12$ and $\Sigma X = 70$. The combined sample has $\Sigma X = 82$ and $n = 10$. The mean for the combined sample is 82/10 = 8.2.

23. Negatively skewed (the mean is displaced toward the tail of the distribution)

25. a. The independent variable is the brand of coffee; the dependent variable is the flavor rating.

b. Nominal

c. Bar graph (coffee brand is a nominal scale)

d.

CHAPTER 4 **VARIABILITY**

1. a. *SS* is the sum of squared deviation scores.

b. Variance is the mean squared deviation.

c. Standard deviation is the square root of the variance. It provides a measure of the standard distance from the mean.

3. *SS* cannot be less than zero because it is computed by adding squared deviations. Squared deviations are always greater than or equal to zero.

5. For a 5-point quiz, the scores range from $X = 0$ to $X = 5$. If the mean is 4.2, then $X = 0$ is the score that is farthest from the mean, and it is only 4.2 points away. If the greatest distance from the mean is only 4.2 points, it is impossible for the standard distance to be 11.4.

7. The standard deviation, σ, can be found by taking the square root of the variance, σ^2. For this example, $\sigma = \sqrt{\sigma^2} = \sqrt{100} = 10$.

9. a. Your score is 5 points above the mean. If the standard deviation is 2, then your score is an extremely high value (out in the tail of the distribution). However, if the standard deviation is 10, then your score is only slightly above average. Thus, you would prefer $\sigma = 2$.

b. If you are located 5 points below the mean, then the situation is reversed. A standard deviation of 2 gives you an extremely low score, but with a standard deviation of 10, you are only slightly below average. Here you would prefer $\sigma = 10$.

11. a. The definitional formula is easier to use when the mean is a whole number and there is a relatively small set of scores.

b. The computational formula is easier to use when the mean is a decimal value and/or there are a large number of scores.

13. $SS = 8$, $s^2 = 8/8 = 1$, $s = 1$

15. a. Sample B covers a wider range.

b. For sample A: $\overline{X} = 9.17$ and $s = 1.72$.
For sample B: $\overline{X} = 9.00$ and $s = 5.66$.

c. Sample A

17. $SS = 20$, $s^2 = 5$, $s = 2.24$

19. a. Range = 12 points; $Q1 = 3.5$ and $Q3 = 8.5$; semi-interquartile range = 2.5; $SS = 120$, $s = 3.30$

b. Adding a constant to every score does not change the variability. The range, semi-interquartile range, and standard deviation are unchanged.

21. a.

b. The mean is $\overline{X} = 50/5 = 10$, and the standard deviation appears to be about 4 or 5.

c. $SS = 100$, $s^2 = 25$, $s = 5$

23. a.

b. The mean is 48/6 = 8, and the standard deviation appears to be about 3 or 4.

c. $SS = 54$, $\sigma^2 = 9$, $\sigma = 3$

CHAPTER 5 *z*-SCORES

1. A *z*-score describes a precise location within a distribution. The sign of the *z*-score tells whether the location is above (+) or below (−) the mean, and the magnitude tells the distance from the mean in terms of the number of standard deviations.

3. a. With $\sigma = 25$, $X = 250$ corresponds to $z = +2.00$.
b. With $\sigma = 100$, $X = 250$ corresponds to $z = +0.50$.

5. a.

X	z	X	z
58	1.00	46	−0.50
34	−2.00	62	1.50
70	2.50	44	−0.75

b.

X	z	X	z
70	2.50	52	0.25
46	−0.50	42	−1.00
38	−1.50	56	0.75

7.

X	z	X	z
63	0.25	54	−0.50
48	−1.00	72	1.00
66	0.50	84	2.00

9. $\sigma = 9$

11. $\mu = 47$

13. The distance between scores is 12 points, which corresponds to 3 standard deviations. Therefore, $\sigma = 4$ and $\mu = 58$.

15. $\sigma = 8$ gives $z = -1.00$
$\sigma = 16$ gives $z = -0.50$ (better score)

17. Tom's CPE score corresponds to $z = 35/50 = 0.70$. Bill's CBT score corresponds to $z = 40/100 = 0.40$. Tom's *z*-score places him higher in the distribution, so he is more likely to be admitted.

19.

Student	z-score	X value
Steve	0.50	95
Sharon	0.90	99
Jill	1.20	102
Ramon	2.00	110

21.

Original Score	Standardized Score
41	50
51	100
43	60
37	30
44	65
46	75

23. a. $\mu = 6$ and $\sigma = 4$
b.

X	z	X	z
12	1.50	3	−0.75
1	−1.25	7	0.25
10	1.00	3	−0.75

25. a. $\mu = 8$ and $\sigma = 4$
b.

X	z
14	1.50
3	−1.25
12	1.00
5	−0.75
9	0.25
5	−0.75

c.

Original X	New X
14	72
3	50
12	68
5	54
9	62
5	54

CHAPTER 6 PROBABILITY

1. a. $p = 60/90 = 0.667$
b. $p = 15/90 = 0.167$
c. $p = 5/90 = 0.056$

3. a. Left = 0.8413, right = 0.1587
b. Left = 0.0668, right = 0.9332
c. Left = 0.5987, right = 0.4013
d. Left = 0.3085, right = 0.6915

5. a. 0.3336
b. 0.0465
c. 0.0885
d. 0.3859

7. a. $z = 0.25$
b. $z = 1.28$
c. $z = -0.84$

9. a. $z = 0.52$, $X = 552$
   b. $z = 0.67$, $X = 567$
   c. $z = -1.28$, $X = 372$

11. a. $z = -1.33$, $p = 0.0918$
   b. $z = -1.67$, $p = 0.9525$
   c. $z = 1.00$, $p = 0.1587$

13. a. 0.1747
   b. 0.6826
   c. 0.4332
   d. 0.7506

15. The portion of the distribution consisting of scores greater than 50 is more than one-half of the whole distribution. The answer must be greater than .50.

17. a. $p(z < 1.22) = 0.8888$
   b. $p(z > -1.67) = 0.9525$
   c. $p(z > 0.56) = 0.2877$

d. $p(z > 2.11) = 0.0174$
   e. $p(-1.33 < z < 1.44) = 0.8333$
   f. $p(-0.33 < z < 0.33) = 0.2586$

19. a. $z = -1.04$, $X = 104.4$
   b. $z = 1.18$, $X = 137.7$
   c. $z = 1.47$, rank $= 92.92\%$
   d. $z = -1.20$, rank $= 11.51\%$
   e. $z = 0$, rank $= 50\%$
   f. semi-interquartile range $= 10.05$ points

21. a. The semi-interquartile range is bounded by $z = 0.67$ and $z = -0.67$. With $\sigma = 10$, the semi-interquartile range is $(0.67)(10) = 6.7$.
   b. With a standard deviation of $\sigma = 20$, the semi-interquartile range is $(0.67)(20) = 13.4$.
   c. In general, the semi-interquartile range is equal to 0.67σ.

CHAPTER 7 **PROBABILITY AND SAMPLES: THE DISTRIBUTION OF SAMPLE MEANS**

1. a. The distribution of sample means is the set of all possible sample means for random samples of a specific size (n) from a specific population.
   b. The expected value of $\overline{X}$ is the mean of the distribution of sample means (μ).
   c. The standard error of $\overline{X}$ is the standard deviation of the distribution of sample means ($\sigma_{\overline{X}} = \sigma/\sqrt{n}$).

3. a. The standard deviation, $\sigma = 30$, measures the standard distance between a score and the population mean.
   b. The standard error, $\sigma_{\overline{X}} = 30/\sqrt{100} = 3$, measures the standard distance between a sample mean and the population mean.

5. a. $n \geq 9$
   b. $n \geq 36$
   c. No. When $n = 1$, the standard error is equal to the standard deviation. For any other sample size, the standard error will be smaller than the standard deviation.

7. a. The standard deviation, $\sigma = 6$, measures the standard distance between X and μ.
   b. With $n = 9$, $\sigma_{\overline{X}} = 2$ points.
   c. With $n = 36$, $\sigma_{\overline{X}} = 1$ point.

9. a. Standard error $= 8$, $z = 0.75$, $p = 0.2266$ (more likely)
   b. Standard error $= 2.67$, $z = 1.12$, $p = 0.1314$

11. a. The distribution is normal with $\mu = 100$ and $\sigma_{\overline{X}} = 4$.
   b. The z-score boundaries are ± 1.96. The $\overline{X}$ boundaries are 92.16 and 107.84.
   c. $\overline{X} = 106$ corresponds to $z = 1.50$. This is not in the extreme 5%.

13. a. The distribution is normal with $\mu = 55$ and $\sigma_{\overline{X}} = 2$.
   b. $z = +2.00$, $p = 0.0228$
   c. $z = +0.50$, $p = 0.6915$
   d. $p(-1.00 < z < +1.00) = 0.6826$

15. a. The distribution is normal with $\mu = 50$ and $\sigma = 10$. The z-score boundaries are ± 0.50, and $p = 0.3830$.
   b. The distribution is normal with $\mu = 50$ and $\sigma_{\overline{X}} = 2$. The z-score boundaries are ± 2.50, and $p = 0.9876$.

17. a. With $n = 4$, the distribution of sample means will not be normal, and you cannot use the unit normal table to find the answer.
   b. With $n = 36$, the distribution of sample means will be normal. $\sigma_{\overline{X}} = 3$, $z = 1.00$, $p = 0.1587$

19. a. $n = 25$
   b. $n = 100$

21. With $n = 4$, $\sigma_{\overline{X}} = 5$ and $p(\overline{X} \geq 150) = p(z \geq 2.00) = 0.0228$.

23. a.

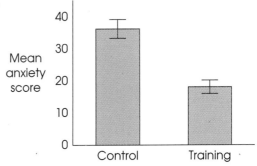

   b. Even considering the standard error for each mean, there is no overlap between the two groups. The relaxation training does seem to have lowered scores more than would be expected by chance.

25.

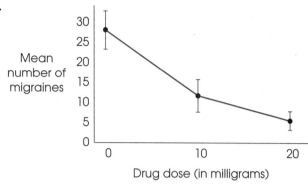

CHAPTER 8 **INTRODUCTION TO HYPOTHESIS TESTING**

1. a. $\overline{X} - \mu$ measures the difference between the sample data and the hypothesized population mean.
   b. A sample mean is not expected to be identical to the population mean. The standard error indicates how much difference between $\overline{X}$ and μ is expected by chance.

3. A smaller alpha level (e.g., .01 rather than .05) has the advantage of reducing the risk of a Type I error. That is, when you reject the null hypothesis, you are more confident that you have made the correct decision. On the other hand, a smaller alpha level makes it more difficult to reject H_0. That is, a small alpha level requires a large discrepancy between the data and the null hypothesis before you can say that a significant difference exists.

5. The analyses are contradictory. The critical region for the two-tailed tests consists of the extreme 2.5% in each tail of the distribution. The two-tailed conclusion indicates that the data were not in this critical region. However, the one-tailed test indicates that the data were in the extreme 1% of one tail. Data cannot be in the extreme 1% and at the same time fail to be in the extreme 2.5%.

7. a. H_0: $\mu = 50$. With $\alpha = .05$, the critical boundaries are $z = \pm 1.96$. For these data, $\sigma_{\overline{X}} = 3$ and $z = -2.33$. Reject H_0.
   b. With $\alpha = .01$, the critical boundaries are $z = \pm 2.58$. The z-score of $z = -2.33$ is not in the critical region. Fail to reject H_0.
   c. As the alpha level decreases, the critical boundaries are moved farther out, and it becomes more difficult to reject H_0.

9. a. The independent variable is presence or absence of the hormone; the dependent variable is weight at 10 weeks of age.
   b. The null hypothesis states that the growth hormone will have no effect on the weight of rats at 10 weeks of age.
   c. H_0: $\mu = 950$. H_1: $\mu \neq 950$.
   d. The critical region consists of z-score values beyond 1.96 or -1.96 in a normal distribution.
   e. For this sample, $\sigma_{\overline{X}} = 6$ and $z = 4.00$.
   f. Reject the null hypothesis, and conclude that the hormone has a significant effect on weight.

11. a. For $n = 25$, $\sigma_{\overline{X}} = 2$ and $z = 1.50$. Fail to reject the null hypothesis. There is no significant difference in vocabulary skills between only children and the general population.

   b. For $n = 100$, $\sigma_{\overline{X}} = 1$ and $z = 3.00$. Reject the null hypothesis, and conclude there is a significant difference in vocabulary skills between only children and the general population.
   c. The larger sample results in a smaller standard error. With $n = 100$, the difference between the data and the null hypothesis is significantly more than chance.

13. For the one-tailed test, the null hypothesis states that there is no memory impairment, H_0: $\mu \geq 50$. The critical region consists of z-scores less than -2.33. For these data, $\sigma_{\overline{X}} = 1.28$ and $z = -2.34$. Reject H_0, and conclude the alcoholics have significantly lower memory scores.

15. a. For the one-tailed test, H_0: $\mu \leq 55$ (there is no increase in depression). With $\alpha = .05$, the critical region consists of z-score values greater than 1.65. The data have $\sigma_{\overline{X}} = 2$ and $z = 2.00$. Reject H_0, and conclude that the elderly subjects are significantly more depressed.
   b. With $\alpha = .01$, the critical region consists of z-score values greater than 2.33. Fail to reject H_0.
   c. Lowering α from .05 to .01 requires more evidence to reject the null hypothesis. The data in this study were sufficient to reject H_0 with $\alpha = .05$ but not with $\alpha = .01$.

17. a. H_0: $\mu = 65$; $\sigma_{\overline{X}} = 4$ and $z = -1.25$; fail to reject H_0.
   b. H_0: $\mu = 55$; $\sigma_{\overline{X}} = 4$ and $z = 1.25$; fail to reject H_0.
   c. With a sample mean of $\overline{X} = 60$, it is reasonable to expect that the population mean has a value near 60. Any value that is reasonably near to 60 will be an acceptable hypothesis for μ. The standard error and the alpha level determine the range of acceptable values.

19. H_0: $\mu \leq 100$ (not above average), and H_0: $\mu > 100$ (above average). The critical region consists of z-score values greater than 2.33. For these data, the standard error is 1.875, and the z-score is $z = 2.61$. Reject H_0.

21. H_0: $\mu = 55$ (patients' scores are not different from the normal population). The critical region consists of z-scores greater than $+2.58$ or less than -2.58. For these data, $\overline{X} = 76.62$, the standard error is 2.62, and $z = 8.25$. Reject the null hypothesis. Scores for depressed patients are significantly different from scores for normal individuals on this test.

CHAPTER 9 INTRODUCTION TO THE t STATISTIC

1. The t statistic is used when the population standard deviation is unknown. You use the sample data to estimate the standard deviation and the standard error.

3. **a.** As variability increases, t becomes smaller (closer to zero).
   b. As sample size increases, the standard error decreases, and the t value increases.
   c. The larger the difference between $\overline{X}$ and μ, the larger the t value.

5. The sample variance (s^2) in the t formula is an estimated value and contributes to the variability.

7. **a.** $\overline{X} = 10$, $s^2 = 36$, $s = 6$
   b. $s_{\overline{X}} = 3$ points

9. The null hypothesis states that self-esteem for the athletes is not different from self-esteem for the general population. In symbols, H_0: $\mu = 70$. The critical boundaries are ± 2.064. For these data, $s_{\overline{X}} = 2$ and $t(24) = 1.50$, which is not in the critical region. Fail to reject the null hypothesis.

11. H_0: $\mu = 10$ (no different from chance). Because $df = 35$ is not in the table, use $df = 30$ to obtain critical values of $t = \pm 2.042$. For these data, the standard error is 2, and $t(35) = 1.75$. Fail to reject H_0.

13. **a.** With $n = 25$, the critical boundaries are ± 2.064. For these data, the standard error is 2.00, and the t statistic is $t(24) = -2.00$. Fail to reject H_0.
    b. Because $df = 399$ is not in the table, use $df = 120$ to obtain critical boundaries of ± 1.98. With $n = 400$, the standard error is 0.50, and the t statistic is $t(399) = -8.00$. Reject H_0.

c. The larger the sample, the smaller the standard error (denominator of the t statistic). If other factors are held constant, a smaller standard error will produce a larger t statistic, which is more likely to be significant.

15. **a.** $\overline{X} = 55$ and $s = 2$. Your sketch should show the scores piled up around a mean value of 55, with most of the scores within a few points of the mean.
    b. A value of $X = 50$ is located outside the sample. It appears that the sample is centered around a location significantly different from $\mu = 50$.
    c. The critical boundaries are $t = \pm 2.131$. For these data, $t(15) = 10.00$. Reject H_0.

17. H_0: $\mu = 21$. Because $df = 99$ is not in the table, use $df = 60$ to obtain critical boundaries of ± 1.980. With a standard error of 0.50, $t(99) = -4.60$. Reject H_0, and conclude that humidity has a significant effect on the rats' eating behavior.

19. H_0: $\mu \leq 60$ (not more pleasant); H_1: $\mu > 60$ (more pleasant). The critical region consists of t values greater than 1.711. For these data, $s^2 = 25$, the standard error is 1.00, and $t(24) = 4.30$. Reject H_0, and conclude that memories for older adults are significantly more pleasant than memories for college students.

21. H_0: $\mu \leq 25$ (not more than $25). The critical region consists of t values greater than 1.833. For these data, $\overline{X} = \$28.00$, $s^2 = 201.11$, the standard error is 4.49, and $t(9) = 0.67$. Fail to reject H_0. The average donation is not significantly greater than $25.00.

CHAPTER 10 HYPOTHESIS TESTS WITH TWO INDEPENDENT SAMPLES

1. An independent-measures study requires a separate sample for each of the treatments or populations being compared. An independent-measures t statistic is appropriate when a researcher has two samples and wants to use the sample mean difference to test hypotheses about the population mean difference.

3. As the difference between the two sample means increases, the value of the t statistic increases. As the variability of the scores increases, the value of t decreases (becomes closer to zero).

5. **a.** The first sample has $s^2 = 7$, and the second has $s^2 = 5$. The pooled variance is $108/18 = 6$ (halfway between).
   b. The first sample has $s^2 = 7$, and the second has $s^2 = 3$. The pooled variance is $108/24 = 4.5$.

7. The null hypothesis states that sex type has no effect on depression, H_0: $\mu_1 - \mu_2 = 0$. For a one-tailed test, the critical value is $t = 1.734$. The pooled variance is 80, the standard error is 4, and $t(18) = 2.00$. Reject H_0, and conclude that the traditional subjects are significantly more depressed than the androgynous subjects.

9. H_0: $(\mu_1 - \mu_2) \leq 0$ (no increase); H_1: $(\mu_1 - \mu_2) > 0$ (increase). Pooled variance = 30 and $t(13) = 3.67$, which is beyond the

one-tailed critical boundary of $t = 1.771$. Reject H_0, and conclude that fatigue has a significant effect.

11. The null hypothesis states that the imagined factors have no effect on persistence, H_0: $\mu_1 - \mu_2 = 0$. With $\alpha = .05$, the critical region consists of t values beyond ± 2.074. The pooled variance is 24, the standard error is 2, and $t(22) = 3.00$. Reject H_0, and conclude that there is a significant difference between the two conditions.

13. **a.** There is a total of 26 subjects in the two samples combined.
    b. With $df = 24$ and $\alpha = .05$, the critical region consists of t values beyond ± 2.064. The t statistic is in the critical region. Reject H_0, and conclude that there is a significant difference.
    c. With $df = 24$ and $\alpha = .01$, the critical region consists of t values beyond ± 2.797. The t statistic is not in the critical region. Fail to reject H_0, and conclude that there is no significant difference.

15. The null hypothesis states that the type of presentation has no effect on learning, H_0: $\mu_1 - \mu_2 = 0$. With $\alpha = .05$, the critical region consists of t values beyond ± 2.101. The pooled variance is 20, the standard error is 2, and $t(18) = 1.50$. Fail to reject H_0, and conclude that there is no significant difference in test scores between the two groups.

17. The null hypothesis states that there is no difference in difficulty between the two problems, H_0: $\mu_1 - \mu_2 = 0$. With $\alpha = .01$, the critical region consists of t values beyond ± 3.355. With the tacks in the box, $\overline{X} = 107.2$ and $SS = 6370.80$. With the tacks and box separate, $\overline{X} = 43.2$ and $SS = 1066.80$. The pooled variance is 929.7, the standard error is 19.28, and $t(8) = 3.32$. Fail to reject H_0. The data are not sufficient to indicate a significant difference between the two conditions.

19. The null hypothesis states that owning a pet has no effect on health, H_0: $\mu_1 - \mu_2 = 0$. The critical boundaries are $t = \pm 2.228$. For the control group, $\overline{X} = 11.14$ and $SS = 56.86$. For the dog owners, $\overline{X} = 6.4$ and $SS = 17.2$. The pooled variance is 7.41, and $t(10) = 2.98$. Reject H_0. The data show a significant difference between the dog owners and the control group.

21. The null hypothesis states that environment has no effect on development, H_0: $\mu_1 - \mu_2 = 0$. With $\alpha = .01$, the critical region consists of t values beyond ± 2.878. For the rich rats, $\overline{X} = 26.0$ and $SS = 214$. For the poor rats, $\overline{X} = 34.2$ and $SS = 313.61$. The pooled variance is 29.31, the standard error is 2.42, and $t(18) = 3.39$. Reject H_0. The data indicate a significant difference between the two environments.

CHAPTER 11 HYPOTHESIS TESTS WITH TWO RELATED SAMPLES

1. a. This is an independent-measures experiment with two separate samples.
   b. This is a repeated-measures design. The same sample is measured twice.
   c. This is a matched-subjects design. The repeated-measures t statistic is appropriate.

3. For a repeated-measures design, the same subjects are used in both treatment conditions. In a matched-subjects design, two different sets of subjects are used. However, in a matched-subjects design each subject in one condition is matched with respect to a specific variable with a subject in the second condition so that the two separate samples are equivalent with respect to the matching variable.

5. The null hypothesis says that there is no change, H_0: $\mu_D = 0$. For these data, $s^2 = 25$, the standard error is 1.25, and $t(15) = 2.56$. The critical region consists of t values beyond ± 2.131. Reject H_0, and conclude that there has been a significant change in the number of cigarettes.

7. a. The histogram for sample A is tightly clustered around a mean of $\overline{D} = 5$ with $SS = 12$. For sample B, the scores in the histogram are widely scattered from well above zero to well below zero. Again, the mean is $\overline{D} = 5$, but now $SS = 1098$.
   b. Sample A does not appear to have come from a population with a mean of zero, but for sample B, it is reasonable that the population mean could be zero.
   c. For sample A, the sample variance is 1.5, the standard error is 0.41, and the t statistic is $t = 12.20$, which is well beyond the critical boundary of 2.306. Reject H_0, and conclude that the population mean is not zero. For sample B, the sample variance is 156.86, the standard error is 4.43, and the t statistic is $t = 1.13$, which is not beyond the critical boundary of 2.365. Fail to reject H_0, and conclude that the population mean is not significantly different from zero.

9. a. For experiment 1, $\overline{D} = 5$ and $s = 0.82$. For experiment 2, $\overline{D} = 5$ and $s = 9.20$.
   b. For experiment 1, the standard error is 0.41, and $t(3) = 12.20$. This is beyond the critical values ($t = \pm 3.182$), so we reject H_0 and conclude that the treatment has a significant effect. For experiment 2, the standard error is 4.60, and $t(3) = 1.09$. This is not in the critical region, so we fail to reject H_0 and conclude that the treatment does not have a significant effect.
   c. The consistent treatment effect in experiment 1 produces small variability and a small standard error. In experiment 2, the inconsistent results produce large variability and a large standard error.

11. The null hypothesis says that stress has no effect on lymphocyte count, H_0: $\mu_D = 0$. With $\alpha = .05$, the critical region consists of t values beyond ± 2.093. For these data, the standard error is 0.018, and $t(19) = 5.00$. Reject H_0, and conclude that the data show a significant change in lymphocyte count.

13. The null hypothesis says that there is no increase in pain tolerance, H_0: $\mu_D \leq 0$. For these data, $s^2 = 64$, the standard error is 2, and $t(15) = 5.25$. For a one-tailed test with $\alpha = .01$, the critical region consists of t values beyond 2.602. Reject H_0, and conclude that there has been a significant increase in pain tolerance.

15. The null hypothesis says that the amount of exercise makes no difference in overall health, H_0: $\mu_D = 0$. With $\alpha = .05$, the critical region consists of t values beyond ± 2.447. For these data, $\overline{D} = 0.43$ with $SS = 33.71$. The standard error is 0.90, and $t(6) = 0.48$. Fail to reject H_0, and conclude that there is no difference between 2 hours and 5 hours of exercise.

17. The null hypothesis says that there is no difference between shots fired during versus between heartbeats, H_0: $\mu_D = 0$. For these data, $\overline{D} = 2.83$, $SS = 34.83$, $s^2 = 6.97$, the standard error is 1.08, and $t(5) = 2.62$. With $\alpha = .05$, the critical region consists of t values beyond ± 2.571. Reject H_0, and conclude that the timing of the shot has a significant effect on the marksmen's scores.

19. The null hypothesis says that there is no difference between the two stores, H_0: $\mu_D = 0$. For these data, $\overline{D} = -0.08$, $SS = .1334$, $s^2 = 0.017$, the standard error is 0.043, and $t(8) = -1.86$. With $\alpha = .05$, the critical region consists of t values beyond ± 2.306. Fail to reject H_0, and conclude that the data show no significant difference in prices between the two stores.

CHAPTER 12 ESTIMATION

1. The general purpose of a hypothesis test is to determine whether or not a treatment effect exists. A hypothesis test always addresses a "yes-no" question. The purpose of estimation is to determine the size of the effect. Estimation addresses a "how much" question.

3. a. The larger the sample, the narrower the interval.
   b. The larger the sample standard deviation (s), the wider the interval.
   c. The higher the percentage of confidence, the wider the interval.

5. a. Use the sample mean, $\overline{X} = 5.1$ hours, as the best point estimate of μ.
   b. The point estimate of change is -0.4 hours.
   c. With a standard error of 0.20 and z-scores ranging from -1.28 to $+1.28$ for 80% confidence, the interval extends from 4.844 hours to 5.356 hours.

7. a. Use $\overline{X} = 820$ for the point estimate.
   b. Using $z = \pm1.28$, the 80% confidence interval extends from 817.44 to 822.56.
   c. Using $z = \pm2.58$, the 99% confidence interval extends from 814.84 to 825.16.

9. a. $s^2 = 8.47$
   b. Use $\overline{X} = 10.09$ for the point estimate. Using $t = \pm2.228$ and a standard error of 0.88, the 95% confidence interval extends from 8.13 to 12.03.

11. Use $\overline{X} = 7.36$ for the point estimate. Using $t = \pm2.228$ and a standard error of 0.43, the 95% confidence interval extends from 6.402 to 8.318.

13. Use the sample mean difference, 5.5, for the point estimate. Using a repeated-measures t statistic with $df = 15$, the standard error is 2, the t-score boundaries for 80% confidence are ±1.341, and the interval extends from 2.818 to 8.182.

15. Use the sample mean difference, 0.72, for the point estimate. Using a repeated-measures t statistic with $df = 24$, the standard error is 0.20, the t-score boundaries for 95% confidence are ±2.064, and the interval extends from 0.3072 to 1.1328.

17. Use the sample mean difference, 11 points, for the point estimate. Using an independent-measures t statistic with $df = 13$, the pooled variance is 30, the standard error is 3, the t-score boundaries for 99% confidence are ±1.771, and the interval extends from 5.687 to 16.313.

19. a. Both experiments show a 10-point mean difference.
    b. The variability is much larger for experiment 2. The SS values are substantially larger.
    c. Experiment 2 would produce a wider interval. The larger variability would produce a larger standard error, which leads to a wider interval.

21. Use the sample mean difference, 4.2, for the point estimate. Using a repeated-measures t statistic with $df = 3$, the standard error is 1, the t-score boundaries for 95% confidence are ±3.182, and the interval extends from 1.018 to 7.382.

CHAPTER 13 ANALYSIS OF VARIANCE

1. When there is no treatment effect, the numerator and the denominator of the F-ratio are both measuring the same sources of variability (individual differences and experimental error). In this case, the F-ratio is balanced and should have a value near 1.00.

3. As the differences between sample means increase, $MS_{between}$ also increases, and the F-ratio increases. Increases in sample variability cause MS_{within} to increase and thereby decrease the F-ratio.

5. Post hoc tests are done after an ANOVA where you reject the null hypothesis with 3 or more treatments. Post hoc tests determine which treatments are significantly different.

7. The means and SS values are

Single	Twin	Triplet
$\overline{X} = 8$	$\overline{X} = 6$	$\overline{X} = 4$
$SS = 10$	$SS = 18$	$SS = 14$

The null hypothesis states that there are no differences in language development among the three groups, H_0: $\mu_1 = \mu_2 = \mu_3$. The critical value for $\alpha = .05$ is 3.88. The analysis of variance produces

Source	SS	df	MS	
Between treatments	40	2	20	$F(2, 12) = 5.71$
Within treatments	42	12	3.5	
Total	82	14		

Reject H_0, and conclude that there are significant differences in language development among the three groups.

9.

Source	SS	df	MS	
Between treatments	10	2	5.00	$F(2, 12) = 3.75$
Within treatments	16	12	1.33	
Total	26	14		

The critical value is $F = 3.88$. Fail to reject H_0. These data do not provide evidence of any differences among the three therapies.

11. a. For the ANOVA

Source	SS	df	MS	
Between treatments	32	1	32	$F(1, 6) = 2.91$
Within treatments	66	6	11	
Total	98	7		

With $\alpha = .05$, the critical value is $F = 5.99$. Fail to reject the null hypothesis.

b. For the independent-measures t, the pooled variance is 11, the standard error is 2.345, and $t(6) = 1.706$. With $\alpha = .05$, the critical value is ± 2.447. Fail to reject the null hypothesis.

c. $(1.706)^2 = 2.91$

13.

Source	SS	df	MS	
Between treatments	45	3	15	$F(3, 36) = 5.00$
Within treatments	108	36	3	
Total	153	39		

15. a. With $df_{between} = 3$, there must be $k = 4$ treatment conditions.

b. Adding the two df values gives $df_{total} = 27$. Therefore, the total number of subjects must be $N = 28$.

17. a.

Source	SS	df	MS	
Between treatments	36	2	18	$F(2, 15) = 9.00$
Within treatments	30	15	4	
Total	66	17		

With a critical value of $F = 6.36$, reject H_0.

b. The greatest change in attitude occurs when there is a moderate discrepancy between the persuasive argument and a person's original opinion.

19. Converting the summarized data into totals (T) and SS values produces

2-year-olds: $T = 42$ and $SS = 32.11$

6-year-olds: $T = 86$ and $SS = 42.75$

10-year-olds: $T = 138$ and $SS = 61.56$

The null hypothesis states that there are no differences among the age groups, H_0: $\mu_1 = \mu_2 = \mu_3$. The critical value for $\alpha = .05$ is 3.17 (using $df = 2, 55$). The analysis of variance produces

Source	SS	df	MS	
Between treatments	230.93	2	115.47	$F(2, 57) = 48.31$
Within treatments	136.42	57	2.39	
Total	367.35	59		

Reject H_0.

21. The means and SS values are

Endomorph	Ectomorph	Mesomorph
$\overline{X} = 22.0$	$\overline{X} = 17.6$	$\overline{X} = 15.0$
$SS = 24.0$	$SS = 23.2$	$SS = 30.0$

The null hypothesis states that there are no differences in personality among the three body types, H_0: $\mu_1 = \mu_2 = \mu_3$. The critical value for $\alpha = .05$ is 3.88. The analysis of variance produces

Source	SS	df	MS	
Between treatments	125.2	2	62.6	$F(2, 12) = 9.74$
Within treatments	77.2	12	6.43	
Total	202.4	14		

Reject H_0, and conclude that there are significant differences in personality among the three groups.

CHAPTER 14 TWO-FACTOR ANALYSIS OF VARIANCE

1. One hypothesis test evaluates mean differences among the rows, and a second test evaluates mean differences among the columns. The test for an interaction evaluates the significance of any mean differences that are not explained by row and/or column differences.

3. a. The graph shows parallel lines with the means for A_1 consistently below the means for A_2. There appear to be main effects for both factors but no interaction.

b. The graph shows crossing lines (the line for A_1 is horizontal, and the line for A_2 slopes sharply). There is a definite interaction and some indication of main effects for both factors.

c. The graph shows parallel, horizontal lines, with A_1 consistently above A_2. There appears to be a main effect for factor A but no effect for factor B and no interaction.

5. a. $df = 1, 36$

b. $df = 1, 36$

c. $df = 1, 36$

7. The null hypotheses state that there is no difference between levels of factor A (H_0: $\mu_{quiet} = \mu_{noisy}$), no difference between levels of factor B (H_0: $\mu_{introvert} = \mu_{extrovert}$), and no interaction. All F-ratios have $df = 1, 16$, and the critical value is $F = 4.49$.

Source	SS	df	MS	
Between treatments	120	3		
A (distraction)	80	1	80	$F(1, 16) = 16$
B (personality)	20	1	20	$F(1, 16) = 4$
A × B	20	1	20	$F(1, 16) = 4$
Within treatments	80	16	5	
Total	200	19		

Distraction has a significant effect on performance. These results do not provide sufficient evidence to conclude that personality affects performance or that personality interacts with distraction.

9.

Source	SS	df	MS	
Between treatments	280	7		
Achievement need	16	1	16	$F(1, 40) = 2.00$
Task difficulty	144	3	48	$F(3, 40) = 6.00$
Interaction	120	3	40	$F(3, 40) = 5.00$
Within treatments	320	40	8	
Total	600	47		

11. The null hypotheses state that there is no difference between genders (H_0: $\mu_{male} = \mu_{female}$), that the amount of the chemical has no effect (H_0: $\mu_{none} = \mu_{small} = \mu_{large}$), and that there is no interaction. For $df = 1, 24$, the critical value is $F = 4.26$. For $df = 2, 24$, the critical value is $F = 3.40$.

Source	SS	df	MS	
Between treatments	70	5		
A	30	1	30	$F(1, 24) = 6$
B	20	2	10	$F(2, 24) = 2$
A × B	20	2	10	$F(2, 24) = 2$
Within treatments	120	24	5	
Total	190	29		

The results indicate that the males are significantly more active than the females. However, the data do not provide sufficient evidence to conclude that the chemical has a significant effect on activity or that the chemical has a different effect on males than on females.

13. a.

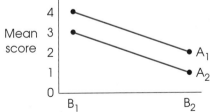

b. The lines are parallel, so there does not appear to be any interaction. There is a small difference between the lines for A_1 and A_2, so there may be a small main effect for factor A. The difference between B_1 and B_2 is larger, so there probably is a main effect for factor B.

c. The null hypotheses state that there is no difference between levels of factor A (H_0: $\mu_{A_1} = \mu_{A_2}$), no difference between

levels of factor B (H_0: $\mu_{B_1} = \mu_{B_2}$), and no interaction. All F-ratios have $df = 1, 36$, and the critical value is $F = 4.11$.

Source	SS	df	MS	
Between treatments	50	3		
A	10	1	10	$F(1, 36) = 2.00$
B	40	1	40	$F(1, 36) = 8.00$
A × B	0	1	0	$F(1, 36) = 0$
Within treatments	180	36	5	
Total	230	39		

The main effect for factor B is significant, but the A-effect and the interaction are not.

15. a. The null hypotheses state that counseling has no effect (H_0: $\mu_{yes} = \mu_{no}$), the drug has no effect (H_0: $\mu_{yes} = \mu_{no}$), and there is no interaction. All F-ratios have $df = 1, 36$, and the critical value is $F = 4.11$.

Source	SS	df	MS	
Between treatments	200	3		
A (counseling)	0	1	0	$F(1, 36) = 0$
B (drug)	160	1	160	$F(1, 36) = 32.00$
A × B	40	1	40	$F(1, 36) = 8.00$
Within treatments	180	36	5	
Total	380	39		

b. The data indicate that overall the drug has a significant effect. The main effect for counseling is not significant. However, there is a significant interaction, meaning that the effect of the drug is significantly affected by counseling. Without counseling, the drug has a minimal effect, but with counseling, the drug has a big effect.

17. a. The treatment means are

	Low	Medium	High
easy	8.00	10.00	12.00
hard	6.00	10.00	8.00

b. The null hypotheses state that task difficulty has no effect (H_0: $\mu_{easy} = \mu_{hard}$), arousal level has no effect (H_0: $\mu_{low} = \mu_{medium} = \mu_{high}$), and there is no interaction. For $df = 1, 54$, the critical value is $F = 4.03$, and for $df = 2, 54$, the critical value is $F = 3.18$ (using $df = 50$ for the denominator because 54 is not in the table).

Source	SS	df	MS	
Between treatments	220	5		
A (difficulty)	60	1	60	$F(1, 54) = 15.00$
B (arousal)	120	2	60	$F(2, 54) = 15.00$
A × B	40	2	20	$F(2, 54) = 5.00$
Within treatments	216	54	4	
Total	436	59		

c. There is a significant main effect for task difficulty, which simply indicates that performance is better on easy tasks than on hard tasks.

d. For easy tasks, increased arousal increases performance until you reach a "high" level of arousal. For hard tasks, the optimum level of arousal is "medium." With higher arousal, performance on hard tasks deteriorates.

19. The null hypotheses state that self-esteem has no effect (H_0: $\mu_{high} = \mu_{low}$), the audience has no effect (H_0: $\mu_{alone} = \mu_{audience}$), and there is no interaction. For $df = 1, 20$, the critical value is $F = 4.35$.

Source	SS	df	MS	
Between treatments	216	3		
A (self-esteem)	96	1	96	$F(1, 20) = 22.33$
B (audience)	96	1	96	$F(1, 20) = 22.33$
A × B	24	1	24	$F(1, 20) = 5.58$
Within treatments	86	20	4.3	
Total	302	23		

The significant interaction indicates that the effect of the audience was different for high self-esteem subjects than for low self-esteem subjects. Specifically, the audience produced an increase in errors for those with low self-esteem and had little effect on those with high self-esteem.

CHAPTER 15 CORRELATION AND REGRESSION

1. Set I, $SP = 6$; Set II, $SP = -16$; Set III, $SP = -4$

3. a.

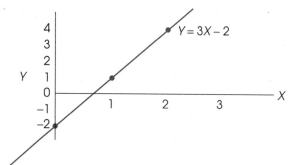

b. Estimate a strong positive correlation, probably $+0.8$ or $+0.9$.

c. $SS_X = 50$, $SS_Y = 8$, $SP = 18$, $r = +0.90$

5. a. The correlation between X and Y is $r = +0.60$.
   b. The correlation between Y and Z is $r = +0.60$.
   c. Based on the answers to part a and part b, you might expect a fairly strong positive correlation between X and Z.
   d. The correlation between X and Z is $r = -0.20$.
   e. Simply because two variables are both related to a third variable does not necessarily imply that they are related to each other.

7. $SS_X = 12.8$, $SS_Y = 22.8$, $SP = 6.2$, r $= +0.36$

9. For these data, $SS_{anxiety} = 18$, $SS_{exam} = 72$, $SP = -32$, and the Pearson correlation is $r = -0.889$.

11. a. $SS_{RT} = 3023.82$, $SS_{errors} = 89.87$, $SP = -402.95$, and the Pearson correlation is $r = -0.773$.
    b. There is a consistent, negative relationship between reaction time and errors. As reaction time gets faster (shorter time), errors increase.

13. The ranked values are as follows:

X ranks	Y ranks	
2	2	
1	1	The Spearman correlation is
3	4	$r_S = +0.90$.
4	3	
5	5	

15. a. $r_S = +0.907$
    b. Yes, there is a strong positive relationship between the grades assigned by the two instructors.

17.

19. $SS_X = 20$, $SS_Y = 24$, $SP = -20$. The regression equation is $\hat{Y} = (-1)X + 10$.

21. $SS_X = 10$, $SS_Y = 67.2$, $SP = 24$, $r = +0.926$. The regression equation is $\hat{Y} = 2.4X + 6.8$.

23. a. $\hat{Y} = 2X + 12.98$
    b. The slope ($b = 2$) indicates an additional salary of $2000 for each additional year of higher education.
    c. The Y-intercept ($a = 12.98$) indicates that the predicted salary for an individual with no higher education ($X = 0$) is 12.98 thousand dollars.

CHAPTER 16 **CHI-SQUARE TESTS**

1. Nonparametric tests make few, if any, assumptions about the populations from which the data are obtained. For example, the populations do not need to form normal distributions, nor is it required that different populations in the same study have equal variances (homogeneity of variance assumption). Parametric tests require data measured on an interval or a ratio scale. For non-parametric tests, any scale of measurement is acceptable.

3. H_0 states that the distribution of automobile accidents is the same as the distribution of registered drivers: 16% under age 20, 28% for age 20 to 29, and 56% for age 30 or older. With $df = 2$, the critical value is 5.99. The expected frequencies for these three categories are 48, 84, and 168. Chi-square = 13.76. Reject H_0, and conclude that the distribution of automobile accidents is not identical to the distribution of registered drivers.

5. The null hypothesis states that there is no preference for one side versus any other: $p = \frac{1}{4}$ for each side. With $df = 3$, the critical value is 7.81. The expected frequencies are $f_e = 12.5$ for all categories, and chi-square = 7.28. Fail to reject H_0, and conclude that there is no significant preference.

7. The null hypothesis states that there is no preference among the three photographs: $p = \frac{1}{3}$ for all categories. The expected frequencies are $f_e = 50$ for all categories, and chi-square = 20.28. With $df = 2$, the critical value is 5.99. Reject H_0, and conclude that there are significant preferences.

9. The null hypothesis states that opinions have not changed since 1970, so the distribution should still be 15% for, 79% against, and 6% no opinion. Chi-square = 2.30. With $df = 2$, the critical value is 5.99. Fail to reject H_0. These data do not provide evidence for a change in opinion.

11. The null hypothesis states that the distribution of preferences is the same for both groups of students (the two variables are independent). With $df = 2$, the critical value is 5.99. The expected frequencies are

	Book 1	Book 2	Book 3
Upper Half	26.0	20.5	13.5
Lower Half	26.0	20.5	13.5

Chi-square = 17.32. Reject H_0.

13. The null hypothesis states that the distribution of perceived importance is the same for adolescents and for college students (the two variables are independent). With $df = 2$, the critical value is 5.99. The expected frequencies are

	Appearance	Popularity	Personality
Adolescents	13.5	9.0	7.5
College	13.5	9.0	7.5

Chi-square = 2.22. Fail to reject H_0.

15. The null hypothesis states that there is no relationship between dream content and gender: The distribution of aggression content should be the same for males and females. The expected frequencies are

	Low	Medium	High
Female	8.8	8.4	6.8
Male	13.2	12.6	10.2

The chi-square statistic is 25.52. The critical value is 9.21. Reject H_0 with $\alpha = .01$ and $df = 2$.

17. The null hypothesis states that there is no relationship between personality and heart disease. The expected frequencies are

	Heart	Vascular	Hypertension	None
Type A	27.20	24.77	27.69	90.34
Type B	28.80	26.23	29.31	95.66

Chi-square = 46.02. With $df = 3$, the critical value is 7.81. Reject H_0.

19. The null hypothesis states that there is no relationship between need for achievement and risk. The expected frequencies are

	Cautious	Moderate	High
High	12.18	15.10	10.72
Low	12.82	15.90	11.28

Chi-square = 17.08. With $df = 2$, the critical value is 5.99. Reject H_0.

STATISTICS ORGANIZER

The following pages present an organized summary of the statistical procedures covered in this book. This organizer is divided into four sections, each of which groups together statistical techniques that serve a common purpose. The four groups are

I. Descriptive Statistics

II. Hypothesis Tests: Inferences About Population Means or Mean Differences

III. Hypothesis Tests: Inferences About Population Proportions and Relative Frequencies

IV. Measures of Relationship Between Two Variables

Each of the four sections begins with a general overview that discusses the purpose for the statistical techniques that follow and points out some common characteristics of the different techniques. Next, there is a decision map that leads you, step by step, through the task of deciding which statistical technique is appropriate for the data you wish to analyze. Finally, there is a brief description of each technique and the necessary formulas.

I DESCRIPTIVE STATISTICS

The purpose of descriptive statistics is to simplify and organize a set of scores. Scores may be organized in a table or graph, or scores may be summarized by computing one or two values that describe the entire set. The most commonly used descriptive techniques are the following.

A. Frequency Distribution Tables and Graphs
A frequency distribution is an organized tabulation of the number of individuals in each category on the scale of measurement. A frequency distribution can be presented as either a table or a graph. The advantage of a frequency distribution is that it presents the entire set of scores, rather than condensing the scores into a single descriptive value. The disadvantage of a frequency distribution is that it can be somewhat complex, especially with large sets of data. Frequency distribution graphs provide visual illustrations of the data set and readily allow one to ascertain the shape of the distribution.

CHOOSING DESCRIPTIVE STATISTICS: A DECISION MAP

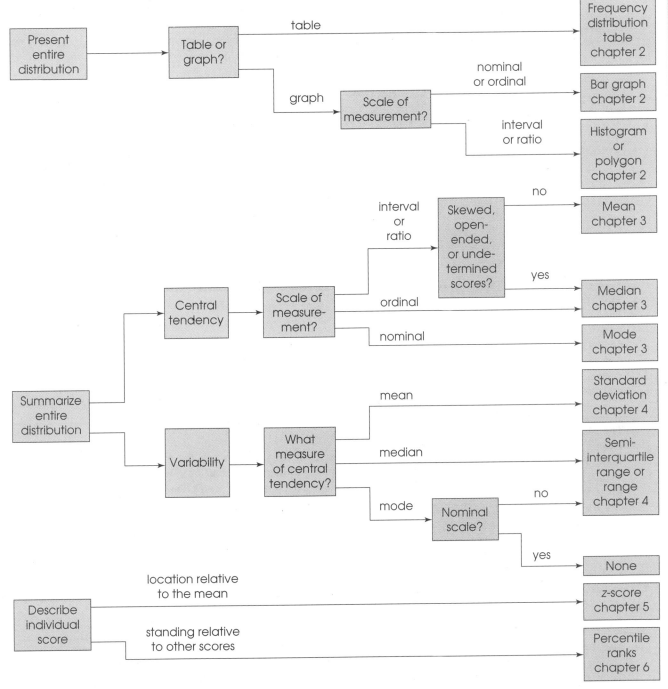

B. Measures of Central Tendency

The purpose of measuring central tendency is to identify a single score that represents an entire data set. The goal is to obtain a single value that is the best example of the average, or most typical, score from the entire set.

Measures of central tendency are used to describe a single data set, and they are the most commonly used measures for comparing two (or more) different sets of data.

C. Measures of Variability

Variability is used to provide a description of how spread out the scores are in a distribution. It also provides a measure of how accurately a single score selected from a distribution represents the entire set.

D. z-Scores

Most descriptive statistics are intended to provide a description of an entire set of scores. However, z-scores are used to describe individual scores within a distribution. The purpose of a z-score is to identify the precise location of an individual within a distribution by using a single number. A z-score describes the position of a raw score (X) relative to the mean of the distribution.

---

1. The Mean (Chapter 3) The mean is the most commonly used measure of central tendency. It is computed by finding the total (ΣX) for the set of scores and then dividing the total by the number of individuals. Conceptually, the mean is the amount each individual will receive if the total is divided equally.	Population: $\mu = \dfrac{\Sigma X}{N}$ Sample: $\overline{X} = \dfrac{\Sigma X}{n}$ (See Demonstration 3.1, page 78.)
2. The Median (Chapter 3) Exactly 50% of the scores in a data set have values less than or equal to the median. The median usually is computed for data sets where the mean cannot be found (undetermined scores, open-ended distribution) or where the mean does not provide a good, representative value (ordinal scale, skewed distribution).	List the scores in order from smallest to largest. a. With an odd number of scores, the median is the middle score. b. With an even number of scores, the median is the average of the middle two scores. c. With several scores tied at the median, use interpolation to find the median. (See Example 3.4, page 64.)
3. The Mode (Chapter 3) The mode is the score with the greatest frequency. The mode is used when the scores consist of measurements on a nominal scale.	No calculation. Simply count the frequency of occurrence for each different score.
4. The Range (Chapter 4) The range is the distance from the lowest to the highest score in a data set. The range is considered to be a relatively crude measure of variability.	Find the upper real limit for the largest score and the lower real limit for the smallest score. The range is the difference between these two real limits.

5. **The Semi-Interquartile Range** (Chapter 4) The semi-interquartile range is one-half of the range covered by the middle 50% of the distribution. The semi-interquartile range is often used to measure variability in situations where the median is used to report central tendency.	Find the first quartile (the X value greater than or equal to 25% of the distribution) and the third quartile (the X value greater than or equal to 75% of the distribution). The semi-interquartile range is one-half of the distance between the two quartiles. (See Example 4.1, page 87.)
6. **Standard Deviation** (Chapter 4) The standard deviation is a measure of the standard distance from the mean. Standard deviation is obtained by first computing SS (the sum of squared deviations) and variance (the mean squared deviation). Standard deviation is the square root of variance.	**Sum of Squares** $\begin{cases} \text{Definitional: } SS = \Sigma(X - \mu)^2 \\ \text{Computational: } SS = \Sigma X^2 - \dfrac{(\Sigma X)^2}{N} \end{cases}$ **Variance** $\begin{cases} \text{Population: } \sigma^2 = \dfrac{SS}{N} \\ \text{Sample: } s^2 = \dfrac{SS}{n-1} \end{cases}$ **Standard Deviation** $\begin{cases} \text{Population: } \sigma = \sqrt{\dfrac{SS}{N}} \\ \text{Sample: } s = \sqrt{\dfrac{SS}{n-1}} \end{cases}$ (See Demonstration 4.1, page 107.)
7. **z-Scores** (Chapter 5) The sign of a z-score indicates whether an individual is above ($+$) or below ($-$) the mean. The numerical value of the z-score indicates how many standard deviations there are between the score and the mean.	$$z = \frac{X - \mu}{\sigma}$$ (See Demonstration 5.1, page 127.)

II HYPOTHESIS TESTS: INFERENCES ABOUT POPULATION MEANS OR MEAN DIFFERENCES

All of the hypothesis tests covered in this section use the means obtained from sample data as the basis for testing hypotheses about population means. Although there are a variety of tests used in a variety of research situations, all use the same basic logic, and all of the test statistics have the same basic structure. In each case, the test statistic (z, t, or F) involves computing a ratio with the following structure:

$$\text{test statistic} = \frac{\text{obtained difference between sample means}}{\text{mean difference expected by chance}}$$

The goal of each test is to determine whether the observed sample mean differences are larger than expected by chance. In general terms, a *significant result* indicates that the results obtained in a research study (sample differences) are not what would be expected by chance. In each case, a large value for the test-statistic ratio indicates a significant result; that is, when the actual difference between sample means (numerator)

CHOOSING A PARAMETRIC TEST: A DECISION MAP FOR MAKING INFERENCES ABOUT POPULATION MEANS OR MEAN DIFFERENCES

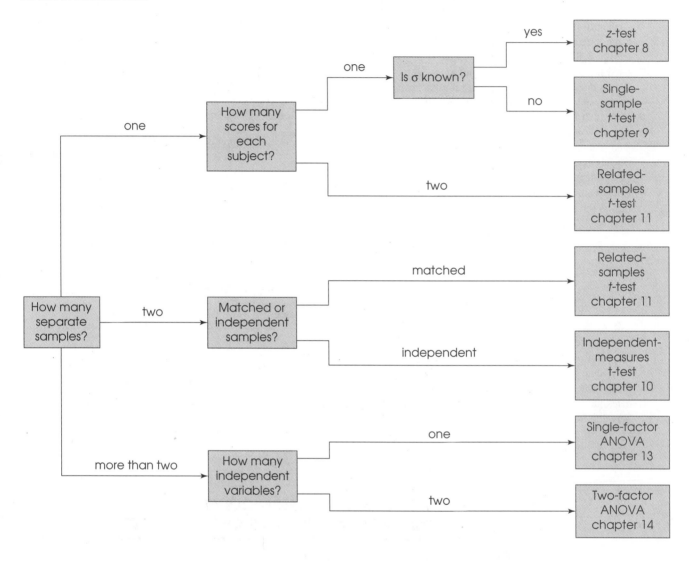

is substantially larger than the difference expected by chance (denominator), you will obtain a large ratio, which indicates that the sample difference is significant.

The actual calculations differ slightly from one test statistic to the next, but all involve the same basic computations.

1. A set of scores (sample) is obtained for each population or each treatment condition.

2. The mean is computed for each set of scores, and some measure of variability (*SS,* standard deviation, or variance) is obtained for each set of scores.

3. The differences between the sample means provide a measure of how much difference exists between treatment conditions. Because these mean differences may be caused by the treatment conditions, they are often called *systematic* or *predicted.* These differences are the numerator of the test statistic.

4. The variability within each set of scores provides a measure of unsystematic or unpredicted differences due to chance. Because the individuals within each treatment condition are treated exactly the same, there is nothing that should cause their scores to be different. Thus, any observed differences (variability) within treatments are assumed to be due to error or chance.

With these considerations in mind, each of the test-statistic ratios can be described as follows:

$$\text{test statistic} = \frac{\text{differences (variability) between treatments}}{\text{differences (variability) within treatments}}$$

The hypothesis tests reviewed in this section apply to three basic research designs:

1. Single-Sample Designs. Data from a single sample are used to test a hypothesis about a single population.
2. Independent-Measures Designs. A separate sample is obtained to represent each individual population or treatment condition.
3. Related-Samples Designs. Related-samples designs include repeated-measures and matched-subjects designs. For a repeated-measures design, there is only one sample, with each individual subject being measured in all of the different treatment conditions. In a matched-subjects design, every individual in one sample is matched with a subject in each of the other samples.

Finally, you should be aware that all of these tests place stringent restrictions on the sample data and the population distributions being considered. First, these tests all require measurements on an interval or a ratio scale (numerical values that allow you to compute means and differences). Second, each test makes assumptions about population distributions and sampling techniques. Consult the appropriate chapter of this book to verify that the specific assumptions are satisfied before proceeding with any hypothesis test.

1. The *z*-Score Test (Chapter 8) The *z*-score test uses the data from a single sample to test a hypothesis about the population mean in situations where the population standard deviation (σ) is known. The null hypothesis states a specific value for the unknown population mean.	$z = \dfrac{\bar{X} - \mu}{\sigma_{\bar{X}}}$ where $\sigma_{\bar{X}} = \dfrac{\sigma}{\sqrt{n}}$ (See Demonstration 8.1, page 214.)
2. The Single-Sample *t* Test (Chapter 9) This test uses the data from a single sample to test a hypothesis about a population mean in situations where the population standard deviation is unknown. The sample variability is used to estimate the unknown population standard deviation. The null hypothesis states a specific value for the unknown population mean.	$t = \dfrac{\bar{X} - \mu}{s_{\bar{X}}}$ where $s_{\bar{X}} = \sqrt{\dfrac{s^2}{n}}$ $df = n - 1$ (See Demonstration 9.1, page 235.)

3. **The Independent-Measures t Test**
(Chapter 10)
The independent-measures t test uses data from two separate samples to test a hypothesis about the difference between two population means. The variability within the two samples is combined to obtain a single (pooled) estimate of population variance. The null hypothesis states that there is no difference between the two population means.

$$t = \frac{(\overline{X}_1 - \overline{X}_2) - (\mu_1 - \mu_2)}{s_{\overline{X}-\overline{X}}} \quad \text{where } s_{\overline{X}-\overline{X}} = \sqrt{\frac{s_p^2}{n_1} + \frac{s_p^2}{n_2}}$$

$$\text{and } s_p^2 = \frac{SS_1 + SS_2}{df_1 + df_2}$$

$$df = df_1 + df_2 = (n_1 - 1) + (n_2 - 1)$$

(See Demonstation 10.1, page 255.)

4. **The Related-Samples t Test** (Chapter 11)
This test evaluates the mean difference between two treatment conditions using the data from a repeated-measures or a matched-subjects experiment. A difference score (D) is obtained for each subject (or each matched pair) by subtracting the score in treatment 1 from the score in treatment 2. The variability of the sample difference scores is used to estimate the population variability. The null hypothesis states that the population mean difference (μ_D) is zero.

$$t = \frac{\overline{D} - \mu_D}{s_{\overline{D}}} \quad \text{where } s_{\overline{D}} = \sqrt{\frac{s^2}{n}}$$

$$df = n - 1$$

(See Demonstration 11.1, page 275.)

5. **Independent-Measures Analysis of Variance**
(Chapter 13)
This test uses data from two or more separate samples to test for mean differences among two or more populations. The null hypothesis states that there are no differences among the population means. The test statistic is an F-ratio that uses the variance between treatment conditions (sample mean differences) as the numerator and variance within treatment conditions (error variability) as the denominator. With only two samples, this test is equivalent to the independent-measures t test.

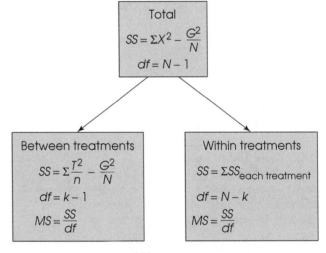

Total
$$SS = \Sigma X^2 - \frac{G^2}{N}$$
$$df = N - 1$$

Between treatments
$$SS = \Sigma \frac{T^2}{n} - \frac{G^2}{N}$$
$$df = k - 1$$
$$MS = \frac{SS}{df}$$

Within treatments
$$SS = \Sigma SS_{\text{each treatment}}$$
$$df = N - k$$
$$MS = \frac{SS}{df}$$

$$F\text{-ratio} = \frac{MS_{\text{between treatments}}}{MS_{\text{within treatments}}}$$

(See Demonstration 13.1, page 346.)

6. Two-Factor, Independent-Measures Analysis of Variance (Chapter 14)

This test is used to evaluate mean differences among populations or treatment conditions using sample data from research designs with two independent variables (factors). The two-factor ANOVA tests three separate hypotheses: mean differences among the levels of factor A (main effect for factor A), mean differences among the levels of factor B (main effect for factor B), and mean differences resulting from specific combinations of the two factors (interaction). Each of the three separate null hypotheses states that there are no population mean differences. Each of the three tests uses an F-ratio as the test statistic, with the variance between samples (sample mean differences) in the numerator and the variance within samples in the denominator.

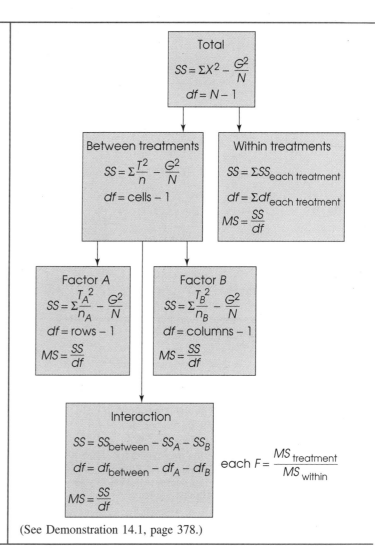

Total
$$SS = \Sigma X^2 - \frac{G^2}{N}$$
$$df = N - 1$$

Between treatments
$$SS = \Sigma \frac{T^2}{n} - \frac{G^2}{N}$$
$$df = \text{cells} - 1$$

Within treatments
$$SS = \Sigma SS_{\text{each treatment}}$$
$$df = \Sigma df_{\text{each treatment}}$$
$$MS = \frac{SS}{df}$$

Factor A
$$SS = \Sigma \frac{T_A^2}{n_A} - \frac{G^2}{N}$$
$$df = \text{rows} - 1$$
$$MS = \frac{SS}{df}$$

Factor B
$$SS = \Sigma \frac{T_B^2}{n_B} - \frac{G^2}{N}$$
$$df = \text{columns} - 1$$
$$MS = \frac{SS}{df}$$

Interaction
$$SS = SS_{\text{between}} - SS_A - SS_B$$
$$df = df_{\text{between}} - df_A - df_B$$
$$MS = \frac{SS}{df}$$

each $F = \dfrac{MS_{\text{treatment}}}{MS_{\text{within}}}$

(See Demonstration 14.1, page 378.)

III HYPOTHESIS TESTS: INFERENCES ABOUT POPULATION PROPORTIONS AND RELATIVE FREQUENCIES

Chi-square tests consist of techniques used to test hypotheses about the relative frequencies (proportions) one expects to find in a frequency distribution. The sample data for these tests consist of the frequencies of observations that fall into various categories of a variable: for example, the frequency of people preferring soft drink A, the frequency preferring soft drink B, and so on. The hypotheses make statements about what proportion of frequencies you should expect for each category for the population. The chi-square test for goodness of fit tests how well the form (or shape) of a sample frequency distribution matches a hypothesized population distribution. The chi-square test for independence also compares sample frequencies to hypothesized population proportions. However, it determines if two variables, consisting of two or more categories apiece, are related.

A DECISION MAP FOR CHI-SQUARE TESTS

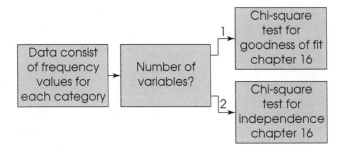

1. The Chi-Square Test for Goodness of Fit (Chapter 16) This chi-square test is used in situations where the measurement procedure results in classifying individuals into distinct categories. The test uses frequency data from a single sample to test a hypothesis about the population distribution. The null hypothesis specifies the proportion or percentage of the population for each category on the scale of measurement.	$$\chi^2 = \Sigma \frac{(f_o - f_e)^2}{f_e}$$ where $f_e = pn$ $df = C - 1$ (See Example 16.1, page 438.)
2. The Chi-Square Test for Independence (Chapter 16) This test uses frequency data to determine whether or not there is a significant relationship between two variables. The null hypothesis states that the two variables are independent. The chi-square test for independence is used when the scale of measurement consists of relatively few categories for both variables and can be used with nominal, ordinal, interval, or ratio scales.	$$\chi^2 = \Sigma \frac{(f_o - f_e)^2}{f_e}$$ where $f_e = \dfrac{(\text{row total})(\text{column total})}{n}$ $df = (R - 1)(C - 1)$ (See Demonstration 16.1, page 452.)

IV MEASURES OF RELATIONSHIP BETWEEN TWO VARIABLES

As we noted in Chapter 1, a major purpose for scientific research is to investigate and establish orderly relationships between variables. The statistical techniques covered in this section all serve the purpose of measuring and describing relationships.

The data for these statistics involve two observations for each individual—one observation for each of the two variables being examined. The goal is to determine whether or not a consistent, predictable relationship exists and to describe the nature of the relationship.

Each of the different statistical methods described in this section is intended to be used with a specific type of data. To determine which method is appropriate, you must first examine your data and identify what type of variable is involved.

CHOOSING A MEASURE OF RELATIONSHIP BETWEEN TWO VARIABLES: A DECISION MAP

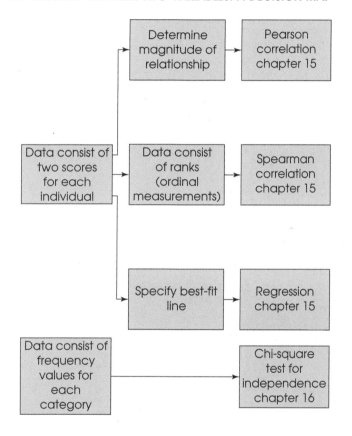

---

1. Pearson Correlation (Chapter 15)
The Pearson correlation measures the degree of linear relationship between two variables. The sign (+ or −) of the correlation indicates the direction of the relationship. The magnitude of the correlation (from 0 to 1) indicates the degree to which the data points fit on a straight line.

$$r = \frac{SP}{\sqrt{SS_X SS_Y}}$$

where $SP = \Sigma(X - \bar{X})(Y - \bar{Y}) = \Sigma XY - \frac{(\Sigma X)(\Sigma Y)}{n}$

(See Demonstration 15.1, page 421.)

2. The Spearman Correlation (Chapter 15)
The Spearman correlation measures the degree to which the relationship between two variables is one-directional, or monotonic. The Spearman correlation is used when both variables, X and Y, are ranks (measured on an ordinal scale).

Use the Pearson formula on the ranked data or the special Spearman formula:

$$r_S = 1 - \frac{6\Sigma D^2}{n(n^2 - 1)}$$

(See Demonstration 15.2, page 424.)

3. Linear Regression (Chapter 15)
The purpose of linear regression is to find the equation for the best-fitting straight line for predicting Y scores from X scores. The regression process determines the linear equation with the least squared error between the actual Y values and the predicted Y values on the line.

$$\hat{Y} = bX + a \quad \text{where } b = \frac{SP}{SS_X} \quad \text{and} \quad a = \overline{Y} - b\overline{X}$$

(See Demonstration 15.1, page 421.)

4. The Chi-Square Test for Independence (Chapter 16)

See Section III, number 2.

REFERENCES

American Psychological Association. (1994). *Publication manual of the American Psychological Association,* 4th ed. Washington, D.C.: Author.

Anderson, C. A., & Anderson, D. C. (1984). Ambient temperature and violent crime: Tests of linear and curvilinear hypothesis. *Journal of Personality and Social Psychology, 46,* 91–97.

Aronson, E., Turner, J. A., & Carlsmith, J. M. (1963). Communicator credibility and communication discrepancy as a determinant of opinion change. *Journal of Abnormal and Social Psychology, 67,* 31–36.

Betz, B. J., & Thomas, C. B. (1979). Individual temperament as a predictor of health or premature disease. *Johns Hopkins Medical Journal, 144,* 81–89.

Bransford, J. D., & Johnson, M. K. (1972). Contextual prerequisites for understanding: Some investigations of comprehension and recall. *Journal of Verbal Learning and Verbal Behavior, 11,* 717–726.

Byrne, D. (1971). *The attraction paradigm.* New York: Academic Press.

Cattell, R. B. (1973, July). Personality pinned down. *Psychology Today, 7,* 40–46.

Cervone, D. (1989). Effects of envisioning future activities on self-efficacy judgments and motivation: An availability heuristic interpretation. *Cognitive Therapy and Research, 13,* 247–261.

Cialdini, R. B., Reno, R. R., & Kallgren, C. A. (1990). A focus theory of normative conduct: Recycling the concept of norms to reduce littering in public places. *Journal of Personality and Social Psychology, 58,* 1015–1026.

Cowles, M., & Davis, C. (1982). On the origins of the .05 level of statistical significance. *American Psychologist, 37,* 553–558.

Craik, F. I. M., & Tulving, E. (1975). Depth of processing and the retention of words in episodic memory. *Journal of Experimental Psychology: General, 104,* 268–294.

Darley, J. M., & Latané, B. (1968). Bystander intervention in emergencies: Diffusion of responsibility. *Journal of Personality and Social Psychology, 8,* 377–383.

Davis, E. A. (1937). *The development of linguistic skills in twins, single twins with siblings, and only children from age 5 to 10 years.* Institute of Child Welfare Series, No. 14. Minneapolis: University of Minnesota Press.

Dollard, J., Miller, N., Doob, L., Mowrer, O. H., & Sears, R. R. (1939). *Frustration and aggression.* New Haven, Conn.: Institute of Human Relations, Yale University Press.

Duncker, K. (1945). On problem-solving. *Psychological Monographs, 58* (No. 270).

Friedman, M., & Rosenman, R. H. (1974). *Type A behavior and your heart.* New York: Knopf.

Gravetter, F. J, & Wallnau, L. B. (1996). *Statistics for the behavioral sciences,* 4th ed. St. Paul, Minn.: West.

Holmes, T. H., & Rahe, R. H. (1967). The social readjustment rating scale. *Journal of Psychosomatic Research, 11,* 213–218.

Hunter, J. E. (1997). Needed: A ban on the significance test. *Psychological Science, 8,* 3–7.

Krech, D., Rozenzweig, M. R., & Bennett, E. L. (1962). Relations between brain chemistry and problem solving among rats raised in enriched and impoverished environments. *Journal of Comparative and Physiological Psychology, 55,* 801–807.

Levine, S. (1960). Stimulation in infancy. *Scientific American, 202,* 80–86.

Loftus, G. R. (1996). Psychology will be a much better science when we change the way we analyze data. *Current Directions in Psychological Science, 5,* 161–171.

McClelland, D. C. (1961). *The achieving society.* Princeton, N.J.: Van Nostrand.

Moore-Ede, M. C., Sulzman, F. M., & Fuller, C. A. (1982). *The clocks that time us.* Cambridge: Harvard University Press.

Pelton, T. (1983). The shootists. *Science83, 4*(4), 84–86.

Reifman, A. S., Larrick, R. P., & Fein, S. (1991). Temper and temperature on the diamond: The heat-aggression relationship in major league baseball. *Personality and Social Psychology Bulletin, 17,* 580–585.

Rosenthal, R. (1963). On the social psychology of the psychological experiment: The experimenter's hypothesis as unintended determinant of experimental results. *American Scientist, 51,* 268–283.

Rosenthal, R., & Fode, K. L. (1963). The effect of experimenter bias on the performance of the albino rat. *Behavioral Science, 8,* 183–189.

Scaife, M. (1976). The response to eye-like shapes by birds. I. The effect of context: A predator and a strange bird. *Animal Behaviour, 24,* 195–199.

Schachter, S. (1968). Obesity and eating. *Science, 161,* 751–756.

Schleifer, S. J., Keller, S. E., Camerino, M., Thornton, J. C., & Stein, M. (1983). Suppression of lymphocyte stimulation following bereavement. *Journal of the American Medical Association, 250,* 374–377.

Sheldon, W. H. (1940). *The varieties of human physique: An introduction to constitutional psychology.* New York: Harper.

Shrauger, J. S. (1972). Self-esteem and reactions to being observed by others. *Journal of Personality and Social Psychology, 23,* 192–200.

Siegel, J. M. (1990). Stressful life events and use of physician services among the elderly: The moderating role of pet ownership. *Journal of Personality and Social Psychology, 58,* 1081–1086.

Tulving, E., & Osler, S. (1968). Effectiveness of retrieval cues in memory for words. *Journal of Experimental Psychology, 77,* 593–601.

Tversky, A., & Kahneman, D. (1974). Judgment under uncertainty: Heuristics and biases. *Science, 185,* 1124–1131.

Volpicelli, J. R., Alterman, A., Hayashida, M., & O'Brien, C. P. (1992). Naltrexone in the treatment of alcohol dependence. *Archives of General Psychiatry, 49,* 881–887.

Winget, C., & Kramer, M. (1979). *Dimensions of dreams.* Gainesville: University Presses of Florida.

INDEX